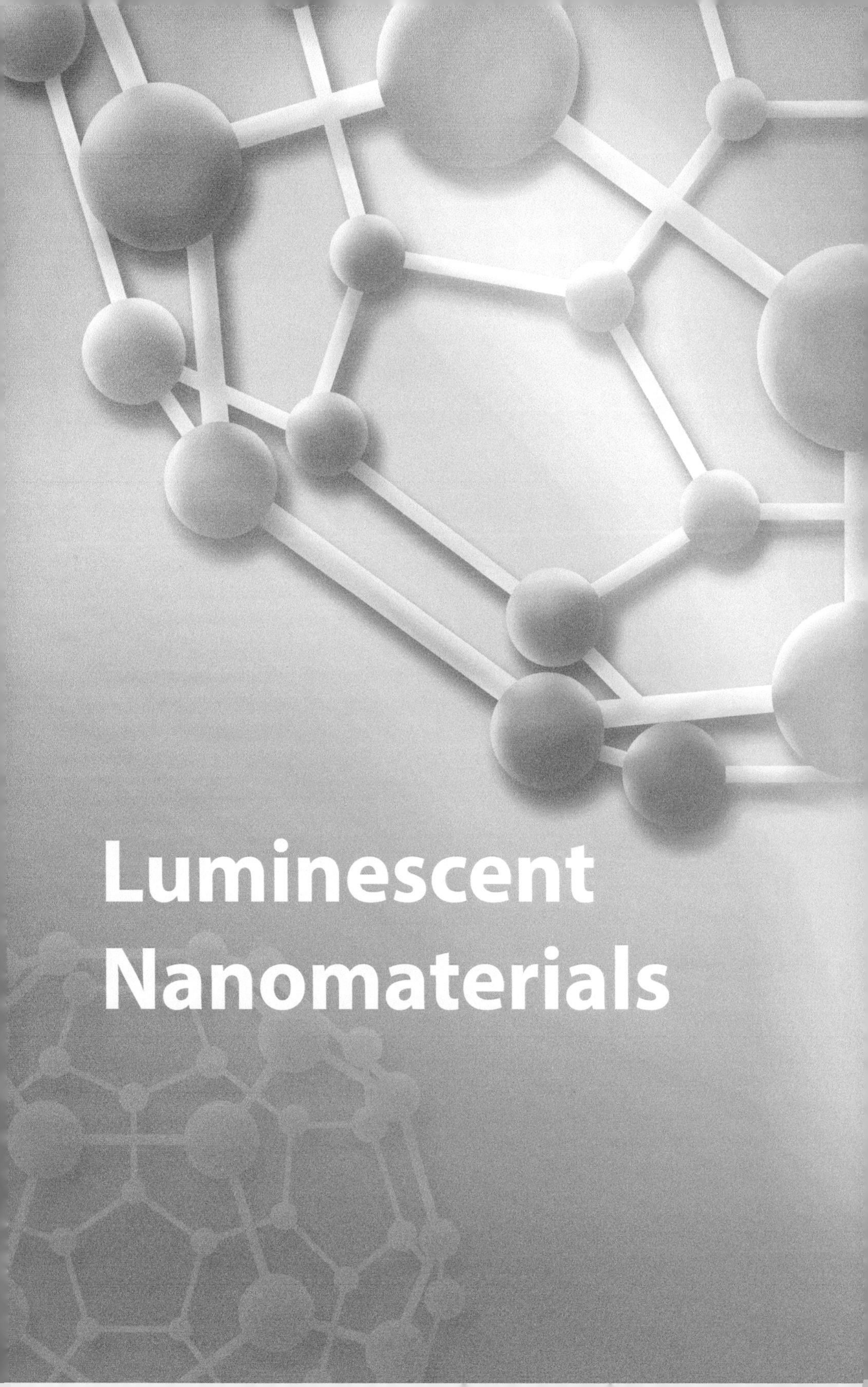

Luminescent Nanomaterials

Luminescent Nanomaterials

edited by

Odireleng Martin Ntwaeaborwa

Jenny Stanford Publishing

Published by

Jenny Stanford Publishing Pte. Ltd.
Level 34, Centennial Tower
3 Temasek Avenue
Singapore 039190

Email: editorial@jennystanford.com
Web: www.jennystanford.com

British Library Cataloguing-in-Publication Data
A catalogue record for this book is available from the British Library.

Luminescent Nanomaterials

ISBN 978-981-4968-11-9 (Hardcover)
ISBN 978-1-003-27738-5 (eBook)

Contents

Preface

Luminescent Nanomaterials provides an extensive review of recent advances in the study of the synthesis and characterization of light-emitting nanomaterials. Today, luminescent nanomaterials find many uses in applications such as displays, solid-state lighting, dosimetry, solar cells, quantum and gas sensing, theft prevention, advertising, and phototherapy. The book covers interesting topics in the study of luminescent nanomaterials including synthesis, characterization, and applications. It provides a comprehensive overview of cutting-edge research in this rapidly developing field. The chapters have been written by leading experts in the field of luminescent nanomaterials from countries across the globe. These chapters reflect the multidisciplinary nature of nanoscience and nanotechnology in the fields of chemistry, materials science, life sciences, and physics. The book serves as a comprehensive reference to early, middle, and advanced researchers in the field of luminescent nanomaterials applications. I hope that the book will provide the readers a stimulating and enriching experience in the field of luminescent nanomaterials and their applications.

Odireleng Martin Ntwaeaborwa
University of the Witwatersrand, South Africa

Chapter 1

Luminescent Diamond: A Platform for Next Generation Nanoscale Optically Driven Quantum Sensors

Nicholas Nunn,[a] Alexander I. Shames,[b] Marco Torelli,[c] Alex I. Smirnov,[a] and Olga Shenderova[c]

[a]*North Carolina State University, Department of Chemistry, Raleigh NC, USA*
[b]*Ben-Gurion University of the Negev, Israel*
[c]*Adamas Nanotechnologies Inc., Raleigh, NC, USA*
nanunn@ncsu.edu

1.1 Introduction

The first observation of luminescence originating from diamond was likely recorded in Robert Boyles 1664 description [1] in *Experiments and Considerations Touching Colours, with Observations on a Diamond that Shines in the Dark*, where he remarks:

> *"Being rubb'd upon my Cloaths, as is usual for the exciting of Amber, Wax, and other Electrical Bodies, it did in the Dark manifestly shine like Rotten Wood, or the Scales of Whitings, or other putrified Fish"*

Luminescent Nanomaterials
Edited by Odireleng Martin Ntwaeaborwa

ISBN 978-981-4968-11-9 (Hardcover), 978-1-003-27738-5 (eBook)
www.jennystanford.com

Interpreting this colorful description, Boyle apparently observed what is now known as triboluminescence. 357 years later, luminescence originating from diamond remains a captivating source of scientific intrigue, albeit at a level of sophistication that Boyle likely could not have imagined. Progress during 20th century in solid-state physics, the understanding of light and matter interactions, and materials science led to unprecedented understanding of not only the origins of luminescence in diamond, but the means to synthesize artificial diamond and tailor its optical properties. Luminescence in diamond originates from a variety of atomic scale defects within the crystalline lattice—many of which are catalogued by Zaitsev [2]. One of the most notable of these optical centers is the nitrogen vacancy (NV). An assessment of literature records demonstrates a significant increase in the number of publications on NV centers since 2000, with the most substantial growth from 2010 onwards (Fig. 1.1). Interest in the NV center stems from the coupling of fluorescence emission and the electronic spin state of the defect. The electronic spins of NV centers can be initialized, manipulated, and read-out optically at room temperature. For these reasons, the NV center continues to be explored as a room temperature quantum memory for quantum information processing (QIP) [3–10]. Additionally, the electronic spins of NV centers are sensitive to magnetic fields, electric fields, temperature, pressure, and strain; meaning making them suitable for applications atomic scale sensors. To date, the NV-based sensors have been demonstrated in nanoscale magnetometry [11, 12], chemical sensing in biological environments [13–15], thermometry [16–21], and pressure/strain sensing [22–24]. It is the opinion of the authors that sensing applications with optical centers in diamond are the most promising, and we discuss some specific examples of the various sensing demonstrations that have appeared in recent years. In addition to the NV center, optical centers based around Group IV (Si, Ge, Sn, Pb) atoms have potentials for use as single photon emission nanophotonic structures in QIP [4, 25] and in nanoscale thermometry [26, 27]. The emergent use of other color centers expands the application portfolio of both bulk and nanoscale diamond.

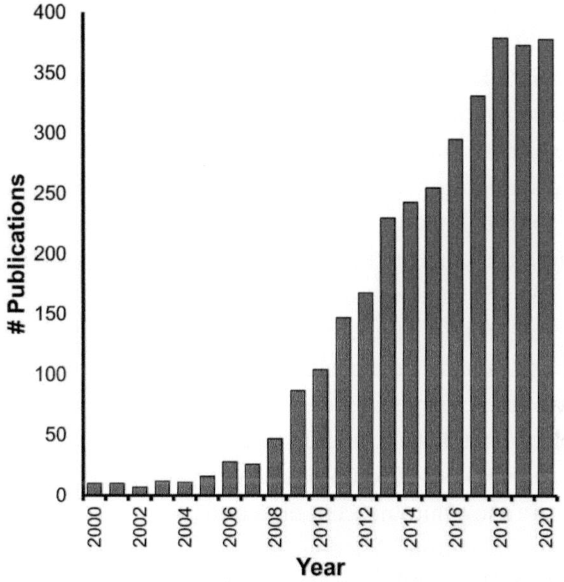

Figure 1.1 Publications per year from 2000 to 2020 based on Web of Science topical search of "NV center" across all databases. A significant increase in the publication volume occurred in the second decade of the 21st century.

Despite these exciting demonstrations, opportunities to improve the quality of the diamond hosting these optical centers remain. These considerations motivated further investigations of correlations between quantum sensing performance and diamond synthesis and processing (*particularly for chemical vapor deposition synthesized diamond*) [28–34]. The ultimate properties and performance of diamond optical centers in applications—be they sensing or QIP—are inherently linked to the quality of the host crystal. Quantum optical centers in diamond are just as sensitive to the many interactions originating from within the diamond lattice as they are to external interactions (*if not more so*). Internal sources of interaction can include: lattice structural defects (e.g., stacking faults, dislocations), magnetic inclusions, unpaired electrons, and non-zero spin nuclei. These additional interactions are particularly undesirable for many sensing methodologies as they add significant noise to the sensor (*the optical center*) and impede overall performance. The only means of minimizing the impact of these unwanted interactions are found at the synthesis and post-synthesis processing stages of luminescent diamond production. Therefore, approaches developed from fundamental materials science and engineering are necessary, as low-quality diamond impairs sensing performance and precludes widespread adoption and use of this promising technology.

Due to the high technological potential of diamond optical centers and the inherent relationship between diamond quality and sensing performance, we seek to: (1) summarize the fundamental methods of luminescent diamond synthesis and post-processing and highlight recent advancements therein; (2) review the unique magneto-optical properties of some prominent optical centers in diamond to establish how they can be employed in sensing; (3) provide an overview of contemporary research (mostly from 2015–2021) in the use of luminescent diamond (both bulk and particulate) in sensing applications. We conclude with an outlook on the future of the technology, with particular regards to what we see as the most important research directions in order to expand the use of the material to broader scientific and engineering communities.

1.2 Synthesis of Luminescent Diamond

Luminescent diamond of technological importance primarily originates from synthetic material as opposed to natural diamond. Synthetic methods afford greater flexibility and precision in controlling crystal structure and impurity atom content and distribution. The impurity content can vary significantly in natural diamond depending on the geological and geographic origin. We therefore limit discussion of luminescent diamond to synthetically produced diamond. However, it is noteworthy that studies of luminescence in diamond predate development of synthetic methods [35].

Not all synthetic processes are viable for the production of luminescent diamond insofar as these methods are currently implemented; however, such processes are documented for completeness. While we restrict discussion primarily to the NV center, these concepts could be applied to other optical centers as well. Minimalism should be the guiding principle in producing high quality diamond for sensing applications. Poor synthesis and excessive post-processing usually lead to compromised properties and performance of NV centers. Examples of how less than ideal synthesis and excessive processing can manifest are: (1) excessive impurity atom content, (2) irreversible lattice damage and distortion, and (3) graphitization, among others. In this section, we discuss the various methods of diamond synthesis, how optical centers are introduced, and emerging processing techniques that can potentially improve optical center performance in sensing.

1.2.1 Methods of Synthetic Diamond Production

Techniques to produce synthetic diamond can be grouped into two broad categories:

(1) Static synthesis
(2) Dynamic synthesis

The distinction between a static or dynamic synthetic process is made with regard to the nature of the growth environment and associated thermodynamic variables (e.g., temperature and pressure). Although no growth processes are truly 'static' in the strictest definition of the word, a static growth process is one in which the growth environment is carefully regulated with near constant temperature and pressure (or carefully regulated gradients of either variable). Examples of static synthesis methods include chemical vapor deposition (CVD) (Fig. 1.2a) and high-pressure high-temperature (HPHT) processes (Fig. 1.2b). In contrast, dynamic synthesis methods involve rapid (e.g. microsecond timescale) and significant changes in temperature and pressure (e.g. $\Delta T \sim 10^3$ K, $\Delta P \sim 10^2$ GPa). Examples of dynamic synthesis include detonation (Fig. 1.2c), shockwave, pulsed laser ablation in liquid (PLAL), and ultrasonic cavitation. The most established industrial methods of diamond synthesis are HPHT, CVD, and detonation for production of so-called detonation nanodiamond (DND).

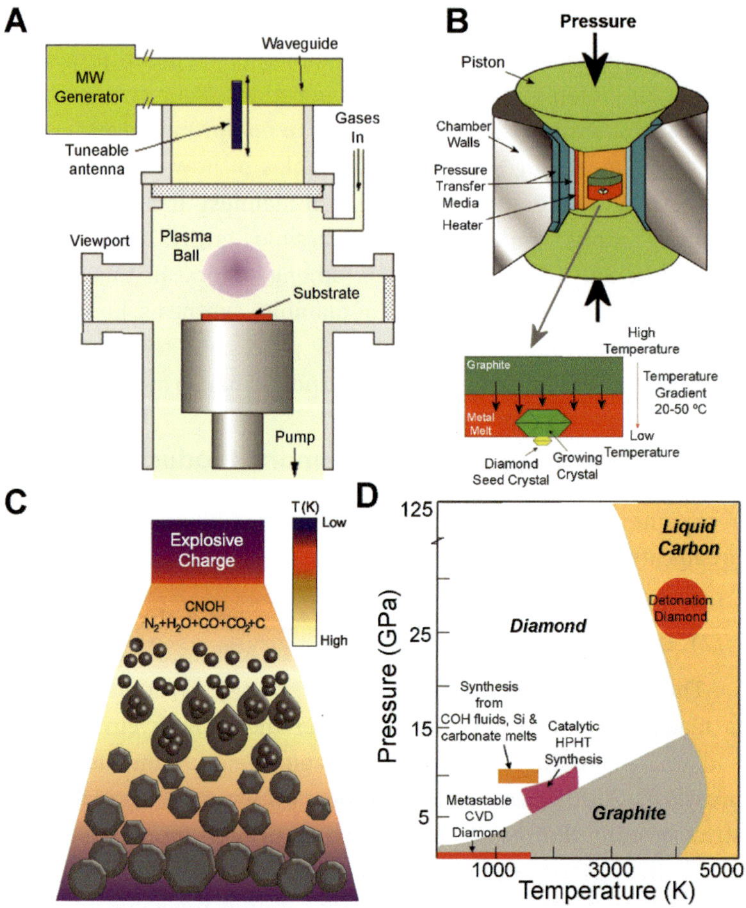

Figure 1.2 Methods for production of synthetic diamond particles and bulk diamond. (a) Microwave assisted chemical vapor deposition (MW-CVD), (b) High pressure high temperature (HPHT) cell, (c) Detonation synthesis, and (d) Pressure vs. temperature phase diagram for diamond, illustrating the relative P-T regimes for each method of synthesis. (a) Reproduced from Ref. [57]. (b) Reproduced with permission from Ref. [58], Copyright 2020, American Chemical Society. (c) Reproduced from Ref. [53]. (d) Adapted from Ref. [59].

Each synthesis method yields diamonds of different size, morphology, and nitrogen content (Table 1.1). Control of the nitrogen content and distribution is of a particular significance if the end goal is the production of luminescent diamond. Lower nitrogen content overall is more desirable for optical centers with greater luminescence (e.g. NV centers) [36]; moreover, nitrogen atoms should be distributed as a substitutional impurity. Because the CVD and HPHT synthesis methods afford the greatest control of nitrogen content and distribution, they are by far the most commonly used procedures for synthesis of luminescent diamond. The nitrogen content is typically too high in detonation diamond, and the nitrogen is primarily incorporated as aggregates. Presently, there are no effective means to reduce the total nitrogen content in DND [37]. However, we note that studies of diamond synthesized from mixtures of graphite and explosives did produce ~40–50 nm particles with viable NV centers [38], and that luminescent diamond nanoparticles were also produced by PLAL [39, 40]. Nevertheless, neither of these approaches have been studied extensively enough to make a conclusive statement on their capability in the production of high-quality luminescent diamond. Aside from an initial paper on ultrasonic cavitation synthesis over a decade ago [41], there have been no significant follow up studies to the best of the knowledge of the authors even though the diamond particles produced were nearly mono-dispersed with sizes of 6 or 9 μm with cubic crystalline morphology.

Table 1.1 Methods of synthetic diamond production along with their respective total nitrogen content and distribution as well as morphology and size of diamond. References pertinent to each synthesis technique or where use of such diamond are listed

Synthesis type	Method	Total nitrogen content (ppm)	Morphology and size	References
Static	HPHT	~100–300 (substitutional nitrogen). High purity type IIa (<1 ppm) possible.	Single crystal and polycrystalline particles from ~20 μm to 1 mm	[42–49]
	CVD	Typically ≤10–20 (substitutional nitrogen). High purity type IIa (<1 ppm) possible.	Primarily single and polycrystalline thin films, CVD grown nanodiamond particles also demonstrated	[50–52]

(Continued)

Table 1.1 (*Continued*)

Synthesis type	Method	Total nitrogen content (ppm)	Morphology and size	References
Dynamic	Detonation	>10,000 (aggregates). Originating from explosives	3–5 nm primary crystals, 100–300 nm aggregates (not disaggregated)	[53]
	PLAL	Not studied	~5 nm to 100 nm crystalline particles (+ hexagonal diamond)	[39, 40, 54, 55]
	Shockwave synthesis*	Not studied	20–100 nm polycrystalline particles	[38, 56]
	Ultrasonic cavitation	Not studied	6–9 μm cubic crystalline particles	[41]

*There are strictly two distinct types of shockwave synthesis: (1) diamond formation from explosive precursors (DND, which has been listed separately as the detonation method due to its relative prominence as compared to the latter two techniques) and (2) explosive shockwave compression of graphite.

1.2.1.1 Detonation synthesis

Detonation synthesis is an ideal method for bottom-up synthesis of nanoscale particulate diamond at large scale; however, it is unfortunately not currently suitable for the production of luminescent diamond hosting optical centers. For this reason, we only briefly overview the technique here and refer the reader to more comprehensive literature on the subject [53, 60, 61].

There are two variants of the method: (1) detonation synthesis from explosives and (2) detonation shockwave compression of graphite to form diamond. Detonation synthesis mostly commonly refers the first method, where diamond forms directly from explosive materials during detonation. Typically, explosives such as

TNT (2,4,6-trinitrotoluene) and RDX (**R**oyal **D**emolition e**X**plosive or cyclonite/hexogen) are mixed such that the carbon content exceeds the oxygen content. Such a mixture is known as a 'negative oxygen balanced' composition. The explosives are detonated in sealed chambers with cooling provided by air, ice, or water. In fractions of microseconds after the detonation, pressures and temperatures are such that the carbon contained in the explosives liquifies (25–30 GPa, ~4000 K) (Fig. 1.2d). The expanding detonation shockwave causes a rapid drop in pressure and temperature, and the liquid carbon crystallizes into 3–5 nm diamond particles (Fig. 1.2c). The process eventually crosses into the regime where graphite is preferentially stabilized (Fig. 1.2d), leading to the growth of a graphitic network between the primary crystalline cores. The resulting detonation soot consists of 200–300 nm aggregates of the 3–5 nm primary particles, along with graphite and metal impurities that require additional purification procedures. Subsequent oxidation, purification, and disaggregation methods have been developed to isolate the 3–5 nm primary particles post synthesis [53, 62, 63]. Nanodiamonds produced via this method are referred to as detonation nanodiamonds (DNDs). The total nitrogen content in DND particles is on the order of 1 at % (10,000 ppm) [64], which is considerably higher than even most naturally occurring diamond. Moreover, nitrogen is distributed non-uniformly, mostly in clusters, along grain boundaries and as aggregates within the diamond lattice [65]. The high nitrogen content and non-uniform distribution are unsuitable in forming NV centers. However, we note that NV centers and their associated fluorescence can be detected in fractions of DND particles [65–67].

In the second approach, graphite is exposed to a shockwave formed from an explosion [56]. The resulting detonation and shockwave provide the necessary temperatures and pressures to facilitate graphite to diamond conversion. This approach allows for a potentially greater control of the composition of the formed diamonds, since high purity graphite can be used as a precursor. Moreover, the only source of nitrogen is from the explosives or the graphite precursor, thus a lower total nitrogen content is possible. Indeed, this approach demonstrated a reduction in nitrogen (<0.5 at %/<5000 ppm) with correspondingly higher NV^- spin quality

(T_2 times approaching 2 μs for graphite shockwave synthesis versus ~0.1 μs for traditional detonation synthesis) [38].

For interested readers, a fascinating account of the shockwave technique was provided by De Carli [68]—who was part of the team that discovered the technique in the 1960s. Earlier recorded work in diamond synthesis dates back as far 1887. Interestingly, some of these earlier experiments employed an apparatus affixed to rifle bores in order to apply shock pressure to carbonaceous materials [69].

1.2.1.2 High pressure high temperature (HPHT) synthesis

HPHT synthesis attempts to mimic the growth conditions of natural diamonds formed deep within the Earth's mantle. Diamonds produced by this method are predominantly utilized in industrial cutting, grinding, and drilling applications. HPHT diamonds are synthesized under high pressure (~5–8 GPa) and temperature (1300–1600 °C) inside of an anvil cell in the presence of a metal alloy catalyst (e.g., Fe-Ni, Fe-Ni-Co) (Fig. 1.2b). Carbon is transported from a high purity graphite source through the metal melt, which has a precisely controlled temperature gradient, to a seed diamond crystal. HPHT synthesis from hydrocarbon precursors (e.g., adamantane [70]) is also possible [71–73]. A potential advantage of hydrocarbon precursors is that they can allow for bottom-up synthesis of nanoscale diamonds with controlled amounts of dopants—originating from the molecular precursors—to generate controlled numbers of optical centers in resulting particles.

The HPHT method yields diamond with nitrogen content of the order of 0.01–0.03 at % (100–300 ppm), with nitrogen primarily distributed in a substitutional arrangement. This nitrogen originates primarily from trapped air contamination during synthesis and is responsible for the yellow coloration exhibited by type Ib diamond. Enhancement of the nitrogen content to upwards of 1000 ppm is possible [74, 75], though undesirable if the end goal is the production NV centers [36]. It is also possible to grow essentially nitrogen free bulk crystals [76] using temperature gradient growth methods [77] with careful addition of nitrogen getters such as Zr, Ti, or Al, leading to colorless diamonds if the nitrogen concentration is below approximately 10^{17} atoms/cm^3 (<0.5–1 ppm) [78, 79]. Precise details of these syntheses processes are closely guarded industrial secrets. Higher nitrogen content (~0.1%) HPHT synthesized diamonds can have a characteristic green to black coloration due to broadening of the nitrogen donor absorption band [36, 74, 80]. The concentration and arrangement of nitrogen in HPHT type Ib diamonds are relatively close to the requirements needed for the production of luminescent diamond particles; however, it would still be desirable to have the nitrogen concentration less than 100 ppm, possibly in the range of 20–50 ppm. In addition to the presence of getter elements previously noted, nitrogen content and incorporation during HPHT synthesis strongly depends on temperature [77, 81], catalysts (especially Ni) [74], growth sector [82, 83], and distance from the seed crystal [81]. Nitrogen preferentially incorporates in {111} growth sectors in HPHT synthesized diamond [82]. For more details on the incorporation of nitrogen in diamond, readers are referred to an excellent recent review by Ashfold and co-workers [58]. In addition to nitrogen, recent studies demonstrate the incorporation of nickel [45], silicon [45], germanium [84], tin [85], and boron [71] during the growth process. An interesting possibility of transmutation of ^{10}B to ^{7}Li defects upon irradiation of diamonds with a thermal neutron flux has also been described [86]. Isotopic enrichment (e.g., ^{12}C, ^{13}C) is possible with a careful selection of precursors, and is of particular importance for quantum applications [87].

The HPHT method typically produces micro- to millimeter scale diamond. HPHT particles below sizes of ~40 μm are typically formed by fragmentation of larger HPHT particles [44]. However, recent reports on using hydrocarbon precursors demonstrated synthesis of nanometer scale particles (<50 nm), though these approaches have not been scaled up beyond research quantities (10s–100s milligrams) thus far [46, 71, 73, 88, 89]. As illustrated in Fig. 1.2d, the growth conditions of HPHT diamond are relatively extreme. As a result, a large energy footprint and associated cost are needed for HPHT synthesis, and this has relegated the technique to primarily large industrial producers (*Hyperion Materials & Technology, Element Six, Sumitomo*), with understandably limited disclosure of information. The primary customer base for these industrial producers has modest requirements on optical properties given the end use in cutting, drilling, etc. Nevertheless, industrial HPHT producers have started to recognize the potential market for diamond with well-controlled optical defect concentrations and quality [36, 45].

A key disadvantage of HPHT diamond is the requirement of metallic catalysts that inevitably become incorporated into the diamond lattice during growth. These metallic inclusions can contribute magnetic noise and lattice distortion due to their size, and it is not possible to remove metallic impurities embedded within the lattice during post synthesis processing. Thus, HPHT produced diamond likely cannot be utilized in extremely demanding quantum applications (e.g., quantum computing and quantum information processing), nevertheless, it still has promise in many sensing applications, provided steps are taken to improve particle-to-particle uniformity, reduce overall nitrogen concentration, and minimize metallic inclusions as much as possible.

1.2.1.3 Chemical vapor deposition (CVD) synthesized diamond

CVD diamond is produced via plasma assisted growth on a heated substrate at either atmospheric or, more commonly, reduced pressure (900–1200°C [90]) using a gas phase carbon precursor in hydrogen (typically CH_4 diluted few mol% in H_2) [58]. The gas mixture is ionized by either a microwave source, hot filament, or arc discharge, and subsequent growth is performed on a substrate material of choice (e.g., silicon, diamond). The crystal habit of the seeding substrate allows for single crystal growth (commonly, (111) and (100) orientations). CVD growth (Fig. 1.2a) is a more accessible technique for synthesis due to its comparably less extreme growth temperature and pressure requirements (Fig. 1.2d) and a smaller infrastructural footprint compared with HPHT synthesis. Several of numerous reviews on the subject of CVD diamond synthesis are provided [50, 51, 91].

The CVD technique can yield high quality, millimeter scale, single crystal diamond [92] (*electronic grade, valuable for quantum optical applications*), gemstone diamond [93], and micro- and nanocrystalline diamond films [94]. Thin film diamond is commonly utilized for coatings (e.g., tungsten carbide cobalt (WC-Co) tooling surfaces [95]). A recent report demonstrated direct CVD synthesis of particulate nanodiamonds [52]; however, in general, the CVD technique is not currently suitable for large scale production (~0.1–1 kg scale) of particulate diamond. Therefore, CVD can be considered primarily a bulk diamond synthesis technique.

High quality CVD grown diamond is ideal for hosting optical centers for sensing or quantum computing applications. This is because carefully grown CVD diamond can facilitate spontaneous preferential [111] alignment of NV centers during synthesis [96–101], leading to greater photon collection and signal-to-noise (SNR) due to uniformly oriented emission dipoles. No metallic catalyst is needed, which eliminates the possibility of additional unwanted magnetic interactions between metal inclusions and NV centers or other optical centers. The nitrogen content in CVD diamond can be precisely controlled by modification of the precursor gas (e.g., addition of N_2, N_2O [102], NH_3 [103]). While the presence of dilute amounts of nitrogen can accelerate the diamond growth [104], excessive nitrogen content in the gas precursor often can lead to poor quality films or even graphitic deposits [58, 105]. Delta doping of nitrogen during growth can form layers of nitrogen at specific depths beneath the surface [106, 107]. Controlled delta doping of nitrogen achieved NV centers formed at depths ranging from 5 nm to 100 nm beneath the surface, with respective spin coherence times (T_2) of >100 μs at 5 nm depth and >600 μs at depths greater than 50 nm [106]. To conclude, the CVD method enables production of high quality of diamond. As the result, many of the exceptional properties of luminescent centers such as the NV and silicon-vacancy (SiV) centers are exploited using CVD synthesized diamond.

1.2.2 Forming and Engineering Luminescent Optical Centers in Diamond

Depending on the method of synthesis, diamond may exhibit in situ luminescence to a greater or lesser degree. Type Ib synthetic diamond may contain a small amount of NV centers incorporated during synthesis as well as other optical centers arising from atomic inclusions such as Ni and Si [45, 108]. CVD diamond can exhibit fluorescence from SiV optical centers also incorporated during synthesis [109, 110]. Although optical centers may form during synthesis, additional processing steps are usually necessary to raise the optical center concentration for practical use. This is usually achieved by one of two means:

(1) Irradiation with high energy particles or ions to form vacancies
(2) Ion implantation of heteroatoms that form optical centers

Subsequent thermal annealing is used to promote diffusion and formation of optical center complexes. Annealing is discussed separately. Approach (1) is widely used, especially for preparation of fluorescent particulate diamond. This method requires dopant heteroatoms to already be present within the lattice—that is, they are incorporated during synthesis. If NV center formation is the desired goal, this dopant atom is nitrogen; however, other heteroatoms such as Ni, Si, Ge, Sn can also be incorporated and used to form optical centers. Approach (2) is suitable for high purity CVD synthesized diamond (type IIa, <1 ppm substitutional nitrogen). We discuss each approach below.

1.2.2.1 Irradiation with high energy particles or ions to form vacancies

Explorations into the effects of irradiation on the optical properties of diamond has been a topic of interest since at least the early 1950s [111, 112], followed by commercial irradiation soon thereafter [113]. Early studies were primarily concerned with bulk natural diamonds within the realms of geology and gemology. With the prospect of quantum computing rising in the early 1980s and the subsequent 'nano-revolution,' a modern resurgence in the study of irradiation effects on the optical properties of diamond followed.

Diamond is traditionally considered a radiation hard material due to the high displacement energy of carbon atoms within the diamond lattice (≈40 eV/atom depending on crystallographic orientation [114]). For comparison, silicon in cubic silicon carbide exhibits a displacement energy of 25 eV (along the [100] orientation) [115]. Irradiation sources such as electron beams (e-beams), proton (H^+) beams, helium ion (He^+) beams, and fast neutrons have been explored. Examples of each of these approaches as reported in the literature are listed in Table 1.2.

Table 1.2 Typical irradiation and annealing conditions reported for various types of irradiation used for formation of optical centers in diamond

Irradiation source	Fluence (cm^{-2})	Energy	Annealing temperature/time	Diamond type	Reference
	1×10^{13}, 1×10^{14}, 5×10^{14}	2 MeV	900°C/1 h	Type Ib 100 μm thick plates	[119]
	3×10^{17}	1.5 MeV	850°C/2 h	Type Ib, 50 nm particles	[120]
	1×10^{17}	1.8 MeV	800°C/2 h	Type Ib, 50 nm particles	[121]
	$0.5\text{-}5\times10^{19}$	3 MeV	850°C/2 h	Type Ib, 15–25 μm particles	[122]
	1×10^{19}	10 MeV	850°C/2 h + 1900°C/5 min	Typa Ib, 150 μm particles	[123]
	5×10^{18}, 2×10^{19}	3 MeV	500–1880°C/5 min	Typa Ib, 40 μm particles	
Electrons	3×10^{18}	1 MeV			
	5×10^{19}	10 MeV	800°C/2 h	Type Ib, 150–190 μm particles	[47]
	3.6×10^{16}–7.2×10^{19}	20 MeV	900°C/1 h	Nitrogen doped CVD films and particles	[124]
		1.5 MeV			
		1.2 MeV			
		800 keV			
		500 keV			

(*Continued*)

Table 1.2 (*Continued*)

Irradiation source	Fluence (cm^{-2})	Energy	Annealing temperature/time	Diamond type	Reference
He^+	1×10^{13}	40 keV	800°C/2 h	Type Ib particles, ~35 nm	[49]
	1×10^{10}–1×10^{13}	30 keV	900°C/1 h	Type Ib and IIa plates	[125]
H^+	5×10^{15}	3 MeV	800°C/2 h	Type Ib particles (~100 nm)	[126]
	1×10^{16}	3 MeV	800°C/2 h	Type Ib (100) and (111) single crystals	[127]
	1×10^{12}–1×10^{17}	2.4 MeV	600–1000°C/1–20 h	Type Ib diamond plates	[128]
Neutrons	2.2×10^{16}	Not Reported	900°C/1 h	Nitrogen doped CVD films and particles	[124]
	2.2×10^{16}–7×10^{17}	~2.5 MeV peak (0.1–7 MeV range)	1000–1700°C/20 h (at 6 GPa)	Type Ib and type IIa diamond	[129]
	0.7×10^{16}–2.8×10^{18}		900°C/1 h		[130]
	1.1×10^{16}–2.8×10^{18}		1000–1700°C/6 h (at 6 GPa)	Type Ib diamond plates	[131]
	-[1]	-	900°C/1 h	Type Ib particles (~50–150 nm)	[132]

[1]Irradiation achieved via neutron activated decay of barium oxide to Li^{7+} and α particles.

While increasing the vacancy concentration is necessary to form optical centers, a naïve assumption would be that maximum vacancy production should be targeted for high densities of optical centers. Excessive irradiation comes at a cost of induced lattice damage and amorphization that can significantly degrade the optical properties (*emission intensity, charge state* [116], *spectral linewidths, fluorescence lifetime*) and electronic spin properties (*spin coherence and relaxation times*) of the luminescent centers [117]. Therefore, optimization of irradiation protocols is necessary to minimize induced damage while maximizing the properties of hosted optical centers. This is a complex multi-parameter space. Factors include: the absorption cross section and vacancy formation rate (*depending on the irradiation source used*); exposure time and dose/fluence (cm^{-2}); beam profile, energy, and current; sample orientation (*if bulk diamond plates are used*) and thickness (*bulk thickness of plates, or powder layer thickness*); temperature (*induced heating from irradiation or intentional heating during irradiation* [118]); and cost. As a result, it is difficult to broadly classify one method of irradiation as significantly better or worse than another, as each has potential advantages and disadvantages. Commercial scale irradiation is expensive, and optimization studies are consequently costly and time consuming. For this reason, commercial vendors of fluorescent diamond optimize conditions for a single source of irradiation. The authors of this chapter have experience primarily with e-beam irradiation, and all data presented herein is from e-beam irradiated diamond. However, insights from e-beam irradiation can be applied to other radiation sources. We therefore attempt to summarize as best as possible the considerations that go into irradiation of diamond with common sources while avoiding broad generalizations.

Electron beam irradiation

Electron beam (e-beam) irradiation is a widely used approach for introducing vacancies in diamond particles. The smaller mass of electrons as compared with neutrons and ions induces less damage for the same energy and fluence. The theoretical minimum electron energy capable of inducing damage is 0.2 MeV [133]. Campbell and Mainwood calculated vacancy production in diamond as a function of depth for e-beam energies of 0.25–10 MeV and arrived at range of 0.74 (vacancies/electron)/cm to 3.42 (vacancies/electron)/cm [134]. The authors of that study note however that their calculated values of vacancy production tend to overestimate actual values. Hunt experimentally determined a production rate of 1.53 (vacancies/electron)/cm with 2 MeV electrons [135]. These production rates are considerably lower than those determined for ion beams (*discussed below*). Due to the small size of the electrons, they have a lower absorption cross section than ions or neutrons. As a result, they can penetrate to greater depths within a sample. 1 MeV electrons can penetrate to a depth of ~1.5 mm, while 2 MeV electrons to a depth of ~3 mm [135]. In practice, the majority of diamond samples are well below 3 mm in thickness; therefore, e-beam energies less than approximately 5 MeV are sufficient in many cases. E-beam energies ranging from 0.5 MeV–3 MeV are common in practice (Table 1.2).

A disadvantage of e-beam irradiation is the high cost due to the necessity to use commercial electron irradiation facilities that are capable of producing beams with a large cross section. Particularly for commercial scale production of luminescent diamond, treatment of batches less than ~100 g of material is not economically feasible as the costs of electron irradiation can easily reach $10k per batch. Additionally, based on the relatively high penetration depths of electrons noted above, it may be challenging to control the depth of NV centers formed in bulk diamond samples unless very low e-beam energies are used (e.g., 200 keV beams common in electron microscopes) or if a nitrogen rich layer is doped into the diamond [136].

The effects of e-beam irradiation and damage manifest in several ways and can be assessed both visually and spectroscopically. As irradiation fluences increase, the color of the diamond particles progresses from an initial yellow color (*if type Ib diamond is used*) to light green, and eventually to a dark *forest green* as a result of increased vacancy concentration (GR1 center [137]) (Fig. 1.3). It should be noted that these color changes are reflective of observations made with ~10–100 μm particles and bulk diamond. Color changes are less pronounced in particle sizes <10 μm, where light scattering leads to mostly a white appearance. After annealing at 800–900 °C for 1–2 h, particles irradiated to higher fluences can exhibit a pale pink to slightly orange hue compared to a deeper purple/violet coloration at lower irradiation fluences [122].

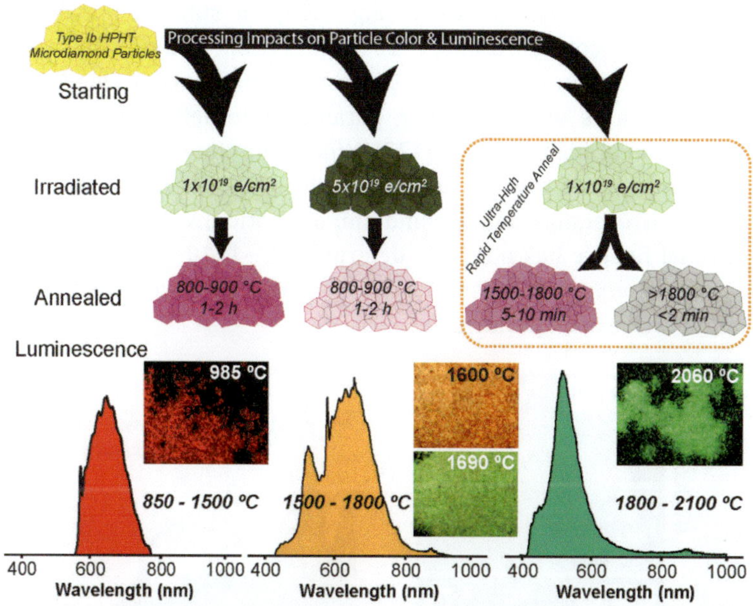

Figure 1.3 Changing visible appearance and fluorescence of diamond particles following e-beam irradiation at lower and higher fluences followed by traditional annealing (800–900 °C, 1–2 h) as well as rapid, ultra-high temperature annealing at temperatures exceeding 1500 °C. Note that the color can only be observed for particles approximately 10 μm and larger in diameter or bulk diamond. Typical luminescence spectra and examples of the observed fluorescence (insets) for 20 μm particles irradiated to a fluence of 1.5×10^{19} e/cm^2 and annealed at specified temperature ranges (850–1500 °C, 1500–1800 °C, and 1800–2000 °C) under 365 nm UV excitation are shown.

The most striking changes induced by irradiation are observed in the luminescence spectra. Irradiation induced TR12 and 3H defects in diamond (Table 1.3, Fig. 1.4b) impart characteristic green luminescence (Fig. 1.4a), while the GR1 (Neutral vacancy, V^0) center leads to a 740 nm peak. Most irradiation induced luminescent defects are removed by annealing at relatively low temperatures (500–700 °C) (see supplementary information in Ref. [138]). However, even with annealing, fluorescence characteristics of the electron irradiated diamond can vary significantly with fluence at a given electron energy. Irradiation induced damage leads to an increase in NV^0 concentration, giving a distinctive orange luminescence instead of the red luminescence associated with the NV^- center (Fig. 1.4c,d). High irradiation fluence eventually degrades overall luminescent intensity (Fig. 1.4e), in spite of apparently higher NV^- concentrations ~10 ppm [122]. Additionally, spin-lattice relaxation times (T_1) also decrease presumably due to the large amounts of surrounding paramagnetic defects induced by irradiation at high fluences [122]. While subsequent annealing can remove some of these defects, excessive irradiation fluences may lead to irreversible damage and amporhization. While a definitive optimal fluence has not been determined, ~1×10^{18}–2×10^{19} e/cm^2 (*with electron irradiation*) tends to be the range at which most studies have been performed.

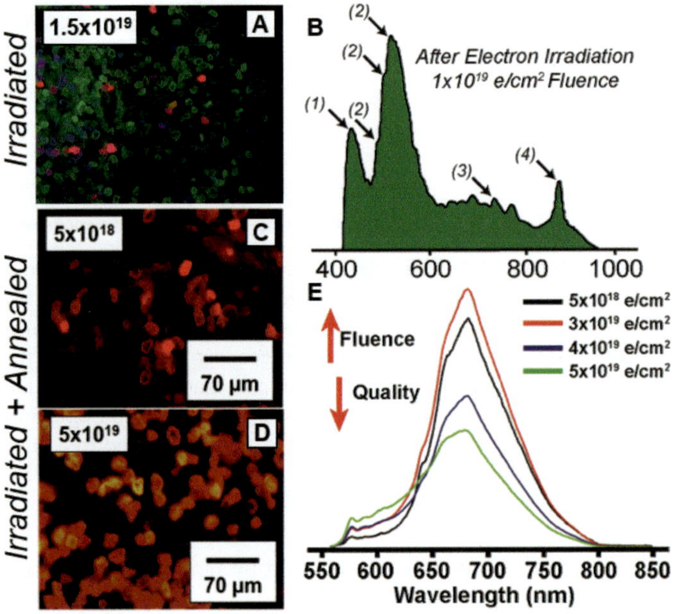

Figure 1.4 Type Ib 20 µm particles following irradiation and annealing. (a) Luminescence under UV excitation following electron irradiation to a fluence of 1.5×10^{19} e/cm^2. (b) The corresponding luminescent spectra of irradiated diamond particles under 365 nm UV excitation. Characteristic irradiation induced defects as well as some additional metal related defect centers are highlighted: (1) TR12 center (420–460 nm), (2) 3H center (triplet peaks at 488, 503, and 516 nm), (3) GR1 center (740 nm), and (4) a nickel related center (879 nm). (c–d) Luminescence under green excitation following fluences of 5×10^{18} and 5×10^{19} e/cm^2 with annealing at 900 °C. (e) Fluence dependent luminescence intensity of NV centers in particles under green excitation. Increasing the irradiation fluence tends to decrease the eventual quality of formed NV centers, both in terms of luminescence intensity and spin properties.

Table 1.3 Some characteristics of common irradiation induced defect centers that exhibit luminescence in type Ib diamond

Defect center designation	Description	Spectral characteristics
TR12	Two vacancies and two interstitial carbons in a hexagonal site	Broad peak centered around 466 nm, a source of green luminescence in as-irradiated diamond
3H	Intrinsic interstitial defect	Triplet peaks at 488, 503, and 518 nm, a source of green luminescence in as-irradiated diamond
GR1	Neutral vacancy (responsible for green coloration of irradiated diamond)	740 nm peak

Proton beam (H^+) irradiation

Protons have a greater vacancy production rate than electrons due to larger mass and collision cross section. The threshold energy for radiation damage is less than 1 keV for protons. 3 MeV protons are calculated to have a penetration depth of 0.05 mm and vacancy generation rate of 13 vacancies/H^+ [126, 127]. The higher vacancy formation efficiency decreases the overall fluence needed compared to electrons. Typical H^+ doses $<10^{17}$ cm^{-2} are sufficient, and this reduces irradiation time and cost. Additionally, because of the lower accelerating energies needed, proton irradiation is a potentially more accessible technique for laboratory scale irradiation [127]. However, the damage induced from proton (*as well as other ions*) can be significant compared to electrons. Doses exceeding 10^{17} cm^{-2} can irreversibly graphitize the diamond [139].

Helium ion (He^+) irradiation

Helium ions induce higher damage than either protons or electrons. A 40 keV He^+ ion has a penetration depth of ~200 nm, and vacancy formation rate of 40 vacancies/He^+ [49]. A 40 keV He^+ beam with fluence on the order of 10^{13} cm^{-2} was used for production of 35 nm fluorescent nanodiamonds (FNDs); however, the brightness of these particles was ~30% less than the same particles irradiated with a 3 MeV proton beam [49]. This decreased fluorescence may be an indication of excessive lattice damage in a similar vein as to what is observed with high fluence e-beam irradiation (Fig. 1.4).

Neutron irradiation

Neutron irradiation of diamond is performed inside research nuclear reactors, where average neutron energies can be ~2–2.5 MeV (*so-called fast neutrons*). The minimum neutron energy necessary to displace carbon in the diamond lattice is ~100 eV [140]. Although is it possible to use neutrons to fabricate NV^- containing diamond [140], heavy neutron irradiation can cause the enhancement of the less desirable NV^0 center in type Ib diamond [130]. Neutrons can also induce radioactivity to the diamond if metallic impurities are present. This is particularly an issue for HPHT synthesized diamond, which can have metallic impurities (e.g., Ni, Cr, Fe) that incorporate into the lattice during synthesis. Upon an irradiation these incorporated impurities can form unstable isotopes such as ^{59}Ni, ^{51}Cr, and ^{59}Fe, among others, which have respective half-lives of 7.6×10^4 years, 27.7 days, and 44.5 days. Radioactive materials typically cannot be released to improperly equipped facilities, thus requiring a waiting period before the diamond can be handled. 2.5 MeV neutrons can also induce the formation of nitrogen B-aggregates (large, even numbered aggregates) at fluences of 7×10^{17} cm^{-2} with 1500°C annealing [131]. This aggregated form of nitrogen is not suitable for NV formation.

The greatest advantage of neutron irradiation is cost. Provided one has access to a research nuclear reactor, the cost of irradiation is significantly lower since the neutrons are generated as a 'free' byproduct, whereas operation of electron beam or accelerators requires significantly greater effort and energy. A promising recent approach demonstrated the use of alpha particle irradiation induced from the neutron activated decay of a boron oxide matrix (^{10}B) into ^7Li and alpha particles for fabrication of NV diamond [132]. The process was demonstrated at ~100 g treatment scale, and led to a greater fraction of luminescent particles at much shorter irradiation times as compared to proton and electron irradiation (3 min of fast neutron exposure vs. 4.5 h in an electron accelerator). The authors of this study [132] reported that the ZPL and ZFS linewidths did not statistically differ from e-beam irradiated diamond; moreover, the T_1 relaxation times were also reported to be similar to e-beam irradiated diamond.

1.2.2.2 Ion implantation of heteroatoms to form optical centers

If nitrogen is not doped into the diamond lattice during synthesis, it may be introduced using ion implantation post synthesis. Ion implantation is accomplished by exposing the diamond substrate to an accelerated beam of ions formed by ionization of a feed gas. A key advantage of ion implantation is the variety of heteroatoms that can be implanted to the lattice. For example, several authors reported diamond implanted with Ge [141] , Sn [142], O, Mg, Ca, F, and P [143] as well as molecular species such as NO^+ [144] and CN [145]. The penetration depth of the ion can be controlled by changing the acceleration voltage, and, depending on the ion and energy, implantation depths can range from several nanometers to microns.

Ion implantation allows for nitrogen to be introduced to diamond with precise spatial and depth control [136, 146–148], including pattering [149]. Subsequent annealing of the implanted samples leads to the formation of NV centers. Ion implantation is not without limitation. Similar to the issues faced with H^+ and He^+ irradiation to form vacancies, ions can induce significant damage due to their size and this can compromise the resulting properties of optical centers. For example, NV centers formed from ^{15}N implanted to CVD diamond showed broadened optical transitions, higher strain, and poorer optical coherence as compared to NV centers formed from native ^{14}N introduced to the diamond during synthesis [150]. The conversion efficiency of implanted nitrogen and the resulting NV centers is strongly dependent on the implantation energy used, ranging from 1–50% for implantation energies 1.5 keV–18 MeV [151]. This is problematic because the ultimate goal would be to achieve a large number single NV centers within a few nanometers of the surface to enable greater sensing capability of the material. Yet, the low conversion efficiency is a roadblock toward these efforts. The use of higher implantation energies can increase the yield of NV centers; however, 'ion-straggling' poses a challenge to precise placement of the ion within the lattice [28].

1.2.2.3 In situ doping and optical center formation

Traditional HPHT synthesis inevitably results in doping with nitrogen originating from atmospheric contamination. The doping level can be varied as discussed in Section 1.2.1.2. However, nitrogen is not the only potential dopant that can be introduced in situ during HPHT synthesis. Microparticles containing Ni and Si related optical centers were recently produced by a commercial scale vendor [45]. HPHT synthesis utilizing hydrocarbon precursors was also demonstrated at a smaller lab scale for the introduction of lanthanides (Eu and Tm) [88] as well as tin (Sn) [85] and germanium (Ge) [84] for production of optical centers in nano and microparticles (though the study in Ref. [88] did not observe incorporation of Eu or Tm optical optical centers upon inspection via cathodoluminescence). Boron can also be introduced for the production of conductive diamond particles for use in microelectronic applications [71], particularly for the growth of low resistivity interfaces [152]. The use of a laser heated diamond anvil cell was also demonstrated to controllably generate NV centers in situ [153]. If color centers could be introduced controllably and reliably during "bottom-up" synthesis of diamond particles, it would be a significant breakthrough since this would essentially eliminate the need for both irradiation and fragmentation. However, these approaches of diamond particles synthesis are currently at early stages. As discussed in Section 1.2.1.3, both high purity and specifically doped CVD diamonds can be produced by controlling the composition of the growth atmosphere, and delta-doping of nitrogen can lead to planar gradients of nitrogen doping throughout the diamond crystal. For quick reference, common low index diamond crystallographic planes are illustrated in Fig. 1.5.

1.2.2.4 Annealing

Typical annealing treatments of diamond materials following irradiation or ion implantation range from 850 to 1000 °C for 1–2 h under inert gas (e.g., N_2) or vacuum to minimize high temperature oxidation of the diamond. This temperature range is sufficient for vacancy diffusion; however, substitutional heteroatoms (e.g., nitrogen) cannot diffuse at these temperatures. Substitutional nitrogen diffusion occurs at temperatures exceeding approximately 1500 °C. The temperature dependent diffusion coefficient of both vacancies and substitutional nitrogen can be roughly modeled as an energy activated process (Eq. 1.1), and the distance can be approximated with Eq. 1.2; where D_0 is the pre-exponential term (0.4 cm^2/s) [154]; E is the energetic barrier (e.g., $E \approx 5$ eV for nitrogen [155] and ≈2.3 eV [156] for vacancy self-diffusion), and k and T are the Boltzmann constant and thermodynamic temperature, respectively.

$$D = D_0 e^{-\frac{E}{kT}} \quad (1.1)$$

$$d = \sqrt{Dt} \quad (1.2)$$

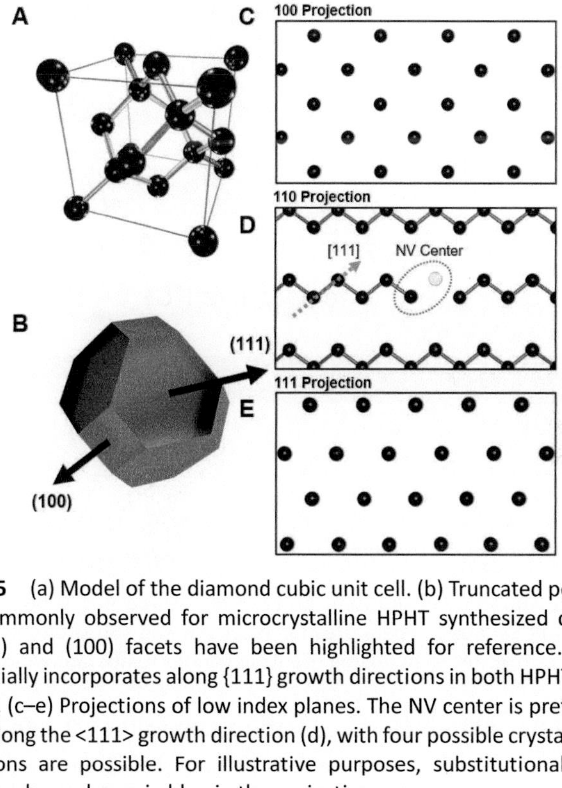

Figure 1.5 (a) Model of the diamond cubic unit cell. (b) Truncated polyhedron shape commonly observed for microcrystalline HPHT synthesized diamonds. The (111) and (100) facets have been highlighted for reference. Nitrogen preferentially incorporates along {111} growth directions in both HPHT and CVD diamond. (c–e) Projections of low index planes. The NV center is preferentially aligned along the <111> growth direction (d), with four possible crystallographic orientations are possible. For illustrative purposes, substitutional nitrogen atoms have been shown in blue in the projections.

Figure 1.6a shows temperature dependence of calculated diffusion distances per second in units of the C-C bond length (*i.e.*, $d(t = 1\text{s})/d_{\text{C-C}}$, where $d_{\text{C-C}} = 0.157$ nm) for V^0 vacancies and substitutional N_s atoms. Some characteristic features of annealing temperature ranges are illustrated in Fig. 1.6b. It should be noted that while instructive, this is a simplified model because the presence of vacancies, impurity metal atoms, and the aggregation state of nitrogen in the lattice can impact energetic barriers for diffusion.

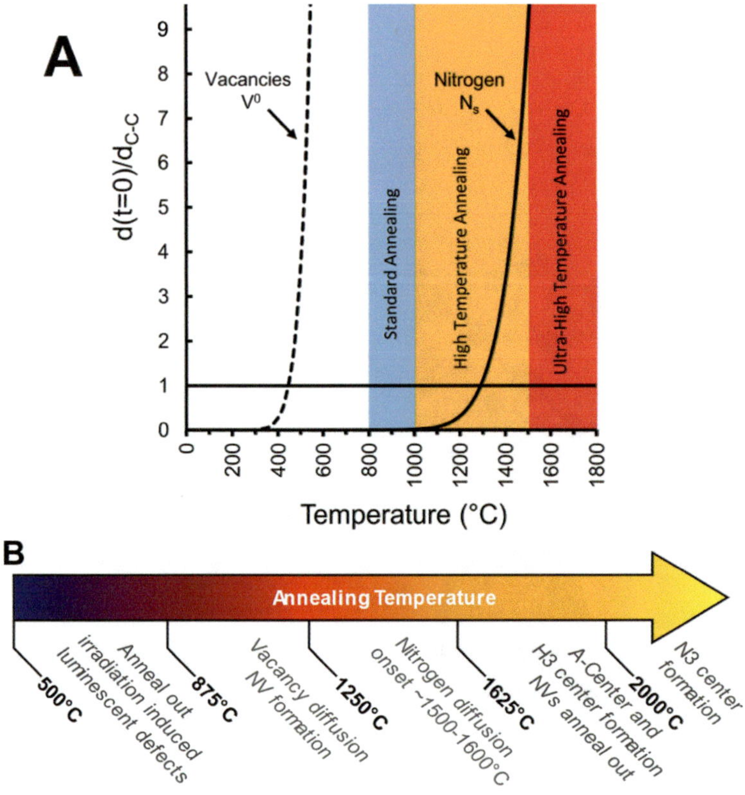

Figure 1.6 (a) Plots of diffusion distance (measured in C-C bonds lengths, 0.157 nm) at various temperatures for vacancies and substitutional nitrogen. (b) Characteristic diffusion related events at various temperature ranges.

Recently, ultra-high temperature annealing at atmospheric pressure under either vacuum or hydrogen (Fig. 1.6a) was shown to produce green and blue fluorescent diamond particles from e-beam irradiated type Ib diamond [138, 157]. Previously, only diamond of natural origin (*which exhibits appreciably higher nitrogen content, primarily distributed as A-centers*) was demonstrated for the production of green fluorescent nanodiamond particles [158]. Annealing temperatures in excess of 1500 °C allows for the formation of various other nitrogen containing defects, such as the H3 (NVN) center, as well as clusters of nitrogen in type Ib diamond [159]. Formation of latter defects leads to dramatic changes in fluorescence spectra (Fig. 1.3). This ultra-high temperature annealing range allows for the production of particles exhibiting yellow-orange fluorescence resulting from combinations of NV and H3 centers within the same particles. The ability to generate a wide range of luminescent color centers with the same material is an exciting prospect for future studies. Finally, high temperature annealing extends the NV^- spin coherence and relaxation times likely as a result of the reduction of paramagnetic species induced from irradiation damage [31, 32].

1.2.2.5 Fragmentation and size fractionation for particulate luminescent diamond

Typically, production of nanoscale fluorescent diamond particles involves fragmentation of larger particles/crystals by ball milling followed by purification and particle separation steps [44, 116]. After milling, a number of washing steps (e.g., acid purification, oxidation) is required to remove contaminations introduced from milling media (typically, metals) as well as amorphous carbon formed during milling [160, 161]. Following purification, centrifugation can be used to separate fractions of the particles with sizes from approximately 10 nm and larger. If smaller particles are desired, the larger particles could be collected and milled again; however, the effects of milling are deleterious to the quality of the final particles due to mechanical crystal damage and surface contamination. Additionally, ball milling, purification, and fractionation steps lead to losses of material, which, based on the authors' experience, could be 10–20% of the starting mass or even higher for unoptimized procedures. Another disadvantage of ball milling is that only a relatively small fraction of the initial diamond would have the desired particle size. Nevertheless, being cost-effective and lacking suitable alternatives, ball milling remains the most widely utilized procedure for fabrication of fluorescent nanodiamonds in both laboratory and industrial settings.

A new approach of fragmentation of microns-sized HPHT particles by detonation was recently demonstrated [162]. The main advantage of this method is in introducing minimal concentrations of impurities vs. a typical milling procedures. Such a fragmentation method could be applied to CVD-sourced diamond with precisely controlled nitrogen content. Currently, the cost of such high-quality CVD diamond is high and conventional milling could negate the advantages offered by such materials by introducing additional impurities and defects. The latter appear to be the main reasons why up to this date only a small scale demonstration of CVD synthesized particulate diamond with color centers was documented in the literature [52]. Note that such particles still required a subsequent irradiation [52].

1.2.3 Characterizing Optical Centers with Magnetic Resonance Techniques

Quantum sensing enabled by optical centers in diamond primarily relies on magnetic interactions associated with the optical center and their surrounding environments. These interactions are magnetic in nature and, therefore, both static and alternating local magnetic fields should be considered. For this reason, reliance on insightful techniques of magnetic resonance (MR) to characterize the optical centers following synthesis and processing is necessary [163].

The phenomenon of MR is the resonant absorption of electromagnetic radiation by spins in the presence of an external static magnetic field. These spins may be unpaired electrons (*S*) or non-zero spin atomic nuclei (*I*). Both a deep theoretical background and application of various magnetic resonance techniques to studies of a large variety of physical objects may be found in the provided Refs. [164–168]. These non-destructive techniques facilitate identification of the unique electron-nuclear structure of various substances containing magnetic species, including gases, organic molecules, complex biological molecules, and solid-state materials (as is the case with diamond).

Many optically active defects and impurities in crystalline diamond samples (including the triplet centers like the NV^-) are non-zero spin systems, making electron paramagnetic resonance (EPR) and related MR techniques adequate for providing useful information on the anisotropy of electronic g tensor; zero-field splitting (ZFS) in species with effective $S > \frac{1}{2}$, and hyperfine interactions with magnetic nuclei of surrounding atoms, etc. [169]. Well-developed theoretical tools in combination with state-of-the-art experimental techniques allows one to obtain detailed information about the electronic structure of specific defects, its surroundings, and the ability to evaluate magnetic interactions in such systems.

With the exception of some specific ODMR experiments, traditional MR measurements report on spin ensembles. A typical sample of synthetic diamonds is an ensemble of nanometer to micrometer sized particles characterized by a distribution of individual sizes and properties. As a result, the MR spectra of such powders report on average properties. This averaging decreases spectral resolution, due to additional line broadening and complicates interpretation of the spectra.

In practice, diamond samples demonstrate a large variety of magnetic properties which descend mainly from the synthesis method as well as any subsequent post-processing treatments. For example, each small diamond particle obtained by micronization (milling) of bulk diamonds inherits magnetic features of their ancestors. This means almost the entire set of paramagnetic defects and potential ferromagnetic inclusions contained in the bulk samples before micronization will be transferred into these descendant diamond particles. However, the micronization process creates new carbon inherited magnetic impurity defects within surface layers of diamond particles and introduces magnetic impurities (e.g., Fe) depending on the milling media used.

A clear discrimination between two types of magnetic defects based on their origin is necessary from the outset:

(1) Inherent or induced defects in the *diamond core* or *surface of the diamond particle* as a result of their synthesis origin (e.g., metal impurity atoms from HPHT catalysts, or hydrogen inclusions in the case of CVD diamond), and
(2) Defects not directly induced on or within the diamond, but stemming from various processing techniques (e.g., impurity contamination from milling media).

Non-impurity related magnetism in diamonds originates from three main sources:

(i) Intrinsic paramagnetic defects inherited from source materials during synthesis (e.g., nitrogen vacancies, metallic impurities from catalysts),

(ii) Lattice defects induced by micronization (e.g., dangling bonds along cleavage planes), and

(iii) Intentionally introduced defects to modify the properties of the initial diamond material (e.g., NV^- centers or other irradiation induced defects).

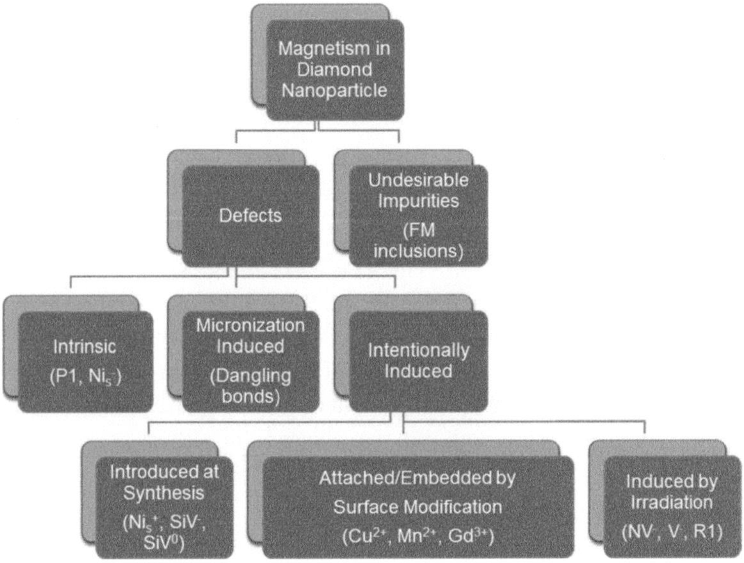

Figure 1.7 Sources of magnetism in diamond.

The latter type (iii) may be sub-classified into: (a) defects attached to the diamond surface (via surface chemistry modification), and (b) defects induced by ionizing irradiation/ion implantation of diamond. Figure 1.7 summarizes this classification scheme of the paramagnetic defects. Generally, lattice defects and paramagnetic impurities are detrimental to optical center luminescence since they will participate in interactions with the optical center electronic spin. Thus, minimizing these unwanted defects is necessary to optimize optical center performance. In this regard, EPR can be used to assess the content of magnetic impurities and track the efficacy of purification procedures [163]. The utility of EPR in assessing optical center quality may be demonstrated with two types of luminescent centers: triplet ($S = 1$) centers and singlet ($S = 1/2$) centers.

1.2.3.1 EPR of luminescent triplet ($S = 1$) centers

Among the variety of optical centers in diamonds, the most studied are color centers having triplet electronic spins with strongly coupled photoluminescent and magnetic properties. These include the negatively charged nitrogen-vacancy (NV^-, denoted as the W15 defect in EPR) and the neutral silicon vacancy (SiV^0, denoted as KUL1 in EPR).

As discussed in Section 1.2.2, NV^- centers can be fabricated in bulk and particulate diamonds containing substitutional nitrogen [58] by high energy irradiation followed by thermal annealing. Irradiation of micrometer-sized diamond crystals by a beam of high-energy particles creates a large number of both neutral and negatively charged vacancies (V^0 and V^-, respectively). In contrast, formation of SiV^0 defects does not require nitrogen or irradiation as they predominantly grow into Si-enriched diamond structures (synthesized by both CVD and HPHT) as complete units, rather than being produced by vacancy migration and trapping at substitutional silicon impurities [45, 170]. EPR spectra of particulate diamonds show intense signals of primary as well as processing induced non-luminescent defects (N_s^0, Ni_s^-, V^- and dangling bonds) grouped around $g = 2.00$ and a less intense groups of satellite lines originating from conventional "allowed" ($\Delta M_s = 1$) transitions between the energy levels of the electronic triplet ground state. In the continuous wave (CW) EPR spectrum, these lines are observed in low- and high-field regions on both sides of the lines due to primary defects (Fig. 1.8a, strong lines around $g = 2.00$ have been omitted for clarity). Splittings between the most notable $\Delta M_s = 1$ triplet lines allow for direct determination of the ZFS parameter of triplet energy levels: 960×10^{-4} cm^{-1} (or 2.87 GHz) for NV^- and 332×10^{-4} cm^{-1} (or 0.94 GHz) for $(SiV)^0$, respectively.

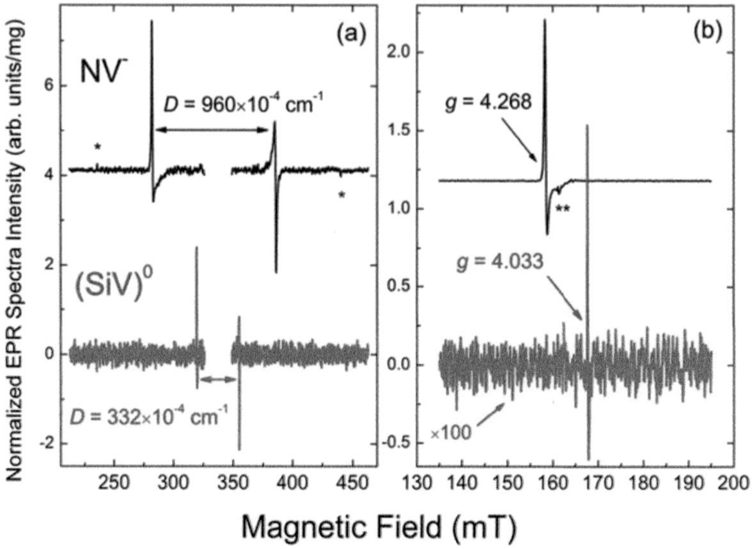

Figure 1.8 RT (T = 295 K) X-band (ν = 9.46 GHz) CW EPR spectra of triplet (S = 1) centers in ensembles of micron-sized diamond particles; black traces – NV^- (W15) centers in e-beam irradiated and annealed Ib type HPHT diamonds, triplet centers content 10.8 ppm, red traces – Si-doped Ib type HPHT diamonds, triplet centers content ~0.06 ppm: (a) x-, y-"allowed" transitions, asterisks indicate weak z-"allowed" transitions separated by 2D, both spectra recorded at partially saturating microwave power P_{MW} = 2 mW. (b) "Forbidden" transitions, NV^- HF spectrum recorded at P_{MW} = 2 mW whereas $(SiV)^0$ one – at P_{MW} = 200 mW and tenfold receiver gain (both resulting in scaling coefficient in the graph), double asterisk indicates "forbidden" line attributed to W16 triplet center. The background/FM impurities signals have been subtracted, intense signals around g = 2.00 are skipped; differences in signal-to-noise ratio originate from the different abundance of triplet centers as well as from different masses of the samples.

However, the most characteristic features of the EPR spectra of polycrystalline diamond samples hosting triplet centers are well observed lines in the half-field region (g = 4.00), attributed to so-called "forbidden" (ΔM_s = 2) transitions—Fig. 1.8b. The observation of these half-field lines reliably indicates the presence of triplet centers. Among them, the X-band g = 4.27[1] line has been

[1]The effective g-value of the "forbidden" line in EPR spectra of polycrystalline samples, containing triplet centers, depends on the actual value of microwave frequency. The real values of the g-factors corresponding to the same "forbidden" lines, being measured at frequencies significantly different from 9.4 GHz, will deviate from the "g = 4.27" value outside the experimental error values.

unambiguously attributed to NV^- centers [171], while the $(SiV)^0$ center is characterized by the half-field line with $g = 4.03–4.04$ [45]. We note that at X-band, high spin rhombic Fe^{3+} would appear at $g = 4.3$; however, it would have a broader line. This is noteworthy, since Fe contamination from the sample or EPR tube could be present. For the same type of paramagnetic centers the spin concentration is proportional to the double integral of the EPR derivative spectrum; therefore, EPR can be used to quantify total spin concentration and hence the optical center concentration. Analysis of triplet related EPR spectra of various diamond samples ascertains the NV^- concentration in particulate HPHT diamonds as a function of irradiation fluence, particle size, and annealing. For example, 150 μm type Ib HPHT particles (black traces in Fig. 1.8), an NV^- concentration of 10.8 ppm was measured. For similarly sized Si-doped HPHT particles (red traces in Fig. 1.8), the SiV^0 concentration does not exceed 0.06 ppm. It is worth mentioning that in most cases, the "allowed" triplet-related lines (Fig. 1.8a) are practically unobservable on a background of much more intense signals originating from primary defects such as substitutional nitrogen (N_S) or other impurities. Therefore, correct EPR-assignment of the remaining spectral features to triplet defects is complicated. Fortunately, the half-field "forbidden" lines remain observable and become the primary EPR signature to indicate the presence of triplet centers in treated polycrystalline diamonds.

For nanodiamonds fabricated by dynamic synthesis techniques [171], changes in the integral intensity of the "forbidden" line strongly correlates with changes in the triplet center content obtained by integration of the entire triplet-related EPR spectrum. The EPR spectrum of the "forbidden" lines ($g = 4$) are a reliable tool for comparative estimations of the triplet centers' content in luminescent nanodiamonds, even in the usual case when "allowed" lines ($g = 2$) are weakly observed. The half-field EPR technique is therefore proposed as a simple and reliable method to identify, attribute, and quantify triplet defects in polycrystalline samples [172].

1.2.3.2 Magnetic resonance of luminescent non-triplet centers

A variety of optically active (luminescent) paramagnetic defects having non-triplet spin multiplicity (e.g., $S = 1/2, 3/2$) are also present in particulate diamond. Representative examples of such defects are the negative silicon vacancy center (SiV^-, denoted as KUL in EPR)

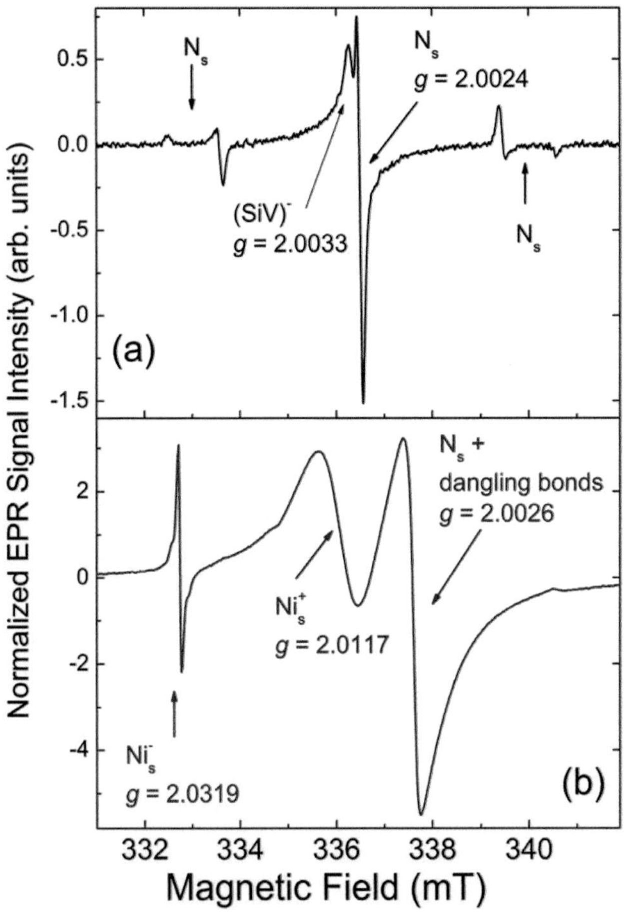

Figure 1.9 X-band CW EPR spectra of polycrystalline samples of micron-sized Ib type HPHT diamonds doped by metals: (a) Si-doped sample, spectrum recorded at $P_{MW} = 2$ μW, RT ($T = 295$ K) and $\nu = 9.430$ GHz; (b) Ni-doped sample, spectrum recorded at $P_{MW} = 2$ mW, $T = 80$ K and 9.463 GHz.

particles [45]. Both centers were fabricated by introducing the corresponding dopants (Si, Ni) during HPHT synthesis [45]. with $S = 1/2$ and the (Ni_S^+, denoted NIRIM-1 in EPR) center with $S = 1/2$. These centers were recently detected in HPHT synthesized Figure 1.9a shows the CW EPR spectrum of the micron-sized Si-doped Ib type HPHT diamond particles recorded around $g = 2.00$ at room temperature at low incident microwave power to avoid saturation effects. The spectrum shows a weak three component signal centered at $g = 2.0024 \pm 0.0002$, attributed to paramagnetic substitutional nitrogen (denoted as N_s, P1) that is always contained in Ib type diamonds (~20 ppm). Additional singlet signals with $g = 2.0033$ indicate the presence of SiV^- centers (~80 ppm). EPR spectra of Ni-doped diamond samples, recorded at room temperature demonstrate only weak signals of N_s centers (~10 ppm) and a relatively broad singlet line with $g = 2.0026$ originating from dangling bonds (not shown). The same spectrum recorded at $T = 80$ K (Fig. 1.9b) reveals two additional well resolved lines: a narrow low field line with $g = 2.0319$ due to negatively charged substitutional Ni centers (Ni_s^-, $S = 3/2$, ~1 ppm) and a broader line with $g = 2.0117$ originating from optically active positively charged substitutional Ni centers (Ni_s^+ or NIRIM-1, $S = 1/2$, ~4 ppm). The latter Ni-related lines may be observed only at temperatures below 150 K due to temperature-dependent electronic relaxation times.

1.3 Quantum Sensing with Diamond Optical Centers

In this section, we outline the magneto-optical properties of the NV^- center that make it an attractive quantum sensor. Examples of the NV^- center in various sensing implementations are presented. Lastly, we outline how other color centers, specifically those consisting of Group IV elements and vacancies, could be potentially utilized for sensing applications.

1.3.1 The Case for Sensing with the NV⁻ Center

1.3.1.1 The NV⁻ center as a shielded electronic spin

It is prudent to conceptually 'separate' the optical centers from the diamond when considering the theoretical basis for their use as sensors. Ultimately, the diamond may be thought of as a host of optical centers such as the NV⁻ center. The NV⁻ center in turn can (in isolation) be thought of as a localized electronic spin system that is generally shielded from external interactions due to its position inside of the diamond lattice and the absence of other nearby spins.

1.3.1.2 The NV⁻ center spin Hamiltonian and electronic structure

The fundamental basis for quantum sensing with optical centers in diamond can be derived by considering energy levels and the corresponding spin Hamiltonian operators. As the NV⁻ center is the most explored optical center, we use it as a primary example. The spin Hamiltonian for the NV⁻ center in the absence of external and external magnetic fields is given by [173]:

$$\hbar^{-1}H = D\left(S_z^2 - \frac{2}{3}\right) + E\left(S_x^2 - S_y^2\right) + \mathbf{S}\bar{\mathbf{A}}\mathbf{I} \tag{1.3}$$

where D is the axial zero-field splitting (ZFS) parameter, which is equal to 2.87 GHz for the ground state and 1.42 GHz for the first excited state [174], E is the non-axial ZFS, S_z, S_x, and S_y are spin projection operators, **S** is the total spin operator, **A** is the hyperfine tensor, and **I** is the nuclear spin operator. We omit electric and quadrupole interactions as they add unnecessary complexity for the present discussion. The full spin Hamiltonian for the NV⁻ center is given in Ref. [175]. In cases where the NV⁻ center exhibits perfect C_{3v} symmetry, the E term vanishes (i.e. $E = 0$). The axial ZFS parameter (D) is perhaps the most important parameter in Eq. 1.3 with regards to sensing, and will be discussed in more detail here.

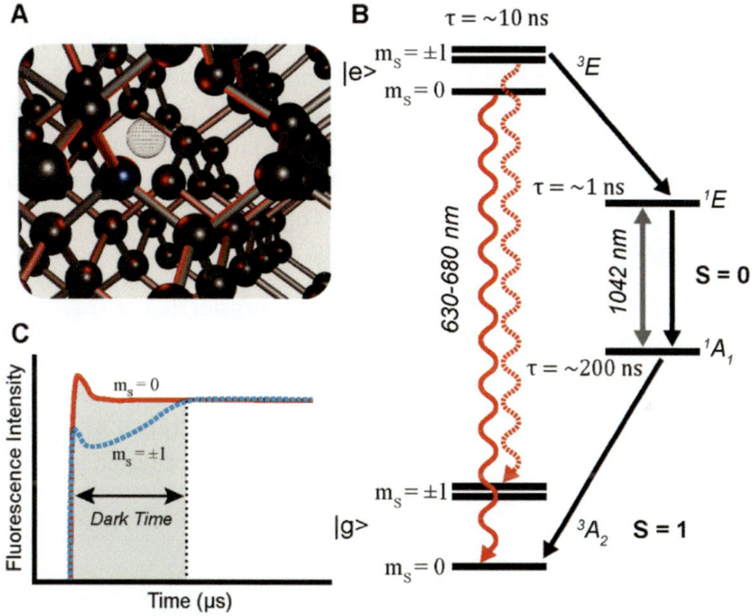

Figure 1.10 (a) Rendering of the NV⁻ center consisting of a nitrogen (blue atom) aligned along a <111> crystallographic axis with a vacancy (wireframe sphere). (b) The electronic structure of the NV⁻ center. In this depiction, solid lines indicate high probability transitions, while dashed lines are low probability transitions. Upon optical pumping (e.g. from 532 nm), spin conserving excitation from the ground state |*g*> to the excited state |*e*> occurs. Preferential relaxation from the excited state $m_s = \pm 1$ sublevels occurs via a non-radiative pathway into singlet states with ^1E and $^1A_1$ symmetry. The singlet-singlet transition can be probed with 1042 nm laser excitation [177]. (c) The long lifetime of the $^1A_1$ singlet state ($\tau \sim 200$ ns) leads to a short 'dark time' in which fluorescence contrast between the $m_s = \pm 1$ and $m_s = 0$ states can be determined. Eventually, preferential relaxation from the $^1A_1$ singlet state to the |*g*> $m_s = 0$ sublevel leads to recovery of the fluorescence.

The NV⁻ center (Fig. 1.10a) is a luminescent triplet state ($S = 1$) defect consisting of six electrons (three from each of the carbon dangling bonds, two from the nitrogen, and one from a nearby donor (not shown in Fig. 1.10a). The electronic structure is a ^3A symmetry ground state, and a ^3E symmetry excited state [173]. The lifetimes of the ^3E excited state, ^1E singlets, and $^1A_1$ singlets are on the order of 10 ns [176], 1 ns [177], and 200 ns [177], respectively (Fig. 1.10b). Optically driven transitions from the $|g\rangle \rightarrow |e\rangle$ states are spin conserving. Optical pumping from the ground electronic |*g*>-state with $m_s = \pm 1$ leads to preferential non-radiative relaxation from the excited |*e*>-state $m_s = \pm 1$ sublevel via non-radiative singlet ($S = 0$) states to the ground state $m_s = 0$ level. The relatively long lifetime of the metastable $^1A_1$ singlet state (~200 ns) leads to a short period of reduced fluorescence called the 'dark time' in which fluorescence contrast between the $m_s = 0$ and $m_s = \pm 1$ sublevels can be measured (with maximum contrast at ≈30%) (Fig. 1.10c).

1.3.1.3 Coupling magnetic and optical properties of the NV⁻ center

When the NV⁻ center is placed in an external magnetic field (B_0), a Zeeman term is added to the spin Hamiltonian (Eq. 1.4). In this equation the non-axial ZFS and hyperfine terms are omitted because their relative contributions are small.

$$\hbar^{-1}H = D\left(S_z^2 - \frac{2}{3}\right) + g\gamma_{NV}\mathbf{B}\cdot\mathbf{S} \tag{1.4}$$

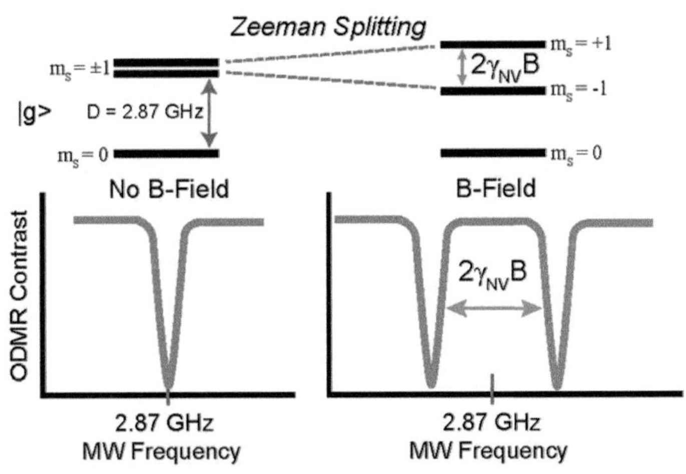

Figure 1.11 Optically detected magnetic resonance (ODMR) of the NV⁻ center in diamond. (left) At zero field, splitting between the |0> and |±1> sublevels is given by the ZFS parameter (D = 2.87 GHz). A drop in fluorescence occurs when a swept microwave (MW) frequency matches the resonance of the transition. (right) When an external magnetic field is applied, the degeneracy of the |±1> sublevels is broken, leading to a splitting in the resonance lines equivalent to $2\gamma_{NV}B$.

In the absence of a magnetic field, the splitting of the ground state $m_s = 0$ and $m_s = \pm 1$ equal to the ZFS parameter $D = 2.87$ GHz (Fig. 1.11, left). If strain effects are present ($E \neq 0$), then the degeneracy of the $m_s = \pm 1$ sublevels may be lifted, but we consider the case when $E = 0$. When an external magnetic field B_0 is applied, the degeneracy of the $m_s = \pm 1$ sublevels is broken, leading to splitting of the $|0\rangle \rightarrow |+1\rangle$ and $|0\rangle \rightarrow |-1\rangle$ transitions to higher and lower frequencies, respectively, with splitting between the transitions given by $2\gamma_{NV}B$ (Fig. 1.11, right). This additional splitting provides the means to probe magnetic fields by ODMR.

1.3.1.4 Generalized sensing schemes with optical centers in diamond

We categorize three general schemes for which optical centers in diamond can be deployed as sensors, summarized in Fig. 1.12. We remark that this categorization is somewhat arbitrary and a matter of opinion; however, it is inclusive of many of the commonly encountered and employed sensing methodologies in the literature. The three general categories are:

(1) Time resolved luminescence intensity measurements;
(2) Optically detected magnetic resonance, including both CW and pulsed methods;
(3) Direct all-optical readout of spectral feature shifts and fluorescence intensity.

The first approach utilizes pulsed laser excitation with time dependent fluorescence readout. Microwave irradiation may or may not be employed. Two time depended fluorescent approaches are optical relaxometry based measurements [13, 178, 179] (Fig. 1.12a) and Ramsey protocol sequences (Fig. 1.12b). Effectively, these two methods can measure spin lattice relaxation (T_1) and spin memory/dephasing times (T_2*), respectively. In T_1 optical relaxometry, the presence of external magnetic species causes alteration in the dark time previously discussed in Section 1.3.1.2 (Fig. 1.10c). Generally, an increased presence of external paramagnetic species shortens the T_1 time and therefore the dark time. Correlation between the relaxation time and concentrations of paramagnetic species is then made possible. The Ramsay protocol is widely used in AC and DC magnetometry measurements with the NV^- center [11, 174–176, 180]. In a typical Ramsay sequence (Fig. 1.12b, inset) (for this

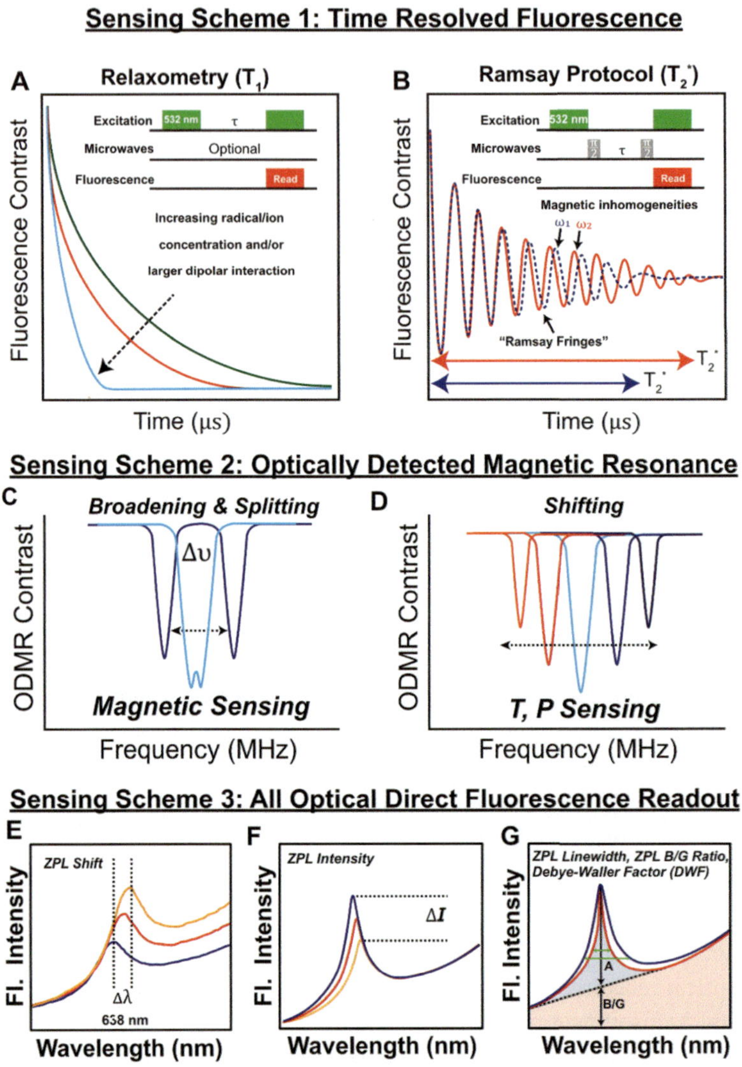

Figure 1.12 Commonly used sensing schemes with NV^- centers and other optical centers in diamond.

example, we consider the NV^- center), a resonant 532 nm laser pulse initializes the NV^- spin(s) to the $m_s = 0$ level. A $\pi/2$ microwave (MW) pulse is applied to create a superposition of the |0> and either the |+1> or |−1> states (note that a static magnetic bias field and tuned MW pulsed is assumed such that either the |0> → |+1> or |0> → |−1>

transition is interrogated). The spins are allowed to process freely for a time τ, before a second $\pi/2$ MW pulse is applied that collapses the superposition. Depending on the phase of the transverse spin magnetization at the time of the second $\pi/2$ pulse, spin populations will map to either the |±1> or |0> state. The former leads to a higher optical contrast while the latter results in a lower one. Continuing this pulse sequence with increasing interrogation time (tau) leads to cycling of the spin populations between the |0> and |±1> states, leading to the characteristic Ramsay fringes. Decay in the oscillating signal results from decoherence of the spins (detected optically by decoherence of the emitted photons). This decay time is T_2 in a perfectly homogenous magnetic field (spin-spin coherence time). In practice the decay is determined by effective T_2* spin-dephasing time ($T_2^* \leq T_2$). Local magnetic field inhomogeneities due to the presence of paramagnetic species, for example, would cause different spins to process with slightly different frequencies, leading to shortening of T_2^*. In this way, the NV^- center can be employed as a sensor of local magnetic interactions. A similar technique introduces an additional π pulse between the two $\pi/2$ pulses is commonly used for AC field sensing, while the Ramsey protocol is used for DC field sensing [181]. Additional discussion of this topic can be found in Refs. [11, 174, 175].

The second scheme (Fig. 1.12c,d) relies on a readout of ODMR resonance line positions or linewidths. ODMR readout can be employed in either CW or pulsed modes, respectively called continuous wave ODMR and pulsed ODMR. For magnetometry applications, CW ODMR is the simplest approach. In CW ODMR, laser excitation, microwaves, and readout of optical signal are all maintained continuously. Optical excitation continuously populates the $m_s = 0$ level (increasing fluorescence signal), while the applied MW equally populates spins from the |0> to the |±1> state (decreasing fluorescence signal at resonance). Fluctuations in the local magnetic environment cause energetic shifts of the resonance conditions of the spin sublevels, leading to changes in the optical contrast which can be measured. Pulsed ODMR shares many features of the Ramsey protocol. First, spins are optically initialized to $m_s = 0$, then over an entire interrogation time τ, a MW π-pulse is applied before optical readout. Magnetic interactions result in detuning of the applied MW frequency from the resonance conditions and an incomplete

π-pulse which manifests in altered fluorescence contrast. Additional discussion of CW and pulsed ODMR in the context of magnetometry can be found in Refs. [11, 174].

The third approach (Fig. 1.12e–g) is the most conceptually straightforward, and relies on assessing changes in the photoluminescent spectra (e.g., intensity changes, zero-phonon line shifts, and peak height ratios or changes in spectral width) in response to external stimuli [21, 182]. This approach does not require laser pulsing sequences or MW irradiation. Methods illustrated in Fig. 1.12e–g are commonly seen in thermometry applications. While shifts in ZPL position and intensity are straightforward, more advanced approaches such as measurement of the Debye–Waller factor (DWF, ratio of area under ZPL peak to total spectral intensity) [182] or ratios of ZPL peak height to background height have been reported in literature [183].

1.3.2 Sensing Demonstrations with NV⁻ Centers in Diamond

1.3.2.1 Detection of paramagnetic ions and radicals

Magnetic interactions between the NV^- center and external fluctuating dipoles can manifest in several ways, including: changes in total fluorescence intensity, changes in the timescales of spin dependent fluorescence recovery, and broadening or splitting of ODMR lines. In this way, external paramagnetic species such as ions or free radicals can be detected and quantified provided the correlation between the concentration of the interacting spins and selected readout is properly calibrated. In absence of fluctuations, dipolar interaction [167, 184] between an NV^- spin (S_{NV}) and an external spin (S_{ext}) leads to an additional term in the spin Hamiltonian described by Eq. 1.5. Where γ_{NV} and γ_{ext} are the gyromagnetic ratios of the NV and external spin, respectively; S_{NV} and S_{ext} are the spin projection operators for the NV^- spin external spin (S_{ext}); r is the distance between the spins; and $\vec{\mathbf{r}}$ is the vector connecting the two spins.

$$\hbar^{-1}H = D\left(S_z^2 - \frac{2}{3}\right) + g^2\gamma_{NV}\gamma_{ext}\left(\frac{S_{NV}\bullet S_{ext}}{r^3} - \frac{3(S_{NV}\bullet\vec{\mathbf{r}})(S_{ext}\bullet\vec{\mathbf{r}})}{r^5}\right) \quad (1.5)$$

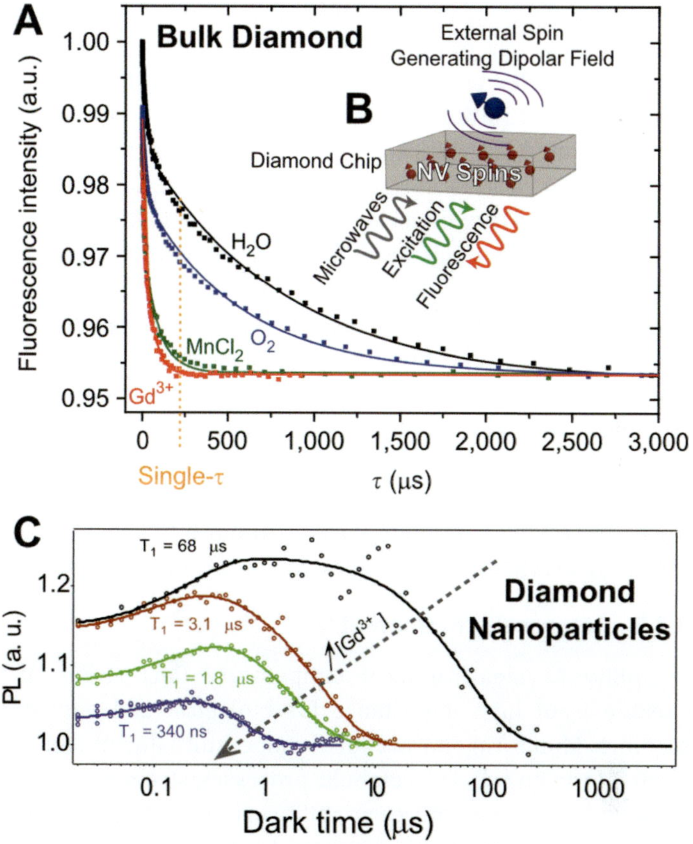

Figure 1.13 Optical relaxometry demonstrations in bulk diamond (a, b) and in particulate diamond (c) in the presence of paramagnetic ions and molecular species. Gd^{3+} is a particularly impactful ion on the T_1 time due to its spin (I = 7/2) and correspondingly large magnetic moment. (a) Reprinted with permission from Ref. [179], Copyright 2013, Nature Publishing Group. (c) Reproduced with permission from Ref. [185], Copyright 2013, American Physical Society.

Dipolar interactions also lead to changes in the spin relaxation times, providing a basis for optical relaxometry sensing. Such protocols were utilized by Schirhagl and co-workers [13, 178] and involved measurement of equilibrium time ('dark time') for fluorescence recovery (T_1) in an all-optical setup. This approach (referred to as optical T_1 relaxometry) does not necessarily require the use of MW radiation (though it can be implemented);

however, precise control of laser pulse sequences is needed. Optical T_1 relaxometry was first demonstrated by Steinhart et al. [179] using ensemble NV centers in high purity ^{15}N implanted type IIa diamond in 2013 (MW radiation was implemented in this specific demonstration) (Fig. 1.13a,b). Subsequently, the technique was extended to single NVs in nanodiamond particles ranging in size from 10 to 100 nm in two specific implementations: (1) the detection magnetic spin noise originating from Gd^{3+} ions [185] (Fig. 1.13c) and (2) for the detection of ferritin non-covalently attached to diamond particle surfaces [186]. More recently, NV relaxometry was utilized in yeast cells to monitor radical concentrations when cells are stressed or when treated with antioxidants [187]. Martínez et al. demonstrated relaxometry-based sensing of UV and peroxide induced hydroxyl radicals [178]. To expand applications utilizing single particle tracking and T_1 relaxometry in cells, efforts geared towards improving particle-to-particle uniformity and NV^- quality, particularly in regards to their quantum properties, are needed.

1.3.2.2 Nanoscale thermometry

The ability to measure local temperature fluctuations at the nanoscale is of high importance for biological and engineering disciplines. Monitoring intracellular temperature fluctuation can be used to inform on cellular metabolic processes, stress response, and pathology. Local temperature variability in integrated circuit devices can inform on failure mechanisms and lead to improved material engineering. In quantum computing, assessment of thermal noise critical to ensure robust control of computation and readout. The NV^- center is an apt temperature sensor because its ZFS parameters D and E are affected by strain fields in the diamond crystal and, as such, are temperature dependent.

The most important parameters of any sensors are the sensitivity and resolution. The sensitivity is the smallest detectable change in temperature that the NV^- center can measure, while the resolution is the degree to which a change can be detected over thermal noise—referred to as a "noise-floor measurement" [27]. The sensitivity (S) can be defined as $S = \delta Q/\delta P$ (where Q is the 'observable quantity,' e.g. splitting between spectral lines, wavelength λ of a spectral peak, etc., and P is the physical parameter to be measured). A relative sensitivity (S_{rel}) can be additionally defined as $S_{rel} = (1/Q)\delta Q/\delta P$, for benchmarking to other measurement methods [26, 27, 188]. The resolution is determined by noise present in the system that is related to its temperature. A nanoscale thermometer experiences thermal fluctuations described by a standard deviation (Eq. 1.6) [189]:

$$\sigma_{\text{th}} = T\sqrt{\frac{k_{\text{b}}}{VC_{\text{V}}}} \tag{1.6}$$

where T is the thermodynamic temperature, k_b is the Boltzmann constant, V is the volume of the nanoparticle, and C_V is the constant volume heat capacity. The nanoparticle sensor experiences these fluctuations on a timescale of τ_{rel}, which is on the order of a^2/α, where a is the radius of the particle and α is the thermal diffusivity of the medium (m^2/s). The noise floor limit can be estimated by Eq. 1.7:

$$\eta_{\text{th, T}} = \sigma_{\text{th}}\sqrt{\tau_{\text{rel}}} \approx T\frac{k_{\text{b}}}{\sqrt{aC_{\text{v}}\alpha}} \tag{1.7}$$

Acosta and co-workers demonstrated that the temperature response of D (in MHz) was—74.2 kHz/K, while the temperature response of E (in MHz) was -1.4×10^{-4} kHz/K (*in bulk CVD diamond with ~10 ppb of* NV^- *centers*) (Fig. 1.14a) [16]. The ODMR resonance line shifts to lower frequencies at higher temperature (Fig. 1.14b,c), and the axial ZFS parameter D is a more sensitive means of assessing temperature response compared to the transverse ZFS parameter E. For this reason, many ODMR based temperature sensing with the NV^- center relies on assessing the shift in the resonance line position with respect to changes in D. The relative sensitivity of the NV^- center can be calculated to be $\approx 2.6\times10^{-5}\ K^{-1}$, with D = 2.87 GHz. In practice, sensitivities of nanodiamonds containing NV centers have been reported in the range from $1.5\times10^{-5} - 1.3\times10^{-2}\ K^{-1}$ [27].

It is possible to simultaneously measure magnetic fields and temperature because the magnetic field induced splitting of the ODMR resonance line, while temperature causes shifting of the line [19]. Numerous studies have demonstrated the temperature sensing capability of the NV^- center, with many relying on ODMR readout techniques [21, 24, 182, 190–197]. Recently, Yukawa and co-workers demonstrated temperature sensing in adipose tissue-derived stem cells (ASCs) over several days using 150 nm fluorescent

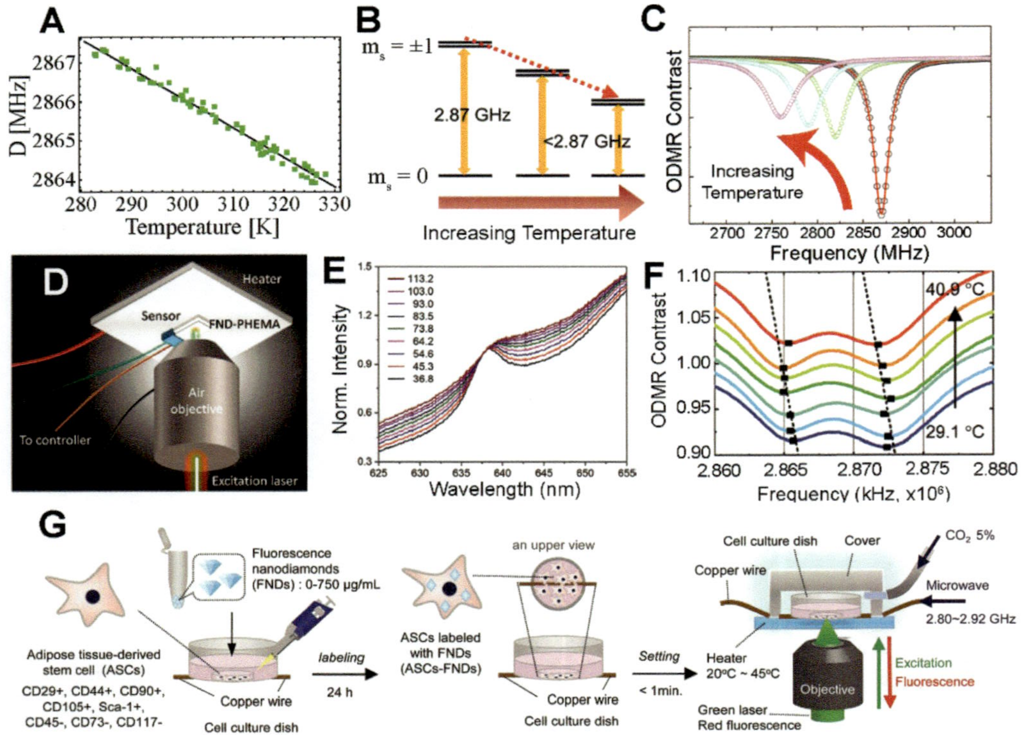

Figure 1.14 Temperature sensing schemes with the NV$^-$ center. (a) Linear dependence of ZFS parameter (D) with respect to temperature. (b–c) Decrease in ZFS parameter leads to ODMR resonance line shifting to lower frequencies at higher temperature. (d–e) All optical temperature measurement using fluorescent nanodiamonds embedded into PHEMA film. Shifts in the zero-phonon line (ZPL) position recorded as a function of temperature from 37°C to 113°C. (f–g) ODMR based technique to assess temperature in adipose derived stem cells. (a) Reproduced from Ref. [16]. (b, c, f, g) Reprinted with permission from Ref. [194], Copyright 2020, the Royal Society of Chemistry. (d, e) Reprinted with permission from Ref. [21], Copyright 2019, American Chemical Society.

nanodiamond particles with a quantum thermometric sensing and analysis system (QTAS) (Fig. 1.14f,g) [194]. A noise floor resolution of 2.2°C/$\sqrt{Hz}$ was achieved. Importantly, in addition to measuring temperature, correlations between cellular temperature and growth factor expression as well as differentiation between adipocytes and osteocytes were assessed. This is an excellent demonstration of utilizing optical centers in diamond to answer a relevant biological question.

Not all temperature sensing modalities rely on ODMR. Hui and co-workers demonstrated an all-optical temperature sensing technique utilizing fluorescent nanodiamonds (FNDs) embedded to poly(2-hydroxyyethyl methacrylate) (PHEMA) films, where the temperature dependence of the zero-phonon line (ZPL) position as well a height ratio method were used to monitor temperature in across the range of 35–120°C (Fig. 1.14d,e) [21]. This study [21] achieved noise floor resolutions of 0.15 K/$\sqrt{Hz}$ at 36.8°C (310.0) and 0.62 K/$\sqrt{Hz}$ at 118°C (391.2 K)—hence, the resolution drops with increasing temperature. The temperature dependence of the longitudinal optical relaxation time (T_1) of single NV^- centers in nanodiamonds at 4 K and 298 K has also been reported [198]; however, it should be noted that objective of this study was to investigate sources of spin noise on the diamond surface rather than to measure temperature.

1.3.2.3 Pressure sensing

The NV^- center is sensitive to external stress, again through changes in the effective D and E parameters as well as off-diagonal terms. These effects can be described by the spin Hamiltonian given in by Eq. 1.8 [24], where α_1, α_2, β_1, and β_2 are stress coefficients and σ values are components of the stress tensor.

$$\begin{aligned}\hbar^{-1}H = {} & D\left(S_z^2 - \frac{2}{3}\right) + E(S_x^2 - S_y^2) + \left[\alpha_1(\sigma_{xx} + \sigma_{yy}) + \beta_1\sigma_{zz}\right]S_z^2 \\ & + [\alpha_2(\sigma_{yy} - \sigma_{xx}) + \beta_2(2\sigma_{xz})](S_y^2 - S_x^2) \\ & + [\alpha_2(2\sigma_{xy}) + \beta_2(2\sigma_{yz})](S_xS_y + S_yS_x) \qquad (1.8)\end{aligned}$$

Both bulk (CVD) and particulate diamond containing NV^- centers have been utilized in pressure sensing. Pressure induces first order shifting and second order splitting of the $m_s = \pm1$ sublevels due to mechanical stress. The shift is characterized by the Π_z parameter (Π_z

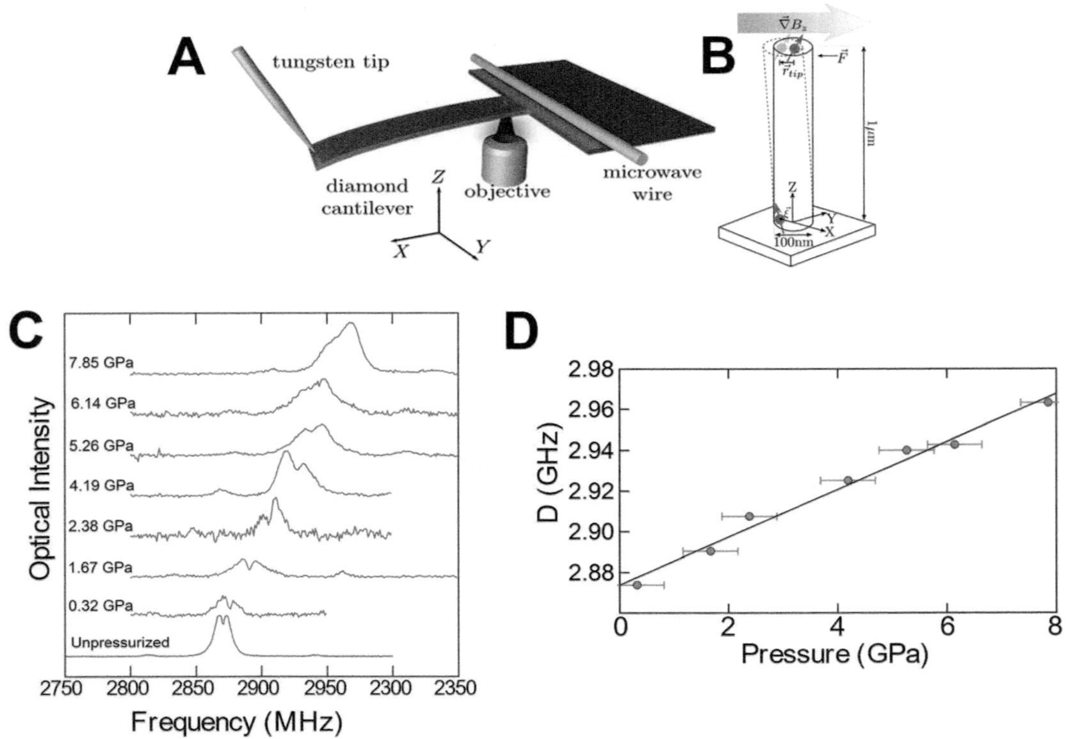

Figure 1.15 Pressure sensing with the NV- center in diamond. (a, b) Proposed diamond nanopillar with NV- center for nanoelectromechanical system (NEMS) for force detection integrated into an atomic force microscope. This could potentially be used for measurement of nanoscale forces in/around cells or for molecular mass assessment. (c) Pressure dependent ODMR readout of 15 μm particles inside diamond anvil cell. (d) Linear dependence of zero-field splitting (ZFS) parameter (D) with pressure. (a, b) Reprinted with permission from Ref. [199], Copyright 2017, American Chemical Society. (c, d) Reprinted with permission from Ref. [23], Copyright 2017, AIP Publishing.

$= \alpha_1(\sigma_{xx} + \sigma_{yy}) + \beta_1\sigma_{zz})$, while pressure induced splitting is described by $\Pi_\perp^2$ ($\Pi_\perp^2 = [\alpha_2(\sigma_{yy} - \sigma_{xx}) + \beta_2(2\sigma_{xz})]^2 + [\alpha_2(2\sigma_{xy}) + \beta_2(2\sigma_{yz})]^2$. Note that in bulk diamond, four crystallographic orientations of the NV^- center lead to a total of eight resonance lines in the ODMR spectrum. Pressure also induces a linear shift in the ZFS parameter (D), with a slope of dD/dP = 14.58(6) MHz/GPa [22] (*calculated relative sensitivity* $\approx 5\times10^{-3}$ GPa^{-1}) (Fig. 1.15d). Hsieh and co-workers demonstrated simultaneous optical readout of pressure and magnetic fields in a diamond anvil cell with ensembles of NV^- centers implanted ~50 nm from the surface of the anvil cutlet [24]. This study [24] achieved a noise floor pressure resolution of {0.023, 0.030, 0.027} GPa/$\sqrt{Hz}$) (*for the hydrostatic, average normal, and average shear stresses, respectively*).

Steele and co-workers demonstrated the use of 15 μm particles containing ~3 ppm of NV^- centers for pressure measurements inside of an anvil cell [23] (Fig 1.15c), and a linear pressure dependence of 11.72 MHz/GPa was measured over the range of 0–8 GPa. The results from this study [23] also indicated that the E parameter was pressure independent—suggesting the local symmetry around the NV^- center was not changed with increasing pressure. Microcantilever force sensors using NV^- centers demonstrated noise floor resolutions of $\approx$ 40 nN/$\sqrt{Hz}$ and are proposed as a means to measure cellular biomechanical forces and potentially single molecule mass in nanoelectromechanical devices (NEMS) (Fig. 1.15 a,b) [199].

1.3.2.4 Nanoscale magnetometry and magnetic resonance spectroscopy

Applications involving the use of the NV^- center in nano-magnetometry, nuclear magnetic resonance (NMR), magnetic resonance imaging (MRI), and electron paramagnetic resonance (EPR) are among the some of the most widely studied applications of luminescent diamond. Literature on the subject is vast and rapidly changing. For the sake of brevity, we refer the reader to several pertinent reviews and original reports on the subject in Table 1.4. Here we and only briefly highlight some key papers and recent developments.

Table 1.4 Several key reviews and original reports on magnetometry with NV^- centers in diamond

Author(s)	Year	Summary	References
Barry et al.	2020	Comprehensive review discussing sensitivity enhancement strategies for NV^- based magnetometers	[11]
Sturner et al.; Kim et al.	2019	Example compact magnetometry devices utilizing NV-diamond sensors	[206, 207]
Hopper et al.	2018	Review discussing spin readout techniques for NV based quantum sensors and devices	[214]
Wolf et al.	2015	Demonstration of sub-picotesla sensitivity with ensemble NV^- centers in bulk diamond	[215]
Rondin et al.	2014	Review discussing fundamentals of NV^- magnetometry	[216]
Simpson et al.	2014	Theoretical paper discussion NV^- magnetometry concepts for single molecule NMR	[217]
Taylor et al.; Balasubramanian et al.	2008	Seminal papers discussing diamond single molecule magnetometry and imaging	[201, 202]

NMR and EPR spectroscopy are mainly utilized to characterize structure and dynamics of molecules and their complexes with applications in chemistry, biology, materials science, and other fields. Typically, the spectroscopic signal is detected by a coil or a resonator that is inductively coupled to large ensemble of nuclear or electronic spins. For example, typical sensitivity of commercial CW EPR spectrometers is about 10^{10} spins/G [200]. NV^- centers in diamond provided researchers with an opportunity to significantly improve on the minimum number of detectable spins by exploring more sensitive optical detection of magnetic resonance down to the single spin level. The first demonstrations of single defect

(NV^-) magnetometers were in 2008 [201, 202]. Since that time, demonstration of single protein NMR and EPR via an optical readout from the NV^- center were shown by Lovchinsky et al. [203] and Shi et al. [204], respectively. 2D homonuclear and heteronuclear correlative spectroscopy (COSY) NMR was recently demonstrated by Smits et al. using a microfluidic diamond sensor (Fig. 1.16 a,b) [205].

The key advantages of an optical readout are (*i*) greatly improved sensitivity and (*ii*) the dimensions of the sensing element can be, in principle, scaled down to a nanoscale-size diamond particle containing a single NV center. These considerations stimulated significant efforts for development of NV-based magnetometers and their integration to compact NMR or NMR on chip devices. Several recent reports described integrated circuit devices based on compact NV^- based magnetometers (Fig. 1.17) [205–207].

In application to electric and magnetic field sensing, bulk diamond containing NV^- centers were utilized to map the current flux through a graphene conductor [209] and diamond particles were used to simultaneously monitor magnetic fields and temperature based on ODMR principles in a transistor [197] (Fig. 1.18). There is growing interest to develop such sensors for biomedical and biophysical applications. Examples of such developments were provided by utilizing the NV^- center to measure the action potentials (*i.e., spatially localized spikes in the membrane electrostatic potentials*) of single neurons [210] and cultured mouse neuronal networks [211].

One of the key challenges with NV^- based magnetic and electric field sensors is the compromise between resolution and sensitivity. Single NV^- centers provide the highest spatial resolution but low optical signal, while ensembles of NV^- increase optical signal and SNR but lack the spatial resolution of single centers. Moreover, fabricating diamond with isolated NV^- centers represents a greater technical challenge vs. that with NV^- ensembles. To date, the highest magnetic sensitivities achieved with NV ensemble-based magnetometers are on the order $pT/\sqrt{Hz}$, whereas competing technologies, such as superconducting quantum interreference devices (SQUIDs) and spin-exchange relaxation-free magnetometers exhibit sensitivities at the $fT/\sqrt{Hz}$ levels or lower [212]; this requires a $\sim 10^6$ greater signal averaging time for ensemble NV magnetometers to achieve the same SNR [11]. For these reasons, significant efforts were focused on developing better approaches for utilizing NV^- ensembles by, for example, implementing intricate pulsing sequences [213].

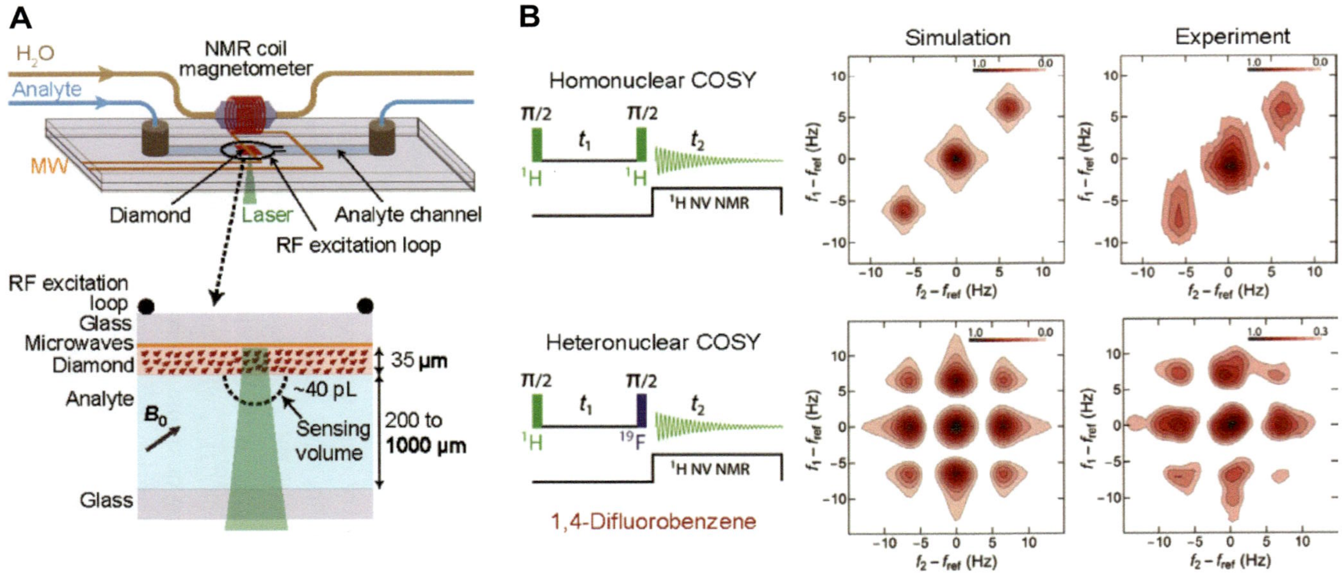

Figure 1.16 2D-NMR COSY using a microfluidic diamond sensor (a). In this example [205], a 13 mT external field was used for homonuclear and heteronuclear COSY, with simulated and experimental spectra shown in (b). Acquisition times were 22 and 25 h, respectively, for the homonuclear and heteronuclear acquisitions. (a, b) Reproduced from Ref. [205] under a Creative Commons (CC-BY) License.

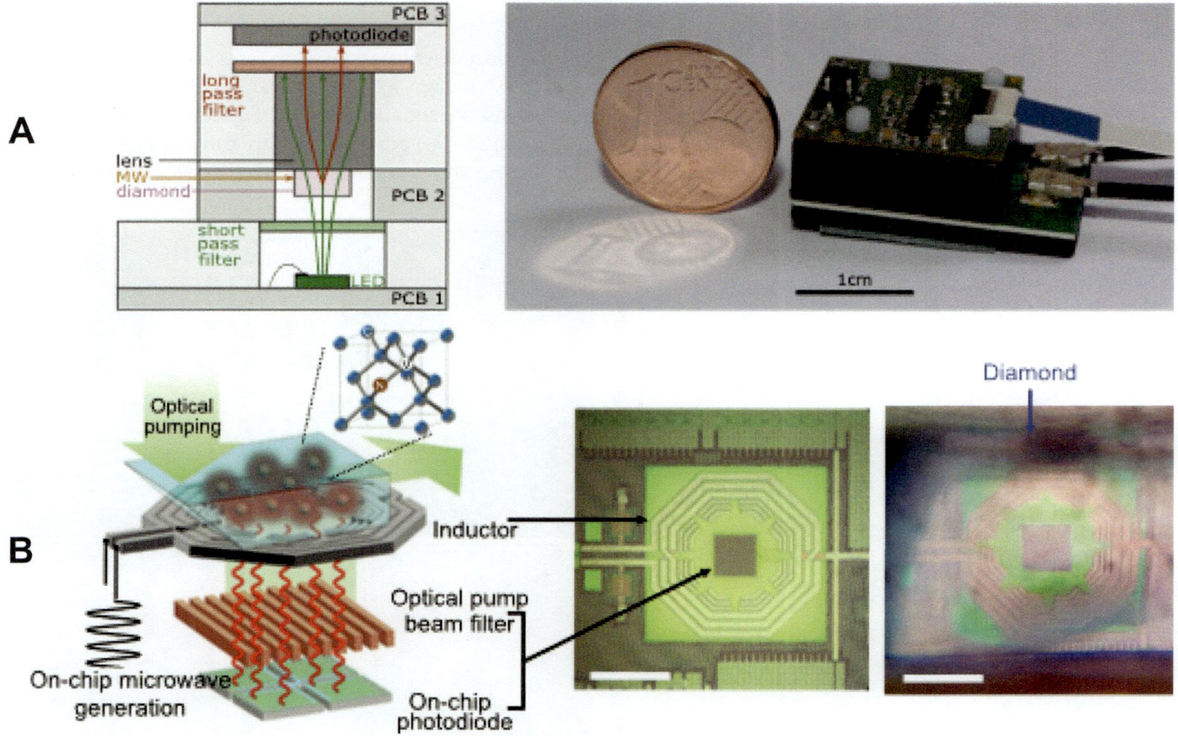

Figure 1.17 (a) Portable compact magnetometer device utilized NV centers in diamond. Reproduced with permission from Ref. [206], Copyright 2019, Elsevier. (b) A CMOS integrated quantum sensor utilizing planar MW generation. Reprinted with permission from Ref. [207], Copyright 2019, Springer Nature.

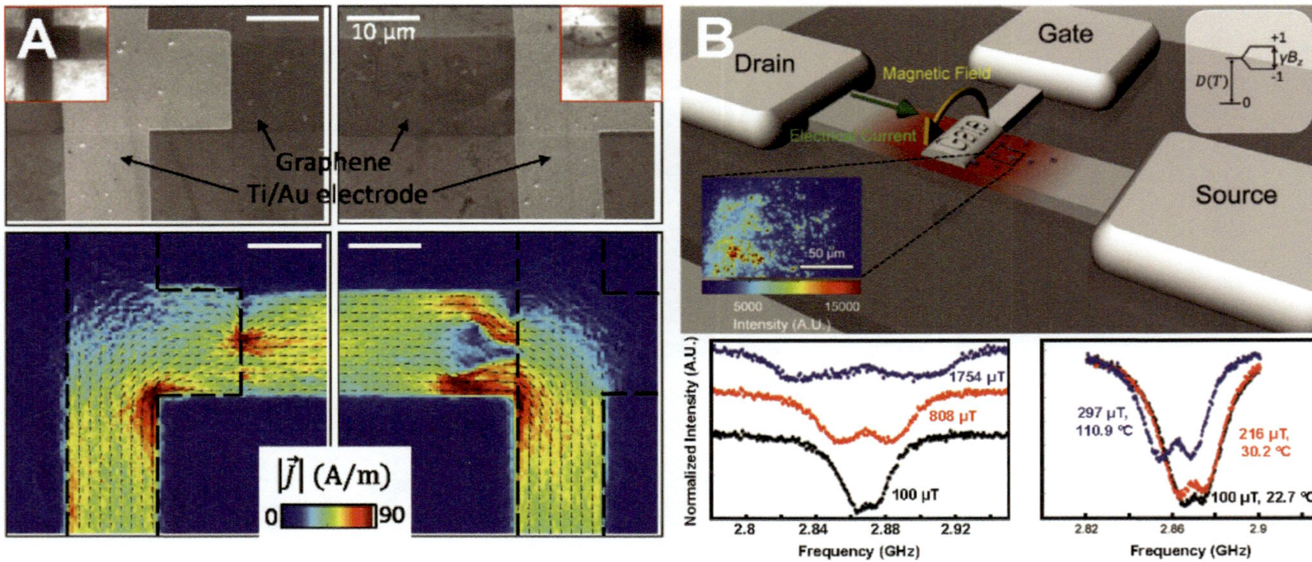

Figure 1.18 (a) NV⁻ centers utilized to map current flux through graphene conductor between two Ti/Au electrodes. Areas of high current density can be sources of failure, thus utilizing the NV⁻ center in diamond to characterize the device can potentially lead to better material and device design and engineering. (b) Simultaneous measurement of magnetic field and temperature induced from current flow through transistor gate using diamond particles with NV⁻ centers with ODMR based readout. (a) Reproduced from Ref. [209] under a Creative Commons (CC-BY) License. (b) Adapted with permission from Ref. [197], Copyright 2020, American Chemical Society.

1.3.2.5 Hyperpolarized nuclear magnetic resonance (NMR) with NV^- centers

Since its discovery nearly 70 years ago, spectroscopic techniques based on magnetic resonance have led to revolutionary developments in chemistry and medicine, and, more recently, in quantum computing. However, a major drawback of all magnetic resonance methods based on traditional inducting based detection techniques is their inherent low sensitivity. In regards to NMR, this sensitivity issue is attributed to a low spin polarization, especially at ambient temperatures. Development of state-of-the-art superconducting magnets with ever higher magnetic fields and superconducting NMR probes significantly improved sensitivity. Unfortunately, the current state of superconductivity technology imposes limits on the achievable magnetic fields and the return on investment for high field magnets is diminishing. For example, the cost of commercial 1.2 GHz (^{1}H resonance frequency) NMR spectrometer installed in 2020 at CERM in Florence, Italy, is $17.8 M [218], and such a massive system requires specialized on-site infrastructure and personnel. Spin polarization could be also increased by decreasing sample temperature T (Eq. 1.9), but this can be impractical in some cases. Additionally, solid-state NMR at cryogenic temperatures has its own challenges. This calls for exploring alternative approaches and, particularly, other means to increase spin polarization without changing magnetic field or bulk sample temperature.

$$P_n = \tanh\left(\frac{\gamma_n \hbar B_0}{2kT}\right) \tag{1.9}$$

Hyperpolarization is a broad term describing any technique in which the spins in a system are initialized to a non-equilibrium state, such that a larger difference in populations, and hence a larger degree of polarization is achieved. This is particularly important for NMR detection of low-γ and low natural abundance nuclei which otherwise require long data acquisition time for acceptable SNR. Over the years, several hyperpolarization techniques have been and continue to be developed. Examples include, but are not limited to: dynamic nuclear polarization (DNP) [219], parahydrogen induced hyperpolarization (PHIP) [220], signal amplification by reversible exchange (SABRE) [220, 221], and spin exchange by optical pumping (SEOP) [222].

A series of studies demonstrated hyperpolarization of natural abundance ^{13}C in diamond by DNP upon mm-wave irradiation of paramagnetic defects (mostly P1 centers) [223–225]. One of these studies [224] investigated the use of nuclear orientation via electron spin-locking (NOVEL), but was largely unsuccessful. However, ^{13}C polarization rates could be improved in diamond with low concentrations of paramagnetic impurities (<5 ppm) utilizing a solid effect DNP mechanism (*continuous microwave pumping of forbidden transitions*). In 2009, the level-anticrossing (LAC) of the NV-center excited state at ~0.05 T was exploited to transfer polarization from the excited state electronic spins of the NV^- center to ^{15}N nuclear spin [226]. This demonstration [226] was performed in high purity type IIa CVD diamond at room temperature and achieved a nuclear spin polarization exceeding 98%—a polarization enhancement of $\approx 6\times10^7$ compared to thermal ^{15}N nuclei at 0.05 T and 300 K. Since that time, a number of studies have investigated the transfer of polarization from the NV^- electronic spin to nearby nuclear spins in bulk diamond [227–235]. Polarization transfer from NV^- centers to ^{13}C in micron sized diamond particles was achieved with diamond particles containing using optical pumping at low field (1–30 mT) with swept microwave frequencies, and achieved a 277-fold NMR signal enhancement versus thermal polarization (Fig. 1.19a,b). Despite these successes, a transfer of polarization from hyperpolarized ^{13}C nuclei within the diamond lattice to external spins has not been demonstrated yet.

Polarization transfer from NV^- to ^{13}C within diamond has been extended to particles as small as approximately 1 μm in size [236]. Even smaller particles would be preferred for increased surface area (particularly those approaching the nanoscale), but currently such particles exhibit poor spin coherence and rapid relaxation processes of NV^- centers due to the surface and other defects present. Spin polarization transfer schemes involve either saturation of electronic transitions or maintaining cross polarization conditions in the rotating frame for sufficient time. Both conditions are difficult to achieve in systems with short electron spin relaxation times. Annealing has been demonstrated as a processing technique of choice to extend both T_1 (spin-lattice) and T_2 (spin-spin) relaxation times of NV^- centers [31, 122, 237]. Recently, it was shown that high temperature annealing enhances ^{13}C NMR signal up to 36-fold (Fig. 1.19c) [238]. We remark that a fundamental limitation of all polarization transfer approaches with ^{13}C in diamond may be the presence of a proton spin bath that would soak up all polarization indued to the ^{13}C, especially if aqueous media are used.

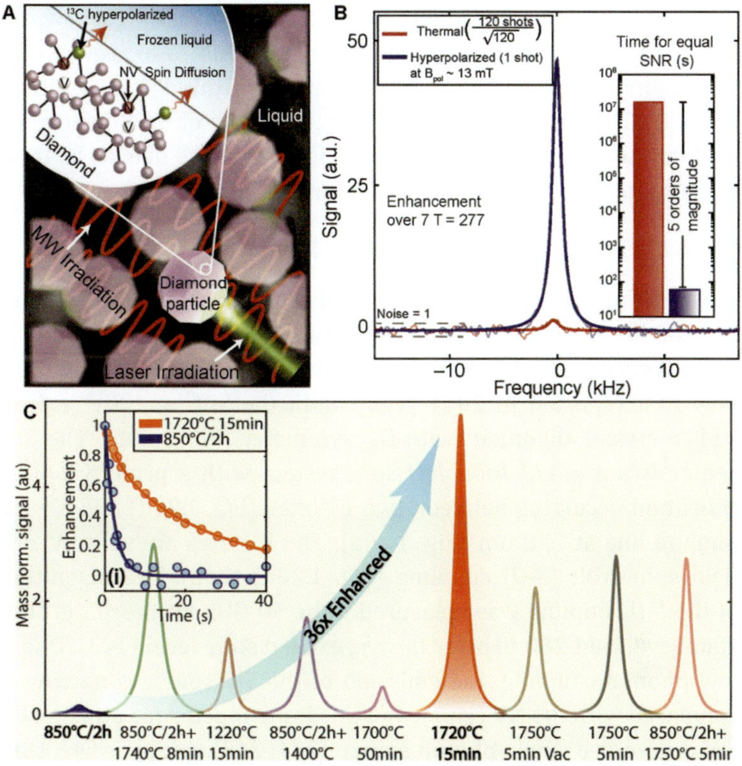

Figure 1.19 (a) Optical polarization of NV centers with polarization transfer to nearby ^{13}C nuclei. (b) Single-shot enhancement of hyperpolarized ^{13}C nuclei NMR signal following optical pumping of nearby NV centers. (c) High temperature annealing enhanced ^{13}C NMR of optically pumped NV centers in 20 μm diamond particles. (a, b) Reproduced from Ref. [239] under a Creative Commons (CC-BY) License. (c) Reproduced from Ref. [238].

1.3.3 Sensing with Group IV-Vacancy (G4V) Optical Centers

While the NV^- center is the most widely studied optical center for sensing applications; other optical centers boast similar capability. Some of these alternatives include Group IV-Vacancy

(G4V) complexes (SiV, GeV, SnV). Many of these defects are finding applications as single photon emitters in quantum information processing and quantum communication systems—topics beyond the scope of this review. With the exception of the SiV center, other G4V centers represent relatively recent discoveries (i.e., within the last six years) and may require further studies to realize their full technological potential. An extensive ab initio computational study by Thiering and Gali [240] on the family of G4V defects is a valuable source of insight into the nature of these defects as more experimental data continues to appear in the literature.

1.3.3.1 The silicon split vacancy (SiV) center

The SiV center was initially discovered in epitaxial thin film diamond in 1980 [241], and can exist in either a negatively charged (SiV^-) or neutral charge state (SiV^0), though the neutral charge variant was only first reported in 2011 [170]. Both the SiV^0 and SiV^- exhibit <111> crystal alignment with D_{3d} symmetry (Fig. 1.20b). The SiV^- center is a $S = 1/2$ *(doublet)* spin system with a primary optical transition occurring between two E states (Fig. 1.20d) and a zero-phonon line at 738 nm (Fig. 1.20a). The E states are split by zero field spin-orbit (*S-O*) coupling (Fig. 1.20d) [242]. The magnitude of the *S-O* coupling was measured to be 50 GHz for the ^2E_g ground state level, and 260 GHz for the ^2E_u excited state level [243]. Due to low phonon coupling, the emission of the SiV center is narrow as compared with the NV center, and this is an impetus for utilizing the SiV center as a single photon source [244, 245]. A higher $^2A_{1g}$ state can act as a 'shelving state' due to its long lifetime [110, 245], where the term 'shelving' refers to the ability to 'store' spins in a long-lived energetic level.

The neutral SiV^0 center is an $S = 1$ *(triplet)* system which exhibits a zero-phonon line (ZPL) at 946 nm (Fig. 1.20c). A proposed electronic structure based on axial stress measurements of the SiV^0 center was suggested by Green and co-workers [246] (Fig. 1.20e). Like the triplet NV^- center, the SiV^0 also exhibits ODMR [247]; however, stabilizing the neutral charge state requires boron doping in the range of 1–3 ppm [245, 248]. Isotopic enrichment with ^{29}Si ($I = 1/2$) leads to additional hyperfine coupling which is reflected in the ODMR spectrum (Fig. 1.20f). The SiV^0 center exhibits a ZFS parameter of 944 MHz (Fig 20e)—compared with 2.87 MHz of the NV^- center. The

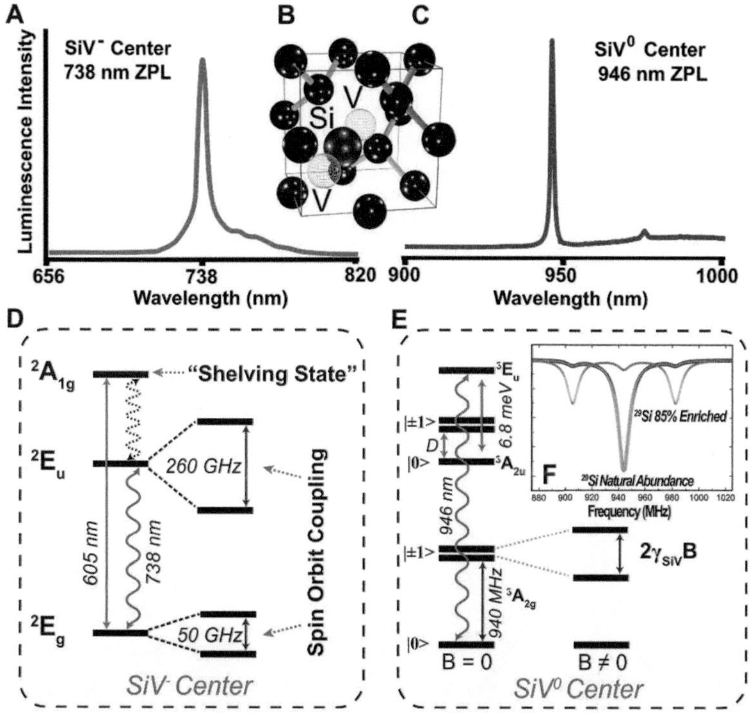

Figure 1.20 (a) Structure of SiV center in diamond lattice, with <111> crystallographic orientation along defect axis. D_{3d} inversion symmetry of the defect eliminates linear Stark effects, protecting the ZPL from spectral shifts [245], providing a basis for many useful quantum applications of the SiV center. (b) Photoluminescence spectra of negatively charged SiV^- center with 738 nm ZPL. (c) Proposed electronic structure of SiV^- center. (d) Photoluminescence spectra of neutral charged SiV^0 center with 946 nm ZPL. (e) Proposed electronic structure of neutral SiV^0 center determined from axial stress (σ), exhibiting a triplet structure with 946 MHz ZFS. (f) $(SiV)^0$ centers in polycrystalline diamond, blue diamonds—spectrum obtained in the magnetic field **B** = 0 and RT, sample contains natural abundant Si, magenta circles - **B** = 0 and RT, sample enriched by 85% ^{29}Si. Spectra shown in (a) and (c) adapted from Refs. [250] and [245], respectively.

SiV^0 ZFS parameter exhibits a larger response to temperature than the NV^- center. For the SiV^0 center, dD/dT = −337 kHz K^{-1} from T = 50 to 150 K and average dD/dT = −202 kHz K^{-1} for T = 50–300 K [246]. This temperature response is about threefold larger than the one observed for the NV^- center (dD/dT= −74 kHz K^{-1}) [16]. An all-optical thermometer based on SiV^- centers in bulk HPHT diamonds achieved 70 mK resolution at room temperature (Fig. 1.21a) while 200 nm particles demonstrated sub-Kelvin resolution [249] (Fig. 1.21b). Nanodiamond thermometers containing SiV centers have reported some of the lowest noise floor of ≈1.8 mK/$\sqrt{Hz}$ [26]. A comprehensive overview of the SiV center is provided in Ref. [245].

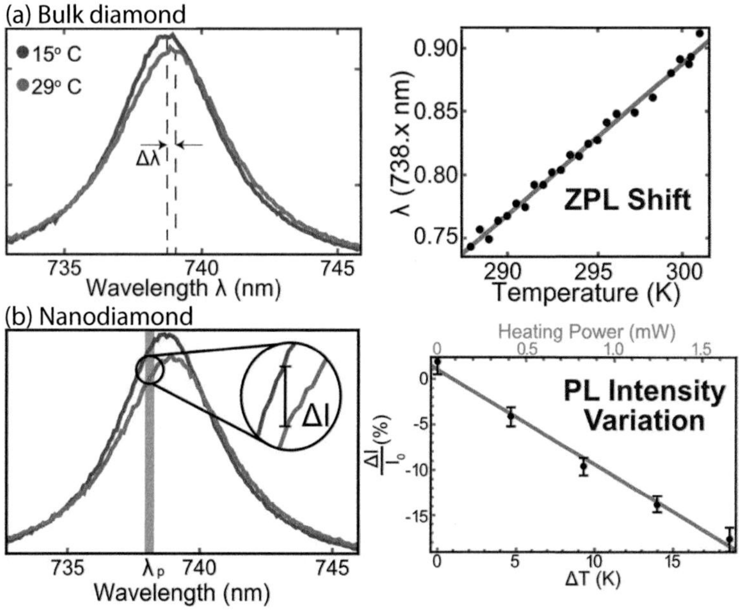

Figure 1.21 All optical temperature sensing with SiV-centers in (a) bulk diamond and (b) 200 nm nanodiamond. For bulk diamond, shifts in ZPL were measured as a function of temperature; for nanodiamonds, changes in fluorescence intensity were measured as function of temperature. Reprinted with permission from Ref. [249], Copyright 2018, AIP Publishing.

1.3.3.2 The germanium vacancy (GeV) center

The first reports on the GeV center appeared in 2015 nearly simultaneously for diamond thin films [251], microcrystalline diamond particles synthesized by HPHT [84, 252], and Ge ion implanted type IIa (100) CVD diamond [141]. The GeV center exhibits a 602 nm ZPL, and is similar in structure to the SiV center, with D_{3d} inversion symmetry and <111> alignment [240, 253] (Fig. 1.22a,b). *S-O* coupling of 181 GHz and 1120 GHz were experimentally determined for the ground and excited states,

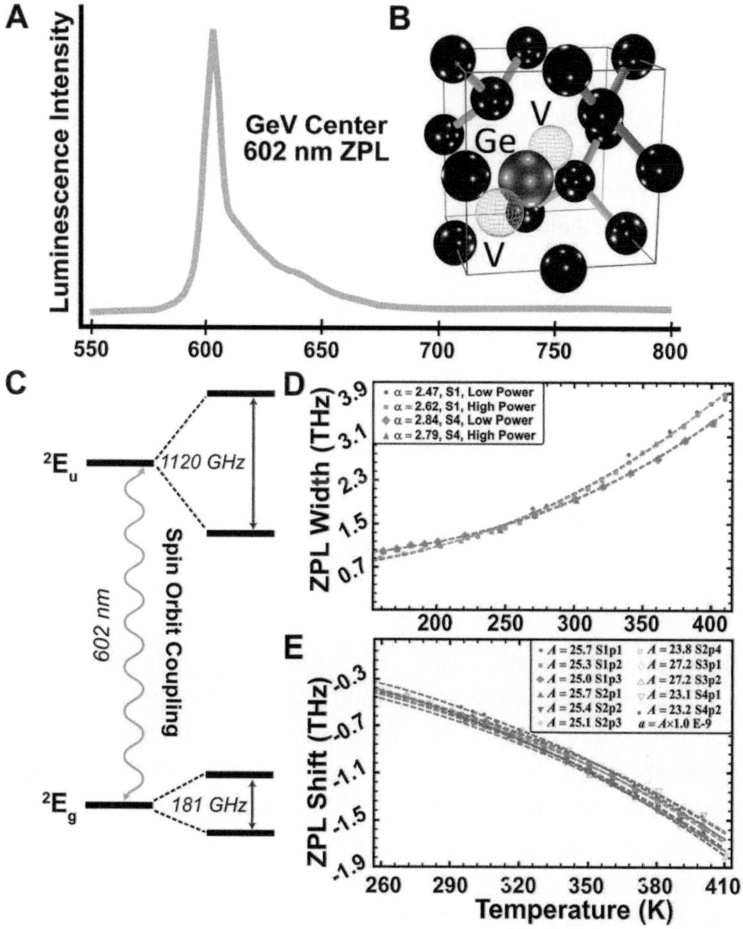

Figure 1.22 Temperature sensing with the GeV center. (a) Photoluminescence spectra of the GeV center exhibiting 602 nm ZPL. (b) The proposed structure of the GeV center. The GeV center is proposed to be similar in structure to the SiV center, with <111> crystal alignment along the defect axis, with D_{3d} symmetry. (c) The proposed electronic structure of the GeV center. (d) Spectral broadening of the ZPL of the GeV center as a function of temperature. (e) Spectral shifting of the ZPL as a function of temperature for the GeV center. (a) Adapted and (d, e) reprinted with permission from Ref. [257], Copyright 2018, American Chemical Society.

respectively (Fig. 1.22c) [84]. Similar to SiV center, the GeV center could be either neutral or negatively charged. Optical parameters (isotopic shifts, ZPL anharmonicity, vibronic peaks) as well as theoretical considerations support the conclusion that the GeV is negatively charged [141, 240, 252, 254]. However, EPR indicates a triplet ($S = 1$) center with ZFS parameter equal to 80.3 mT [255, 256] in HPHT synthesized crystals (~2–3 mm). The authors of the EPR study [256] concluded a neutrally charged GeV center based on shared symmetry and EPR spectral characteristics to that of the SiV^0 center. Nevertheless, the optical signature of the neutral GeV^0 has not been identified [240].

Thermometry based on the GeV center was demonstrated based on all-optical protocol which correlated the shifts in ZPL and spectral broadening with temperature (Fig. 1.22d,e) [257, 258]. Miller and co-workers demonstrated temperature sensing with nanodiamonds containing GeV centers synthesized from an adamantane ($C_{10}H_{16}$) mixture with tetraphenylgermane ($C_{24}H_{20}Ge$) [259]. Tran and co-workers demonstrated temperature sensing with the GeV center (as well as the NV and SiV) utilizing anti-Stokes excitation [26].

1.3.3.3 Tin vacancy (SnV) center

Continuing down the Group IV carbon family, the SnV center was discovered in 2017 in diamond implanted with Sn ions [142, 260]. Subsequently in 2018, SnV centers were identified in HPHT synthesized microcrystalline diamond particles [85]. The SnV center exhibits a 620 nm ZPL (Fig. 1.23a), and like the other G4V centers, is believed to also exhibit D_{3d} inversion symmetry, with the Sn atom occupying an interstitial site situated between two vacancies along the <111> crystallographic axis (Fig. 1.23a, inset) [240, 261]. The SnV center is negatively charged (SnV^-) based on ab initio calculations [240], and is $S = 1/2$ spin system [261]. In the absence of an external magnetic field, *S-O* coupling leads to a ground state splitting of 850 GHz and an excited state splitting of 3 THz [261] (Fig. 1.23b). An all-optical thermometry was also demonstrated based on measurements of ZPL shift with respect to temperature (Fig. 1.23c,d) with sensitivity ($\Delta\lambda/\Delta T = 0.0142$ nm/K [262]. Comparably, the SiV^0 and GeV^- centers exhibit 0.012 nm/K [249] and 0.009 nm/K, respectively [257].

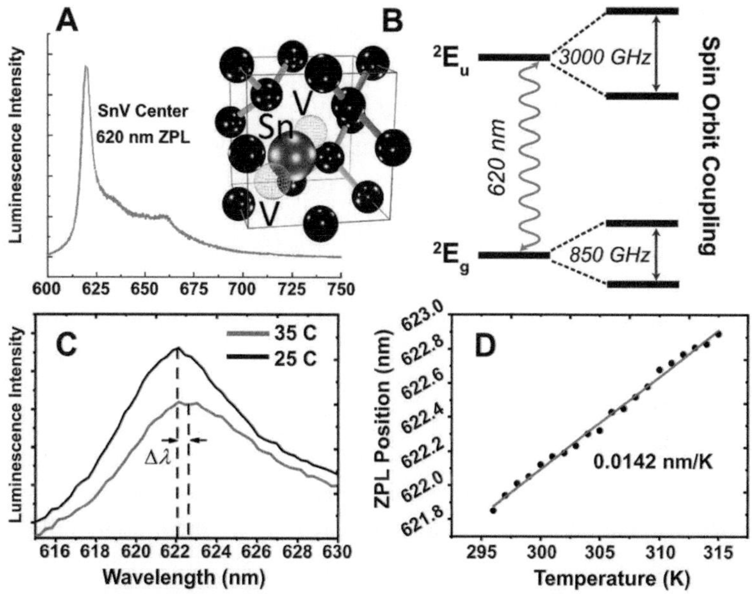

Figure 1.23 (a) Photoluminescence spectra of SnV center with 620 nm ZPL. Inset model of the SnV center with D_{3d} inversion symmetry along the <111> axis. (b) Electronic structure of SnV center in the absence of an external magnetic field. Spin-orbit coupling leads to splitting of the S = 1/2 ground and excited states by 850 GHz and 3 THz, respectively. (c) All optical thermometry with SnV centers in an ion implanted (100) diamond crystal with readout based on shift in the ZPL of the optical center. (d) Linear red-shift of the SnV ZPL with respect to temperature. (a) Adapted from Ref. [263] and (c,d) Reprinted with permission from Ref. [262], Copyright 2018, AIP Publishing.

1.4 Conclusion and Outlook

Optical centers in diamond are exceptional candidates for next generation solid-state quantum sensors. Significant technical advancements in the past 25 years in the implementation of diamond optical centers facilitated significant growth in the use and interest in this material and also laid a solid foundation for future technology development. However, the current progress cannot be sustained without further fundamental research and, particularly, advancing diamond fabrication methods. Many of the issues encountered with the use of optical centers for sensing ultimately stem from the quality of the host diamond crystal. This quality issue is mostly in regards to particulate diamond, as exceptional quality bulk CVD diamond with NV centers is now commercially available as of 2020 (Element Six's DNV-B1™ series [264]) for use in quantum computing and sensing applications. While bulk diamond is suitable for use in compact devices (e.g., compact magnetometers, radiation detectors), particulate diamond is required for sensing localized interactions within environments where bulk diamond cannot be introduced (e.g., intracellular sensing). The CVD method is not suitable for mass production of particulate diamond due to high production cost of CVD single crystal plates and absence of a developed approach for fragmentation of the bulk crystals into micron- and nano-sized particles. HPHT synthesis remains the major source of particulate diamond thus far. However, particulate HPHT diamond suffers from magnetic impurities in the as-grown particles and a variety of issues for particles fragmented by milling to smaller sizes, such as inhomogeneity in size and shape, lattice structural defects, inhomogeneous optical center content, and extensive, costly processing of milling and purification [116, 265]. Particulate diamond is unlikely to reach the quality standards of bulk CVD diamond; nevertheless, insights on quality improvements from CVD diamond can be applied to particulate diamond. The following recommendations to improve the quality of particulate diamond to enhance the quantum properties of hosted optical centers can be considered:

(1) **Reducing nitrogen content**: High nitrogen content is associated with poorer luminescence intensity of NV centers [36] and can also shorten spin-memory/dephasing T_2* times due to magnetic interactions with NV centers. A lower nitrogen content should also increase the average distance between neighboring NV centers, thereby reducing magnetic interactions between nearby NV centers. A substitutional nitrogen content of 10–20 ppm can be considered as a means to significantly reduce magnetic noise in particulate diamond.

(2) **Isotopic enrichment**: Efforts towards hyperpolarization with NV centers seek to increase the ^{13}C concentration; however, for sensing applications, the ^{12}C enrichment would be preferred

since it is a magnetically silent isotope. At 1.1% (11,000 ppm) natural abundance, ^{13}C could contribute significant magnetic noise to NV centers which are typically at a level of 1–3 ppm. ^{13}C enrichment with HPHT synthesized diamond has recently been demonstrated [87], and ^{12}C enrichment is also possible. ^{12}C enrichment has already been shown to significantly enhance coherence times of NV centers in bulk diamond [266, 267].

(3) **Further optimization of irradiation and annealing**: Irradiation optimization studies are costly, but more insight into the relationships between irradiation parameters (e.g., electron flux) and optical center spin quality is needed so that local maxima in the processing and performance parameter space can be more precisely identified. Annealing provides a straightforward and cost-effective way to improve optical center quantum properties [31, 32], and the use of ultra-high temperature annealing [138] to form various different color centers and reduce irradiation induced lattice damage can also reduce magnetic noise within the diamond lattice.

(4) **Optimized surface chemistry**: The surface presents perhaps the greatest challenge for particulate diamond. Interactions between near surface NVs and surface functional groups is inevitable, and they can impact quantum coherence [34]. Because many target applications for particulate diamond is in biology, it is necessary that the functional chemistry can be modified to facilitate targeting while not compromising sensing performance.

While challenges still remain, there are paths available to further improve the quality of diamond for quantum sensing. This is a critically important endeavor based on the wide scope of potential applications, some of which we have highlighted in this chapter. Bulk diamond has already established itself as a valuable material in quantum computing and magnetometry, and upon improving material properties, greater adoption of particulate diamond in the biological and biophysical disciplines will occur. The future of luminescent diamond as next generation quantum sensors is bright.

References

1. Boyle, R. (1664). *Experiments and Considerations Touching Colours.*
2. Zaitsev, A. M. (2001). *Optical Properties of Diamond: A Data Handbook* (1st ed.). Springer-Verlag Berlin Heidelberg.
3. Dutt, M. V. G., Childress, L., Jiang, L., Togan, E., Maze, J., Jelezko, F., Zibrov, A. S., Hemmer, P. R., & Lukin, M. D. (2007). Quantum register based on individual electronic and nuclear spin qubits in diamond. *Science, 316*(5829), 1312–1316. https://doi.org/10.1126/science.1139831
4. Shim, J. H., Niemeyer, I., Zhang, J., & Suter, D. (2013). Room-temperature high-speed nuclear-spin quantum memory in diamond. *Physical Review A, 87*(1), 012301. https://doi.org/10.1103/PhysRevA.87.012301
5. Aharonovich, I., & Prawer, S. (2014). Chapter 13 - Promising directions in diamond technologies for quantum information processing (QIP) and sensing. In S. Prawer & I. Aharonovich (eds.), *Quantum Information Processing with Diamond* (pp. 307–317). Woodhead Publishing. https://doi.org/https://doi.org/10.1533/9780857096685.3.307
6. Grezes, C., Kubo, Y., Julsgaard, B., Umeda, T., Isoya, J., Sumiya, H., Abe, H., Onoda, S., Ohshima, T., Nakamura, K., Diniz, I., Auffeves, A., Jacques, V., Roch, J.-F., Vion, D., Esteve, D., Moelmer, K., & Bertet, P. (2016). Towards a spin-ensemble quantum memory for superconducting qubits. *Comptes Rendus Physique, 17*(7), 693–704. https://doi.org/https://doi.org/10.1016/j.crhy.2016.07.006
7. Pfender, M., Aslam, N., Simon, P., Antonov, D., Thiering, G., Burk, S., Fávaro de Oliveira, F., Denisenko, A., Fedder, H., Meijer, J., Garrido, J. A., Gali, A., Teraji, T., Isoya, J., Doherty, M. W., Alkauskas, A., Gallo, A., Grüneis, A., Neumann, P., & Wrachtrup, J. (2017). Protecting a diamond quantum memory by charge state control. *Nano Letters, 17*(10), 5931–5937. https://doi.org/10.1021/acs.nanolett.7b01796
8. Awschalom, D. D., Hanson, R., Wrachtrup, J., & Zhou, B. B. (2018). Quantum technologies with optically interfaced solid-state spins. *Nature Photonics, 12*(9), 516–527. https://doi.org/10.1038/s41566-018-0232-2
9. Lai, Y.-Y., Lin, G.-D., Twamley, J., & Goan, H.-S. (2018). Single-nitrogen-vacancy-center quantum memory for a superconducting flux qubit mediated by a ferromagnet. *Physical Review A, 97*(5), 052303. https://doi.org/10.1103/PhysRevA.97.052303
10. Kalb, N., Reiserer, A. A., Humphreys, P. C., Bakermans, J. J. W., Kamerling, S. J., Nickerson, N. H., Benjamin, S. C., Twitchen, D. J., Markham, M., & Hanson, R. (2017). Entanglement distillation between solid-state

quantum network nodes. *Science, 356*(6341), 928. https://doi.org/10.1126/science.aan0070

11. Barry, J. F., Schloss, J. M., Bauch, E., Turner, M. J., Hart, C. A., Pham, L. M., & Walsworth, R. L. (2020). Sensitivity optimization for NV-diamond magnetometry. *Reviews of Modern Physics, 92*(1), 015004. https://doi.org/10.1103/RevModPhys.92.015004
12. Bucher, D. B., Aude Craik, D. P. L., Backlund, M. P., Turner, M. J., Ben Dor, O., Glenn, D. R., & Walsworth, R. L. (2019). Quantum diamond spectrometer for nanoscale NMR and ESR spectroscopy. *Nature Protocols, 14*(9), 2707–2747. https://doi.org/10.1038/s41596-019-0201-3
13. Schirhagl, R., Chang, K., Loretz, M., & Degen, C. L. (2014). Nitrogen-vacancy centers in diamond: nanoscale sensors for physics and biology. *Annual Review of Physical Chemistry, 65,* 83–105. https://doi.org/10.1146/annurev-physchem-040513-103659
14. Sigaeva, A., Hamoh, T., Perona, F., & Schirhagl, R. (2018). Fluorescent nanodiamonds: potential free radical detectors in live cells. *Free Radical Biology and Medicine, 120,* S87–S87. https://doi.org/10.1016/j.freeradbiomed.2018.04.288
15. Sigaeva, A., Ong, Y., Damle, V. G., Morita, A., van der Laan, K. J., & Schirhagl, R. (2019). Optical detection of intracellular quantities using nanoscale technologies. *Accounts of Chemical Research, 52*(7), 1739–1749. https://doi.org/10.1021/acs.accounts.9b00102
16. Acosta, V. M., Bauch, E., Ledbetter, M. P., Waxman, A., Bouchard, L. S., & Budker, D. (2010). Temperature dependence of the nitrogen-vacancy magnetic resonance in diamond. *Physical Review Letters, 104*(7), 070801. https://doi.org/10.1103/PhysRevLett.104.070801
17. Kucsko, G., Maurer, P. C., Yao, N. Y., Kubo, M., Noh, H. J., Lo, P. K., Park, H., & Lukin, M. D. (2013). Nanometre-scale thermometry in a living cell. *Nature, 500*(7460), 54–58. https://doi.org/10.1038/nature12373
18. Neumann, P., Jakobi, I., Dolde, F., Burk, C., Reuter, R., Waldherr, G., Honert, J., Wolf, T., Brunner, A., Shim, J. H., Suter, D., Sumiya, H., Isoya, J., & Wrachtrup, J. (2013). High-precision nanoscale temperature sensing using single defects in diamond. *Nano Letters, 13*(6), 2738–2742. https://doi.org/10.1021/nl401216y
19. Toyli, D. M., de las Casas, C. F., Christle, D. J., Dobrovitski, V. V., & Awschalom, D. D. (2013). Fluorescence thermometry enhanced by the quantum coherence of single spins in diamond. *Proceedings of the National Academy of Sciences of the United States of America, 110*(21), 8417–8421.http://www.jstor.org.prox.lib.ncsu.edu/stable/42656737

20. Simpson, D. A., Morrisroe, E., McCoey, J. M., Lombard, A. H., Mendis, D. C., Treussart, F., Hall, L. T., Petrou, S., & Hollenberg, L. C. L. (2017). Non-neurotoxic nanodiamond probes for intraneuronal temperature mapping. *ACS Nano, 11*(12), 12077–12086. https://doi.org/10.1021/acsnano.7b04850

21. Hui, Y. Y., Chen, O. Y., Azuma, T., Chang, B.-M., Hsieh, F.-J., & Chang, H.-C. (2019). All-optical thermometry with nitrogen-vacancy centers in nanodiamond-embedded polymer films. *The Journal of Physical Chemistry C, 123*(24), 15366–15374. https://doi.org/10.1021/acs.jpcc.9b04496

22. Doherty, M. W., Struzhkin, V. V., Simpson, D. A., McGuinness, L. P., Meng, Y., Stacey, A., Karle, T. J., Hemley, R. J., Manson, N. B., Hollenberg, L. C. L., & Prawer, S. (2014). Electronic properties and metrology applications of the diamond ${\mathrm{NV}}^{\ensuremath{-}}$ center under pressure. *Physical Review Letters, 112*(4), 047601. https://doi.org/10.1103/PhysRevLett.112.047601

23. Steele, L. G., Lawson, M., Onyszczak, M., Bush, B. T., Mei, Z., Dioguardi, A. P., King, J., Parker, A., Pines, A., Weir, S. T., Evans, W., Visbeck, K., Vohra, Y. K., & Curro, N. J. (2017). Optically detected magnetic resonance of nitrogen vacancies in a diamond anvil cell using designer diamond anvils. *Applied Physics Letters, 111*(22), 221903. https://doi.org/10.1063/1.5004153

24. Hsieh, S., Bhattacharyya, P., Zu, C., Mittiga, T., Smart, T. J., Machado, F., Kobrin, B., Höhn, T. O., Rui, N. Z., Kamrani, M., Chatterjee, S., Choi, S., Zaletel, M., Struzhkin, V. V., Moore, J. E., Levitas, V. I., Jeanloz, R., & Yao, N. Y. (2019). Imaging stress and magnetism at high pressures using a nanoscale quantum sensor. *Science, 366*(6471), 1349. https://doi.org/10.1126/science.aaw4352

25. Bradac, C., Gao, W., Forneris, J., Trusheim, M. E., & Aharonovich, I. (2019). Quantum nanophotonics with group IV defects in diamond. *Nature Communications, 10*(1), 5625. https://doi.org/10.1038/s41467-019-13332-w

26. Tran, T. T., Regan, B., Ekimov, E. A., Mu, Z., Zhou, Y., Gao, W.-B., Narang, P., Solntsev, A. S., Toth, M., Aharonovich, I., & Bradac, C. (2019). Anti-stokes excitation of solid-state quantum emitters for nanoscale thermometry. *Science Advances, 5*(5), eaav9180. https://doi.org/10.1126/sciadv.aav9180

27. Bradac, C., Lim, S. F., Chang, H.-C., & Aharonovich, I. (2020). Optical nanoscale thermometry: from fundamental mechanisms to emerging practical applications [https://doi.org/10.1002/adom.202000183].

Advanced Optical Materials, *8*(15), 2000183. https://doi.org/https://doi.org/10.1002/adom.202000183

28. Smith, J. M., Meynell, S. A., Bleszynski Jayich, A. C., & Meijer, J. (2019). Colour centre generation in diamond for quantum technologies. *Nanophotonics*, *8*(11), 1889–1906. https://doi.org/https://doi.org/10.1515/nanoph-2019-0196
29. Achard, J., Jacques, V., & Tallaire, A. (2020). Chemical vapour deposition diamond single crystals with nitrogen-vacancy centres: a review of material synthesis and technology for quantum sensing applications. *Journal of Physics D: Applied Physics*, *53*, 313001. https://doi.org/10.1088/1361-6463/ab81d1
30. Meynell, S. A., McLellan, C. A., Hughes, L. B., Wang, W., Mates, T. E., Mukherjee, K., & Bleszynski Jayich, A. C. (2020). Engineering quantum-coherent defects: The role of substrate miscut in chemical vapor deposition diamond growth. *Applied Physics Letters*, *117*(19), 194001. https://doi.org/10.1063/5.0029715
31. Naydenov, B., Reinhard, F., Lämmle, A., Richter, V., Kalish, R., D'Haenens-Johansson, U. F. S., Newton, M., Jelezko, F., & Wrachtrup, J. (2010). Increasing the coherence time of single electron spins in diamond by high temperature annealing. *Applied Physics Letters*, *97*(24), 242511. https://doi.org/10.1063/1.3527975
32. Yamamoto, T., Umeda, T., Watanabe, K., Onoda, S., Markham, M. L., Twitchen, D. J., Naydenov, B., McGuinness, L. P., Teraji, T., Koizumi, S., Dolde, F., Fedder, H., Honert, J., Wrachtrup, J., Ohshima, T., Jelezko, F., & Isoya, J. (2013). Extending spin coherence times of diamond qubits by high-temperature annealing. *Physical Review B*, *88*(7), 075206. https://doi.org/10.1103/PhysRevB.88.075206
33. Nagl, A., Hemelaar, S. R., & Schirhagl, R. (2015). Improving surface and defect center chemistry of fluorescent nanodiamonds for imaging purposes: a review. *Analytical and Bioanalytical Chemistry*, *407*(25), 7521–7536. https://doi.org/10.1007/s00216-015-8849-1
34. Ryan, R. G., Stacey, A., O'Donnell, K. M., Ohshima, T., Johnson, B. C., Hollenberg, L. C. L., Mulvaney, P., & Simpson, D. A. (2018). Impact of surface functionalization on the quantum coherence of nitrogen-vacancy centers in nanodiamonds. *ACS Applied Materials & Interfaces*, *10*(15), 13143–13149. https://doi.org/10.1021/acsami.7b19238
35. Bull, C., & Garlick, G. F. J. (1950). The luminescence of diamonds. *Proceedings of the Physical Society. Section A*, *63*(11), 1283–1291. https://doi.org/10.1088/0370-1298/63/11/312

36. Shames, A. I., Osipov, V. Y., Bogdanov, K. V., Baranov, A. V., Zhukovskaya, M. V., Dalis, A., Vagarali, S. S., & Rampersaud, A. (2017). Does progressive nitrogen doping intensify negatively charged nitrogen vacancy emission from e-beam-irradiated Ib type high-pressure–high-temperature diamonds? *The Journal of Physical Chemistry C, 121*(9), 5232–5240. https://doi.org/10.1021/acs.jpcc.6b12827

37. Dolmatov, V. Y. (2020). On the possibility of obtaining nitrogen-free detonation nanodiamonds: effect of covalently bound nitrogen in explosive molecules on the yield of nanodiamonds. *Journal of Superhard Materials, 42*(4), 223–228. https://doi.org/10.3103/S1063457620040036

38. Shenderova, O. A., Vlasov, I. I., Turner, S., Van Tendeloo, G., Orlinskii, S. B., Shiryaev, A. A., Khomich, A. A., Sulyanov, S. N., Jelezko, F., & Wrachtrup, J. (2011). Nitrogen control in nanodiamond produced by detonation shock-wave-assisted synthesis. *The Journal of Physical Chemistry C, 115*(29), 14014–14024. https://doi.org/10.1021/jp202057q

39. Basso, L., Bazzanella, N., Cazzanelli, M., & Miotello, A. (2019). On the route towards a facile fluorescent nanodiamonds laser-synthesis. *Carbon, 153,* 148–155. https://doi.org/https://doi.org/10.1016/j.carbon.2019.07.025

40. Basso, L., Gorrini, F., Cazzanelli, M., Bazzanella, N., Bifone, A., & Miotello, A. (2018). An all-optical single-step process for production of nanometric-sized fluorescent diamonds [10.1039/C7NR08791H]. *Nanoscale, 10*(12), 5738–5744. https://doi.org/10.1039/C7NR08791H

41. Khachatryan, A., Aloyan, S., May, P. W., Sargsyan, R., Khachatryan, V., & Baghdasaryan, V. (2008). Graphite-to-diamond transformation induced by ultrasound cavitation. *Diamond and Related Materials, 17,* 931–936.

42. Bovenkerk, H. P., Bundy, F. P., Hall, H. T., Strong, H. M., & Wentorf, R. H. (1959). Preparation of diamond. *Nature, 184*(4693), 1094–1098. https://doi.org/10.1038/1841094a0

43. Bundy, F. P., Hall, H. T., Strong, H. M., & Wentorfjun, R. H. (1955). Man-made diamonds. *Nature, 176*(4471), 51–55. https://doi.org/10.1038/176051a0

44. Shenderova, O. A., Shames, A. I., Nunn, N. A., Torelli, M. D., Vlasov, I., & Zaitsev, A. (2019). Review article: synthesis, properties, and applications of fluorescent diamond particles. *Journal of Vacuum Science & Technology B, 37*(3), 030802. https://doi.org/10.1116/1.5089898

45. Shames, A. I., Dalis, A., Greentree, A. D., Gibson, B. C., Abe, H., Ohshima, T., Shenderova, O., Zaitsev, A., & Reineck, P. (2020). Near-infrared fluorescence from silicon- and nickel-based color centers in high-pressure high-temperature diamond micro- and nanoparticles [https://doi.org/10.1002/adom.202001047]. *Advanced Optical Materials, 8*(23), 2001047. https://doi.org/https://doi.org/10.1002/adom.202001047

46. Ekimov, E. A., Kudryavtsev, O. S., Mordvinova, N. E., Lebedev, O. I., & Vlasov, I. I. (2018). High-pressure synthesis of nanodiamonds from adamantane: myth or reality? *ChemNanoMat, 4*(3), 269–273. https://doi.org/10.1002/cnma.201700349

47. Boudou, J. P., Tisler, J., Reuter, R., Thorel, A., Curmi, P. A., Jelezko, F., & Wrachtrup, J. (2013). Fluorescent nanodiamonds derived from HPHT with a size of less than 10 nm [Article]. *Diamond and Related Materials, 37*, 80–86. https://doi.org/10.1016/j.diamond.2013.05.006

48. Boudou, J. P., Curmi, P. A., Jelezko, F., Wrachtrup, J., Aubert, P., Sennour, M., Balasubramanian, G., Reuter, R., Thorel, A., & Gaffet, E. (2009). High yield fabrication of fluorescent nanodiamonds [Article]. *Nanotechnology, 20*(23), 11, Article 235602. https://doi.org/10.1088/0957-4484/20/23/235602

49. Chang, Y. R., Lee, H. Y., Chen, K., Chang, C. C., Tsai, D. S., Fu, C. C., Lim, T. S., Tzeng, Y. K., Fang, C. Y., Han, C. C., Chang, H. C., & Fann, W. (2008). Mass production and dynamic imaging of fluorescent nanodiamonds. *Nature Nanotechnology, 3*(5), 284–288. https://doi.org/10.1038/nnano.2008.99

50. Liu, H., & Dandy, D. S. (1995). Chapter 3 - Diamond CVD techniques. In H. Liu & D. S. Dandy (eds.), *Diamond Chemical Vapor Deposition* (pp. 14–45). William Andrew Publishing. https://doi.org/https://doi.org/10.1016/B978-081551380-3.50004-9

51. Tallaire, A., Achard, J., Silva, F., Brinza, O., & Gicquel, A. (2013). Growth of large size diamond single crystals by plasma assisted chemical vapour deposition: recent achievements and remaining challenges. *Comptes Rendus Physique, 14*(2), 169–184. https://doi.org/https://doi.org/10.1016/j.crhy.2012.10.008

52. Tallaire, A., Brinza, O., De Feudis, M., Ferrier, A., Touati, N., Binet, L., Nicolas, L., Delord, T., Hétet, G., Herzig, T., Pezzagna, S., Goldner, P., & Achard, J. (2019). Synthesis of loose nanodiamonds containing nitrogen-vacancy centers for magnetic and thermal sensing. *ACS Applied Nano Materials, 2*(9), 5952–5962. https://doi.org/10.1021/acsanm.9b01395

53. Mochalin, V. N., Shenderova, O., Ho, D., & Gogotsi, Y. (2011). The properties and applications of nanodiamonds. *Nature Nanotechnology, 7*(1), 11–23. https://doi.org/10.1038/nnano.2011.209
54. Panich, A. M., Shames, A. I., Zousman, B., & Levinson, O. (2012). Magnetic resonance study of nanodiamonds prepared by laser-assisted technique. *Diamond and Related Materials, 23*, 150–153. https://doi.org/https://doi.org/10.1016/j.diamond.2011.12.047
55. Yang, L., May, P. W., Yin, L., Smith, J. A., & Rosser, K. N. (2007). Growth of diamond nanocrystals by pulsed laser ablation of graphite in liquid. *Diamond and Related Materials, 16*(4), 725–729. https://doi.org/https://doi.org/10.1016/j.diamond.2006.11.010
56. DeCarli, P. S., & Jamieson, J. C. (1961). Formation of diamond by explosive shock. *Science, 134*(3472), 92. https://doi.org/10.1126/science.134.3472.92
57. Leeds, S. M. (1999). *Characterisation of the gas-phase environment in a microwave plasma enhanced diamond chemical vapour deposition reactor using molecular beam mass spectrometry*. http://inis.iaea.org/search/search.aspx?orig_q=RN:31010675
58. Ashfold, M. N. R., Goss, J. P., Green, B. L., May, P. W., Newton, M. E., & Peaker, C. V. (2020). Nitrogen in diamond. *Chemical Reviews, 120*(12), 5745–5794. https://doi.org/10.1021/acs.chemrev.9b00518
59. Dobrzhinetskaya, L. F. (2012). Microdiamonds — Frontier of ultrahigh-pressure metamorphism: a review. *Gondwana Research, 21*(1), 207-223. https://doi.org/https://doi.org/10.1016/j.gr.2011.07.014
60. Dolmatov, V. Y., Ozerin, A. N., Kulakova, I. I., Bochechka, O. O., Lapchuk, N. M., Myllymäki, V., & Vehanen, A. (2020). Detonation nanodiamonds: new aspects in the theory and practice of synthesis, properties and applications. *Russian Chemical Reviews, 89*(12), 1428–1462. https://doi.org/10.1070/rcr4924
61. Vul', A. Y., & Shenderova, O. A. (2013). *Detonation Nanodiamonds: Science and Applications*. CRC Press.
62. Krüger, A., Kataoka, F., Ozawa, M., Fujino, T., Suzuki, Y., Aleksenskii, A. E., Vul', A. Y., & Ōsawa, E. (2005). Unusually tight aggregation in detonation nanodiamond: Identification and disintegration. *Carbon, 43*(8), 1722–1730. https://doi.org/https://doi.org/10.1016/j.carbon.2005.02.020
63. Osswald, S., Yushin, G., Mochalin, V., Kucheyev, S. O., & Gogotsi, Y. (2006). Control of sp2/sp3 carbon ratio and surface chemistry of nanodiamond powders by selective oxidation in air. *Journal of the American Chemical Society, 128*(35), 11635–11642. https://doi.org/10.1021/ja063303n

64. Turner, S., Lebedev, O., Shenderova, O., Vlasov, I., Verbeeck, J., & van Tendeloo, G. (2009). Determination of size, morphology, and nitrogen impurity location in treated detonation nanodiamond by transmission electron microscopy. *Advanced Functional Materials, 19*, 2116–2124. https://doi.org/10.1002/adfm.200801872

65. Vlasov, I. I., Shenderova, O., Turner, S., Lebedev, O. I., Basov, A. A., Sildos, I., Rähn, M., Shiryaev, A. A., & Van Tendeloo, G. (2010). Nitrogen and luminescent nitrogen-vacancy defects in detonation nanodiamond [https://doi.org/10.1002/smll.200901587]. *Small, 6*(5), 687–694. https://doi.org/https://doi.org/10.1002/smll.200901587

66. Reineck, P., Capelli, M., Lau, D. W. M., Jeske, J., Field, M. R., Ohshima, T., Greentree, A. D., & Gibson, B. C. (2017). Bright and photostable nitrogen-vacancy fluorescence from unprocessed detonation nanodiamond [10.1039/C6NR07834F]. *Nanoscale, 9*(2), 497–502. https://doi.org/10.1039/C6NR07834F

67. Terada, D., Segawa, T. F., Shames, A. I., Onoda, S., Ohshima, T., Ōsawa, E., Igarashi, R., & Shirakawa, M. (2019). Monodisperse five-nanometer-sized detonation nanodiamonds enriched in nitrogen-vacancy centers. *ACS Nano, 13*(6), 6461–6468. https://doi.org/10.1021/acsnano.8b09383

68. Decarli, P. S. (1995). Shock wave synthesis of diamond and other phases. *MRS Proceedings, 383*, 21, Article 21. https://doi.org/10.1557/PROC-383-21

69. Parsons, C. A. (1920). III. Bakerian lecture, 1918.—Experiments on the artificial production of diamond. *Philosophical Transactions of the Royal Society of London. Series A, Containing Papers of a Mathematical or Physical Character, 220*(571–581), 67–107. https://doi.org/10.1098/rsta.1920.0003

70. Kudryavtsev, O. S., Ekimov, E. A., Romshin, A. M., Pasternak, D. G., & Vlasov, I. I. (2018). Structure and luminescence properties of nanonodiamonds produced from adamantane. *Physica Status Solidi (A), 215*(22), 1800252. https://doi.org/10.1002/pssa.201800252

71. Ekimov, E. A., Kudryavtsev, O. S., Khomich, A. A., Lebedev, O. I., Dolenko, T. A., & Vlasov, II. (2015). High-pressure synthesis of boron-doped ultrasmall diamonds from an organic compound. *Advanced Materials (Deerfield Beach, Fla.), 27*(37), 5518–5522. https://doi.org/10.1002/adma.201502672

72. Ekimov, E. A., & Kondrin, M. V. (2020). Chapter 6 - High-pressure, high-temperature synthesis and doping of nanodiamonds. In C. E. Nebel, I. Aharonovich, N. Mizuochi, & M. Hatano (eds.), *Semiconductors and*

Semimetals (Vol. 103, pp. 161–199). Elsevier. https://doi.org/https://doi.org/10.1016/bs.semsem.2020.03.006

73. Alkahtani, M., Lang, J., Naydenov, B., Jelezko, F., & Hemmer, P. (2019). Growth of High-purity low-strain fluorescent nanodiamonds. *ACS Photonics, 6*(5), 1266–1271. https://doi.org/10.1021/acsphotonics.9b00224
74. Liang, Z. Z., Kanda, H., Jia, X., Ma, H. A., Zhu, P. W., Guan, Q.-F., & Zang, C. Y. (2006). Synthesis of diamond with high nitrogen concentration from powder catalyst-C-additive NaN3 by HPHT. *Carbon, 44*(5), 913–917. https://doi.org/https://doi.org/10.1016/j.carbon.2005.10.018
75. Chen, L., Shen, W., Fang, C., Zhang, Y., Mu, P., Zhou, G., Wang, Q., Zhang, Z., & Jia, X. (2020). High pressure and high temperature annealing of Ni-containing, nitrogen-rich synthetic diamonds and the formation of NE8 centers. *Crystal Growth & Design, 20*(5), 3257–3263. https://doi.org/10.1021/acs.cgd.0c00080
76. Sumiya, H., Toda, N., & Satoh, S. (2005). Development of high-quality large-size synthetic diamond crystals. *SEI Technical Review*, 10–16.
77. Tian, Y., Jia, X., Zang, C., Li, S., Xiao, H., Zhang, Y., Huang, G., Li, R., Han, Q., Ma, L., Li, Y., Chen, X., Zhang, C., & Ma, H. (2009). Dependence of nitrogen concentration in type Ib diamonds on synthesis temperature. *Chinese Science Bulletin, 54*(9), 1459–1462. https://doi.org/10.1007/s11434-009-0211-6
78. Wentorf, R. H. (1971). Diamond growth rates. *The Journal of Physical Chemistry, 75*(12), 1833–1837. https://doi.org/10.1021/j100681a013
79. Strong, H. M., & Chrenko, R. M. (1971). Diamond growth rates and physical properties of laboratory-made diamond. *The Journal of Physical Chemistry, 75*(12), 1838–1843. https://doi.org/10.1021/j100681a014
80. Nassau, K. (1993). In R. F. Davis (ed.), *Diamond Films and Coatings* (p. 31). Noyes Publishers.
81. Babich, Y. V., & Feigelson, B. N. (2006). Distribution of nitrogen-related defects in diamond single crystals grown under nonisothermal conditions. *Inorganic Materials, 42*(9), 971–975. https://doi.org/10.1134/S0020168506090081
82. Burns, R. C., Cvetkovic, V., Dodge, C. N., Evans, D. J. F., Rooney, M.-L. T., Spear, P. M., & Welbourn, C. M. (1990). Growth-sector dependence of optical features in large synthetic diamonds. *Journal of Crystal Growth, 104*(2), 257–279. https://doi.org/https://doi.org/10.1016/0022-0248(90)90126-6

83. Boyd, S. R., Pillinger, C. T., Milledge, H. J., Mendelssohn, M. J., & Seal, M. (1988). Fractionation of nitrogen isotopes in a synthetic diamond of mixed crystal habit. *Nature, 331*(6157), 604–607. https://doi.org/10.1038/331604a0

84. Ekimov, E. A., Lyapin, S. G., Boldyrev, K. N., Kondrin, M. V., Khmelnitskiy, R., Gavva, V. A., Kotereva, T. V., & Popova, M. N. (2015). Germanium–vacancy color center in isotopically enriched diamonds synthesized at high pressures. *JETP Letters, 102*(11), 701–706. https://doi.org/10.1134/S0021364015230034

85. Ekimov, E. A., Lyapin, S. G., & Kondrin, M. V. (2018). Tin-vacancy color centers in micro- and polycrystalline diamonds synthesized at high pressures. *Diamond and Related Materials, 87,* 223–227. https://doi.org/https://doi.org/10.1016/j.diamond.2018.06.014

86. Jagannadham, K., Verghese, K., & Butler, J. E. (2001). Neutron transmutation of 10B isotope-doped diamond. *Applied Physics Letters, 78*(4), 446–447. https://doi.org/10.1063/1.1342207

87. Mindarava, Y., Blinder, R., Liu, Y., Scheuer, J., Lang, J., Agafonov, V., Davydov, V. A., Laube, C., Knolle, W., Abel, B., Naydenov, B., & Jelezko, F. (2020). Synthesis and coherent properties of ^{13}C-enriched sub-micron diamond particles with nitrogen vacancy color centers. *Carbon, 165,* 395–403. https://doi.org/https://doi.org/10.1016/j.carbon.2020.04.071

88. Ekimov, E. A., Zibrov, I. P., Malykhin, S. A., Khmel'nitskii, R. A., & Vlasov, I. I. (2017). Luminescence properties of diamond prepared in the presence of rare-earth elements. *Inorganic Materials, 53*(8), 809–815. https://doi.org/10.1134/S0020168517080039

89. Ekimov, E. A., Zibrov, I. P., Malykhin, S. A., Khmelnitskiy, R. A., & Vlasov, I. I. (2017). Synthesis of diamond in double carbon-rare earth element systems. *Materials Letters, 193,* 130–132. https://doi.org/https://doi.org/10.1016/j.matlet.2017.01.110

90. Linares, R. (2013). *CVD-Grown Synthetic Diamonds, Part 1: History.* https://www.gia.edu/news-research-CVD-grown-part1#:~:text=To%20produce%20a%20gem%2Dquality,900%20to%201200%C2%B0C.

91. Lee, S. T., Lin, Z., & Jiang, X. (1999). CVD diamond films: nucleation and growth. *Materials Science and Engineering: R: Reports, 25*(4), 123–154. https://doi.org/https://doi.org/10.1016/S0927-796X(99)00003-0

92. Liang, Q., Yan, C.-s., Meng, Y., Lai, J., Krasnicki, S., Mao, H.-k., & Hemley, R. J. (2009). Recent advances in high-growth rate single-crystal CVD

diamond. *Diamond and Related Materials, 18*(5), 698–703. https://doi.org/https://doi.org/10.1016/j.diamond.2008.12.002

93. Eaton-Manana, S., & Shigley, J. E. (2016). Observations on CVD-grown synthetic diamonds: a review. *Gems & Gemology, 52*(3).
94. Thompson, J. M. T., & May, P. W. (2000). Diamond thin films: a 21st-century material. *Philosophical Transactions of the Royal Society of London. Series A: Mathematical, Physical and Engineering Sciences, 358*(1766), 473–495. https://doi.org/10.1098/rsta.2000.0542
95. Raghuveer, M. S., Yoganand, S. N., Jagannadham, K., Lemaster, R. L., & Bailey, J. (2002). Improved CVD diamond coatings on WC–Co tool substrates. *Wear, 253*(11), 1194–1206. https://doi.org/https://doi.org/10.1016/S0043-1648(02)00244-2
96. Ishiwata, H., Nakajima, M., Tahara, K., Ozawa, H., Iwasaki, T., & Hatano, M. (2017). Perfectly aligned shallow ensemble nitrogen-vacancy centers in (111) diamond. *Applied Physics Letters, 111*(4), 043103. https://doi.org/10.1063/1.4993160
97. Ozawa, H., Tahara, K., Ishiwata, H., Hatano, M., & Iwasaki, T. (2017). Formation of perfectly aligned nitrogen-vacancy-center ensembles in chemical-vapor-deposition-grown diamond (111). *Applied Physics Express, 10*(4), 045501. https://doi.org/10.7567/apex.10.045501
98. Fukui, T., Doi, Y., Miyazaki, T., Miyamoto, Y., Kato, H., Matsumoto, T., Makino, T., Yamasaki, S., Morimoto, R., Tokuda, N., Hatano, M., Sakagawa, Y., Morishita, H., Tashima, T., Miwa, S., Suzuki, Y., & Mizuochi, N. (2014). Perfect selective alignment of nitrogen-vacancy centers in diamond. *Applied Physics Express, 7*(5), 055201. https://doi.org/10.7567/apex.7.055201
99. Lesik, M., Tetienne, J. P., Tallaire, A., Achard, J., Mille, V., Gicquel, A., Roch, J. F., & Jacques, V. (2014). Perfect preferential orientation of nitrogen-vacancy defects in a synthetic diamond sample. *Applied Physics Letters, 104*(11), 113107. https://doi.org/10.1063/1.4869103
100. Edmonds, A. M., D'Haenens-Johansson, U. F. S., Cruddace, R. J., Newton, M. E., Fu, K. M. C., Santori, C., Beausoleil, R. G., Twitchen, D. J., & Markham, M. L. (2012). Production of oriented nitrogen-vacancy color centers in synthetic diamond. *Physical Review B, 86*(3), 035201. https://doi.org/10.1103/PhysRevB.86.035201
101. Michl, J., Teraji, T., Zaiser, S., Jakobi, I., Waldherr, G., Dolde, F., Neumann, P., Doherty, M. W., Manson, N. B., Isoya, J., & Wrachtrup, J. (2014). Perfect alignment and preferential orientation of nitrogen-vacancy centers during chemical vapor deposition diamond growth on (111)

surfaces. *Applied Physics Letters, 104*(10), 102407. https://doi.org/10.1063/1.4868128

102. Tallaire, A., Mayer, L., Brinza, O., Pinault-Thaury, M. A., Debuisschert, T., & Achard, J. (2017). Highly photostable NV centre ensembles in CVD diamond produced by using N_2O as the doping gas. *Applied Physics Letters, 111*(14), 143101. https://doi.org/10.1063/1.5004106

103. Othman, M. Z., May, P. W., Fox, N. A., & Heard, P. J. (2014). Incorporation of lithium and nitrogen into CVD diamond thin films. *Diamond and Related Materials, 44*, 1–7. https://doi.org/https://doi.org/10.1016/j.diamond.2014.02.001

104. Dunst, S., Sternschulte, H., & Schreck, M. (2009). Growth rate enhancement by nitrogen in diamond chemical vapor deposition—a catalytic effect. *Applied Physics Letters, 94*(22), 224101. https://doi.org/10.1063/1.3143631

105. Ashkinazi, E. E., Khmelnitskii, R. A., Sedov, V. S., Khomich, A. A., Khomich, A. V., & Ralchenko, V. G. (2017). Morphology of diamond layers grown on different facets of single crystal diamond substrates by a microwave plasma CVD in CH4-H2-N2 gas mixtures. *Crystals, 7*(6). https://doi.org/10.3390/cryst7060166

106. Ohno, K., Joseph Heremans, F., Bassett, L. C., Myers, B. A., Toyli, D. M., Bleszynski Jayich, A. C., Palmstrøm, C. J., & Awschalom, D. D. (2012). Engineering shallow spins in diamond with nitrogen delta-doping. *Applied Physics Letters, 101*(8), 082413. https://doi.org/10.1063/1.4748280

107. Lobaev, M. A., Gorbachev, A. M., Bogdanov, S. A., Vikharev, A. L., Radishev, D. B., Isaev, V. A., Chernov, V. V., Drozdov, M. N., & Yunin, P. A. (2017). Influence of CVD diamond growth conditions and misorientation angle on nitrogen incorporation [10.1051/epjconf/201714902003]. *EPJ Web of Conference, 149*. https://doi.org/10.1051/epjconf/201714902003

108. D'Haenenes-Johansson, U. F. S., Katrusha, A., Moe, K. S., Johnson, P., & Wang, W. (2015). Large Colorless HPHT-grown synthetic gem diamonds from new diamond technology, Russia. *Gems & Gemology, 51*(3).

109. Gatto Monticone, D., Traina, P., Moreva, E., Forneris, J., Olivero, P., Degiovanni, I. P., Taccetti, F., Giuntini, L., Brida, G., Amato, G., & Genovese, M. (2014). Native NIR-emitting single colour centres in CVD diamond. *New Journal of Physics, 16*(5), 053005. https://doi.org/10.1088/1367-2630/16/5/053005

110. Neu, E., Steinmetz, D., Riedrich-Möller, J., Gsell, S., Fischer, M., Schreck, M., & Becher, C. (2011). Single photon emission from silicon-vacancy

colour centres in chemical vapour deposition nano-diamonds on iridium. *New Journal of Physics, 13*(2), 025012. https://doi.org/10.1088/1367-2630/13/2/025012

111. Hamilton, J. G., Putnam, T. M., & Ehrmann, M. L. (1952). Effect of heavy charged particle and fast neutron irradiation on diamonds. *American Mineralogist, 37*(11–12), 941–949.

112. Benny, A. H., & Champion, F. C. (1954). Some effects of neutron irradiation of diamond. *Nature, 173*(4414), 1087–1087. https://doi.org/10.1038/1731087a0

113. Commercial Irradiation of Diamonds. (1956). *Nature, 178*(4529), 348–348. https://doi.org/10.1038/178348c0

114. Koike, J., Parkin, D., & Mitchell, T. (1992). Displacement threshold energy for type IIa diamond. *Applied Physics Letters, 60*, 1450–1452. https://doi.org/10.1063/1.107267

115. Lefèvre, J., Costantini, J.-M., Esnouf, S., & Petite, G. (2009). Silicon threshold displacement energy determined by photoluminescence in electron-irradiated cubic silicon carbide. *Journal of Applied Physics, 105*(2), 023520. https://doi.org/10.1063/1.3072681

116. Shenderova, O., Nunn, N., Oeckinghaus, T., Torelli, M., McGuire, G., Smith, K., Danilov, E., Reuter, R., Wrachtrup, J., Shames, A., Filonova, D., & Kinev, A. (2017). Commercial quantities of ultrasmall fluorescent nanodiamonds containing color centers. *SPIE Procedding, 10118*. https://doi.org/10.1117/12.2256800

117. Laube, C., Oeckinghaus, T., Lehnert, J., Griebel, J., Knolle, W., Denisenko, A., Kahnt, A., Meijer, J., Wrachtrup, J., & Abel, B. (2019). Controlling the fluorescence properties of nitrogen vacancy centers in nanodiamonds [10.1039/C8NR07828A]. *Nanoscale, 11*(4), 1770–1783. https://doi.org/10.1039/C8NR07828A

118. Capelli, M., Heffernan, A. H., Ohshima, T., Abe, H., Jeske, J., Hope, A., Greentree, A. D., Reineck, P., & Gibson, B. C. (2019). Increased nitrogen-vacancy centre creation yield in diamond through electron beam irradiation at high temperature. *Carbon, 143*, 714–719. https://doi.org/https://doi.org/10.1016/j.carbon.2018.11.051

119. Gruber, A., Dräbenstedt, A., Tietz, C., Fleury, L., Wrachtrup, J., & Borczyskowski, C. v. (1997). Scanning confocal optical microscopy and magnetic resonance on single defect centers. *Science, 276*(5321), 2012. https://doi.org/10.1126/science.276.5321.2012

120. Beveratos, A., Kühn, S., Brouri, R., Gacoin, T., Poizat, J. P., & Grangier, P. (2002). Room temperature stable single-photon source. *The European*

Physical Journal D - Atomic, Molecular, Optical and Plasma Physics, 18(2), 191–196. https://doi.org/10.1140/epjd/e20020023

121. Treussart, F., Jacques, V., Wu, E., Gacoin, T., Grangier, P., & Roch, J. F. (2006). Photoluminescence of single colour defects in 50 nm diamond nanocrystals. *Physica B: Condensed Matter, 376–377*, 926–929. https://doi.org/https://doi.org/10.1016/j.physb.2005.12.232

122. Shames, A. I., Smirnov, A. I., Milikisiyants, S., Danilov, E. O., Nunn, N., McGuire, G., Torelli, M. D., & Shenderova, O. (2017). Fluence-dependent evolution of paramagnetic triplet centers in e-beam irradiated microcrystalline Ib type HPHT diamond. *The Journal of Physical Chemistry C, 121*(40), 22335–22346. https://doi.org/10.1021/acs.jpcc.7b06514

123. Shenderova, O. A., Nunn, N. A., Torelli, M. D., McGuire, G. E., Shames, A. I., & Zaitsev, A. M. (2020). Toward production of diamond particles with improved fluorescence uniformity. *Physica B: Condensed Matter, 579*, 411868. https://doi.org/https://doi.org/10.1016/j.physb.2019.411868

124. Yokota, Y., Kotsuka, H., Sogi, T., Ma, J. S., Hiraki, A., Kawarada, H., Matsuda, K., & Hatada, M. (1992). Formation of optical centers in CVD diamond by electron and neutron irradiation. *Diamond and Related Materials, 1*(5), 470–477. https://doi.org/https://doi.org/10.1016/0925-9635(92)90147-G

125. Huang, Z., Li, W. D., Santori, C., Acosta, V. M., Faraon, A., Ishikawa, T., Wu, W., Winston, D., Williams, R. S., & Beausoleil, R. G. (2013). Diamond nitrogen-vacancy centers created by scanning focused helium ion beam and annealing. *Applied Physics Letters, 103*(8), 081906. https://doi.org/10.1063/1.4819339

126. Yu, S. J., Kang, M. W., Chang, H. C., Chen, K. M., & Yu, Y. C. (2005). Bright fluorescent nanodiamonds: no photobleaching and low cytotoxicity. *Journal of the American Chemical Society, 127*(50), 17604–17605. https://doi.org/10.1021/ja0567081

127. Wee, T.-L., Tzeng, Y.-K., Han, C.-C., Chang, H.-C., Fann, W., Hsu, J.-H., Chen, K.-M., & Yu, Y.-C. (2007). Two-photon excited fluorescence of nitrogen-vacancy centers in proton-irradiated type Ib diamond. *The Journal of Physical Chemistry A, 111*(38), 9379–9386. https://doi.org/10.1021/jp073938o

128. Botsoa, J., Sauvage, T., Adam, M. P., Desgardin, P., Leoni, E., Courtois, B., Treussart, F., & Barthe, M. F. (2011). Optimal conditions for $N{V}^{\ensuremath{-}}$ center formation in type-1b diamond studied using photoluminescence and positron annihilation spectroscopies.

Physical Review B, 84(12), 125209. https://doi.org/10.1103/PhysRevB.84.125209

129. Mita, Y., Kanehara, H., & Nisida, Y. (1997). The 0.545 eV center in neutron irradiated and annealed type Ib diamond. *Diamond and Related Materials, 6*(11), 1722–1725. https://doi.org/https://doi.org/10.1016/S0925-9635(97)00132-5

130. Mita, Y. (1996). Change of absorption spectra in type-Ib diamond with heavy neutron irradiation. *Physical Review B, 53*(17), 11360–11364. https://doi.org/10.1103/PhysRevB.53.11360

131. Mita, Y., Nisida, Y., & Okada, M. (2018). Formation of the nitrogen aggregates in annealed diamond by neutron irradiation. *AIP Advances, 8*(2), 025106. https://doi.org/10.1063/1.5011262

132. Havlik, J., Petrakova, V., Kucka, J., Raabova, H., Panek, D., Stepan, V., Zlamalova Cilova, Z., Reineck, P., Stursa, J., Kucera, J., Hruby, M., & Cigler, P. (2018). Extremely rapid isotropic irradiation of nanoparticles with ions generated in situ by a nuclear reaction. *Nature Communications, 9*(1), 4467. https://doi.org/10.1038/s41467-018-06789-8

133. Avalos, V., & Dannefaer, S. (2005). Positron annihilation studies of γ-irradiated type Ib and IIa diamonds. *Diamond and Related Materials, 14*(2), 155–159. https://doi.org/https://doi.org/10.1016/j.diamond.2004.09.006

134. Campbell, B., & Mainwood, A. (2000). Radiation damage of diamond by electron and gamma irradiation [https://doi.org/10.1002/1521-396X(200009)181:1<99::AID-PSSA99>3.0.CO;2-5]. *Physica Status Solidi (A), 181*(1), 99–107. https://doi.org/https://doi.org/10.1002/1521-396X(200009)181:1<99::AID-PSSA99>3.0.CO;2-5

135. Hunt, D. C., Twitchen, D. J., Newton, M. E., Baker, J. M., Anthony, T. R., Banholzer, W. F., & Vagarali, S. S. (2000). Identification of the neutral carbon $\langle 100 \rangle$-split interstitial in diamond. *Physical Review B, 61*(6), 3863–3876. https://doi.org/10.1103/PhysRevB.61.3863

136. McLellan, C. A., Myers, B. A., Kraemer, S., Ohno, K., Awschalom, D. D., & Bleszynski Jayich, A. C. (2016). Patterned formation of highly coherent nitrogen-vacancy centers using a focused electron irradiation technique. *Nano Letters, 16*(4), 2450–2454. https://doi.org/10.1021/acs.nanolett.5b05304

137. Breeding, C. M., Eaton-Manana, S., & Shigley, J. E. (2018). Natural-color green diamonds: a beutiful conundrum. *Gems & Gemology, 54*(1).

138. Dei Cas, L., Zeldin, S., Nunn, N., Torelli, M., Shames, A. I., Zaitsev, A. M., & Shenderova, O. (2019). From fancy blue to red: controlled production

of a vibrant color spectrum of fluorescent diamond particles. *Advanced Functional Materials, 0*(0), 1808362. https://doi.org/10.1002/adfm.201808362

139. McCloskey, D., Fox, D., O'Hara, N., Usov, V., Scanlan, D., McEvoy, N., Duesberg, G. S., Cross, G. L. W., Zhang, H. Z., & Donegan, J. F. (2014). Helium ion microscope generated nitrogen-vacancy centres in type Ib diamond. *Applied Physics Letters, 104*(3), 031109. https://doi.org/10.1063/1.4862331

140. Nöbauer, T., Buczak, K., Angerer, A., Putz, S., Steinhauser, G., Akbarzadeh, J., Peterlik, H., Majer, J., Schmiedmayer, J., & Trupke, M. (2013). Creation of ensembles of nitrogen-vacancy centers in diamond by neutron and electron irradiation. https://ui.adsabs.harvard.edu/abs/2013arXiv1309.0453N

141. Iwasaki, T., Ishibashi, F., Miyamoto, Y., Doi, Y., Kobayashi, S., Miyazaki, T., Tahara, K., Jahnke, K. D., Rogers, L. J., Naydenov, B., Jelezko, F., Yamasaki, S., Nagamachi, S., Inubushi, T., Mizuochi, N., & Hatano, M. (2015). Germanium-vacancy single color centers in diamond. *Scientific Reports, 5*(1), 12882. https://doi.org/10.1038/srep12882

142. Tchernij, S. D., Herzig, T., Forneris, J., Küpper, J., Pezzagna, S., Traina, P., Moreva, E., Degiovanni, I. P., Brida, G., Skukan, N., Genovese, M., Jakšić, M., Meijer, J., & Olivero, P. (2017). Single-photon-emitting optical centers in diamond fabricated upon sn implantation. *ACS Photonics, 4*(10), 2580–2586. https://doi.org/10.1021/acsphotonics.7b00904

143. Lühmann, T., Raatz, N., John, R., Lesik, M., Rödiger, J., Portail, M., Wildanger, D., Kleißler, F., Nordlund, K., Zaitsev, A., Roch, J.-F., Tallaire, A., Meijer, J., & Pezzagna, S. (2018). Screening and engineering of colour centres in diamond. *Journal of Physics D: Applied Physics, 51*(48), 483002. https://doi.org/10.1088/1361-6463/aadfab

144. Matindi, T. B., Naidoo, S. R., & Ntwaeaborwa, O. M. (2019). Luminescence induced by NO ion implantation into diamond. *Diamond and Related Materials, 96,* 11–19. https://doi.org/https://doi.org/10.1016/j.diamond.2019.04.024

145. Spinicelli, P., Dréau, A., Rondin, L., Silva, F., Achard, J., Xavier, S., Bansropun, S., Debuisschert, T., Pezzagna, S., Meijer, J., Jacques, V., & Roch, J. F. (2011). Engineered arrays of nitrogen-vacancy color centers in diamond based on implantation of CN−molecules through nanoapertures. *New Journal of Physics, 13*(2), 025014. https://doi.org/10.1088/1367-2630/13/2/025014

146. Meijer, J., Burchard, B., Domhan, M., Wittmann, C., Gaebel, T., Popa, I., Jelezko, F., & Wrachtrup, J. (2005). Generation of single color centers

by focused nitrogen implantation. *Applied Physics Letters, 87*(26), 261909. https://doi.org/10.1063/1.2103389

147. Rabeau, J. R., Reichart, P., Tamanyan, G., Jamieson, D. N., Prawer, S., Jelezko, F., Gaebel, T., Popa, I., Domhan, M., & Wrachtrup, J. (2006). Implantation of labelled single nitrogen vacancy centers in diamond using N15. *Applied Physics Letters, 88*(2), 023113. https://doi.org/10.1063/1.2158700

148. Meijer, J., Pezzagna, S., Vogel, T., Burchard, B., Bukow, H. H., Rangelow, I. W., Sarov, Y., Wiggers, H., Plümel, I., Jelezko, F., Wrachtrup, J., Schmidt-Kaler, F., Schnitzler, W., & Singer, K. (2008). Towards the implanting of ions and positioning of nanoparticles with nm spatial resolution. *Applied Physics A, 91*(4), 567–571. https://doi.org/10.1007/s00339-008-4515-1

149. Fukuda, R., Balasubramanian, P., Higashimata, I., Koike, G., Okada, T., Kagami, R., Teraji, T., Onoda, S., Haruyama, M., Yamada, K., Inaba, M., Yamano, H., Stürner, F. M., Schmitt, S., McGuinness, L. P., Jelezko, F., Ohshima, T., Shinada, T., Kawarada, H., Kada, W., Hanaizumi, O., Tanii, T., & Isoya, J. (2018). Lithographically engineered shallow nitrogen-vacancy centers in diamond for external nuclear spin sensing. *New Journal of Physics, 20*(8), 083029. https://doi.org/10.1088/1367-2630/aad997

150. van Dam, S. B., Walsh, M., Degen, M. J., Bersin, E., Mouradian, S. L., Galiullin, A., Ruf, M., Ijspeert, M., Taminiau, T. H., Hanson, R., & Englund, D. R. (2019). Optical coherence of diamond nitrogen-vacancy centers formed by ion implantation and annealing. *Physical Review B, 99*(16), 161203. https://doi.org/10.1103/PhysRevB.99.161203

151. Pezzagna, S., Naydenov, B., Jelezko, F., Wrachtrup, J., & Meijer, J. (2010). Creation efficiency of nitrogen-vacancy centres in diamond. *New Journal of Physics, 12*(6), 065017. https://doi.org/10.1088/1367-2630/12/6/065017

152. Tsigkourakos, M., Hantschel, T., Simon, D. K., Nuytten, T., Verhulst, A. S., Douhard, B., & Vandervorst, W. (2014). On the local conductivity of individual diamond seeds and their impact on the interfacial resistance of boron-doped diamond films. *Carbon, 79*, 103–112. https://doi.org/https://doi.org/10.1016/j.carbon.2014.07.048

153. Crane, M. J., Smith, B. E., Meisenheimer, P. B., Zhou, X., Stroud, R. M., James Davis, E., & Pauzauskie, P. J. (2018). Photothermal effects during nanodiamond synthesis from a carbon aerogel in a laser-heated diamond anvil cell. *Diamond and Related Materials, 87*, 134–142. https://doi.org/https://doi.org/10.1016/j.diamond.2018.05.013

154. Koga, K. T., Walter, M. J., Nakamura, E., & Kobayashi, K. (2005). Carbon self-diffusion in a natural diamond. *Physical Review B, 72*(2), 024108. https://doi.org/10.1103/PhysRevB.72.024108

155. Haque, A., & Sumaiya, S. (2017). An overview on the formation and processing of nitrogen-vacancy photonic centers in diamond by ion implantation. *Journal of Manufacturing and Materials Processing, 1*(1), 6. https://www.mdpi.com/2504-4494/1/1/6

156. Davies, G., Lawson, S. C., Collins, A. T., Mainwood, A., & Sharp, S. J. (1992). Vacancy-related centers in diamond. *Physical Review B, 46*(20), 13157–13170. https://doi.org/10.1103/PhysRevB.46.13157

157. Nunn, N., Prabhakar, N., Reineck, P., Magidson, V., Kamiya, E., Heinz, W. F., Torelli, M. D., Rosenholm, J., Zaitsev, A., & Shenderova, O. (2019). Brilliant blue, green, yellow, and red fluorescent diamond particles: synthesis, characterization, and multiplex imaging demonstrations [10.1039/C9NR02593F]. *Nanoscale, 11*(24), 11584–11595. https://doi.org/10.1039/C9NR02593F

158. Wee, T.-L., Mau, Y.-W., Fang, C.-Y., Hsu, H.-L., Han, C.-C., & Chang, H.-C. (2009). Preparation and characterization of green fluorescent nanodiamonds for biological applications. *Diamond and Related Materials, 18*(2), 567–573. https://doi.org/https://doi.org/10.1016/j.diamond.2008.08.012

159. Kazuchits, N. M., Rusetsky, M. S., Kazuchits, V. N., & Zaitsev, A. M. (2016). Aggregation of nitrogen in synthetic diamonds annealed at high temperature without stabilizing pressure. *Diamond and Related Materials, 64*, 202–207. https://doi.org/https://doi.org/10.1016/j.diamond.2016.03.002

160. Narulkar, R., Bukkapatnam, S., Raff, L. M., & Komanduri, R. (2009). Graphitization as a precursor to wear of diamond in machining pure iron: a molecular dynamics investigation. *Computational Materials Science, 45*(2), 358–366. https://doi.org/https://doi.org/10.1016/j.commatsci.2008.10.007

161. Niwase, K., Tanaka, T., Kakimoto, Y., Ishihara, K. N., & Shingu, P. H. (1995). Raman spectra of graphite and diamond mechanically milled with agate or stainless steel ball-mill. *Materials Transactions, JIM, 36*(2), 282–288. https://doi.org/10.2320/matertrans1989.36.282

162. Abdullahi, I. M., Langenderfer, M., Shenderova, O., Nunn, N., Torelli, M. D., Johnson, C. E., & Mochalin, V. N. (2020). Explosive fragmentation of luminescent diamond particles. *Carbon, 164*, 442–450. https://doi.org/https://doi.org/10.1016/j.carbon.2020.03.057

163. Shames, A. I., Osipov, V. Y., Boudou, J. P., Panich, A. M., Bardeleben, H. J. v., Treussart, F., & Vul', A. Y. (2015). Magnetic resonance tracking of fluorescent nanodiamond fabrication. *Journal of Physics D: Applied Physics, 48*(15), 155302. http://stacks.iop.org/0022-3727/48/i=15/a=155302

164. Weil, J. A., & Bolton, J. R. (2007). *Electron Paramagnetic Resonance: Elementary Theory and Practical Applications.* Wiley.

165. Abragam, A., & Bleaney, B. (2012). *Electron Paramagnetic Resonance of Transition Ions,* reprinted ed. Oxford University Press.

166. Baranov, P. G., von Bardeleben, H. J., Jelezko, F., & Wrachtrup, J. (2017). *Magnetic Resonance of Semiconductors and Their Nanostructures.* Springer-Verlag Wien.

167. Levitt, M. H. (2008). *Spin Dynamics Basiscs of Nuclear Magnetic Resonance* (2nd ed.). John Wiley & Sons.

168. Duer, M. J. (2001). *Solid State NMR Spectroscopy: Principles and Applications.* Wiley-Blackwell.

169. Newton, M. E. (2007). Chapter 6 - EPR, ENDOR and EPR imaging of defects in diamond. *Electron Paramagnetic Resonance* (Vol. 20, pp. 131–156). The Royal Society of Chemistry. https://doi.org/10.1039/9781847557568-00131

170. D'Haenens-Johansson, U. F. S., Edmonds, A. M., Green, B. L., Newton, M. E., Davies, G., Martineau, P. M., Khan, R. U. A., & Twitchen, D. J. (2011). Optical properties of the neutral silicon split-vacancy center in diamond. *Physical Review B, 84*(24), 245208. https://doi.org/10.1103/PhysRevB.84.245208

171. Shames, A. I., Osipov, V. Y., von Bardeleben, H. J., Boudou, J. P., Treussart, F., & Vul', A. Y. (2014). Native and induced triplet nitrogen-vacancy centers in nano- and micro-diamonds: half-field electron paramagnetic resonance fingerprint. *Applied Physics Letters, 104*(6), 063107. https://doi.org/10.1063/1.4865205

172. Segawa, T. F., & Shames, A. I. (2020). How to identify, attribute, and quantify triplet defects in ensembles of small nanoparticles. *The Journal of Physical Chemistry Letters, 11*(17), 7438–7442. https://doi.org/10.1021/acs.jpclett.0c01922

173. Jelezko, F., & Wrachtrup, J. (2006). Single defect centres in diamond: a review. *physica status solidi (a), 203*(13), 3207–3225. https://doi.org/https://doi.org/10.1002/pssa.200671403

174. Abe, E., & Sasaki, K. (2018). Tutorial: Magnetic resonance with nitrogen-vacancy centers in diamond—microwave engineering,

materials science, and magnetometry. *Journal of Applied Physics, 123*(16), 161101. https://doi.org/10.1063/1.5011231

175. Rembold, P., Oshnik, N., Müller, M. M., Montangero, S., Calarco, T., & Neu, E. (2020). Introduction to quantum optimal control for quantum sensing with nitrogen-vacancy centers in diamond. *AVS Quantum Science, 2*(2), 024701. https://doi.org/10.1116/5.0006785

176. Batalov, A., Zierl, C., Gaebel, T., Neumann, P., Chan, I. Y., Balasubramanian, G., Hemmer, P. R., Jelezko, F., & Wrachtrup, J. (2008). Temporal coherence of photons emitted by single nitrogen-vacancy defect centers in diamond using optical rabi-oscillations. *Physical Review Letters, 100*(7), 077401. https://doi.org/10.1103/PhysRevLett.100.077401

177. Acosta, V. M., Jarmola, A., Bauch, E., & Budker, D. (2010). Optical properties of the nitrogen-vacancy singlet levels in diamond. *Physical Review B, 82*(20), 201202. https://doi.org/10.1103/PhysRevB.82.201202

178. Perona Martínez, F., Nusantara, A. C., Chipaux, M., Padamati, S. K., & Schirhagl, R. (2020). Nanodiamond relaxometry-based detection of free-radical species when produced in chemical reactions in biologically relevant conditions. *ACS Sensors, 5*(12), 3862–3869. https://doi.org/10.1021/acssensors.0c01037

179. Steinert, S., Ziem, F., Hall, L. T., Zappe, A., Schweikert, M., Götz, N., Aird, A., Balasubramanian, G., Hollenberg, L., & Wrachtrup, J. (2013). Magnetic spin imaging under ambient conditions with sub-cellular resolution. *Nature Communications, 4*(1), 1607. https://doi.org/10.1038/ncomms2588

180. Balasubramanian, P., Osterkamp, C., Chen, Y., Chen, X., Teraji, T., Wu, E., Naydenov, B., & Jelezko, F. (2019). dc Magnetometry with engineered nitrogen-vacancy spin ensembles in diamond. *Nano Letters, 19*(9), 6681–6686. https://doi.org/10.1021/acs.nanolett.9b02993

181. Hong, S., Grinolds, M. S., Pham, L. M., Le Sage, D., Luan, L., Walsworth, R. L., & Yacoby, A. (2013). Nanoscale magnetometry with NV centers in diamond. *MRS Bulletin, 38*(2), 155–161. https://doi.org/10.1557/mrs.2013.23

182. Plakhotnik, T., Doherty, M. W., Cole, J. H., Chapman, R., & Manson, N. B. (2014). All-optical thermometry and thermal properties of the optically detected spin resonances of the NV– center in nanodiamond. *Nano Letters, 14*(9), 4989–4996. https://doi.org/10.1021/nl501841d

183. Plakhotnik, T., Aman, H., & Chang, H. C. (2015). All-optical single-nanoparticle ratiometric thermometry with a noise floor of

0.3 K Hz(-1/2). *Nanotechnology, 26*(24), 245501. https://doi.org/10.1088/0957-4484/26/24/245501

184. Slichter, C. P. (1992). *Principles of Magnetic Resonance.* Springer-Verlag.

185. Tetienne, J. P., Hingant, T., Rondin, L., Cavaillès, A., Mayer, L., Dantelle, G., Gacoin, T., Wrachtrup, J., Roch, J. F., & Jacques, V. (2013). Spin relaxometry of single nitrogen-vacancy defects in diamond nanocrystals for magnetic noise sensing. *Physical Review B, 87*(23), 235436. https://doi.org/10.1103/PhysRevB.87.235436

186. Ermakova, A., Pramanik, G., Cai, J. M., Algara-Siller, G., Kaiser, U., Weil, T., Tzeng, Y. K., Chang, H. C., McGuinness, L. P., Plenio, M. B., Naydenov, B., & Jelezko, F. (2013). Detection of a few metallo-protein molecules using color centers in nanodiamonds. *Nano Letters, 13*(7), 3305–3309. https://doi.org/10.1021/nl4015233

187. Morita, A., Nusantara, A. C., Perona Martinez, F. P., Hamoh, T., Damle, V. G., van der Laan, K. J., Sigaeva, A., Vedelaar, T., Chang, M., Chipaux, M., & Schirhagl, R. (2020). Quantum monitoring the metabolism of individual yeast mutant strain cells when aged, stressed or treated with antioxidant. https://ui.adsabs.harvard.edu/abs/2020arXiv200716130M

188. Brites, C. D. S., Lima, P. P., Silva, N. J. O., Millán, A., Amaral, V. S., Palacio, F., & Carlos, L. D. (2012). Thermometry at the nanoscale [10.1039/C2NR30663H]. *Nanoscale, 4*(16), 4799–4829. https://doi.org/10.1039/C2NR30663H

189. Choi, S., Agafonov, V. N., Davydov, V. A., & Plakhotnik, T. (2019). Ultrasensitive all-optical thermometry using nanodiamonds with a high concentration of silicon-vacancy centers and multiparametric data analysis. *ACS Photonics, 6*(6), 1387–1392. https://doi.org/10.1021/acsphotonics.9b00468

190. Kucsko, G., Maurer, P. C., Yao, N. Y., Kubo, M., Noh, H. J., Lo, P. K., Park, H., & Lukin, M. D. (2013). Nanometre-scale thermometry in a living cell. *Nature, 500*(7460), 54–58. https://doi.org/10.1038/nature12373

191. Alkahtani, M., Jiang, L., Brick, R., Hemmer, P., & Scully, M. (2017). Nanometer-scale luminescent thermometry in bovine embryos. *Optics Letters, 42*(23), 4812-4815. https://doi.org/10.1364/OL.42.004812

192. Alkahtani, M. H., Alghannam, F., Jiang, L., Rampersaud, A. A., Brick, R., Gomes, C. L., Scully, M. O., & Hemmer, P. R. (2018). Fluorescent nanodiamonds for luminescent thermometry in the biological transparency window. *Optics Letters, 43*(14), 3317–3320. https://doi.org/10.1364/OL.43.003317

193. Choe, S., Yoon, J., Lee, M., Oh, J., Lee, D., Kang, H., Lee, C.-H., & Lee, D. (2018). Precise temperature sensing with nanoscale thermal sensors based on diamond NV centers. *Current Applied Physics, 18*(9), 1066–1070. https://doi.org/https://doi.org/10.1016/j.cap.2018.06.002

194. Yukawa, H., Fujiwara, M., Kobayashi, K., Kumon, Y., Miyaji, K., Nishimura, Y., Oshimi, K., Umehara, Y., Teki, Y., Iwasaki, T., Hatano, M., Hashimoto, H., & Baba, Y. (2020). A quantum thermometric sensing and analysis system using fluorescent nanodiamonds for the evaluation of living stem cell functions according to intracellular temperature [10.1039/D0NA00146E]. *Nanoscale Advances, 2*(5), 1859–1868. https://doi.org/10.1039/D0NA00146E

195. Sekiguchi, T., Sotoma, S., & Harada, Y. (2018). Fluorescent nanodiamonds as a robust temperature sensor inside a single cell. *Biophys Physicobiol, 15*, 229–234. https://doi.org/10.2142/biophysico.15.0_229

196. Hayashi, K., Matsuzaki, Y., Taniguchi, T., Shimo-Oka, T., Nakamura, I., Onoda, S., Ohshima, T., Morishita, H., Fujiwara, M., Saito, S., & Mizuochi, N. (2018). Optimization of temperature sensitivity using the optically detected magnetic-resonance spectrum of a nitrogen-vacancy center ensemble. *Physical Review Applied, 10*(3), 034009. https://doi.org/10.1103/PhysRevApplied.10.034009

197. Foy, C., Zhang, L., Trusheim, M. E., Bagnall, K. R., Walsh, M., Wang, E. N., & Englund, D. R. (2020). Wide-field magnetic field and temperature imaging using nanoscale quantum sensors. *ACS Applied Materials & Interfaces, 12*(23), 26525–26533. https://doi.org/10.1021/acsami.0c01545

198. de Guillebon, T., Vindolet, B., Roch, J. F., Jacques, V., & Rondin, L. (2020). Temperature dependence of the longitudinal spin relaxation time T_1 of single nitrogen-vacancy centers in nanodiamonds. *Physical Review B, 102*(16), 165427. https://doi.org/10.1103/PhysRevB.102.165427

199. Barson, M. S. J., Peddibhotla, P., Ovartchaiyapong, P., Ganesan, K., Taylor, R. L., Gebert, M., Mielens, Z., Koslowski, B., Simpson, D. A., McGuinness, L. P., McCallum, J., Prawer, S., Onoda, S., Ohshima, T., Bleszynski Jayich, A. C., Jelezko, F., Manson, N. B., & Doherty, M. W. (2017). Nanomechanical sensing using spins in diamond. *Nano Letters, 17*(3), 1496–1503. https://doi.org/10.1021/acs.nanolett.6b04544

200. Blank, A., Twig, Y., & Ishay, Y. (2017). Recent trends in high spin sensitivity magnetic resonance. *Journal of Magnetic Resonance, 280*, 20–29. https://doi.org/https://doi.org/10.1016/j.jmr.2017.02.019

201. Taylor, J. M., Cappellaro, P., Childress, L., Jiang, L., Budker, D., Hemmer, P. R., Yacoby, A., Walsworth, R., & Lukin, M. D. (2008). High-sensitivity diamond magnetometer with nanoscale resolution. *Nature Physics, 4*(10), 810–816. https://doi.org/10.1038/nphys1075

202. Balasubramanian, G., Chan, I. Y., Kolesov, R., Al-Hmoud, M., Tisler, J., Shin, C., Kim, C., Wojcik, A., Hemmer, P. R., Krueger, A., Hanke, T., Leitenstorfer, A., Bratschitsch, R., Jelezko, F., & Wrachtrup, J. (2008). Nanoscale imaging magnetometry with diamond spins under ambient conditions. *Nature, 455*(7213), 648–651. https://doi.org/10.1038/nature07278

203. Lovchinsky, I., Sushkov, A. O., Urbach, E., de Leon, N. P., Choi, S., De Greve, K., Evans, R., Gertner, R., Bersin, E., Muller, C., McGuinness, L., Jelezko, F., Walsworth, R. L., Park, H., & Lukin, M. D. (2016). Nuclear magnetic resonance detection and spectroscopy of single proteins using quantum logic. *Science, 351*(6275), 836–841. https://doi.org/10.1126/science.aad8022

204. Shi, F., Zhang, Q., Wang, P., Sun, H., Wang, J., Rong, X., Chen, M., Ju, C., Reinhard, F., Chen, H., Wrachtrup, J., Wang, J., & Du, J. (2015). Single-protein spin resonance spectroscopy under ambient conditions. *Science, 347*(6226), 1135. https://doi.org/10.1126/science.aaa2253

205. Smits, J., Damron, J. T., Kehayias, P., McDowell, A. F., Mosavian, N., Fescenko, I., Ristoff, N., Laraoui, A., Jarmola, A., & Acosta, V. M. (2019). Two-dimensional nuclear magnetic resonance spectroscopy with a microfluidic diamond quantum sensor. *Science Advances, 5*(7), eaaw7895. https://doi.org/10.1126/sciadv.aaw7895

206. Stürner, F. M., Brenneis, A., Kassel, J., Wostradowski, U., Rölver, R., Fuchs, T., Nakamura, K., Sumiya, H., Onoda, S., Isoya, J., & Jelezko, F. (2019). Compact integrated magnetometer based on nitrogen-vacancy centres in diamond. *Diamond and Related Materials, 93*, 59–65. https://doi.org/https://doi.org/10.1016/j.diamond.2019.01.008

207. Kim, D., Ibrahim, M. I., Foy, C., Trusheim, M. E., Han, R., & Englund, D. R. (2019). A CMOS-integrated quantum sensor based on nitrogen–vacancy centres. *Nature Electronics, 2*(7), 284–289. https://doi.org/10.1038/s41928-019-0275-5

208. Stürner, F. M., Brenneis, A., Buck, T., Kassel, J., Rölver, R., Fuchs, T., Savitsky, A., Suter, D., Grimmel, J., Hengesbach, S., Förtsch, M., Nakamura, K., Sumiya, H., Onoda, S., Isoya, J., & Jelezko, F. (2020). Integrated and portable magnetometer based on nitrogen-vacancy ensembles in diamond. https://ui.adsabs.harvard.edu/abs/2020arXiv201201053S

209. Tetienne, J.-P., Dontschuk, N., Broadway, D. A., Stacey, A., Simpson, D. A., & Hollenberg, L. C. L. (2017). Quantum imaging of current flow in graphene. *Science Advances, 3*(4), e1602429. https://doi.org/10.1126/sciadv.1602429

210. Barry, J. F., Turner, M. J., Schloss, J. M., Glenn, D. R., Song, Y., Lukin, M. D., Park, H., & Walsworth, R. L. (2016). Optical magnetic detection of single-neuron action potentials using quantum defects in diamond. *Proceedings of the National Academy of Sciences, 113*(49), 14133. https://doi.org/10.1073/pnas.1601513113

211. Price, J. C., Mesquita-Ribeiro, R., Dajas-Bailador, F., & Mather, M. L. (2020). Widefield, spatiotemporal mapping of spontaneous activity of mouse cultured neuronal networks using quantum diamond sensors [Original Research]. *Frontiers in Physics, 8*(255). https://doi.org/10.3389/fphy.2020.00255

212. Kitching, J. (2018). Chip-scale atomic devices. *Applied Physics Reviews, 5*(3), 031302. https://doi.org/10.1063/1.5026238

213. Hart, C. A., Schloss, J. M., Turner, M. J., Scheidegger, P. J., Bauch, E., & Walsworth, R. L. (2020). NV-diamond magnetic microscopy using a double quantum 4-Ramsey protocol. https://ui.adsabs.harvard.edu/abs/2020arXiv200902371H

214. Hopper, D. A., Shulevitz, H. J., & Bassett, L. C. (2018). Spin readout techniques of the nitrogen-vacancy center in diamond. *Micromachines, 9*(9). https://doi.org/10.3390/mi9090437

215. Wolf, T., Neumann, P., Nakamura, K., Sumiya, H., Ohshima, T., Isoya, J., & Wrachtrup, J. (2015). Subpicotesla diamond magnetometry. *Physical Review X, 5*(4), 041001. https://doi.org/10.1103/PhysRevX.5.041001

216. Rondin, L., Tetienne, J. P., Hingant, T., Roch, J. F., Maletinsky, P., & Jacques, V. (2014). Magnetometry with nitrogen-vacancy defects in diamond. *Reports on Progress in Physics, 77*(5), 056503. https://doi.org/10.1088/0034-4885/77/5/056503

217. Perunicic, V. S., Hall, L. T., Simpson, D. A., Hill, C. D., & Hollenberg, L. C. L. (2014). Towards single-molecule NMR detection and spectroscopy using single spins in diamond. *Physical Review B, 89*(5), 054432. https://doi.org/10.1103/PhysRevB.89.054432

218. Bettenhausen, C. A. (2020, May 15, 2020). Bruker installs world's first 1.2 GHz NMR. *Chemical and Engineering News, 98*(19). https://cen.acs.org/business/instrumentation/Bruker-installs-12-GHz-NMR/98/i19

219. Lilly Thankamony, A. S., Wittmann, J. J., Kaushik, M., & Corzilius, B. (2017). Dynamic nuclear polarization for sensitivity enhancement

in modern solid-state NMR. *Progress in Nuclear Magnetic Resonance Spectroscopy, 102–103,* 120–195. https://doi.org/https://doi.org/10.1016/j.pnmrs.2017.06.002

220. Kovtunov, K. V., Koptyug, I. V., Fekete, M., Duckett, S. B., Theis, T., Joalland, B., & Chekmenev, E. Y. (2020). Parahydrogen-induced hyperpolarization of gases [https://doi.org/10.1002/anie.201915306]. *Angewandte Chemie International Edition, 59*(41), 17788–17797. https://doi.org/https://doi.org/10.1002/anie.201915306

221. Mewis, R. E. (2015). Developments and advances concerning the hyperpolarisation technique SABRE [https://doi.org/10.1002/mrc.4280]. *Magnetic Resonance in Chemistry, 53*(10), 789–800. https://doi.org/https://doi.org/10.1002/mrc.4280

222. Barskiy, D. A., Coffey, A. M., Nikolaou, P., Mikhaylov, D. M., Goodson, B. M., Branca, R. T., Lu, G. J., Shapiro, M. G., Telkki, V.-V., Zhivonitko, V. V., Koptyug, I. V., Salnikov, O. G., Kovtunov, K. V., Bukhtiyarov, V. I., Rosen, M. S., Barlow, M. J., Safavi, S., Hall, I. P., Schröder, L., & Chekmenev, E. Y. (2017). NMR hyperpolarization techniques of gases [https://doi.org/10.1002/chem.201603884]. *Chemistry – A European Journal, 23*(4), 725–751. https://doi.org/https://doi.org/10.1002/chem.201603884

223. Nevzorov, A. A., Milikisiyants, S., Marek, A. N., & Smirnov, A. I. (2018). Multi-resonant photonic band-gap/saddle coil DNP probehead for static solid state NMR of microliter volume samples. *Journal of Magnetic Resonance, 297,* 113–123. https://doi.org/https://doi.org/10.1016/j.jmr.2018.10.010

224. Reynhardt, E. C., & High, G. L. (1998). Dynamic nuclear polarization of diamond. II. Nuclear orientation via electron spin-locking. *The Journal of Chemical Physics, 109*(10), 4100–4107. https://doi.org/10.1063/1.477010

225. Reynhardt, E. C., & High, G. L. (2000). Dynamic nuclear polarization of diamond. III. Paramagnetic electron relaxation times from enhanced 13C nuclear magnetic resonance signals. *The Journal of Chemical Physics, 113*(2), 744–750. https://doi.org/10.1063/1.481849

226. Jacques, V., Neumann, P., Beck, J., Markham, M., Twitchen, D., Meijer, J., Kaiser, F., Balasubramanian, G., Jelezko, F., & Wrachtrup, J. (2009). Dynamic polarization of single nuclear spins by optical pumping of nitrogen-vacancy color centers in diamond at room temperature. *Physical Review Letters, 102*(5), 057403. https://doi.org/10.1103/PhysRevLett.102.057403

227. King, J. P., Coles, P. J., & Reimer, J. A. (2010). Optical polarization of ^{13}C nuclei in diamond through nitrogen vacancy centers. *Physical Review B, 81*(7), 073201. https://doi.org/10.1103/PhysRevB.81.073201

228. Fischer, R., Jarmola, A., Kehayias, P., & Budker, D. (2013). Optical polarization of nuclear ensembles in diamond. *Physical Review B, 87*(12), 125207. https://doi.org/10.1103/PhysRevB.87.125207

229. Henshaw, J., Pagliero, D., Zangara, P. R., Franzoni, M. B., Ajoy, A., Acosta, R. H., Reimer, J. A., Pines, A., & Meriles, C. A. (2019). Carbon-13 dynamic nuclear polarization in diamond via a microwave-free integrated cross effect. *Proceedings of the National Academy of Sciences, 116*(37), 18334. https://doi.org/10.1073/pnas.1908780116

230. Schwartz, I., Scheuer, J., Tratzmiller, B., Müller, S., Chen, Q., Dhand, I., Wang, Z.-Y., Müller, C., Naydenov, B., Jelezko, F., & Plenio, M. B. (2018). Robust optical polarization of nuclear spin baths using Hamiltonian engineering of nitrogen-vacancy center quantum dynamics. *Science Advances, 4*(8), eaat8978. https://doi.org/10.1126/sciadv.aat8978

231. Loretz, M., Takahashi, H., Segawa, T. F., Boss, J. M., & Degen, C. L. (2017). Optical hyperpolarization of nitrogen donor spins in bulk diamond. *Physical Review B, 95*(6), 064413. https://doi.org/10.1103/PhysRevB.95.064413

232. Wunderlich, R., Kohlrautz, J., Abel, B., Haase, J., & Meijer, J. (2017). Optically induced cross relaxation via nitrogen-related defects for bulk diamond $^{13}\mathrm{C}$ hyperpolarization. *Physical Review B, 96*(22), 220407. https://doi.org/10.1103/PhysRevB.96.220407

233. Fernández-Acebal, P., Rosolio, O., Scheuer, J., Müller, C., Müller, S., Schmitt, S., McGuinness, L. P., Schwarz, I., Chen, Q., Retzker, A., Naydenov, B., Jelezko, F., & Plenio, M. B. (2018). Toward hyperpolarization of oil molecules via single nitrogen vacancy centers in diamond. *Nano Letters, 18*(3), 1882–1887. https://doi.org/10.1021/acs.nanolett.7b05175

234. Parker, A. J., Jeong, K., Avalos, C. E., Hausmann, B. J. M., Vassiliou, C. C., Pines, A., & King, J. P. (2019). Optically pumped dynamic nuclear hyperpolarization in $^{13}\mathrm{C}$-enriched diamond. *Physical Review B, 100*(4), 041203. https://doi.org/10.1103/PhysRevB.100.041203

235. Bucher, D. B., Glenn, D. R., Park, H., Lukin, M. D., & Walsworth, R. L. (2020). Hyperpolarization-enhanced NMR spectroscopy with femtomole sensitivity using quantum defects in diamond. *Physical Review X, 10*(2), 021053. https://doi.org/10.1103/PhysRevX.10.021053

236. Zangara, P. R., Dhomkar, S., Ajoy, A., Liu, K., Nazaryan, R., Pagliero, D., Suter, D., Reimer, J. A., Pines, A., & Meriles, C. A. (2019). Dynamics of frequency-swept nuclear spin optical pumping in powdered diamond at low magnetic fields. *Proceedings of the National Academy of Sciences, 116*(7), 2512. https://doi.org/10.1073/pnas.1811994116

237. Torelli, M. D., Nunn, N. A., Jones, Z. R., Vedelaar, T., Padamati, S. K., Schirhagl, R., Hamers, R. J., Shames, A. I., Danilov, E. O., Zaitsev, A., & Shenderova, O. A. (2020). High temperature treatment of diamond particles toward enhancement of their quantum properties [Original Research]. *Frontiers in Physics, 8*(205). https://doi.org/10.3389/fphy.2020.00205

238. Gierth, M., Krespach, V., Shames, A. I., Raghavan, P., Druga, E., Nunn, N., Torelli, M., Nirodi, R., Le, S., Zhao, R., Aguilar, A., Lv, X., Shen, M., Meriles, C. A., Reimer, J. A., Zaitsev, A., Pines, A., Shenderova, O., & Ajoy, A. (2020). Enhanced optical ^{13}C hyperpolarization in diamond treated by high-temperature rapid thermal annealing. *Advanced Quantum Technologies, 3*(10), 2000050. https://doi.org/https://doi.org/10.1002/qute.202000050

239. Ajoy, A., Liu, K., Nazaryan, R., Lv, X., Zangara, P. R., Safvati, B., Wang, G., Arnold, D., Li, G., Lin, A., Raghavan, P., Druga, E., Dhomkar, S., Pagliero, D., Reimer, J. A., Suter, D., Meriles, C. A., & Pines, A. (2018). Orientation-independent room temperature optical ^{13}C hyperpolarization in powdered diamond. *Science Advances, 4*(5), eaar5492. https://doi.org/10.1126/sciadv.aar5492

240. Thiering, G., & Gali, A. (2018). Ab Initio Magneto-optical spectrum of group-iv vacancy color centers in diamond. *Physical Review X, 8*(2), 021063. https://doi.org/10.1103/PhysRevX.8.021063

241. Vavilov, V., & Gippius, A. (1980). Investigation of the cathodoluminescence of epitaxial diamond films. *Sov. Phys. Semicond., 14*(9), 1078–1079.

242. Rogers, L. J., Jahnke, K. D., Doherty, M. W., Dietrich, A., McGuinness, L. P., Müller, C., Teraji, T., Sumiya, H., Isoya, J., Manson, N. B., & Jelezko, F. (2014). Electronic structure of the negatively charged silicon-vacancy center in diamond. *Physical Review B, 89*(23), 235101. https://doi.org/10.1103/PhysRevB.89.235101

243. Hepp, C., Müller, T., Waselowski, V., Becker, J. N., Pingault, B., Sternschulte, H., Steinmüller-Nethl, D., Gali, A., Maze, J. R., Atatüre, M., & Becher, C. (2014). Electronic structure of the silicon vacancy color center in diamond. *Physical Review Letters, 112*(3), 036405. https://doi.org/10.1103/PhysRevLett.112.036405

244. Wang, C., Kurtsiefer, C., Weinfurter, H., & Burchard, B. (2005). Single photon emission from SiV centres in diamond produced by ion implantation. *Journal of Physics B: Atomic, Molecular and Optical Physics, 39*(1), 37–41. https://doi.org/10.1088/0953-4075/39/1/005

245. Becker, J. N., & Neu, E. (2020). Chapter 7 - The silicon vacancy center in diamond. In C. E. Nebel, I. Aharonovich, N. Mizuochi, & M. Hatano (eds.), *Semiconductors and Semimetals* (Vol. 103, pp. 201–235). Elsevier. https://doi.org/https://doi.org/10.1016/bs.semsem.2020.04.001

246. Green, B. L., Mottishaw, S., Breeze, B. G., Edmonds, A. M., D'Haenens-Johansson, U. F. S., Doherty, M. W., Williams, S. D., Twitchen, D. J., & Newton, M. E. (2017). Neutral silicon-vacancy center in diamond: spin polarization and lifetimes. *Physical Review Letters, 119*(9), 096402. https://doi.org/10.1103/PhysRevLett.119.096402

247. Zhang, Z.-H., Stevenson, P., Thiering, G., Rose, B. C., Huang, D., Edmonds, A. M., Markham, M. L., Lyon, S. A., Gali, A., & de Leon, N. P. (2020). Optically detected magnetic resonance in neutral silicon vacancy centers in diamond via bound exciton states. *Physical Review Letters, 125*(23), 237402. https://doi.org/10.1103/PhysRevLett.125.237402

248. Rose, B. C., Huang, D., Zhang, Z.-H., Stevenson, P., Tyryshkin, A. M., Sangtawesin, S., Srinivasan, S., Loudin, L., Markham, M. L., Edmonds, A. M., Twitchen, D. J., Lyon, S. A., & de Leon, N. P. (2018). Observation of an environmentally insensitive solid-state spin defect in diamond. *Science, 361*(6397), 60. https://doi.org/10.1126/science.aao0290

249. Nguyen, C. T., Evans, R. E., Sipahigil, A., Bhaskar, M. K., Sukachev, D. D., Agafonov, V. N., Davydov, V. A., Kulikova, L. F., Jelezko, F., & Lukin, M. D. (2018). All-optical nanoscale thermometry with silicon-vacancy centers in diamond. *Applied Physics Letters, 112*(20), 203102. https://doi.org/10.1063/1.5029904

250. Yang, B., Li, J., Guo, L., Huang, N., Liu, L., Zhai, Z., Long, W., & Jiang, X. (2018). Fabrication of silicon-vacancy color centers in diamond films: tetramethylsilane as a new dopant source [10.1039/C7CE02181J]. *CrystEngComm, 20*(8), 1158–1167. https://doi.org/10.1039/C7CE02181J

251. Ralchenko, V. G., Sedov, V. S., Khomich, A. A., Krivobok, V. S., Nikolaev, S. N., Savin, S. S., Vlasov, I. I., & Konov, V. I. (2015). Observation of the Ge-vacancy color center in microcrystalline diamond films. *Bulletin of the Lebedev Physics Institute, 42*(6), 165–168. https://doi.org/10.3103/S1068335615060020

252. Palyanov, Y. N., Kupriyanov, I. N., Borzdov, Y. M., & Surovtsev, N. V. (2015). Germanium: a new catalyst for diamond synthesis and a new

optically active impurity in diamond. *Scientific Reports*, *5*(1), 14789. https://doi.org/10.1038/srep14789

253. Siyushev, P., Metsch, M. H., Ijaz, A., Binder, J. M., Bhaskar, M. K., Sukachev, D. D., Sipahigil, A., Evans, R. E., Nguyen, C. T., Lukin, M. D., Hemmer, P. R., Palyanov, Y. N., Kupriyanov, I. N., Borzdov, Y. M., Rogers, L. J., & Jelezko, F. (2017). Optical and microwave control of germanium-vacancy center spins in diamond. *Physical Review B*, *96*(8), 081201. https://doi.org/10.1103/PhysRevB.96.081201

254. Ekimov, E. A., Krivobok, V. S., Lyapin, S. G., Sherin, P. S., Gavva, V. A., & Kondrin, M. V. (2017). Anharmonicity effects in impurity-vacancy centers in diamond revealed by isotopic shifts and optical measurements. *Physical Review B*, *95*(9), 094113. https://doi.org/10.1103/PhysRevB.95.094113

255. Komarovskikh, A., Dmitriev, A., Nadolinny, V., & Palyanov, Y. (2017). A DFT calculation of EPR parameters of a germanium-vacancy defect in diamond. *Diamond and Related Materials*, *76*, 86–89. https://doi.org/https://doi.org/10.1016/j.diamond.2017.04.013

256. Nadolinny, V. A., Komarovskikh, A. Y., Palyanov, Y. N., Kupriyanov, I. N., Borzdov, Y. M., Rakhmanova, M. I., Yuryeva, O. P., & Veber, S. L. (2016). EPR study of germanium-vacancy defects in diamonds. *Journal of Structural Chemistry*, *57*(5), 1041–1043. https://doi.org/10.1134/S0022476616050280

257. Fan, J.-W., Cojocaru, I., Becker, J., Fedotov, I. V., Alkahtani, M. H. A., Alajlan, A., Blakley, S., Rezaee, M., Lyamkina, A., Palyanov, Y. N., Borzdov, Y. M., Yang, Y.-P., Zheltikov, A., Hemmer, P., & Akimov, A. V. (2018). Germanium-vacancy color center in diamond as a temperature sensor. *ACS Photonics*, *5*(3), 765–770. https://doi.org/10.1021/acsphotonics.7b01465

258. Blakley, S., Liu, X., Fedotov, I., Cojocaru, I., Vincent, C., Alkahtani, M., Becker, J., Kieschnick, M., Lühman, T., Meijer, J., Hemmer, P., Akimov, A., Scully, M., & Zheltikov, A. (2019). Fiber-optic quantum thermometry with germanium-vacancy centers in diamond. *ACS Photonics*, *6*(7), 1690–1693. https://doi.org/10.1021/acsphotonics.9b00206

259. Miller, C., Puust, L., Ekimov, E., Vlasov, I., Vanetsev, A., Vinogradova, E., Orlovskii, Y., Treshchalov, A., & Sildos, I. (2020). Toward performance and applications of large area optical thermometry based on the luminescence of germanium-vacancy defects in diamond nanocrystals [https://doi.org/10.1002/pssa.202000217]. *Physica Status Solidi (A)*, *n/a*(n/a), 2000217. https://doi.org/https://doi.org/10.1002/pssa.202000217

260. Iwasaki, T., Miyamoto, Y., Taniguchi, T., Siyushev, P., Metsch, M. H., Jelezko, F., & Hatano, M. (2017). Tin-vacancy quantum emitters in diamond. *Physical Review Letters, 119*(25), 253601. https://doi.org/10.1103/PhysRevLett.119.253601

261. Rugar, A. E., Dory, C., Sun, S., & Vučković, J. (2019). Characterization of optical and spin properties of single tin-vacancy centers in diamond nanopillars. *Physical Review B, 99*(20), 205417. https://doi.org/10.1103/PhysRevB.99.205417

262. Alkahtani, M., Cojocaru, I., Liu, X., Herzig, T., Meijer, J., Küpper, J., Lühmann, T., Akimov, A. V., & Hemmer, P. R. (2018). Tin-vacancy in diamonds for luminescent thermometry. *Applied Physics Letters, 112*(24), 241902. https://doi.org/10.1063/1.5037053

263. Görlitz, J., Herrmann, D., Thiering, G., Fuchs, P., Gandil, M., Iwasaki, T., Taniguchi, T., Kieschnick, M., Meijer, J., Hatano, M., Gali, A., & Becher, C. (2020). Spectroscopic investigations of negatively charged tin-vacancy centres in diamond. *New Journal of Physics, 22*(1), 013048. https://doi.org/10.1088/1367-2630/ab6631

264. Element Six. (2020). Element Six launches DNV-B1™ - its first commercially available, general purpose quantum diamond.

265. Reineck, P., Trindade, L. F., Havlik, J., Stursa, J., Heffernan, A., Elbourne, A., Orth, A., Capelli, M., Cigler, P., Simpson, D. A., & Gibson, B. C. (2019). Not all fluorescent nanodiamonds are created equal: a comparative study [https://doi.org/10.1002/ppsc.201900009]. *Particle & Particle Systems Characterization, 36*(3), 1900009. https://doi.org/https://doi.org/10.1002/ppsc.201900009

266. Jahnke, K. D., Naydenov, B., Teraji, T., Koizumi, S., Umeda, T., Isoya, J., & Jelezko, F. (2012). Long coherence time of spin qubits in ^{12}C enriched polycrystalline chemical vapor deposition diamond. *Applied Physics Letters, 101*(1), 012405. https://doi.org/10.1063/1.4731778

267. Herbschleb, E. D., Kato, H., Maruyama, Y., Danjo, T., Makino, T., Yamasaki, S., Ohki, I., Hayashi, K., Morishita, H., Fujiwara, M., & Mizuochi, N. (2019). Ultra-long coherence times amongst room-temperature solid-state spins. *Nature Communications, 10*(1), 3766. https://doi.org/10.1038/s41467-019-11776-8

Chapter 2

TL, OSL, and PL Properties of Zinc Oxide Nanocrystals

Alvaro de Farias Soares,[a] Sonia Hatsue Tatumi,[a,b] and Lilia Coronato Courrol[c]

[a]*Polytechnic School, University of São Paulo,*
158 Professor Luciano Gualberto Av., 05508-010 São Paulo/SP, Brazil
[b]*Federal University of São Paulo, Campus Baixada Santista,*
144 Carvalho de Mendonca St., 11070-100 Santos/SP, Brazil
[c]*Federal University of São Paulo, Campus Diadema,*
275 Artur Riedel St., 09972-270 Diadema/SP, Brazil
alvarodefarias@usp.br, sonia.tatumi@unifesp.br

2.1 Introduction

In recent years, there has been a growth in studies addressing nanomaterials, due to the properties found only in materials from this category [1]. Nanomaterials is a classification to the set of materials made up of a set of substances, where there is at least a dimension less than about 100 nm, which includes nanotubes, nanofilms, quantum dots, nanopowders, among others [2, 3]. Since such materials have a high superficial area by volume ratio, this

Luminescent Nanomaterials
Edited by Odireleng Martin Ntwaeaborwa

ISBN 978-981-4968-11-9 (Hardcover), 978-1-003-27738-5 (eBook)
www.jennystanford.com

leads to a high percentage of atoms on surface, allowing greater interaction with other atoms, and providing unique properties [2].

Within the group of nanomaterials, they can be divided in a subclassification, according to the composition: into organics, which include for example fullerenes and nanofibers; inorganic, such as metals, metallic oxides and quantum dots; in addition, there are the hybrids, that allow combinations of different organic and inorganic materials [4].

In this chapter, some luminescent properties of a nanomaterial from inorganic category, the zinc oxide (ZnO), will be discussed. ZnO is a metal oxide, semiconductor, of the group II–IV, with applications in several areas, such as photocatalytic activity [5], batteries [6], magnetic properties [7], antibacterial potential [7, 8], optical biosensors [9], detectors [10], solar cells [11], among others. This wide versatility is attributed to its properties, such as a characteristic direct band gap of approximately 3.4 eV, large exciton binding energy (60 meV), overlapping the band gap of more traditional materials as GaN, easy synthesis and crystal growth, diversity of synthesis methods and low production cost [12].

However, due to the different application possibilities, not just restricted to luminescence, few studies in the last decades have addressed the thermoluminescence (TL) and optically stimulated luminescence (OSL) emission for non-doped ZnO. In addition, considering the various methods of synthesis, precursor materials, heat treatments, crystalline phases, and other possible production configurations, it makes possible to obtain ZnO with different properties [12].

In the next sections, typical TL, OSL and photoluminescence (PL) curves of ZnO nanocrystals, synthesized by the coprecipitation method will be presented, with hexagonal Wurtzite structure, followed by heat treatment in atmospheric air (calcination) at 900 and 1000°C for 2 h, developing and detailing the discussions of the material properties, which was already started in previous work [13].

2.2 Thermoluminescence of ZnO Nanocrystals

TL emission for ZnO has been reported since the 1960s, when peaks at low negative temperature (<0°C) were detected in a commercial material sample, after exposure to light [14], and since then, several studies have evaluated the behavior from different types ionizing irradiation doses [15–17], in addition to different methods of synthesis, which allowed obtaining nanostructured samples [18].

Despite the knowledge of the potential application of ZnO as a TL material, few studies have been done about TL response for non-doped ones. A hypothesis pointed out by Cruz- Vázquez et al. (2005) [15], is that, as ZnO has several applications, such as in the areas of optoelectronics, catalysis, etc., together with the low TL response observed in the first studies, they may have discouraged such a study. However, this has changed in recent years, and more studies have been indexed.

In Table 2.1, there is a summary of some TL emission studies already reported, for non-doped ZnO.

Nowadays, in addition to studies on TL of pure ZnO, which attested the potential for application in the dosimetry area, new works have been dedicated to assessing how doping with elements such as Mg, Yb, Gd, Tb [26–29], among others, can improve the TL response.

Forward, TL emission characteristics of ZnO nanocrystals synthesized by the coprecipitation method will be presented. In these analyses, irradiations were performed with β-rays in a dose range of 0.08–40.6 Gy, from a $^{90}Sr/^{90}Y$ source, with a dose rate of 0.081 Gy/s, and carry out in Risoe TL/OSL model DA/20 reader from DTU Nutech, using Schott BG39 optical filter, allowing the transmission on visible region.

Figure 2.1 illustrates images obtained by scanning electron microscopy (SEM), using FEG-SEM Supra 35 VP model equipment (from Carl Zeiss, Germany), of the ZnO nanocrystals, which the luminescent will be approached in the next sections. It can be noted that, disregarding the heat treatment adopted, all samples supplied particles in nanoscale dimension, proving to be a nanocrystal scale. Due the increasing in heat treatment temperature, a conformation in "balloon" format is preferred, as it can be seen in ZnO calcined at 1000°C (Fig. 2.1c) compared to these ones without calcination (Fig. 2.1a) and calcined at 900°C (Fig. 2.1b).

Table 2.1 Summary of some TL emission studies for ZnO non-doped

TL peak(s) reported	Synthesis method	Crystalline structure	Irradiation type and dose interval	References
Peaks at −161, −144, −129, −117 and −101°C	Pure powder (commercial)	—	Woods' light 30 min	Muer and Vorst (1968) [14]
Peaks at −123 and −91°C	(commercial)	—	Ultraviolet (UV) and visible	Seitz, Pinter and Hirthe (1971) [19]
A broad set of peaks centered at 250°C approximately	Thermal annealing of ZnS powders	Hexagonal Zincite	β-rays 0.15–10.5 kGy	Cruz-Vázquez et al. (2005) [15]
Peaks at 150 and 222°C	Glycol mediated chemical synthesis	Hexagonal Wurtzite	β-rays 100 Gy	Pal et al. (2008) [20]
Peaks at 116 and 207°C	Chemical coprecipitation method	Hexagonal	X-ray 0.2–1.2 Gy	Manam, Das and Isaac (2009) [21]
Single peak at 180°C	Sol-gel	Hexagonal Wurtzite	X-ray 10 kGy	Srivastava et al. (2009) [22]
Peaks at 118, 164, 210, 266 and 326°C (if synthesized by pulsed mode) and peaks at 126, 166, 230, and 288°C (if continuous mode)	Sonification method using pulsed and continuous mode	Hexagonal Wurtzite	β-rays 10–200 Gy	Sahu et al. (2010) [23]
A broad peak centered at 343°C	Low temperature solution combustion method	Hexagonal Wurtzite	γ-rays 10–50 Gy	Reddy et al. (2011) [17]
Peaks at 139, 257 and 327°C for one sintered at 900°C; peaks at 157, 297, and 367°C for one sintered at 950°C	Controlled chemical reaction	Hexagonal	β-rays 25–800 Gy	Borbón-Nuñez et al. (2014) [24]
Peaks at 97, 147 and 207°C (pulsed electron beam evaporation) and at 97, 127 and 217–247°C	Pulsed electron beam evaporation/solar physical vapor deposition (SPVD)	Hexagonal	X-ray 1.5–66 Gy	Grigorjeva et al. (2017) [25]

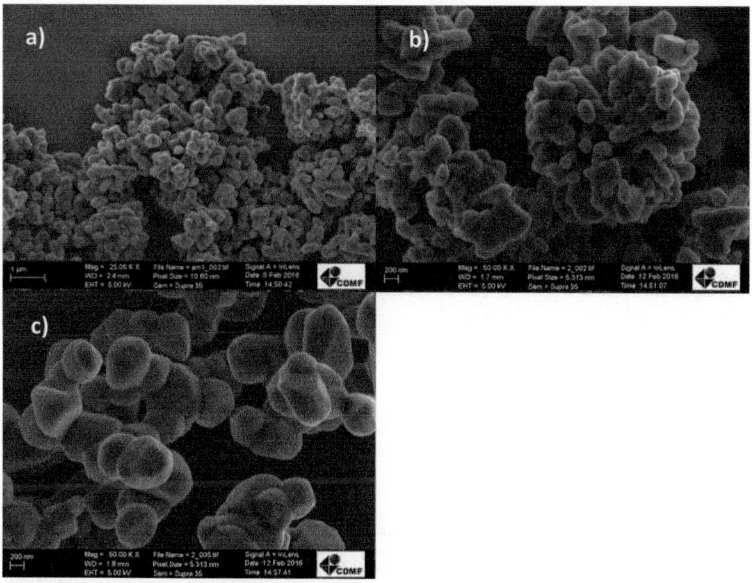

Figure 2.1 SEM images of ZnO synthesized by coprecipitation method, and (a) without calcination, (b) calcined at 900°C, and (c) calcined at 1000°C.

2.2.1 Heat Treatment Effect on TL Emission of ZnO Nanocrystals

Using heat treatment (as calcination) after ZnO synthesis allows modifying its properties, including luminescent properties, since these changes can alter the atomic arrangement and the crystalline structure, as well as to increase or create new defects, which are related with the TL and OSL emission of the material.

Figure 2.2 shows TL glow curves for ZnO nanocrystals synthesized by coprecipitation method, followed by different heat treatments at 900 and 1000°C for 2 h. After irradiation with a dose of 40.6 Gy of β-rays, the non-calcined ZnO did not present any TL emission in the visible region of the spectrum between 0–450°C. When calcined at 900°C, ZnO presents a broad emission peak centered at 358°C, with high intensity, indicating the existence of deep traps. And the ZnO calcined at 1000°C shows a very different TL behavior: several emission peaks, being the most prominent centered at approximately 120 and 256°C, and the less intense one, and deeper, at 409°C. In addition, the intensity of the main peak is more than double that obtained for the sample calcined at 900°C.

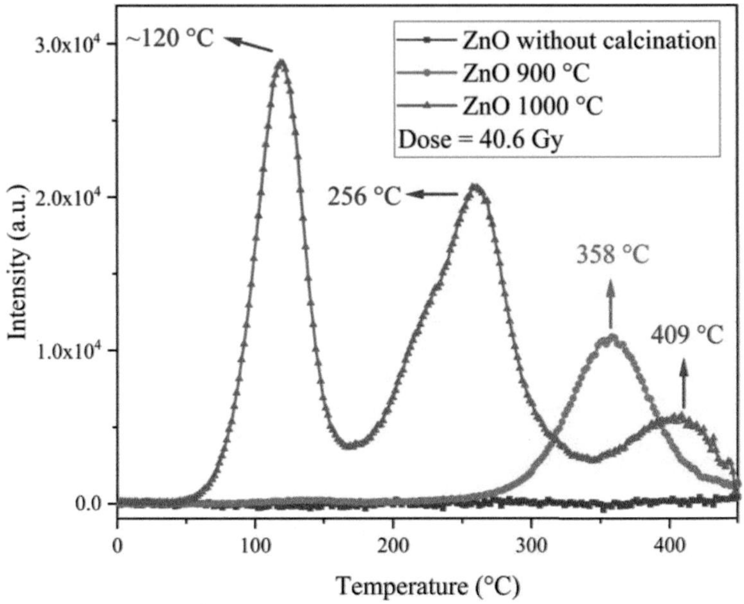

Figure 2.2 TL glow curves for ZnO nanocrystals synthesized by coprecipitation method and calcined at different temperatures, after a dose = 40.6 Gy of β-rays.

These findings indicate, that among the effects arising from calcination, the defects creation in the crystalline structure, since the non-calcined sample did not present TL emission in comparison to those heat treated. In addition, among the calcined samples, there was a difference in the number and peaks position, indicating that the energy levels of the created defects are different, according to the adopted temperature.

The behavior of increasing the ZnO TL intensity with heat treatment was also observed in the study Borbón-Nuñez et al. (2014) [24], which synthesized ZnO phosphors by controlled chemical reaction method, followed by sintering at 850, 900 and 950°C. The sample sintered at 950°C for 24 h presented the highest TL intensity, with peaks located at 157, 297, and 367°C, being the last one the most intense.

A possible explanation for these observations is that the heat treatment would be responsible for the increase in oxygen vacancies in the material [30–32]. As these vacancies are identified as one of the main reasons of luminescence in the green region, together with zinc vacancies, or zinc and oxygen interstitials, the TL glow growth is related to these defects [17].

Improvement on luminescent response with heat treatment is not restricted to ZnO nanocrystals, but has already been observed in other oxides, such as hafnium oxide [33], beryllium oxide [34], yttrium oxide [35], among others, proving to be an important mechanism for obtaining better luminescent properties.

2.2.2 TL Dose Response of ZnO Nanocrystals

To be applied as a dosimetric material, among the desired properties, it is expected that the material has a linear TL response with the dose, in a wide range of doses [36].

Figure 2.3a shows that ZnO nanocrystals calcined at 900°C has a TL intensity that increases with the irradiation dose, supporting that higher doses promote greater electron trapping, which are later released by heating, emitting visible photons from recombination in trapped holes at luminescence centers. However, for doses below 0.8 Gy, it was not possible to verify significant emission, therefore, in this case, low doses were not enough to perform the electrons trapping.

Figure 2.3b depicts that the dose response has a linear behavior in the range of 0.8–20.3 Gy, afterwards, a sublinear behavior is observed, which may be indicative of saturation—according to dose response curve, considering the total area under the curve throughout the TL region.

The sample calcined at 1000°C supplied a TL emission since lower doses, compared to calcined at 900°C (Fig. 2.4a), reinforcing that calcination has a significant role on luminescent properties, as shown in the previous section. For the dose range between 0.08–0.49 Gy, all peaks show growth with dose, being the first one the most prominent.

The analysis of dose-response (Fig. 2.4b), shows that the first set of peaks, located between 0–200°C, has a supralinear behavior, followed by a sublinear between 0.08–0.41 Gy. On the other hand, the peak located between 200–350°C shows a linear behavior over the entire interval.

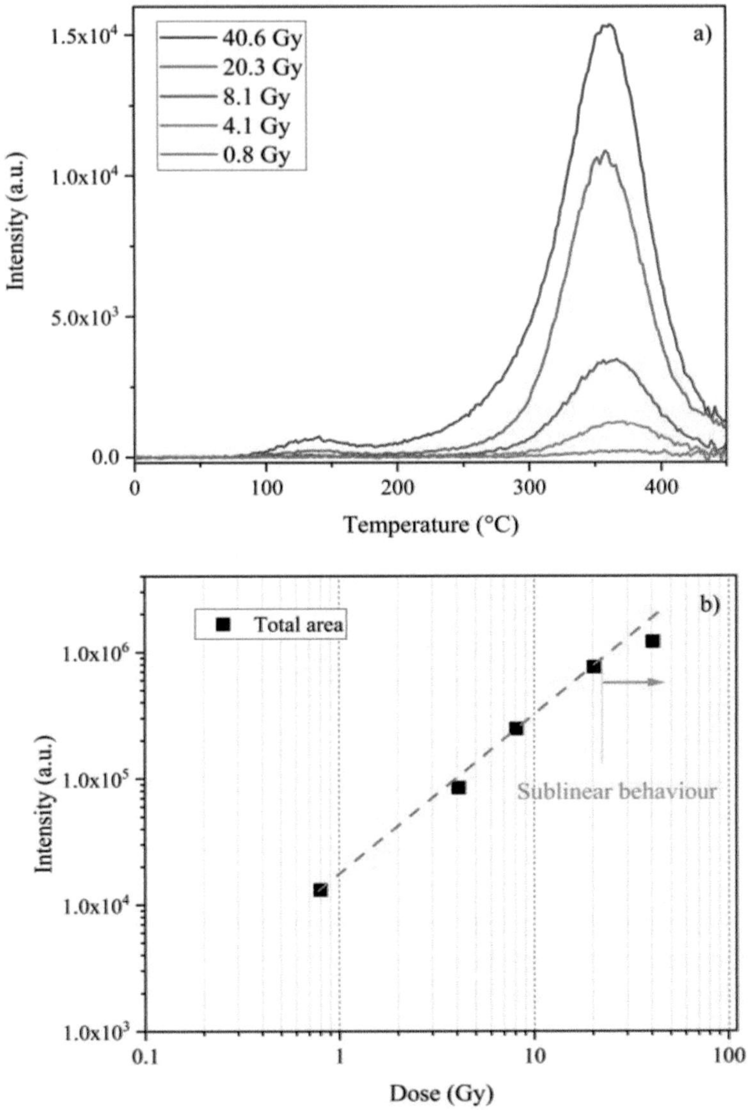

Figure 2.3 (a) TL glow curves for ZnO nanocrystals synthesized by coprecipitation method and calcined at 900°C, after doses between 0.8–40.6 Gy of β-rays, and (b) respective dose response curve.

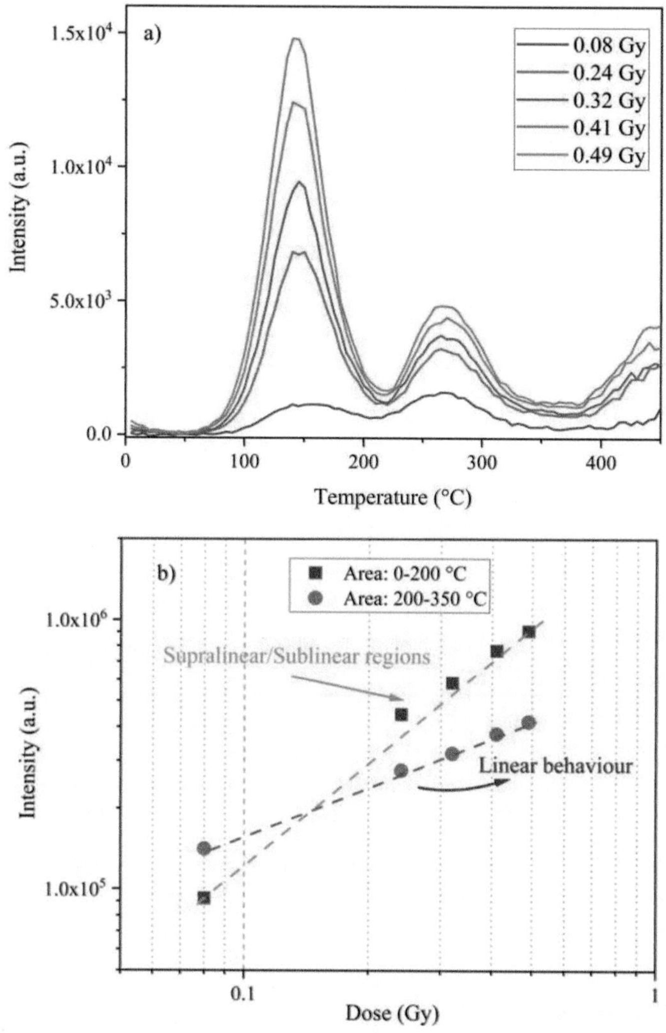

Figure 2.4 (a) TL glow curves for ZnO nanocrystals synthesized by coprecipitation method and calcined at 1000°C, after doses between 0.08–0.49 Gy of *β*-rays, and (b) respective dose response curve.

Analyzing the intensity in a range of higher doses, between 0.8–40.6 Gy, the increasing with dose is still observed, and the peaks located between 200–350°C show an intensity of a magnitude order similar to that located at lower temperatures (Fig. 2.5a). The dose-response curve (Fig. 2.5b) for both 0–200 and 200–350°C regions, depicts a quite similar behavior, which is linear up to a dose of 8.1 Gy, followed by a region of supralinearity.

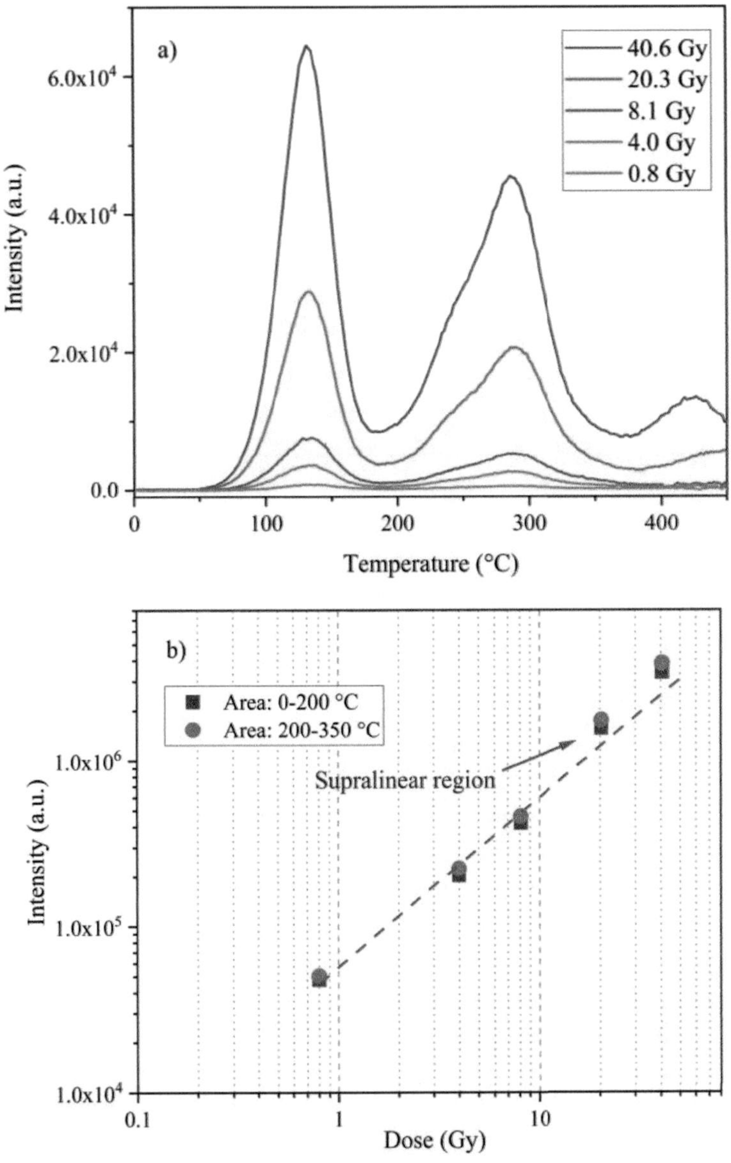

Figure 2.5 (a) TL glow curves for ZnO nanocrystals synthesized by coprecipitation method and calcined at 1000°C, after doses between 0.8–40.6 Gy of β-rays, and (b) respective dose response curve.

Results obtained for the ZnO nanocrystals synthesized by coprecipitation method followed by thermal treatment are different from those found by other studies, indicating that the adopted synthesis method, as well as the thermal treatment, promote changes in luminescent properties.

Reddy et al. (2011) [17], for example, evaluated the TL emission of ZnO nanopowders synthesized by chemical combustion, after irradiation by γ-rays in the range of 10–50 Gy, observed a single wide peak located at 343°C. This emission was attributed to oxygen defects, mainly to oxygen interstitials. When synthesized by the sonification method, the TL obtained after irradiation by beta particles between 10–200 Gy, was observed with two sets of peaks, the first and most intense being close to 120°C [23].

Cruz-Vázquez et al. (2005) [15], also using beta irradiation (0.15–10.5 kGy), observed a single and wide set of peaks, between 100–400°C, centered at 220°C, which showed an exponential response dose in the interval analyzed, for ZnO nanophosphors synthesized from thermal annealing of ZnS.

2.2.3 Deconvolution Study for TL of ZnO Nanocrystals

In order to verify which peaks are present in the TL curve, as well as to estimate the parameters of each peak, such as kinetic order (b), activation energy (E_a), frequency factor (s), it is used the general orders model in the computerized glow curve deconvolution (CGCD), according to literature [27, 38]. For this, the follow equation (Eq. 2.1) proposed by Kitis, Gomez-Ros, and Tuyn [39] is used:

$$I(T) = I_m \cdot \frac{b}{b^{b-1}} \exp\left(\frac{E}{kT}\frac{T-T_m}{T_m}\right) \cdot [(b-1)(1-\Delta)\frac{T^2}{T_m^2}\exp\left(\frac{E}{kT}\frac{T-T_m}{T_m}\right) + Z_m]^{\frac{b}{b-1}} \quad (2.1)$$

where:

$$\Delta = \frac{2kT}{E}, \Delta_m = \frac{2kT_m}{E}, Z_m = 1 + (b-1)\Delta_m \quad (2.2)$$

where $I(T)$ is the TL intensity as function of temperature (T), E (or E_a) is the activation energy (eV), b is the kinetic order, k is the

Boltzmann constant, and T_m is the peak position (maximal). In addition, to check the adjustment, it is used the FOM equation (Eq. 2.3) [40]:

$$\mathrm{FOM} = \sum_i \frac{|Y_{Exp} - Y_{Fit}|}{A} \tag{2.3}$$

where A is the area below the fit curve, Y_{Exp} are the experimental curve data, and Y_{Fit} are the fit curve values. Lower the FOM, better the adjust.

From CGCD study, it was observed that the sample calcined at 900°C shows a TL emission after 20.3 Gy of beta dose composed of three peaks, at approximately 141, 247 and 357°C, being the last, the main one (Fig. 2.6). Values of E_a varied between 0.7–1.5 eV, b assumed values higher than 1.5, and the FOM obtained was less than 4% (3.64%), attesting the fitting quality (Table 2.2).

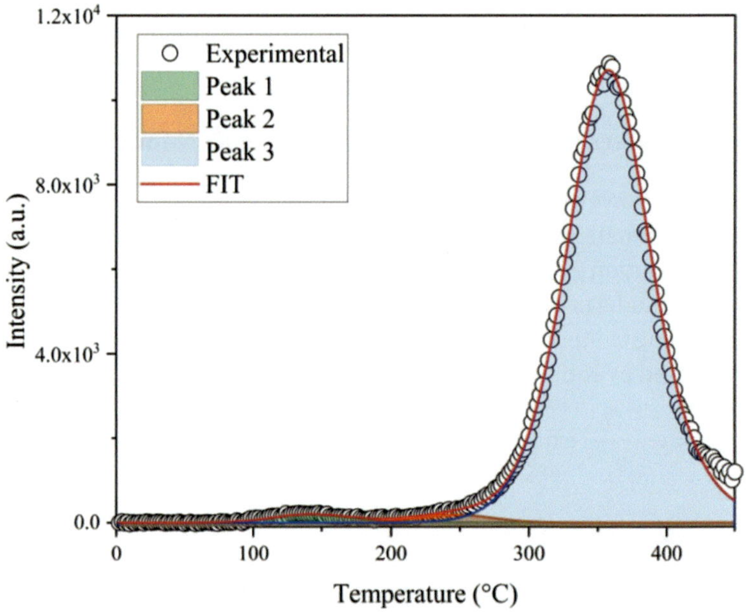

Figure 2.6 CGCD study to TL glow curve of ZnO synthesized by coprecipitation method, calcined at 900°C, for a dose of 20.3 Gy of β-rays.

Table 2.2 CGCD parameters to TL glow curve of ZnO synthesized by coprecipitation method

Sample			**ZnO – 900°C - Dose = 20.3 Gy**		
Peak number	I_M **(a.u.)**	T_M **(°C)**	E_a **(eV)**	**b**	s **(s⁻¹)**
1	200	141	0.7	2	7.1491×10^{7}
2	190	247	1	1.5	1.1011×10^{9}
3	10700	357	1.5	1.82	2.0726×10^{11}
FOM			3.64%		
Sample			**ZnO – 1000°C - Dose = 0.41 Gy**		
Peak number	I_M **(a.u.)**	T_M **(°C)**	E_a **(eV)**	**b**	s **(s^{-1})**
1	12100	141	0.9	2	2.5538×10^{10}
2	850	177	1.10	1.1	4.5058×10^{10}
3	4100	268	1.11	2	4.7442×10^{9}
4	510	342	1.2	1.5	1.2033×10^{9}
5	3550	450	1.4	1.1	8.8491×10^{8}
FOM			4.51%		
Sample			**ZnO – 1000°C - Dose = 20.3Gy**		
Peak number	I_M **(a.u.)**	T_M **(°C)**	E_a **(eV)**	**b**	s **(s^{-1})**
1	28746	132	0.9	1.5	4.8570×10^{10}
2	1950	191	1.01	1.65	2.4196×10^{10}
3	5950	238	1.1	1.0001	1.6235×10^{10}
4	20400	287	1.35	1.8	3.3370×10^{11}
5	1610	357	1.48	1.5	1.4437×10^{11}
6	5760	447	1.55	1.4	1.1900×10^{10}
FOM			2.51%		

As previously demonstrated, ZnO nanocrystals, when calcined at 1000°C, present a TL emission after β irradiation composed of several peaks. When a dose of 0.41 Gy is applied, five peaks make up this curve (Fig. 2.7a), the most intense is centered at 141°C, and others are at 177, 268, 342, and 450°C. The values of E_a increasing with the peak temperature, and spanning between 0.9–1.4 eV, an interval similar to that found for the sample calcined at 900°C (Table 2.2). The kinetic order b varied between 1.1–2, and the FOM was 4.51%.

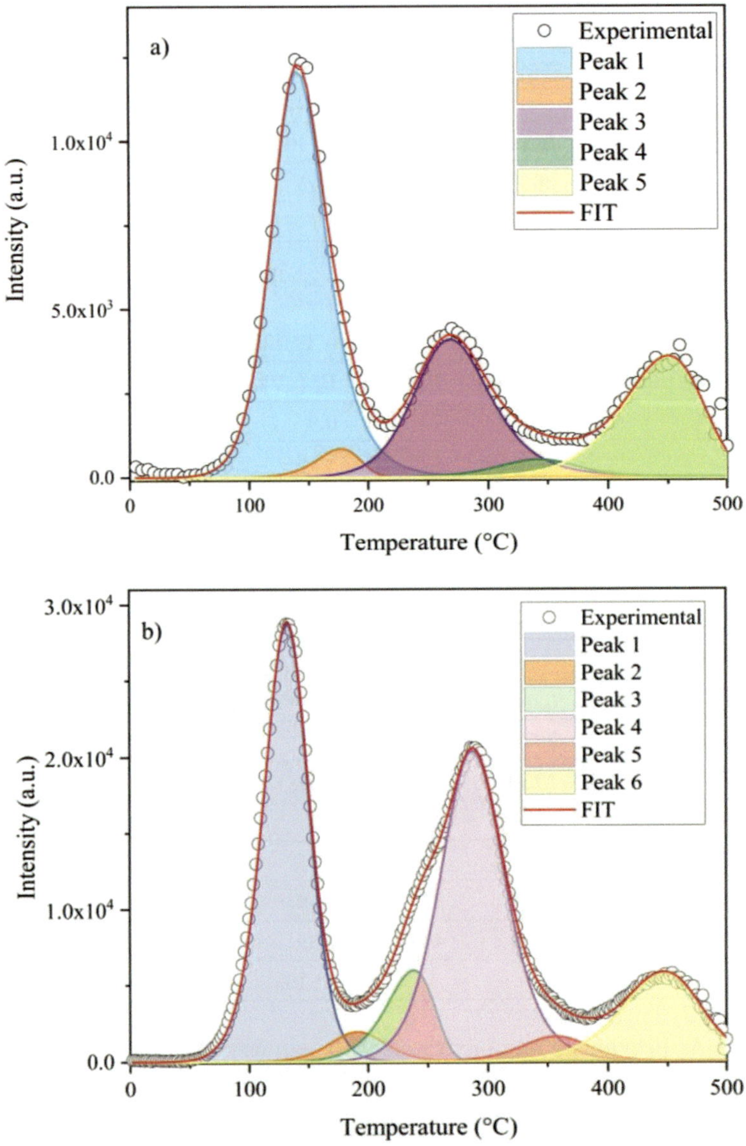

Figure 2.7 CGCD study to TL glow curves of ZnO synthesized by coprecipitation method, calcined at 1000°C, for a dose of (a) 0.41 and (b) 20.3 Gy of β-rays.

When the same analysis is performed for a higher dose (20.3 Gy), although the curve shows a similar behavior, the fitting by

CGCD is performed with six peaks, showing that higher doses of irradiation can modify the distribution of trapped electrons in the material (Fig. 2.7b). The main peak is also the shallower, therefore, more unstable, as a lower temperature is enough for emptying, and it is located at 132°C. The first two peaks show a small shifting to higher temperatures, compared to the analysis made for TL curve obtained with a dose of 0.41 Gy, while the last two ones have similar characteristics (Table 2. 2).

The values of E_a varied between 0.9–1.55 eV, and b between 1.0001–1.8. The fitting quality is confirmed by the low value of the FOM, which was 2.51%.

2.3 Optically Stimulated Luminescence of ZnO Nanocrystals

Further, OSL results will be presented for using constant (CW) and linearly modulated (LM) mode for ZnO nanocrystals. The measurements were carried out with a blue LED excitation centered at 470 nm, with a Hoya U-340 filter, whose transmissibility is given mainly in the UV region.

2.3.1 CW-OSL of ZnO Nanocrystals

In addition to the emission in the visible region observed for TL, ZnO nanocrystals also supplied emission in the UV region, observed on CW-OSL decay curves. The non- calcined ZnO, or calcined at 900°C, did not present CW-OSL emission in the dose range analyzed (0.81–40.6 Gy), therefore, the material was not sensitive to doses below 40.6 Gy. However, ZnO treated at 1000°C, showed decay curves, which response raises with the low doses (Fig. 2.8a).

From dose-response curve, considering the intensity of the first point, it is possible to verify a linear behavior for doses above 4 Gy (Fig. 2.8b), similar to that observed when the total area under the curve was used. One hypothesis for the first point (0.81 Gy) does not coincide with the adjust line, is that with a lower dose, the OSL intensity presents a higher noisy aspect, with great contribution from background.

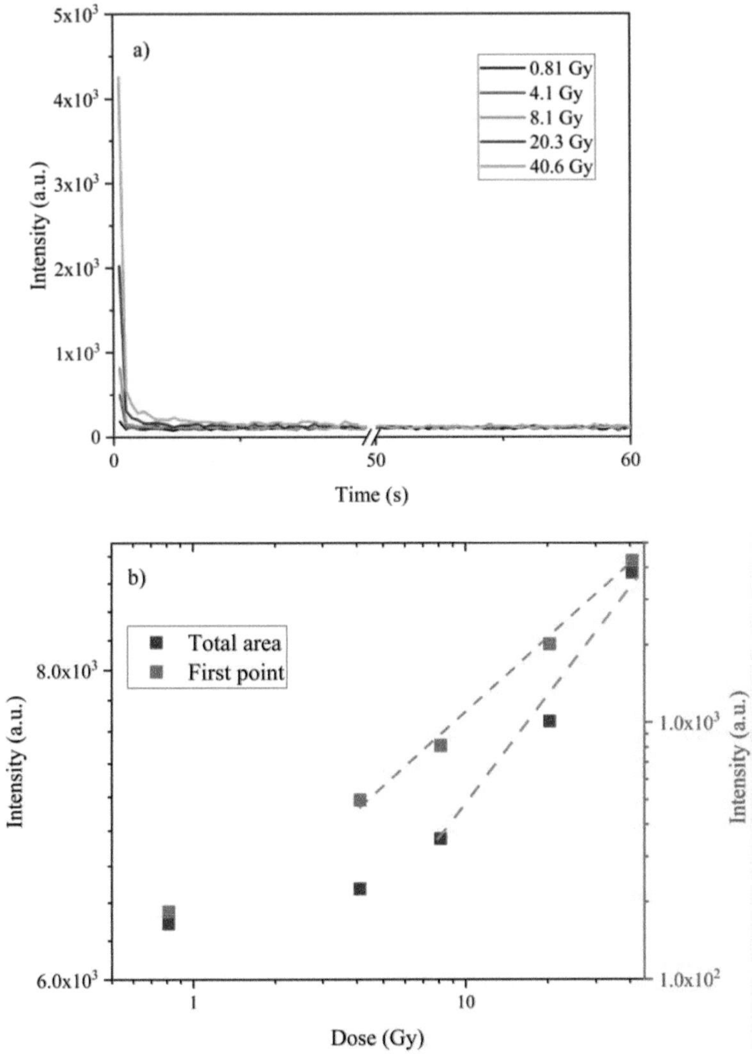

Figure 2.8 (a) CW-OSL decay curves for ZnO nanocrystals synthesized by coprecipitation method and calcined at 1000°C, after doses between 0.81–40.6 Gy of β-rays, and (b) respective dose response curve.

These results may indicate that the material could be applied for OSL dosimetry for high doses, provided that the linearity of the response remains in a wide range. Another important aspect is that, as observed for the TL emission, the heat treatment improved the OSL intensity increase, creating defects related to the UV emission.

Few studies in the literature have evaluated the OSL emission of nanostructured ZnO. Perhaps, because TL is a technique that has been used for a long time, and previously established and also, since the material can present one or the other emission, instead of both, considering that the detection region is different. Guckan et al. [41] (2020), verified the CW-OSL of pure and Eu-doped ZnO synthesized by the sol-gel and precipitation methods, after β irradiation, verifying that the Eu doping improved the intensity. In other studies, a preliminary one presented the OSL response of ZnO synthesized by thermal annealing of ZnS [42], and also the influence of doping with Ce ions [43]. Thus, this response observed for β-rays is a promising indication of material as a OSL dosimeter.

2.3.2 LM-OSL of ZnO Nanocrystals

The LM-OSL differs from the CW mode about the stimulus: while in CW the excitation power is constant during the measurement, in LM, it grows linearly [44]. This type of analysis is interesting, because instead of a simple decay curve, it can observe peaks, which correspond to the different levels of traps present in the material.

As with CW, ZnO also presents LM-OSL, however, it is also significant for doses above 4 Gy (Fig. 2.9a). The curve indicates that the maximum intensity is reached before 5 s, followed by a rapid decay up to 20–30 s.

The dose-response analysis (Fig. 2.9b) shows that at high doses, higher than 20 Gy, the OSL response behaves linearly with the dose. In smaller doses, perhaps, due to the low intensity along with the noise of the measurement, the response still does not meet this behavior.

The maximum intensity in the CW mode, for the same dose, was more than 10 times higher than the maximum found in the LM mode, and this is explained, because in the first case, with the maximum power adopted, the electrons are readily released from the traps, while in the second one, the release increases with the power of the stimulus.

Studies on the undoped ZnO LM-OSL are quite scarce. Such results were reported in our previous study [13], which are being expanded here.

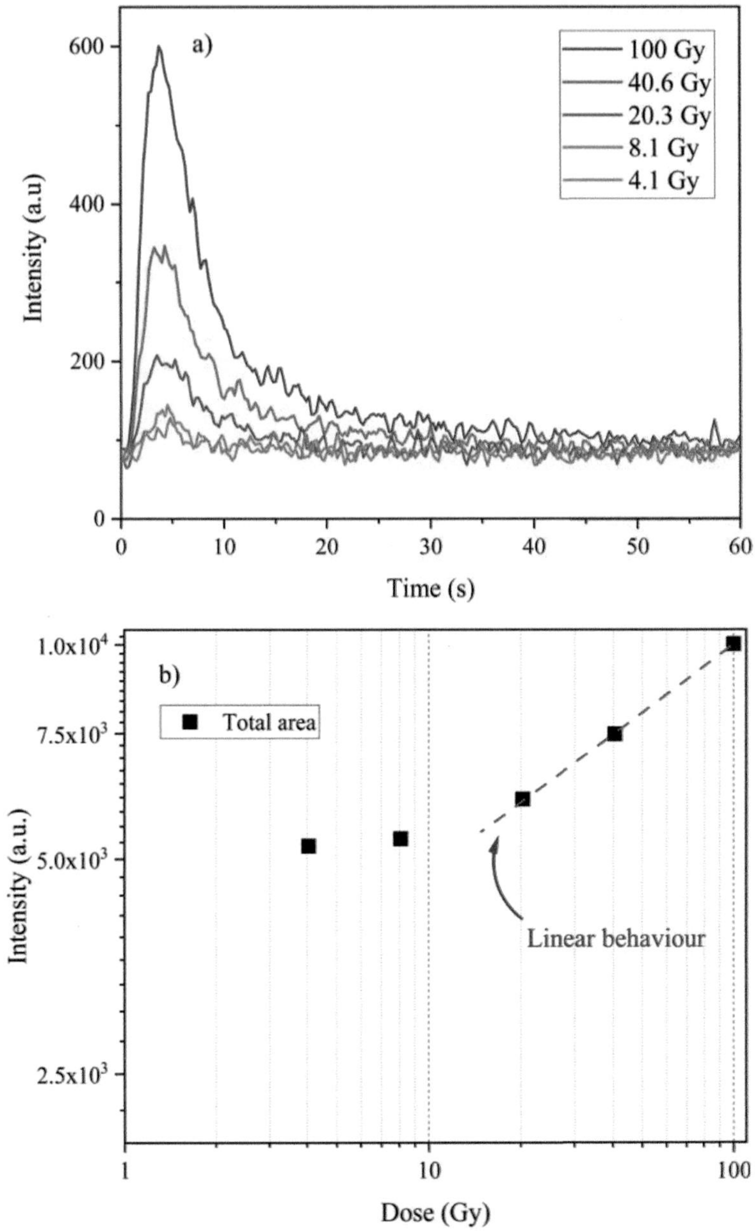

Figure 2.9 (a) LM-OSL curves for ZnO nanocrystals synthesized by coprecipitation method and calcined at 1000°C, after doses between 4.1–100 Gy of β-rays, and (b) respective dose response curve.

2.3.3 Deconvolution Study for OSL of ZnO Nanocrystals

In order to understand which components are present in the OSL curves, the CGCD study can also be applied. For the CW-OSL, it is possible to check the presence or absence of the fast, medium and long components, as well as their kinetic order values (b), and decay time (τ). For this, the adjustment is made by Eq. 2.4 [37, 38]:

$$I(t) = I_0 \left(1 + (b-1)\frac{t}{\tau}\right)^{\frac{b}{b-1}} \quad (2.4)$$

Where $I(t)$ is the intensity as function of time, I_0 is the initial intensity, b is the kinetic order, and τ is lifetime. The FOM (Eq. 2.3) was also used to verify the fitting quality.

When irradiated with a dose of 4.1 Gy, the ZnO calcined at 1000°C presents a CW-OSL decay curve composed only with fast component (Fig. 2.10a), whose b is close to 1, and has a decay time of 0.28 s. There is still a background of 108 a.u., and the FOM was 4.53% (Table 2.3).

In turn, when a higher dose of 40.6 Gy is applied, in addition to the rapid component, whose parameters are maintained, except for I_0, there is also the presence of a medium component (Fig. 2.10b), whose decay is slower, of 8.8 s, and with b = 1.1 (Table 2.3).

Table 2.3 CGCD parameters to CW-OSL decay curves of ZnO synthesized by coprecipitation method

	ZnO – Dose of 4.1 Gy			ZnO – Dose of 40.6 Gy		
Parameters	**Component**			**Component**		
	Fast	**Medium**	**Slow**	**Fast**	**Medium**	**Slow**
I_o (a.u.)	900		—	7000	72	—
b	1.0000001	—	—	1.0000001	1.1	—
τ(s)	0.28	—	—	0.35	8.8	—
BGK		108			123	
FOM		4.53%			8.78%	

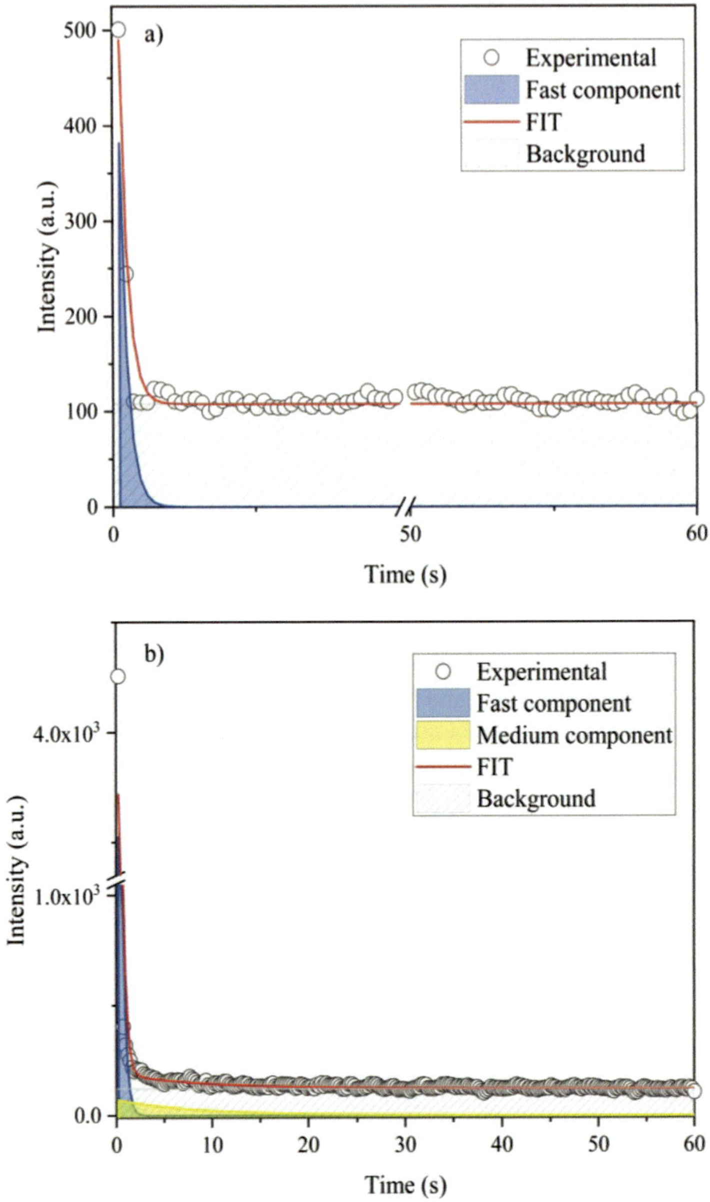

Figure 2.10 CGCD study to CW-OSL decay curves of ZnO synthesized by coprecipitation method, calcined at 1000°C, for a dose of (a) 4.1 and (b) 40.6 Gy of β-rays.

The results show that a higher dose of β irradiation has played an important role in trapping electron in deeper traps, as evidenced by the presence of the medium component. However, with a dose below 4.1 Gy, in turn, only trapped electrons at shallower levels in the material, so only a quick component was observed.

Similar to the CW mode, the CGCD study is also applicable to check which components are present. To the fitting, the following equation is used [37, 38, 45]:

$$I_{\mathrm{LMOSL}}(t) = I_m \frac{t}{t_m}\left[\frac{b-1}{2b}\frac{t^2}{t_m^2} + \frac{b+1}{2b}\right]^{\frac{b}{1-b}} \tag{2.5}$$

Where I_{LMOSL} is the intensity, I_m is the maximum peak intensity, t_m is the time at I_m, t is time, and b is the kinetic order. Effectiveness was also measured by FOM.

In the CGCD study, it was verified that the LM-OSL curve, obtained after an irradiation of 100 Gy, is composed of three components, located in t_m of 3.75, 10 and 35 s, and with respective orders b of 1, 1 and 2 (Fig. 2.11; Table 2.4). The value of the FOM was close to 5%, ensuring an adequate adjustment.

Table 2.4 CGCD parameters to LM-OSL curve of ZnO synthesized by coprecipitation method

Parameter	Component 1	Component 2	Component 3
I_{LMOSL} (a.u.)	463	124	113
t_m (s)	3.75	10	35
b	1.00	1.00	2.00
FOM		5.11%	

In addition to presenting emission in UV region, observed by the OSL measurements, the study by CGCD allowed us to understand that these emissions are not simple, that is, they are not composed of just one component or peak. Conversely, the CW-OSL emission curve is composed of up to two components, depending on the applied radiation dose, and the LM-OSL by three peaks.

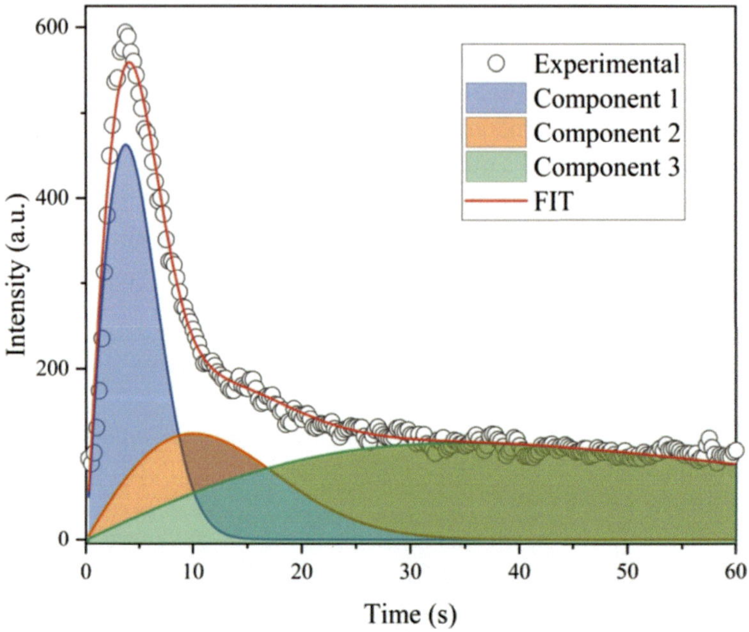

Figure 2.11 CGCD study to LM-OSL curve of ZnO synthesized by coprecipitation method, calcined at 1000°C, for a dose of 100 Gy of β-rays.

2.4 Photoluminescence of ZnO Nanocrystals

PL is a phenomenon that refers to photon excitation from the incidence light (between UV and infrared), which is applied to the material, while the luminescent response is measured simultaneously, at different wavelengths [46]. Thereby, it is possible to check the emission bands for a given excitation wavelength.

In Fig. 2.12, it is possible to observe that ZnO, when excited at 360 nm, has several emission bands, spanning from violet to yellow, and that it varies according to the heat treatment adopted.

Using multiple Gaussian curves fitting, it was observed that the non-calcined ZnO has a broad emission band centered in the green region, at approximately 558 nm, other band in the blue region at 463 nm, and three bands in the violet region, at 390, 413, and 439 nm (Fig. 2.12a). Considering the ZnO nanocrystals treated at 900°C, there are still five bands in regions very close to those found in the non-calcined sample, however, the band in the green region is shifted to a shorter wavelength, being centered at 502 nm, in addition, the intensity of this band is lower, indicating that the emission in green, although majority, becomes lower (Fig. 2.12b). About the ZnO treated at 1000°C, it preserves the three peaks in the violet region, as well as the one in the blue region (Fig. 2.12c), moreover, the band in the green region remained in a similar position to that of the sample calcined at 900°C (498 nm), but with an intensity about 1.5 times higher, indicating that the increase in the heat treatment temperature improved the emission in that region. Finally, a new band in the yellow region, centered at 582 nm, is also observed.

Plotting these results on Chromaticity Diagram (Fig. 2.12d), it can observe what is the main emission region related to the whole curve. It is verified that the non-calcined sample shows white luminescence ($x = 0.32374659, y = 0.34456177$), which is changed to a bluish when calcined in 900°C ($x = 0.24762038, y = 0.26625979$), and it is changed again to a blue-greenish when calcined at 1000°C ($x = 0.26911568, y = 0.32795841$).

Emission in the visible region for nanostructured ZnO has already been reported by other studies. Mazhdi et al. (2013) [18], evaluated the PL of nanocrystalline ZnO synthesized by the micelle reverse method, and observed four bands when excited at 330 nm, located in the violet region, between 395–400 nm (main band), in the blue (~450 nm), blue-green (~490 nm), and a weaker one in green (~530 nm). These bands are located in regions similar to those shown here, however, the intensity relationship between them is different, since the PL presented previously has the main band in the green and non-violet region. This indicates how the synthesis method and type of heat treatment modifies the PL properties.

Sahu et al. (2010) [23], also reported six emission bands for the material synthesized by sonification method with continuous mode, at 412, 437, 468, 544, 570, and 618 nm, after an excitation at 365 nm, however, it differs about the main bands, which are centered at 437 and 570 nm. In another study, Reddy et al. (2011) [17] showed that nanocrystalline ZnO synthesized by the low combustion method has a PL, after excitation at 254 nm, composed of peaks in violet (a ~402 nm), blue (at 421, 437, and 485 nm) and a low one in the green at 525 nm, therefore, unlike the ZnO presented in this chapter, whose main band is in the green region.

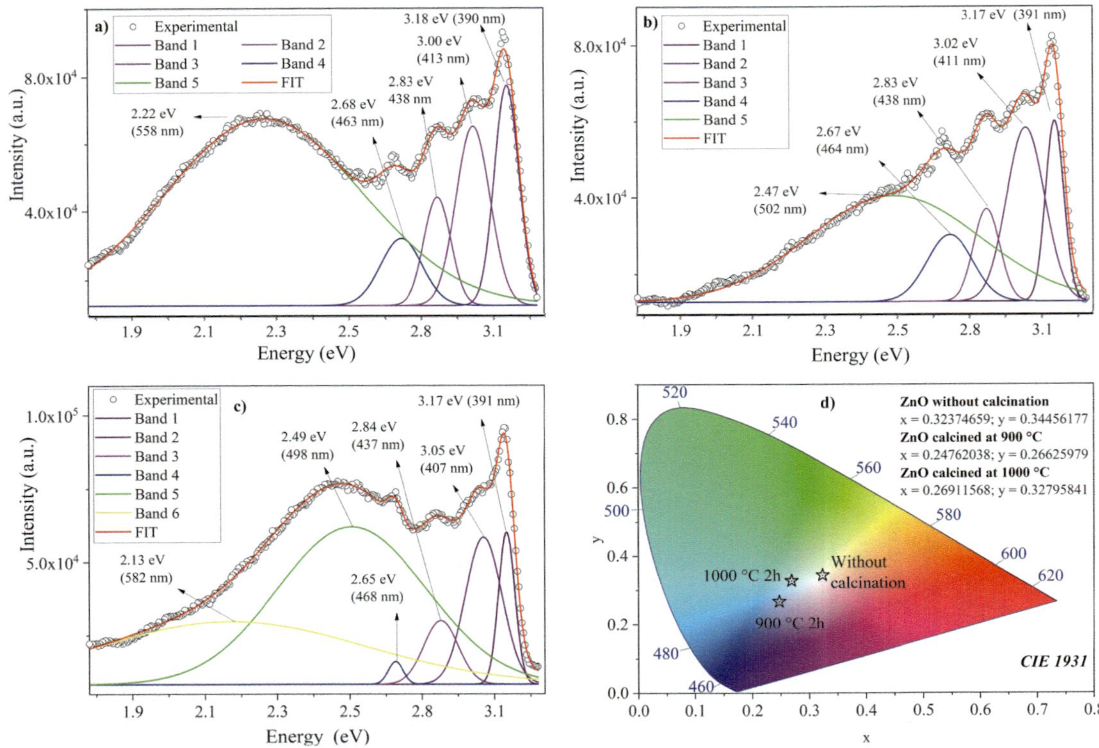

Figure 2.12 Photoluminescence emission of ZnO nanocrystals synthesized by coprecipitation method (a) without calcination, (b) calcined at 900°C, (c) calcined at 1000°C, with a λ_{EXC} = 365 nm, and (d) the respective chromaticity diagram.

The defects responsible for each emission band are not yet fully elucidated, but studies have pointed out that the emission in violet would be related to excitonic recombination referent to the band close to the edge transition, and that the visible one, related to surface defects [47]. The green emission would be related, mainly to oxygen vacancies [18, 48], or even, it may be related to other defects such as zinc vacancies, or zinc and oxygen interstitials [17, 30]. For the emission bands in blue region, it has been reported that it occurs due to defects in the ZnO surface, such as oxygen vacancies, and zinc interstitials [17].

The PL bands shown indicate strong emission in the visible region, and considering that the Schott BG39 filter has a transmission between 300–600 nm [49], this explains the important emission observed in the TL curves, while for LOE, the intensity was lower.

2.5 Summary

In this chapter, some luminescent properties of ZnO nanocrystals synthesized by coprecipitation method, of hexagonal Wurtzite phase, were studied by TL, OSL and PL measurements.

TL glow curves showed that ZnO supplied several peaks between 0–500°C, in the visible region, after irradiation with β-rays, which the behavior depends to the heat treatment adopted. ZnO nanocrystals without calcination, did not present a TL glow curve. While the one calcined at 900°C has two set of peaks, located at 120 and 350°C approximately, being the dose response, considering the whole curve, with a linear behavior up to 20.3 Gy. ZnO calcined at 1000°C was more sensitive sample to the irradiation, showing TL glow curves since low doses, as 0.08 Gy. In addition, the peak positions were different, with two sets between 100–200°C and 200–350°C. The dose response analyses pointed out linear, sub and supralinear regions.

CGCD study for TL allowed to verify that ZnO calcined at 900°C presents 3 peaks at 141, 247, and 357°C, with b spanning 1.5–2, and the E_a between 0.7–1.5 eV. Whilst ZnO calcined at 1000°C has 5 peaks for low doses (of 0.41 Gy, for example), at 141, 177, 268, 342, and 450°C, which have little changes when a higher dose is applied (as 20.3 Gy), being located at 132, 191, 238, 357, and 447°C.

In both cases. The b values varied between 1–2, and the E_a between 0.9–1.55 eV.

CW and LM-OSL curves were only supplied for ZnO calcined at 1000°C, showing the importance of heat treatment in luminescent properties, especially in ultraviolet region, for the OSL case specifically. The CGCD study depicted of those for a low dose, just a fast component is present in the CW-OSL, with b close to 1, and decay time (τ) of 0.28 s. After irradiation with a dose of 40.6 Gy, a medium component is also observed, with $b = 1.1$ and $\tau = 8.8$ s. In the LM-OSL case, 3 components at 3.75, 10, and 35 s were noted, with respective values of b of 1, 1, and 2.

Photoluminescence curves indicated that ZnO after calcinations has a strong emission band in green region, and other in violet and blue region. The bands positions shifted according the heat treatment. ZnO without calcination has 5 bands, being the main one at 558 nm; ZnO calcined at 900°C also supplied 5 bands, with a main one at 502 nm; and the ZnO calcined at 1000°C presented 6 bands, with the most intense band at 498 nm. The Chromaticity Diagram showed how heat treatment shifted the luminescence on white region to a bluish or blue-greenish one.

The findings approached in this chapter pointed out that ZnO nanocrystals have an important potential in application to luminescence area, including dosimetry. In addition, showed how heat treatment plays a significative role in luminescence materials properties, such as ZnO.

References

1. M. Rafique, S. Hajra, M.B. Tahir, T.I. Awan, A. Bashir, A. Tehseen, Chapter 4 - Properties of nanomaterials, eds.: T.I. Awan, A. Bashir, A. Tehseen, In: *Chemistry of Nanomaterials*, Elsevier, 2020, pp. 89–117. ISBN: 9780128189085. doi: 10.1016/B978-0-12-818908-5.00004-4
2. A.B. Asha, R. Narain, Chapter 15 - Nanomaterials properties, eds.: Ravin Narain, In: *Polymer Science and Nanotechnology*, Elsevier, 2020, pp. 343–359. ISBN: 9780128168066. doi: 10.1016/B978-0-12-816806-6.00015-7
3. S.P. Patil, V.V. Burungale, Chapter 2 - Physical and chemical properties of nanomaterials, eds.: N.D. Thorat, J.Bauer, In: *Micro and Nano Technologies, Nanomedicines for Breast Cancer Theranostics*, Elsevier,

2020, pp. 17–31. ISBN: 9780128200162. doi: 10.1016/B978-0-12-820016-2.00002-1

4. M. Rizwan, A. Shoukat, A. Ayub, B. Razzaq, M.B. Tahir, Chapter 3 - Types and classification of nanomaterials, eds.: M.B. Tahir, M. Sagir, A.M. Asiri, In: *Micro and Nano Technologies, Nanomaterials: Synthesis, Characterization, Hazards and Safety*, Elsevier, 2021, pp. 31–54. ISBN: 9780128238233. doi: 10.1016/B978-0-12-823823-3.00001-X
5. S. Ameen, M.S. Akhtar, H.-K. Seo, H.S. Shin, Advanced ZnO–graphene oxide nanohybrid and its photocatalytic applications, *Mater. Lett.* 100 (2013) 261–264. doi: 10.1016/j.matlet.2013.03.012
6. E. Quartarone, V. Dall'Asta, A. Resmini, C. Tealdi, I.G. Tredici, U.A. Tamburini, P. Mustarelli, Graphite-coated ZnO nanosheets as high-capacity, highly stable, and binder-free anodes for lithium-ion batteries, *J. Power Sources.* 320 (2016) 314–321. doi: 10.1016/j.jpowsour.2016.04.107
7. N. Mala, K. Ravichandran, S. Pandiarajan, N. Srinivasan, B. Ravikumar, K.C.S. Pushpa, K. Swaminathan, T. Arun, Formation of hexagonal plate shaped ZnO microparticles – A study on antibacterial and magnetic properties, *Ceram. Int.* 42 (2016) 7336–7346. doi: 10.1016/j.ceramint.2016.01.132
8. F.S. Jebel, H. Almasi, Morphological, physical, antimicrobial and release properties of ZnO nanoparticles-loaded bacterial cellulose films, *Carbohydr. Polym.* 149 (2016) 8–19. doi: 10.1016/j.carbpol.2016.04.089
9. A. Tereshchenko, M. Bechelany, R. Viter, V. Khranovskyy, V. Smyntyna, N. Starodub, R. Yakimova, Optical biosensors based on ZnO nanostructures: advantages and perspectives. A review, *Sens. Actuators, B.* 229 (2016) 664–677. doi: 10.1016/j.snb.2016.01.099
10. P. Wang, D. Wang, M. Zhang, Y. Zhu, Y. Xu, X. Ma, X. Wang, ZnO nanosheets/graphene oxide nanocomposites for highly effective acetone vapor detection, *Sens. Actuators, B.* 230 (2016) 477–484. doi: 10.1016/j.snb.2016.02.056
11. D.K. Sharma, S. Shukla, K.K. Sharma, V. Kumar, A review on ZnO: Fundamental properties and applications, *Mater. Today Proc.*, In press. 2020. doi: 10.1016/j.matpr.2020.10.238
12. Ü. Özgür, Y.I. Alivov, C. Liu, A. Teke, M.A. Reshchikov, S. Doğan, V. Avrutin, S.-J. Cho, and H. Morkoç, A comprehensive review of ZnO materials and devices, *J. Appl. Phys.* 98 (2005) 041301. doi: 10.1063/1.1992666
13. A.F. Soares, S.H. Tatumi, T.M. Mazzo, R.R. Rocca, L.C. Courrol, Study of morphological and luminescent properties (TL and OSL) of ZnO

nanocrystals synthetized by coprecipitation method, *J. Lumin.* 186 (2017) 135–143. doi: 10.1016/j.jlumin.2017.02.028

14. D. Muer, W.M.-V. Vorst, Thermoluminescence of ZnO powder, *Physica.* 39 (1968) 123–132. doi: 10.1016/0031-8914(68)90053-0
15. C. Cruz-Vázquez, R. Bernal, S.E. Burruel-Ibarra, H. Grijalva-Monteverde, M. Barboza-Flores, Thermoluminescence properties of new ZnO nanophosphors exposed to beta irradiation, *Opt. Mat.* 27 (2005) 1235–1239. doi: 10.1016/j.optmat.2004.11.016
16. L. Grigorjeva, A. Zolotarjovsa, S.Y. Sokovnin, D. Millers, K. Smits, V.G. Il`ves, Radioluminescence, thermoluminescence and dosimetric properties of ZnO ceramics, *Ceram. Int.* 43 (2017) 6187–6191. doi: 10.1016/j.ceramint.2017.02.016
17. A.J. Reddy, M.K. Kokila, H. Nagabhushana, J.L. Rao, C. Shivakumara, B.M. Nagabhushana, R.P.S. Chakradhar, EPR, thermo and photoluminescence properties of ZnO nanopowders, *Spectrochim. Acta, Part A.* 81 (2011) 59–63. doi: 10.1016/j.saa.2011.06.048
18. M. Mazhdi, J. Saydi, M. Karimi, J. Seidi, F. Mazhdi, A study on optical, photoluminescence and thermoluminescence properties of ZnO and Mn doped-ZnO nanocrystalline particles, *Optik.* 124 (2013) 4128–4133. doi: 10.1016/j.ijleo.2012.12.068
19. M.A. Seitz, W.F. Pinter, W.M. Hirthe, Thermoluminescence of ZnO, *Mater. Res. Bull.* 6 (1971) 275–282. doi: 10.1016/0025-5408(71)90041-9
20. U. Pal, R. Meléndrez, V. Chernov, M. Barboza-Flores, Effect of Yb doping on the afterglow and thermoluminescent properties of ZnO nanophosphors, *J. Nanosci. Nanotechnol.* 8 (2008) 1–6. doi:10.1166/jnn.2008.020
21. J. Manam, S. Das, A. Isaac, Preparation, characterization and thermally stimulated luminescence of ZnO nanophosphor. *Indian J. Phys.* 83 (2009) 1407–1419. doi: 10.1007/s12648-009-0129-5
22. A.K. Srivastava, K. Ninagawa, S. Toyoda, B.R. Chakraborty, S. Chandra, Induced thermoluminescence of X-ray irradiated nanostructured zinc oxide, *Opt. Mat.* (2009) 410– 413. doi: 10.1016/j.optmat.2009.09.007
23. D. Sahu, B.S. Acharya, B.P. Bag, Th. Basanta Singh, R.K. Gartia, Probing the surface states in nano ZnO powder synthesized by sonication method: Photo and thermo-luminescence studies, *J. Lumin.* 130 (2010) 1371–1378. doi: 10.1016/j.jlumin.2010.02.049
24. H.A. Borbón-Nuñez, C. Cruz-Vázquez, R. Bernal, G. Kitis, C. Furetta, V.M. Castaño, Thermoluminescence properties of sintered ZnO. *Opt. Mat.* (2014) 398–403. doi: 10.1016/j.optmat.2014.06.034

25. L. Grigorjeva, A. Zolotarjovsa, S.Y. Sokovnin, D. Millers, K. Smits, V.G. Il`ves, Radioluminescence, thermoluminescence and dosimetric properties of ZnO ceramics, *Ceram. Int.* 43 (2017) 6187–6191. doi: 10.1016/j.ceramint.2017.02.016

26. N. Pushpa, M.K. Kokila, K.R Nagabhushana, Thermoluminescence studies of γ- irradiated ZnO:Mg^{2+} nanoparticles, *Nucl. Instrum. Methods Phys. Res., Sect. B*, 379 (2016) 62–68. doi: 10.1016/j.nimb.2016.04.042

27. M. Isik, N.M. Gasanly, The defect state of Yb-doped ZnO nanoparticles using thermoluminescence study, *Mater. Sci. Semicond. Process.* 100 (2019) 29–34. doi: 10.1016/j.mssp.2019.04.030

28. M. Isik, N.M. Gasanly, Gd-doped ZnO nanoparticles: Synthesis, structural and thermoluminescence properties, *J. Lumin.* 207 (2019) 220–225. 10.1016/j.jlumin.2018.11.022

29. P.P. Pal, J. Manam, Photoluminescence and thermoluminescence studies of Tb^{3+} doped ZnO nanorods, *Mater. Sci. Eng., B.* 178 (2013) 400–408.

30. K.S. Babu, A.R. Reddy, Ch. Sujatha, K.V.G. Reddy, A.N. Mallika, Annealing effects on photoluminescence of ZnO nanoparticles, *Mater. Lett.* 110 (2013) 10–12. doi: 10.1016/j.matlet.2013.07.114

31. S. Wei, J. Lian, H. Wu, Annealing effect on the photoluminescence properties of ZnO nanorod array prepared by a PLD-assistant wet chemical method, *Mater. Charact.* 61 (2010) 1239–1244. doi: 10.1016/j.matchar.2010.08.002

32. Y. Zhang, Y. Liu, L. Wu, H. Li, L. Han, B. Wang, E. Xie, Effect of annealing atmosphere on the photoluminescence of ZnO nanospheres. *Appl. Surf. Sci.* 255 (2009) 4801–4805.

33. A.F. Soares, S.H. Tatumi, R.R. Rocca, Influence of heat treatment on luminescent properties of the HfO_2, synthesized by precipitation method, *Radiat. Phys. Chem.* 180 (2021) 109292. doi: 10.1016/j.radphyschem.2020.109292

34. V. Altunal, V. Guckan, A. Ozdemir, Z. Yegingil, A calcination study on BeO ceramics for radiation dosimetry, *Mater. Res. Bull.* 130 (2020) 110921. doi: 10.1016/j.materresbull.2020.110921

35. R.H. Krishna, B.M. Nagabhushana, H. Nagabhushana, N.S. Murthy, S.C. Sharma, C. Shivakumara, R.P.S. Chakradhar, Effect of calcination temperature on structural, photoluminescence, and thermoluminescence properties of Y_2O_3:Eu^{3+} nanophosphor. *Phys. Chem. C.* 117 (2013) 1915–1924. doi: 10.1021/jp309684b

36. S.W.S. McKeever, *Thermoluminescence of Solids*, 1st edition, Cambridge University Press, Cambridge, 1985. doi:10.1017/CBO9780511564994
37. D. Afouxenidis, G.S. Polymeris, N.C. Tsirliganis, G. Kitis, Computerised curve deconvolution of TL/OSL curves using a popular spreadsheet program, *Radiat. Prot. Dosimetry*. 149 (2012) 363–370. doi:10.1093/rpd/ncr315
38. A. Soni, B.C. Bhatt, M.S. Kulkarni, D.R. Mishra, G.S. Polymeris, OSL and thermally assisted OSL response in dental enamel for its possible application in retrospective dosimetry, *Radiat. Environ. Biophys*. 53 (2014) 763–774. doi:10.1007/s00411-014-0554-5
39. G. Kitis, J.M. Gomez-Ros, J.W.N. Tuyn, Thermoluminescence glow-curve deconvolution functions for first, second and general orders of kinetics, *J. Phys. D. Appl. Phys*. 31 (1998) 2636–2641. doi:10.1088/0022-3727/31/19/037
40. H.G. Balian, N.W. Eddy, Figure-of-merit (FOM), an improved criterion over the normalized chi-squared test for assessing goodness-of-fit of gamma-ray spectral peaks, *Nucl. Instruments Methods*. 145 (1977) 389–395. doi:10.1016/0029-554X(77)90437-2
41. V. Guckan, V. Altunal, A. Ozdemir, V. Tsiumra, Y. Zhydachevskyy, Z. Yegingil, Calcination effects on europium doped zinc oxide as a luminescent material synthesized via sol-gel and precipitation methods, *J. Alloys Compd*. 823 (2020) 153878.
42. C. Cruz-Vázquez, S.E. Burruel-Ibarra, H. Grijalva-Monteverde, V. Chernov, R. Bernal, Thermally and optically stimulated luminescence of new ZnO nanophosphors exposed to beta particle irradiation, *Radiat. Eff. Defects Solids*. 162 (2007) 737–743. doi: 10.1080/10420150701482089
43. N.R. Panda, B.S. Acharya, Th. B. Singh, R.K. Gartia, Luminescence properties and decay kinetics of nano ZnO powder doped with cerium ions, *J. Lumin*. 136 (2013) 369–377. doi: 10.1016/j.jlumin.2012.12.002
44. E.G. Yukihara, S.W.S. McKeever, *Optically Stimulated Luminescence*, John Wiley & Sons Ltd Registered, West Sussex, 2011.
45. G. Kitis, V. Pagonis, Computerized curve deconvolution analysis for LM-OSL, *Radiat. Meas*. 43 (2008) 737–741. doi:10.1016/j.radmeas.2007.12.055
46. B. Zhang, Chapter 8 - Optical properties of nanomaterials, eds.: B. Zhang, In: *Micro and Nano Technologies, Physical Fundamentals of Nanomaterials*, William Andrew Publishing, 2018, pp. 291–335, ISBN 9780124104174, https://doi.org/10.1016/B978-0-12-410417-4.00008-3

47. Z.-J. Mo, Z.-H. Hao, H.-Z Wu, Q. Yang, P. Zhuo, H. Yang, J.-P. Xu, X.-S. Zhang, L. Li, Synthetic and effect of annealing on the luminescent properties of ZnO nanowire, *J. Lumin.* 175 (2016) 232–236. doi: 10.1016/j.jlumin.2016.02.010

48. N. Tu, H. V. Bui, D.Q. Trung, A.-T. Duong, D.M. Thuy, D.H. Nguyen, K.T. Nguyen, P.T. Huy, Surface oxygen vacancies of ZnO: A facile fabrication method and their contribution to the photoluminescence, *J. Alloys Compd.* 791 (2019) 722–729. doi: 10.1016/j.jallcom.2019.03.395

49. SCHOTT. Datasheet SHOTT filters. Disponible on: https://shop.schott.com/advanced_optics/en/Matt-Filter-Plates/BG39/c/matt-plate/glass-BG39 Accessed on: 03/20/21.

Chapter 3

Lanthanide Luminescence Enhancement in Nanostructures by Coordination Chemistry

Miki Hasegawa[a,b]
[a]*Department of Chemistry and Biological Science,*
College of Science and Engineering, and
[b]*Mirai Molecular Materials Design Institute,*
Aoyama/Future Gakuin University, 5-10-1 Fuchinobe,
Chuo-ku, Sagamihra, Kanagawa 252-5258, Japan
hasemiki@chem.aoyama.ac.jp

3.1 Introduction

Lanthanide ions exhibit very high color purity originating from the ff transitions localized on f-orbitals [1–5] The ff emission in hybridized compounds with organic chromophores can be sensitized through the photo-excitation energy transfer via the excited $\pi\pi^*$ transitions of the ligand. The complexation of lanthanide and organic ligands recently has much possibility to apply for industrial luminescent

Luminescent Nanomaterials
Edited by Odireleng Martin Ntwaeaborwa

ISBN 978-981-4968-11-9 (Hardcover), 978-1-003-27738-5 (eBook)
www.jennystanford.com

materials such as diodes, sensors for bio-materials [6–17]. It also remains scientific agenda to explain the electronic behaviors of f-elements due to the electric configuration of a series of lanthanide ions has the inner-transition metal configuration and f-electrons do not directly coordinate to the neighbor atoms. Then, it takes the coordination number surrounding lanthanide ions from six to twelve.

Coordination chemistry is one of the key approaches to control the physicochemical properties of lanthanide ions with organic ligands, that can act as spacers between two lanthanide ions of neighbor molecules, media, or photo-antenna to enhance the ff-emission. Here, the nanostructured compounds consisting of the lanthanide ion and organic species are focused to enhance the luminescence phenomena.

3.2 Photo-Antenna Effect in Lanthanide Complexes

Energy transfer in bichromophore has been established in 1960's [18]. For instance, a system consisting of naphthalene and anthracene shows energy transfer from the excited state of naphthalene to the ground state of anthracene [19]. The energy transfer accelerates via the superimposition of the fluorescence band of naphthalene and phosphorescence of anthracene, as an energy donor and an acceptor, respectively. This energy transfer system is known as Förster mechanism [20] by the Coulomb interaction. The energy transfer rate k_{EET} depends on the distance R_0 between energy donor and acceptor, $k_{EET} \propto 1/R_0{}^6$. Then the linked components of naphthalene and anthracene were developed for the evaluation [21].

While, another energy transfer mechanism is Dexter type, appears by the electron exchange [22]. Both cases should consider the spin multiplicity of before/after the energy transfer.

In the case of luminescent lanthanide complexes, above energy transfer mechanisms can be applied to sensitize the emission or to design the organic ligands. Energy transfer in a combination of lanthanide ion and organic chromophore is useful to enhance the ff emission of the metals [23]. Figure 3.1 shows the energy transfer diagram of europium complex.

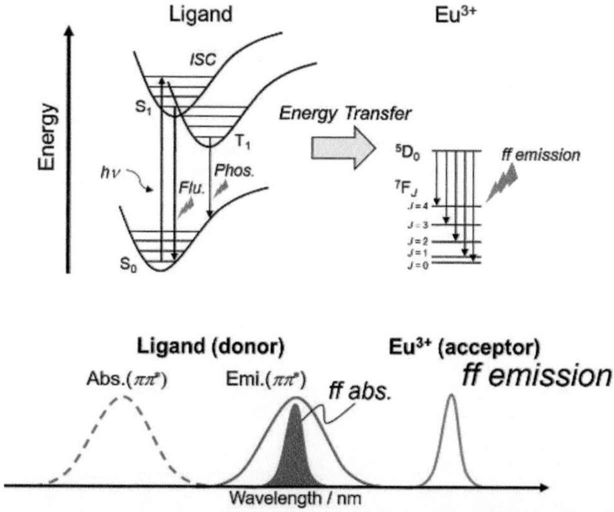

Figure 3.1 Photo-antenna effect: Model of energy transfer mechanism for the sensitizing of ff emission in the europium complexes with aromatic ligands. Superimposition between the ligand-emission and ff-absorption bands accelerates the energy transfer.

The split f-orbitals was assigned by using UPS (helium photoelectron spectroscopy) measurements of a series of lanthanide oxides (Fig. 3.2), and the diagram is known as "Dieke's diagram" [24–26]. The energy levels are unique in each elements and ions of lanthanide. These levels exist at most similarly even in different crystalline fields due to their electron configuration with the shielding effect surrounding 5d- and 6s-orbitals. The stability of f-level splitting becomes priority to design the energy donor level of photo-antenna ligands and to occur the lanthanide-lanthanide energy transfer (Ln-Ln EnT or MMEnT) toward to down-/ upconversions.

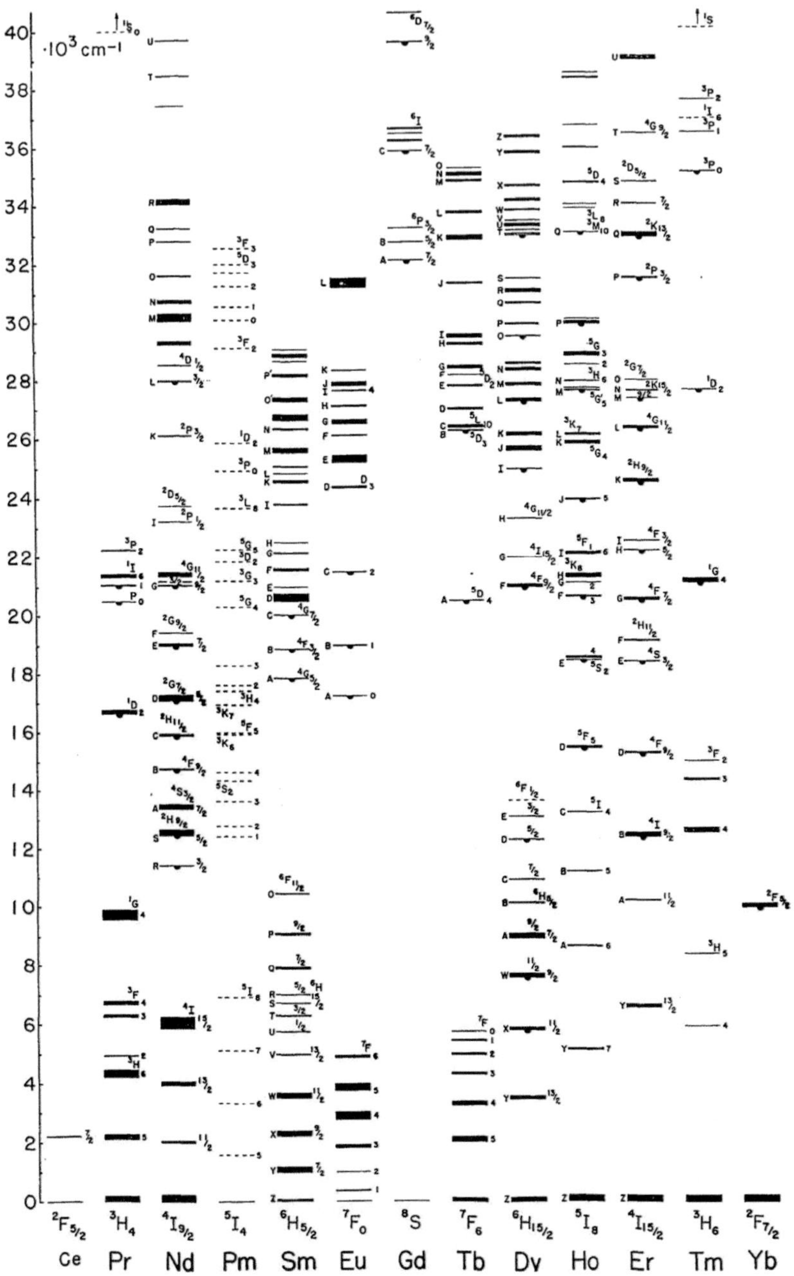

Figure 3.2 Classical Dieke's diagram from the observed results for a series of trivalent lanthanide salts [25].

3.3 Electronic Absorption and Luminescence Spectra of Lanthanide

The bands of ff transitions in lanthanide ions are Laporte forbidden, therefore the absorption coefficients are quite low from few to several hundred in $M^{-1}cm^{-1}$ [27]. The shape of each band sharply appears since the ff transition occurs in the isolated inner-core of lanthanide ion. Similarly, the ff emission bands are sharply observed and their Stokes shift is quite low due to the stable split energy levels in the f-orbital. The sharp luminescence band means highly pure luminescence color depending on the sorts of element in any compounds [28, 29]. Such characteristic luminescence behaviors are already used for lasers, information transportation through optical fibers and security inks [30–32].

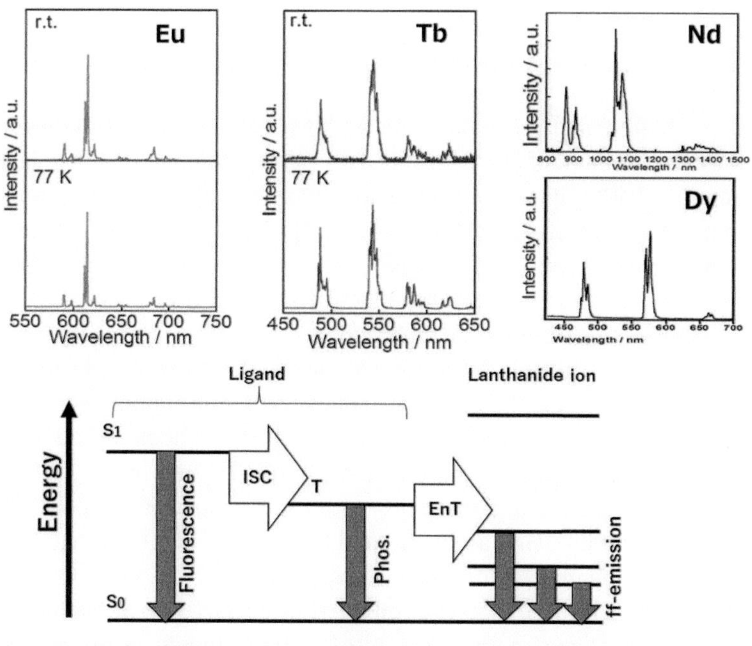

Figure 3.3 (Top) Luminescence spectra of Eu, Tb, Nd, and Dy complexes with L (λ_{ex} = 310 nm) [28, 29] and (bottom) the representation of intramolecular energy transfer from the ligand to lanthanide ion. ISC and EnT are intersystem crossing and energy transfer, respectively.

Figure 3.3 shows typical luminescence spectra of Eu, Tb, Nd, and Dy complexes in the solid state and the energy diagram to enhance photo-antenna effect. Each band position (wavelength) appears almost at same position in any complexes, because the transition energy is stable according to the Dieke's diagram. Thus, Eu and Tb luminescence in red and green, respectively. Nd shows luminescence bands in near infrared region. While, the luminescence intensities depend on the energy transfer efficiency, the ligand-field symmetry and/or non-radiative energy relaxation caused by the atmosphere. The luminescence quantum yields of Tb complexes increases at low temperature than at room temperature drastically because of the possibility of thermal equilibrium between energy donor (ligand) and acceptor (lanthanide ion) [30–32].

3.4 Lanthanide Luminescence in Nanostructures

3.4.1 Modification of Nanoparticles with Lanthanides

Core-shell structured nanoparticles have been widely studied for the enhancement of photo-physical properties compared with single nanoparticles. Of course, there are huge examples of inorganic-inorganic core-shell systems [33, 34]. Here, inorganic-organic core-shell systems of lanthanide compounds are forced which were considered the antenna effects and Dieke's diagram for the enhancement of lanthanide emission.

The base as core in the system are used silica, metal oxide, metal halide, metal phosphate, metal sulfate, and non-metals, etc. The surface of nanoparticles in core-shell systems can be modified with organic chromophore through covalent bond as shown in Fig. 3.4 [35, 36]. The core-shell nanoparticles with organic species are possible to composite by using ambient condition without high voltage and vacuum atmosphere. This type of materials can be modified to the procedure of layered materials on the rigid surface (see Section 3.5).

The strategic components to obtain highly luminescent lanthanide-based core-shell nanoparticles for inorganic compounds are well summarized as shown in Table 3.1 which is the modified one based on the review literature by Martin and Jaque [37]. This table shows the photo-antenna ability to obtain photoluminescence

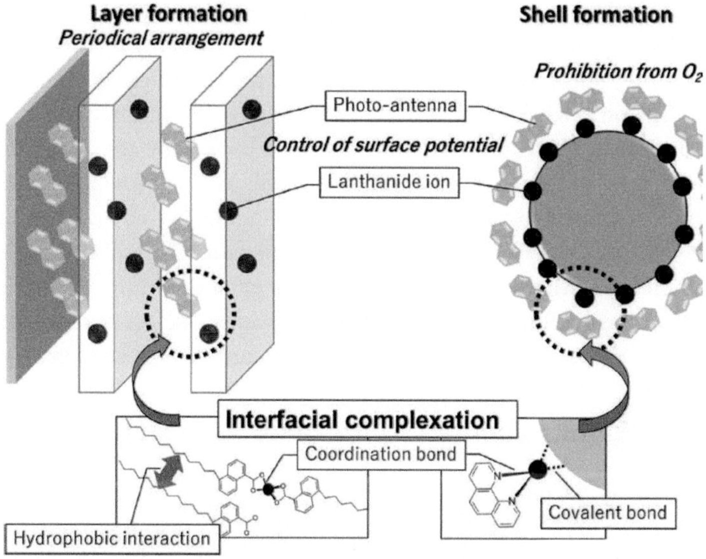

Figure 3.4 Concept of lanthanide-based layered materials (left: layer formation and right: core-shell formation) [36].

Table 3.1 Summary of the strategies for highly luminescent lanthanide-base core-shell nanoparticles based on the table in the reference [37]

Strategy	Photo-antenna	Absorption cross section	Absorption tuning possibility	Photo-stability
Multiple core-shell inorganic NP	Ln^{3+}	10^{21}–10^{20} cm^2	Difficult	High
Co-doping with other metal ions	$Mn^{2+/4+}$, Cr^{3+}, Ce^{3+}...	10^{20}–10^{18} cm^2	Medium	High
Organic dye–Ln hybridization	The π-electronic systems	10^{17}–10^{16} cm^2	Flexible	Enough

originated from ff-transition in nanoparticles. Thus, the components consisting of core and shell materials should be considered not only sorts of lanthanides and organic moieties, but also the absorption cross section, absorption, absorption tuning possibility, photostability of photo-antenna moieties. Especially, the absorption coefficient of aromatic compounds such as 1,10-phenanthroline (phen) is much higher than the inorganic species, and will induce the efficient energy transfer and ff-emission. The solid state–Eu complexes with phen show the luminescence quantum yields over 88% at room temperature [38]. Organic shells also act as shields for the prohibition of metal oxidation and as contourable surface.

3.4.2 Lanthanide-Based Core-Shell Silica Nanoparticles

Silica, SiO_2, is most high abundant metal oxide in the crust, and known as an optical transparent compound. The size control of silica nanoparticles has been established in several methods such as Stöber-van method, reverse microemulsion or water-in-oil method, and the synthesis of silica-core/PEG-shell nanoparticles (Fig. 3.5) [39]. Size distribution of silica nanoparticles shows sharp thresholds and reproducibility under the same condition. Chemical treated silica nanoparticles by surfactants differs their surface potentials. For instance, if the surface of silica nanoparticles takes hydrophobicity by using alkyl chains, the luminescence band of intercalated chromophore appears at different wavelength in various concentrations due to the intermolecular interactions on the surface. While, in the case of hydrophilic surface on the silica nanoparticle by hydroxy, amine or carboxyl groups, other metal ions and cation can reach for the formation of multiple core-shell nanoparticles.

Efficient luminescent resonance energy transfer from the Tb complex (N,N,N^1,N^1-[4′-phenyl-2,2′:6′,2′-terpyridine-6,6′-diyl] bis(methylene-nitrilo)tetrakis(acetate)-Tb^{3+}) to CTMR (5-carboxy-tetramethylrhodamine) in the core-shell nanostructures composed of them covalently bonded silica nanoparticles [40]. The Tb complex act as an energy donor and CTMR as acceptor in the system. The energy transfer on the silica nanoparticles was examined the color changes from green of Tb^{3+} to red of CTMR and evaluated its long-lived luminescence by the coexistence of two luminophores (Fig. 3.6).

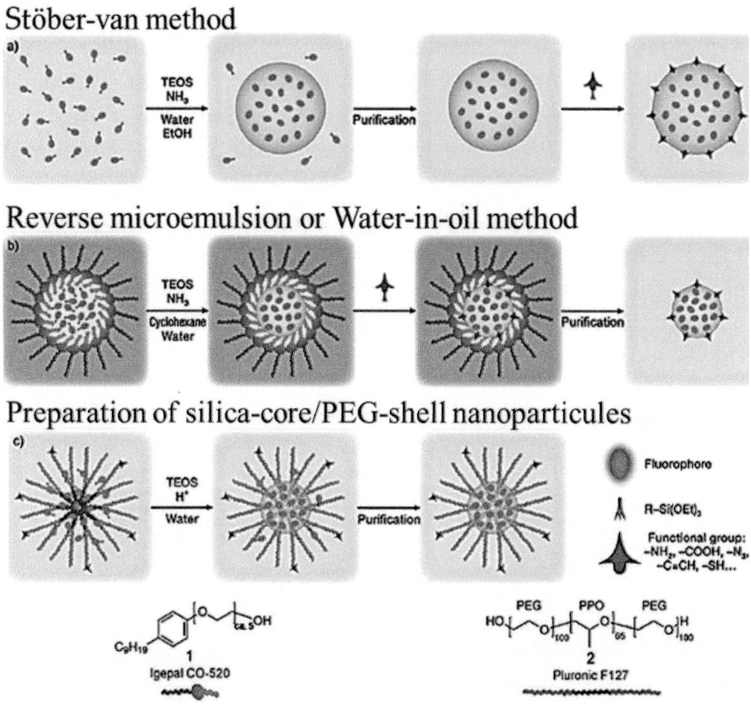

Figure 3.5 Strategies for the preparation of silica nanoparticles. TEOS and PEG are monomeric tetrathoysilane and polyethyleneglycol, respectively. Modified figure of Ref. [39].

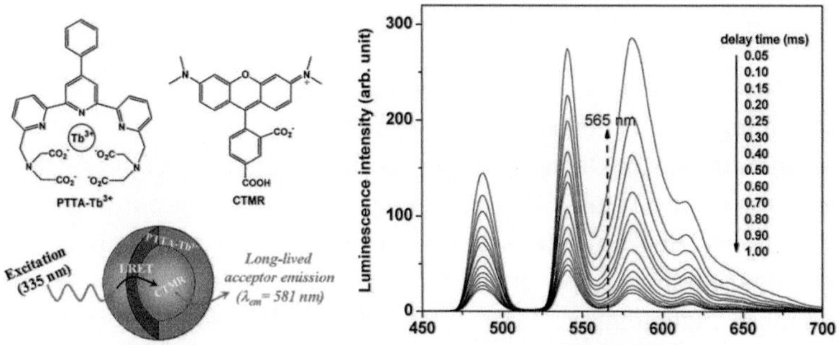

Figure 3.6 Energy transfer system from Tb complex to rhodamine chromophore on the silica nanoparticle surface. Luminescence lifetime of CTMR via the energy transfer was elongated. Figure was modified from Ref. [40].

During the preparation of silica nanoparticles with TEOS, dye components can dissolve into the solution resulting in the composition having concentration gradient [41]. As shown in Fig. 3.7, three types of silica nanoparticles including rhodamine were prepared. From the luminescence spectra and brightness measurements, the brightness of luminescence drastically increased in the expanded core-shell particle. The enhancement of the radiative rate is independent of particle architecture and is greater than the single dye. Thus, optical property of silica nanoparticle affected to enhance the composed fluorophores' brightness. This case also uses the optical behavior of silica for the brightness to atmosphere.

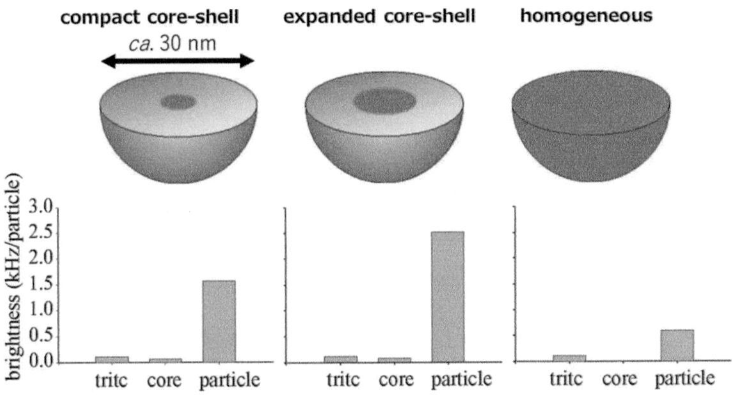

Figure 3.7 Strategy of various core-shell silica nanoparticles with organic dye in various gradients (top) and the brightness for each synthesis stage. λ_{ex}: 1.2 mW at 900 nm. Modified figure of Ref. [41].

The covering of silica nanoparticles with organic compound via interface complexations also functions as antioxidants for lanthanides [42]. Divalent Eu ion is much unstable than trivalent ion, and does not exist in nature. To prepare luminescent materials containing divalent state of Eu, it is necessary to reduce the trivalent Eu in an appropriate matrix, using a reducing atmosphere at quite high temperatures above 1000°C [43–48]. For instance, Matsuzawa et al., succeeded to develop persistent luminescence compound, LumiNova®, which can keep the divalent Eu ion in strontium alumina [43], and has been used for emergency lights in the buildings, airplanes, subways, etc. in the world. The Eu ion is reduced from trivalent to divalent in the defect of strontium after the ceramitaion. Their modification has expanded to the ceramic mechanoluminescent compounds [49]. Although some reports show that this reduction occurs even in air, high temperatures are still necessary, along with a rigid inorganic crystal structure as a host [50–52].

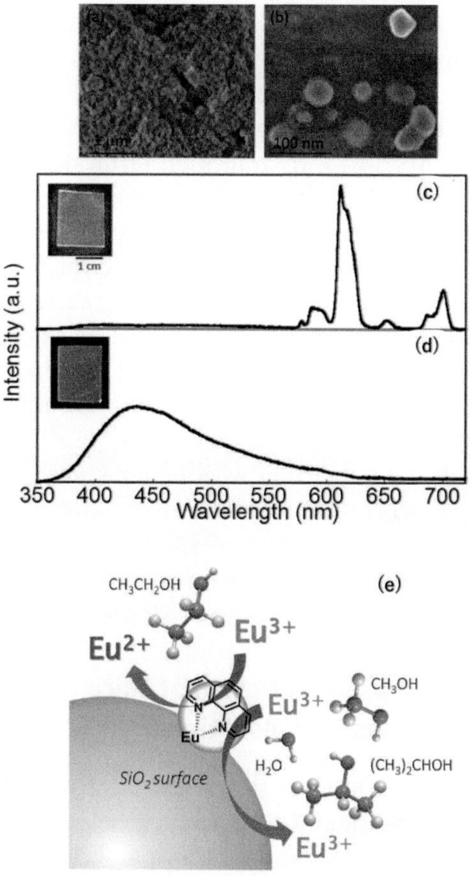

Figure 3.8 (a) Low and (b) high magnification SEM images of silica nanoparticles covered with Eu ions. Luminesce spectra of (c) as-prepared and (d) sintered Eu-silica nanoparticles covered with 1,10-phenanthroline (λ_{ex} = 280 nm). (e) Possible redox mechanism of ethanol on the nanoparticle for the reduction of Eu. These figures are modified from Refs. [42, 53].

Figure 3.8 shows the core-shell silica nanoparticle doubly covered with Eu and 1,10-phenanthroline, which shows the interfacial reduction of a Eu [42, 53]. The size of nanoparticle does not change after the complexation by the measurement of microscope and synchrotron XRD [42]. The as-prepared Eu-covered silica nanoparticle with 1,10-phenanthroline exhibits sharp luminescence bands at 578.2, 589.5, 611.1, 651.1, and 700.5 nm, assigned to the $^5D_0 \rightarrow {}^7F_0$, $^5D_0 \rightarrow {}^7F_1$, $^5D_0 \rightarrow {}^7F_2$, $^5D_0 \rightarrow {}^7F_3$, and $^5D_0 \rightarrow {}^7F_4$ transitions, respectively, for trivalent Eu ion. These bands are sensitized by the photo-excitation of 1,10-phenanthroline at 280 nm as the photo-antenna effect. While the broad luminescence at 434 nm of the sintered nanoparticles is the result of the allowed electric dipole $4f^6\ 5d \rightarrow 4f^7$ transition of divalent Eu ion, via an allowed transition of trivalent Eu ion rather than energy transfer from the photo-antenna. The color of luminescence of Eu changes from red of the trivalent state to blue of the divalent by a thermally activated reaction at relatively low temperature around 200°C in air. Additionally, the existence of ethanol on silica surface from the solvent of 1,10-phenanthroline during immersion with silica nanoparticle [53], since the oxidation energy of ethanol (–3.83 eV [53]) corresponds to the reduction of Eu^{3+}/Eu^{2+} (–4.0 eV [54]). Normally, such redox reaction with luminescence color changing of Eu complexes were discussed with electro chemical experiments [55–57]. Therefore, the surface of silica nanoparticles exhibits a characteristic reaction field for the redox reaction of Eu ions through the interface complexation of Eu and 1,10-phenanthroline.

3.4.3 Surface Potential Design of Silica Nanoparticle for Lanthanides Complexation

Tetraalchoxysilane is used for the formation of sol-gel silica grasses and surface modification of silica nanoparticles [58–60] or quartz substrates [61, 62] based on their transparency after the chemical reaction to form covalent bond on silica surfaces for lanthanide luminescence. Binnemans summarized a series of alchoxysilane agents for the hybridization of lanthanide ions and silica matrixes in his review article [58]. Figure 3.9 shows molecular structures of hydolizable precursors. These precursors make covalent attachment of lanthanide complexes to silica matrix.

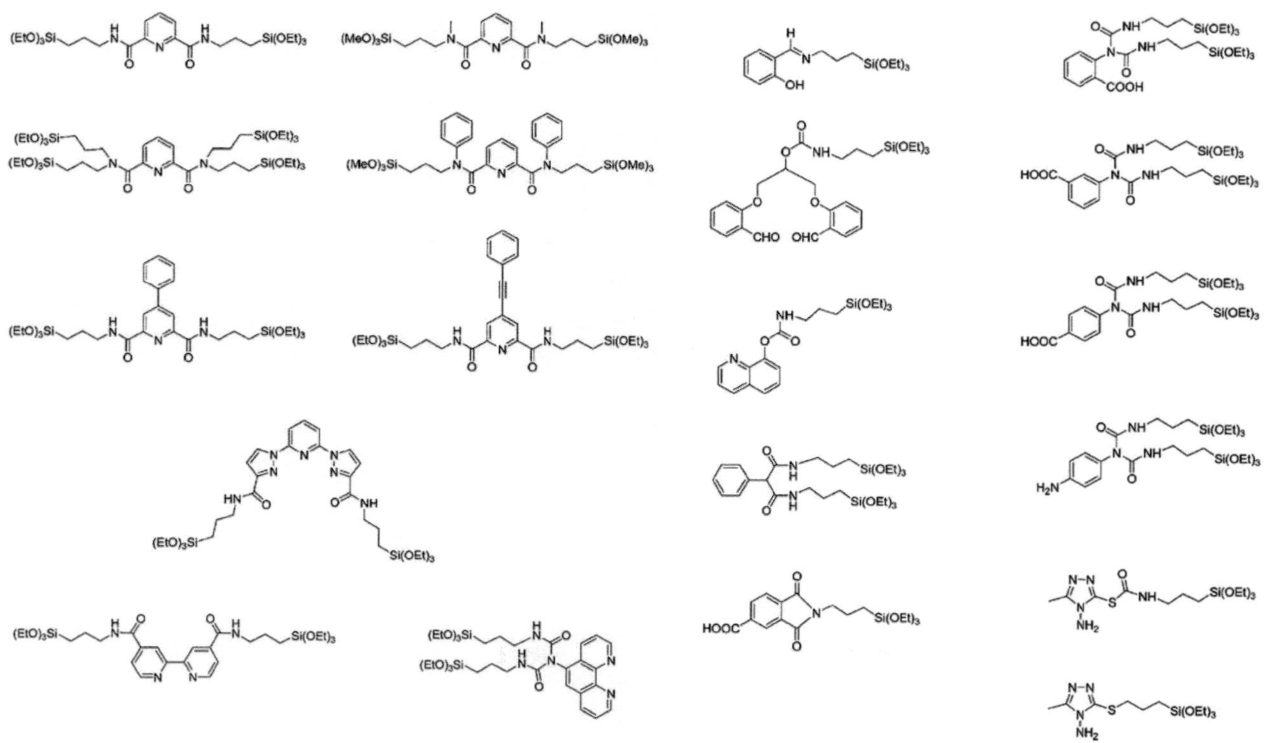

Figure 3.9 Hydrolyzable precursors covalently linked to silica hybrid materials. Nitrogen, oxygen or sulfur atoms will coordinate lanthanide ions. Modified from Ref. [58 and their-in].

An example of surface treatment for carboxylate-modified quartz is shown in Fig. 3.10 [63]. Fresh hydrolyzed quartz plates were dipped in ethanol solution of 3-aminopro-pyltrihydroxysilane and heated at 60°C for a few hours. Followed by immersion around 10 h, in glutaric anhydride dimethylformamide-solution at room temperature. Finally, the quartz plate was dipped in an aqueous solution of NaOH for 2 h and rinsed with ultrapure water. Such method is possible to apply for both silica nanoparticles and quartz substrates for their modification with different surface potentials.

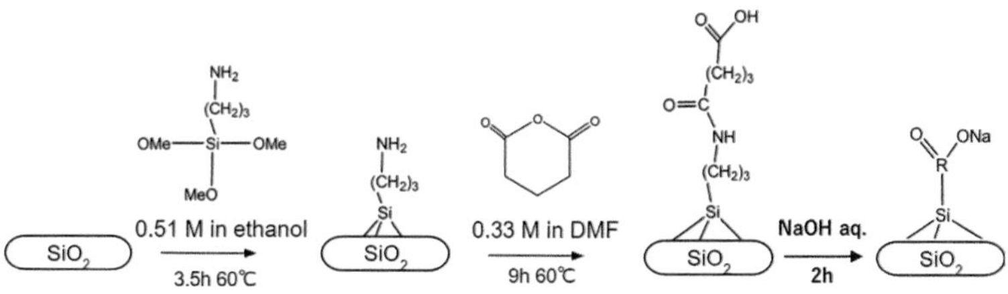

Figure 3.10 Surface treatment by solution system [63].

Silica-coated nanocomposits of magnetic nanoparticles or nanostructures were reported [64–66]. For instance, the magnetic behavior of such magnetic silica nanoparticles is enhanced by the magnetite [67, 68]. Lanthanides' luminescence is quenched often by the magnetic interaction, however, the silica-coating surrounding magnetite acts as a substrate for lanthanide-dye-modification through chemical bonding with their luminescence. Furthermore, the chain-complex of lanthanide can also elongate on the surface of silica-nanoparticles meaning that the distance between magnetite-center and the luminophore of lanthanide can periodically control in molecular level [69]. The chain-complex is composed of helicate lanthanide complexes and 1,4-benzendicarboxylate (bdc) [28, 70], and anchored on the magnetic silica nanoparticles [69] as shown in Fig. 3.11. The photo-antenna of Eu/Tb complexes is an organic chromophore, two bipyridine moiety bridged with ethylenediamine [28, 71, 72], and helically coordinates to lanthanide ion as hexadentate ligand. Two nitrate ions on the helicate complexes were replaced bdc to elongate the chain-complexes [70]. The initial

Tb complex luminesces negligibly weak, but the chain-complex formation drastically induced the luminescence even on the magneto silica nanoparticles. From TEM images, the surface of magnetic silica nanoparticles is clearly covered silica and is found suitable for the hierarchical chain-complexation linking. Their luminescence and magnetic properties can be recognized by naked eyes, since the particles in the suspension were drawn into the magnet with strong luminescence under UV light.

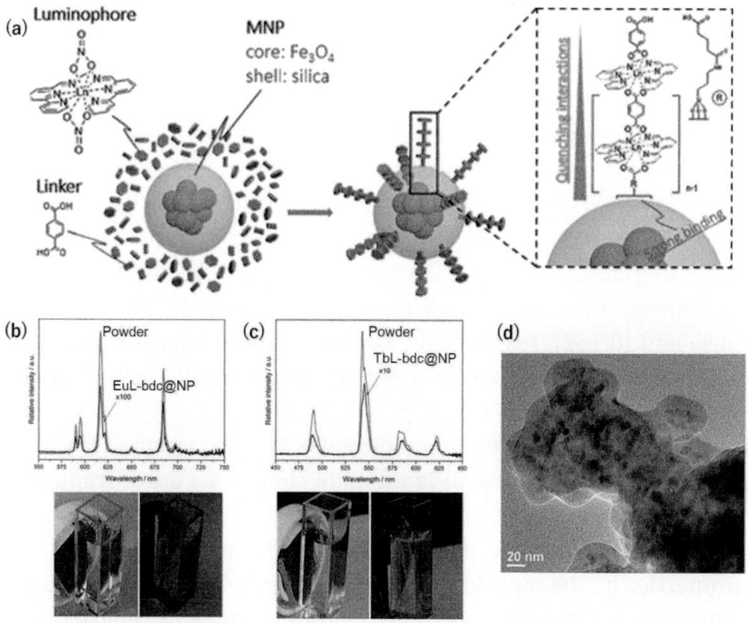

Figure 3.11 Magneto-luminescence lanthanide complexation on silica nanoparticles. (a) Strategic representation of magneto silica nanoparticles covering chain complexes with helicate Eu/Tb complex as luminophore and 1,4-benzenedicarboxylate (bdc) as linker. Luminescence spectra of Eu(b) and Tb(c) system (λ_{ex} = 335 nm) and pictures of luminescence of nanoparticles. (d) TEM image of EuL-bdc@magnetic silica nanoparticles. Figures are modified from Ref. [63].

This system was also applied for the quartz substrate [63] and formed the film of oligomeric-brush lanthanide complexes. Details will be described in Section 3.5.

3.5 Nanostructures of Lanthanide Complexes

3.5.1 Effect of Molecular Aggregation on the Lanthanide Luminescence

Molecular aggregation in molecules also enhances their luminescence properties [73] without/with matrix such as silica, zeolite, clusters, etc. Circularly polarized luminescence (CPL) is one of key phenomena for the next devices [74, 75]. In lanthanide complexes, it is known that lanthanide complexes coordinating chiral organic ligands exhibit CPL in solutions. The molecular chirality surrounding lanthanide ion sensitized CPL originated from ff-transitions [76]. The luminescence dissymmetry factor (g_{lum}) of organic molecules, a measure of degree of chirality in CPL, is typically in a range of 10^{-3} to 10^{-5} [77]. While it reveals that the aggregation of organic species significantly induces CPL phenomena because of the integration of molecular helicity in own 1D-stacking [78], and the organic cylinder in toluene takes the high quantum yield and significant g_{lum} value, 80% and |0.152|, respectively. Similarly, supramolecular gels of an achiral components exhibit strong CPL, g_{lum} = |0.105| more than that in solution g_{lum} =|0.003| [79].

Luminescence bands of organic molecules are assigned to the $\pi\pi^*$ transitions. While those of lanthanide complexes are ff transitions of metal ion via excitation energy transfer from organic ligands, but surprisingly there are few reports the magnification of CPL properties by the molecular aggregations [80, 81]. For example, the spiral Eu coordination polymers of Hasegawa were reported with CPL as shown in Fig. 3.12. Here, the CPL signals (I_{CPL}) are given by the differences between the left-handed (I_{L}) and right-handed (I_{R}) circularly polarized luminescence intensities (Eq. 3.1).

$$I_{\text{CPL}} = I_{\text{L}} - I_{\text{R}} \tag{3.1}$$

The spiral coordination polymer named $[\text{Eu}(+/-\text{ftc})_3(\text{dpbp})]_n$ in the solid state shows strong CPL property with high g_{lum} value, |0.17| at 586 nm compared with that in monomer, g_{lum} = |0.09| by the following Eq. 3.2.

$$g_{\text{lum}} = \frac{2(I_{\text{L}} - I_{\text{R}})}{(I_{\text{L}} + I_{\text{R}})} \tag{3.2}$$

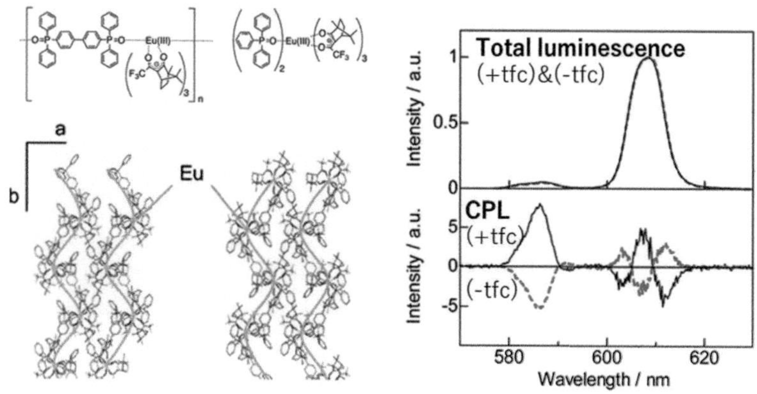

Figure 3.12 (Left) Molecular structure of Eu-coordination polymer and its component complex with +/–tfc and dpbp. Packing of their chiral crystals take mirror image with spirally linking. (Right) Total luminescence and CPL spectra of Eu complexes with +/–tfc and dpbp in the solid state (λ_{ex} = 380 nm). Modified figures of Ref. [80].

Kaizaki and Mullar reported pioneering work on the chirotopical properties of the tandem type Eu-Cs complexes with heptafluorobuthyryl camphorate [81–83], exhibiting the highest g_{lum} value, 1.38 in solution. This molecules as a building block helically aggregates to form chiral micro pieces under the control by Yuasa and Kawai [84], and performed the magnification of CPL phenomena with g_{lum} = 1.45 (Fig. 3.13). The performance also clarified by using a series of alkaline metal ions such as Na^+, K^+, Rb^+. Thus, molecular aggregation is effective for the magnify the CPL of lanthanide complexes.

3.5.2 Enhancement of Polarized Luminescence Phenomena of Lanthanide Complexes in Nano-thickness Films

From the viewpoints of interfacial complexation of lanthanides on rigid substrates, the molecular films are attracted much attention for the surface modification to develop efficient luminescence devices or sensors. As described above, materials with the solid-state CPL are difficult to design and development [78, 85], and there are only several literatures published. Recently, novel evaluation systems of

CPL which may be suitable for the luminescent lanthanide complex films are reported.

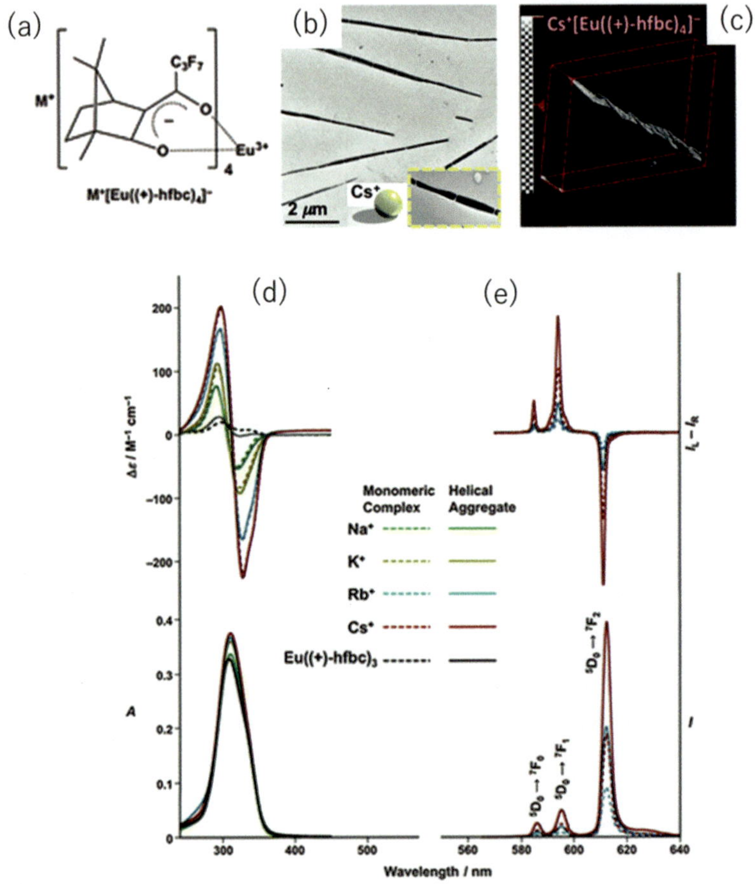

Figure 3.13 Helicate molecular aggregation-induced high CPL performance of Eu complexes with Cs with +/–hfbc (a). (b) and (c) are TEM and its topographic image of Cs[Eu((+)–hfbc)], respectively and inset of (b) is the high resolution TEM image. (d) and (e) are CD and CPL spectra of the Eu complexes with various alkaline metal ions, respectively, bottoms are total spectra.

Nozaki and Iwamura et al. developed the microscopic CPL imaging system and successfully observed the CPL map on the surface of molecular aggregation system consisting of Eu complex with 1,10-phenanthroline derivative and chiral amino acid in agar gels [86]. As shown in Fig. 3.14, each digit significantly indicates the phase of CPL of chiral amino acid because of the high resolution based on the microfocus system. The luminescence intensities depending on the concentration of complexation between the Eu-phenanthroline and amino acid in gel were corrected by the equilibrium constant. Relating detecting system of CPL were also newly developed [87]. These systems are possible for further applicable usage and evaluation for nanoparticles of luminescent lanthanide complexes.

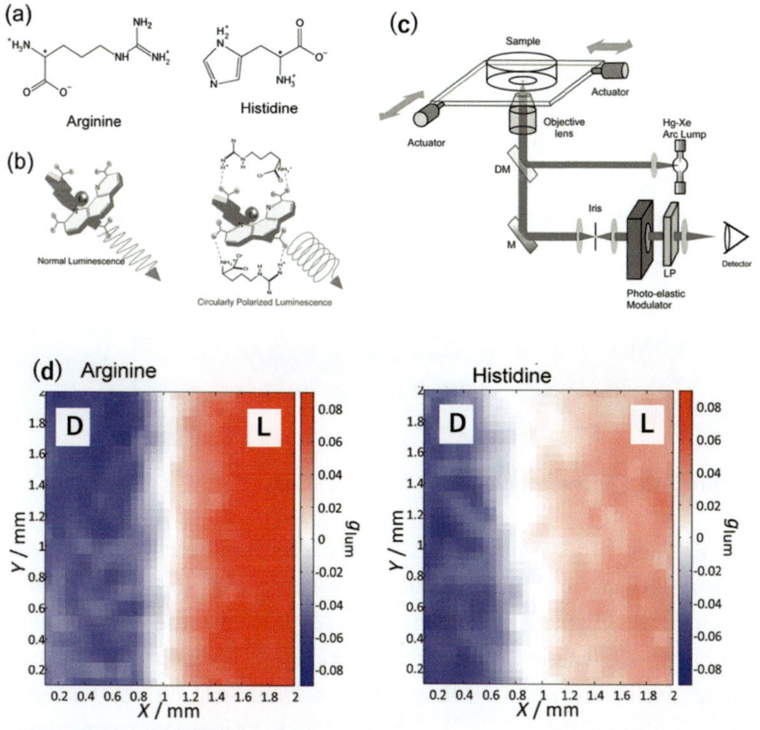

Figure 3.14 (a) Target chiral molecules, and (b) induced-CPL through intermolecular interaction between Eu complex with amino acid. (c) Schematic diagram of the microscopic CPL measurement system and (d) its image for g_{lum} at 591.4 nm spatial dispersion of agar gel containing Eu complex and amino acid, arginine (left) and histidine (right). λ_{ex} = 340–400 nm.

Luminescence quantum yields of solid compounds such as crystals, films, nanoparticles can be estimated by experimentally observed results [88–91]. The apparatus mainly uses an integrating sphere as a sample folder and absolute luminescence quantum yields' calculation approaches are suggested. It is shifted that the relative estimation of luminescence quantum yields in solutions with their concentration, reflective index of solvents compared to those of authentic samples changes to both usage such relative values and absolute ones without authentic information. Thus, absolute luminescence quantum yield estimation is widely used for various compounds.

Langmuir-Blodgett (LB) films and layer-by-layer (LbL) ones are major wet-procedure to periodically construct nano-thickness molecular aggregation with different packing from that in crystalline states. For the LB film preparation, surfactant molecules spread at interface of water/air are deposited on the substrates under the surface pressure (Fig. 3.15). The π-value is estimated from the pressure-area isotherm and adopted to keep the Langmuir film [92–94].

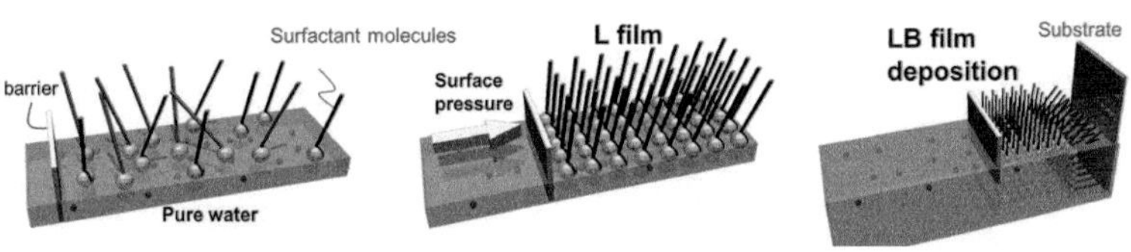

Figure 3.15 Representation of LB film formation.

Kitchen et al. reported LB films of lanthanide complexes with chiral-surfactant ligand [95, 96]. The complexes form the LB film on quartz substrate under the surface pressure and shows the ff emission of trivalent Eu ion in the films (Fig. 3.16). Each luminescence band also clearly shows CPL with g_{lum} = *ca.* |0.161| and *ca.* |0.068| in the LB film for the $^5D_0 \rightarrow {}^7F_1$ and $^5D_0 \rightarrow {}^7F_2$ transition, respectively. Corresponding g_{lum} values in solution are g_{lum} = *ca.* |0.177| and *ca.* |0.104|, respectively. Their slight differences were discussed that it means the molecular packing in the LB films.

Linearly polarized luminescence (LPL) of lanthanide complex in the LB films was also reported [97, 98]. The periodically deposited molecular films were clarified by the synchrotron XRD measurement, and it takes *ca.* 50 Å intervals as the inter-layer distances (Fig. 3.17).

The XRD pattern of the LB film prepared at the surface pressure of 30 mM/m shows single components, and that of 40 mN/m does at least double components due to the appearance of shoulder accompanied to each peak. Their nanostructure induces the efficient luminescence quantum yield (5.8%) of Eu complex in the film more than that in solid state (less than 0.1%), and LPL via energy transfer. The photo-antenna moiety, naphthoic acid derivative, has strong transition dipole moment. Then, the dipole moment localized on the ligand will interpret to Eu ion through the excitation energy transfer in the LB films. This suggestion corresponds to the assumption already published the literature of LPL in polymer films including Eu ion [99].

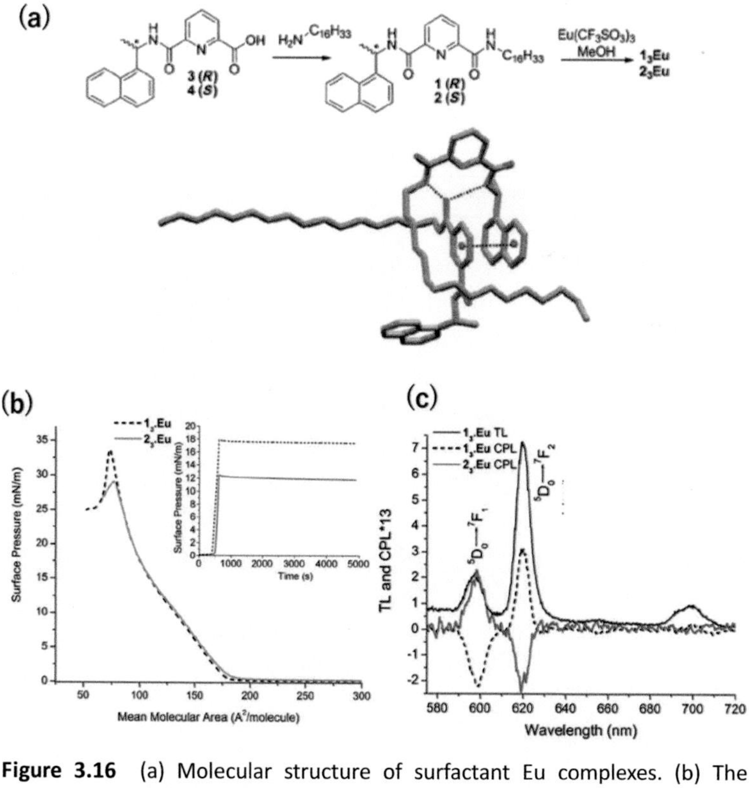

Figure 3.16 (a) Molecular structure of surfactant Eu complexes. (b) The surface-area isotherm of the complexes. (c) CPL and total luminescence spectra of the LB films. Modified figures of Ref. [95].

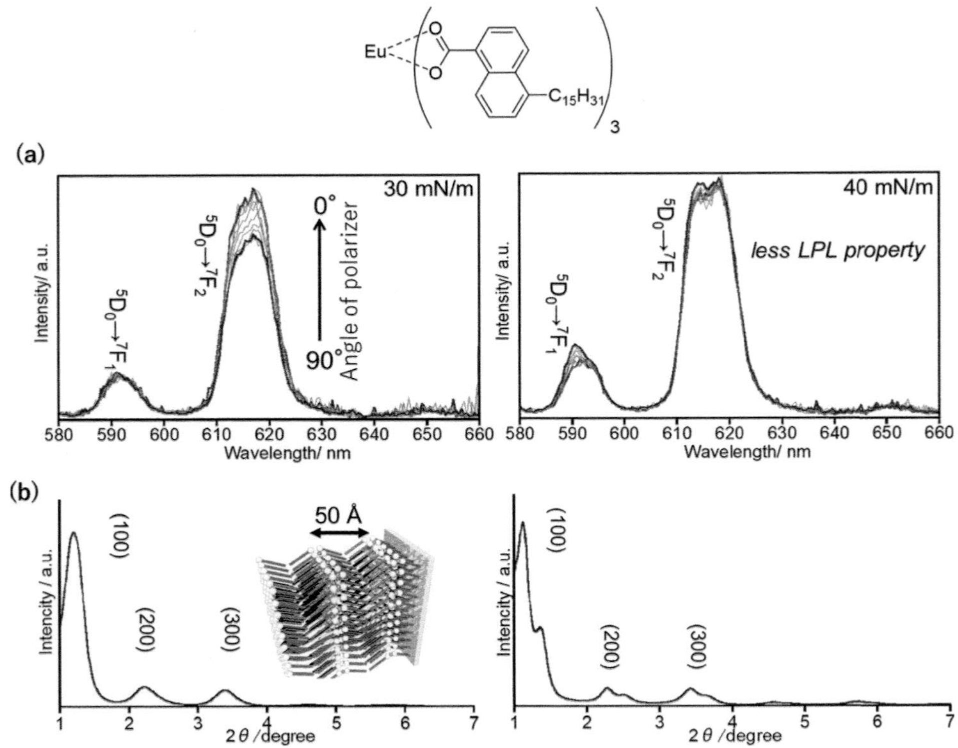

Figure 3.17 (Top) Molecular structure of the Eu complex with a surfactant/photo-antenna moiety. (a) Linearly polarized luminescence and (b) synchrotron XRD of the LB films of Eu complex with naphthoic acid prepared at the surface pressure of 30 (left) and 40 (right) mN/m.

The LbL films [63] of above chain lanthanide complexes [70] on the quartz substrate also brightly luminesces as well as those on the magnetic silica nanoparticles. In the case of LbL films of chain lanthanide complexes, the unit lanthanide complex was linked with bdc until fifth-generation. The luminescence is brightly induced by UV excitation and appear the transitions of Eu or Tb, but no LPL phenomena was observed, because of the chains aggregated each other to form spherical domes. The chain conformations lost the periodic inter-metal distances in inter-/intra-chains. XPS reveals the consequent complex depositions, and halo patterns by XRD measurements of the films supported the stable chain structure and their flexible aggregation on the substrate. Such soft crystalline behavior will have possibility to understand phase transitions [100].

3.6 Conclusion and Outlook

In this chapter, significant efforts for the synthesis of lanthanide-based core-shell nanoparticles have been reported with some cases of nanostructure compounds on the substrate considering molecular aggregation. Such unleash nano-structures will support to improve luminescence properties of lanthanide compounds. Furthermore, we found the possibility that interfacial complexation and active development of surface science lead to the creation of new photo-functions.

For instance, these also include upconversion emission [33, 34, 101–104], and the application to biological science [9, 13, 15, 16, 105] by using NIR luminescent lanthanide nanoparticles. Especially, enough amount of lanthanide complexes for bio-sensing is quite small, and their toxicity is known negligibly low [71]. Some water-soluble-Gd complexes are already applied as sensitizer of MRI observation [106–109] because of their less toxicity.

The technique and scientific approach of the modification of nanoparticles with lanthanide complexes are hybridized to the thin film systems. Here, quartz substrates were mainly mentioned as media for photo-functional lanthanides. Of course, there are many options for matrixes such as various lanthanide oxides films, electrodes [110, 111]. The molecular design of aromatic ligands adopting the variety of surfactant, the Ln complexes can also keep maintain the amphiphilic properties to facilitate the construction of self-assembled molecular films. It is useful to adsorb on the unique surface potentials to retain the lanthanide luminescence.

Finally, from the above-summarized examples that exhibit characteristic luminescence properties of lanthanide with core-shell silica nanoparticles or in their films, it can be hybridized to other unique surfaces. These materials were produced at ambient or relatively low risk conditions and wet-processes. These strategic lanthanide nano-structures will be applicable for the development of future devices such as bio-monitoring proves, security optical filter, photo-energy storages, condensed matters by environmentally friend procedures.

Acknowledgments

This work is partly supported by KAKENHI MEXT (17H06374).

References

1. De Betten Court-Dias, A. (2014). *Luminescence of Lanthanide Ions in Coordination Compounds and Nanomaterials* (John Wiley and Sons, Ltd).
2. Pöttgen, R., Jüstel, T., Strassert, C.A. (2020). *Rare Earth Chemistry* (De Gruyter).
3. Bünzli, J.-C.G., Piguet, C. (2005). Taking advantage of luminescent lanthanide ions, *Chem. Soc. Rev.*, **34**, pp. 1048–1077.
4. Bünzli, J.-C.G., Eliseeva, S.V. (2013). Intriguing aspects of lanthanide luminescence, *Chem. Sci.*, **4**, pp. 1939–1949.
5. Bünzli, J.-C.G. (2015). Intriguing aspects of lanthanide luminescence, *Coord. Chem. Rev.*, **294**, pp. 19–47.
6. Kido, J., Okamoto, Y. (2002). Organo lanthanide metal complexes for electroluminescent materials, *Chem. Rev.*, **102**, pp. 2357–2368.
7. Hemmilä, I., Laitala, V. (2005). Progress in lanthanides as luminescent probes, *J. Fluorescence*, **15**, pp. 529–542.
8. Eliseeva, S.V., Bünzli, J.-C.G. (2010). Lanthanide luminescence for functional materials and bio-sciences, *Chem. Soc. Rev.*, **39**, pp. 189–227.
9. Bünzli, J.-C.G. (2009). Lanthanide luminescent bioprobes (LLBs), *Chem. Lett.*, **38**, pp. 104–109

10. Alpha, B., Lehn, J.-M., Mathis, G. (1987). Energy transfer luminescence of europium(III) and terbium(III) cryptates of macrobicyclic polypyridine ligands, *Angew. Chem. Int. Ed.*, **26**, pp. 266–267.

11. Ashokkumar, S., Ravi, S., Kathiravan, V., Velmurugan, S. (2017). Retraction notice to "synthesis of silver nanoparticles using A. indicum leaf extract and their antibacterial activity", *Spectrochim. Acta A: Mol. Biomol. Spectrosc.*, **171**, p. 526.

12. Goetz, J., Nonat, A., Diallo, A., Sy, M., Sera, I., Lecointre, A., Lefevre, C., Chan, C.F., Wong, K.-L., Charbonnière, L.J. (2016). Ultrabright lanthanide nanoparticles, *ChemPlusChem*, **81**, pp. 526–534.

13. Law, G.-L., Wong, K.-L., Man, C.W.-Y., Wong, W.-T., Tsao, S.-W., Lam, M.H.-W., Lam, P.K.-S. (2008). Emissive terbium probe for multiphoton in vitro cell imaging, *J. Am. Chem. Soc.*, **130**, pp. 3714–3715.

14. Petoud, S., Cohen, S.M., Bünzli, J.-C.G., Raymond, K.N. (2003). Stable lanthanide luminescence agents highly emissive in aqueous solution: multidentate 2-hydroxyisophthalamide complexes of Sm^{3+}, Eu^{3+}, Tb^{3+}, Dy^{3+}, *J. Am. Chem. Soc.*, **125**, pp. 13324–13325.

15. Shinoda, S., Tsukube, H. (2011). Luminescent lanthanide complexes as analytical tools in anion sensing, pH indication and protein recognition, *Analyst*, **136**, pp. 431–435.

16. Tsukube, H., Shinoda, S. (2002). Lanthanide complexes in molecular recognition and chirality sensing of biological substrates, *Chem. Rev.*, **102**, 2389–2404.

17. Butler, S.J., Parker, D. (2013). Anion binding in water at lanthanide centres: from structure and selectivity to signalling and sensing, *Chem. Soc. Rev.*, **42**, pp. 1652–1666.

18. Turro, N.J., Ramamurthy, V., Scaiano, J.C. (2010). *Modern Molecular Photochemistry of Organic Molecules* (University Science Books).

19. Berlman, I.B. (1973). *Energy Transfer Parameters of Aromatic Compounds* (Academic Press, New York).

20. Förster, T. (1950). *Fluoreszenz organischer Verbindungen* (Vandenhoeck & Ruprecht, Göttingen).

21. Hasegawa, M., Enomoto, S., Hoshi, T., Igarashi, K., Yamazaki, T., Nishimura, Y., Speiser, S., Yamazaki, I. (2002). Intramolecular excitation energy transfer in bichromophoric compounds in stretched polymer films, *J. Phys. Chem. B*, **106**, pp. 4925–4932.

22. Dexter, D.L. (1953). A theory of sensitized luminescence in solids, *J. Chem. Phys.*, **21**, pp. 836–850.

23. Binnemans, K. (2015). Interpretation of europium(III) spectra, *Coord. Chem. Rev.*, **295**, pp. 1–45.
24. Dieke, G.H., Crosswhite, H.M., Dunn, B. (1961). Emission spectra of the doubly and triply ionized rare earths, *J. Opt. Soc. Ame.*, **51**, pp. 820–827.
25. Dieke, G.H., Crosswhite, H.M. (1963). The spectra of the doubly and triply ionized rare earths, *Appl. Opt.*, 2, pp. 675–686.
26. Tegh, R.T., Meijerink, A., Lamminmäki, R.-J., Hölsä (2000). Extending Dieke's diagram, *J. Lumin.*, **87–89**, pp. 1002–1004.
27. Hasegawa, M., Nakao, A., Masui, M., Tamura, T., Suzuki, D., Linert, W., Fukuda, Y., Hoshi, T. (2001). Higher excited state fluorescence of mixed ligand lanthanide(III) complexes with acetylacetonato, nitrate and terpyridine, *Chem. Phys.*, **269**, pp. 323–337.
28. Hasegawa, M., Ohtsu, H., Kodama, D., Kasai, T., Sakurai, S., Ishii, A., Suzuki, K. (2014). Luminescence behaviour in acetonitrile and in the solid state of a series of lanthanide complexes with a single helical ligand, *New J. Chem.*, **38**, pp. 1225–1234.
29. Wada, H., Ooka, S., Iwasawa, D., Hasegawa, M., Kajiwara, T. (2016). Slow magnetic relaxation of lanthanide(III) complexes with a helical ligand, *Magnetochemistry*, **2**, pp. 43–55.
30. Kagagiri, S., Tsukahara, Y., Hasegawa, Y., Wada, Y. (2007). Energy-transfer mechanism in photoluminescent terbium(III) complexes causing their temperature-dependence, *Bull. Chem. Soc. Jpn.*, **80**, pp. 1492–1503.
31. Samuel, A.P.S., Moore, E.G., Melchior, M., Xu, J., Raymond, K.N. (2008). Water-soluble 2-hydroxyisophthalamides for sensitization of lanthanide luminescence, *Inorg. Chem.*, **47**, pp. 7535–7544.
32. Starck, M., Kadjane, P., Bois, E., Darbouret, B., Incamps, A., Ziessel, R., Charbonnière, L.J. (2011). Towards libraries of luminescent lanthanide complexes and labels from generic synthons, *Chem. Eur. J.*, **17**, pp. 9164–9179.
33. Kömpe, K., Borchert, H., Storz, J., Lobo, A., Adam, S., Möller, T., Haase, M., (2003). Green-emitting $CePO_4$:Tb/$LaPO_4$ core-shell nanoparticles with 70% photoluminescence quantum yield, *Angew. Chem., Int. Ed.*, **42**, pp. 5513–5516.
34. Kar, A., Patra, A. (2012). Impacts of core-shell structures on properties of lanthanide-based nanocrystals: crystal phase, lattice, downconversion, upconversion and energy transfer, *Nanoscale*, **4**, pp. 3608–3619.

35. Chaudhuri, R.G., Paria, S. (2012). Core/shell nanoparticles: classes, properties, synthesis mechanisms, characterization, and applications, *Chem. Rev.*, **112**, pp. 2373–2433.

36. Hasegawa, M., Ishii, A. (2020). Thin-film formation for promoting the potential of luminescent lanthanide coordination complexes, *Coord. Chem. Rev.*, **241**, 213458 (pp. 1–17).

37. Martin, R., Jaque D. (2021), Doping lanthanide ions in colloidal semiconductor nanocrystals for brighter photoluminescence, *Chem. Rev.*, online published.

38. Ogata, S., Ishii, A., Lu, C.L., Kondo, T., Yajima, N., Hasegawa, M. (2017). Polymorphism-based luminescence of lanthanide complexes with a deuterated 1,10-phenanthroline, *J. Photochem. Photobiol. A*, **334**, pp. 55–60.

39. Bonacchim S., Genovese, D., Juris, R., Montalti, M., Prodi, L., Rampazzo, E., Zaccheroni, N. (2010). Luminescent silica nanoparticles: extending the frontires of brightness, *Angew. Chem. Int. Ed.*, **50**, pp. 4056–4066.

40. Song, C., Ye, Z., Wang, G., Yuan, J. Guan, Y. (2010). Core–shell nanoarchitectures: a strategy to improve the efficiency of luminescence resonance energy transfer, *ACS Nano*, **4**, pp. 5389–5397.

41. Larson, R.L., Ow, H., Vishwasrao, H. D., Heikal, A.A., Wiesner, U, Webb, W.W. (2008). Silica nanoparticle architecture determines radiative properties of encapsulated fluorophores, *Chem. Mater.*, **20**, pp. 2677–2684.

42. Ishii, A., Hasegawa, M. (2015). An interfacial europium complex on SiO_2 nanoparticles: reduction-induced blue emission system, *Sci. Rep.*, **5**, 11714 (pp. 1–8).

43. Matsuzawa, T., Aoki, Y., Takeuchi, N., Murayama, Y. (1996). A new long phosphorescent phosphor with high brightness, $SrAl_2O_4$:Eu^{2+}, Dy^{3+}, *J. Electrochem. Soc.*, **143**, pp. 2670–2673.

44. Suriyamurthy, N., Panigrahi, B.S. (2008). Effects of non-stoichiometry and substitution on photoluminescence and afterglow luminescence of $Sr_4Al_{14}O_{25}$:Eu^{2+}, Dy^{3+} phosphor, *J. Lumin.*, **128**, pp. 1809–1814.

45. Qiao, X., Wang, J., Huang, Y., Cai, P., Kim, S.I., Seo, H.J. (2014). Synthesis and luminescence properties of blue-emitting phosphor Eu^{2+}-doped zinc fluoro-phosphate $Zn_2[PO_4]$ F, *J. Am. Ceram. Soc.*, **97**, pp. 3561–3567.

46. Dutczak, D., Ronda, C., Jüstel, T., Meijerink, A. (2014). Anomalous trapped exciton and d-f emission in $Sr_4Al_{14}O_{25}$:Eu^{2+}, *J. Phys. Chem. A*, **118**, pp. 1617–1621.

47. Li, K., Geng, D., Shang, M., Zhang, Y., Lian, H., Lin, J. (2014). Color-tunable luminescence and energy transfer properties of $Ca_9Mg(PO_4)_6F_2$:Eu^{2+}, Mn^{2+} phosphors for UV-LEDs, *J. Chem. Phys. C*, **118**, pp. 11026–11034.

48. Liu, B., Wang, Y., Zhou, J., Zhang, F., Wang, Z. (2009). The reduction of Eu3+ to Eu^{2+} in $BaMgAl_{10}O_{17}$:Eu and the photoluminescence properties of $BaMgAl_{10}O_{17}$:Eu^{2+} phosphor, *J. Appl. Phys.*, **106**, p. 053102–5.

49. Zhang, J.-C., Wang, X., Marriott, G., Xu, C.-N. (2019). Trap-controlled mechanoluminescent materials, *Prog. Mater. Sci.*, **103**, pp. 678–742.

50. Xie, H. Lu, J., Guan, Y., Huang, Y., Wei, D., Seo, H.J. (2014). Abnormal reduction, $Eu^{3+} \rightarrow Eu^{2+}$, and defect centers in Eu^{3+}-doped pollucite, $CsAlSi_2O_6$, prepared in an oxidizing atmosphere, *Inorg. Chem.*, **53**, pp. 827–834.

51. Lian, Z., Wang, J., Lv, Y., Wang, S., Su, Q. (2007). The reduction of Eu^{3+} to Eu^{2+} in air and luminescence properties of Eu^{2+} activated $ZnO–B_2O_3–P_2O_5$ glasses, *J. Alloys Compd.*, **430**, pp. 257–261.

52. Peng, M., Hong, G., (2007). Reduction from Eu^{3+} to Eu^{2+} in $BaAl_2O_4$:Eu phosphor prepared in an oxidizing atmosphere and luminescent properties of $BaAl_2O_4$:Eu, *J. Lumin.*, **127**, pp. 735–740.

53. Ishii, A., Hasegawa, M. (2016). The ethanol-induced interfacial reduction of a europium complex on SiO_2 nanoparticles, *Chem. Lett.*, **45**, pp. 1265–1267.

54. Evans ,W.J. (2000). Perspectives in reductive lanthanide chemistry, *Coord. Chem. Rev.*, **206–207**, pp. 263–283.

55. Adachi, G., Tomokiyo, K., Sorita, K., Shiokawa, J., (1980). Luminescence of divalent europium complexes with crown ethers and polyethylene glycols, *J. Chem. Soc., Chem. Commn.*, **19**, pp. 914–915.

56. Jiang, J., Higashiyama, N., Machida, K., Adachi, G., (1998). The luminescence properties of divalent europium complexes of crown ethers and cryptands, *Coord. Chem. Rev.*, **170**, pp. 1–29.

57. Sabbatini, N., Ciano, M., Dllonte, S., Bonazzi, A., Bolletta, F., Balzani, V., (1984). Photophysical properties of europium(II) cryptates, *J. Phys. Chem.*, **88**, pp. 1534–1537.

58. Binnemans, K. (2009). Lanthanide-based luminescent hybrid materials, *Chem. Rev.*, **109**, pp. 4283–4374.

59. Bekiari, V., Pistolis, G., Lianos, P. (1998). Improvement of the emission properties of sol-gel silica matrices containing Eu^{3+} in the presence of poly(ethylene glycol)-200, *J. Non-Cryst. Solids*, **226**, p. 200.

60. Bekiari, V., Pistolis, G., Lianos, P. (1999). *Chem. Mater.*, **11**, p. 3189.

61. Cras, J.J., Rowe-Taitt, C.A., Nivens, D.A., Ligler, F.S. (1999). Comparison of chemical cleaning methods of glass in preparation for silanization, *Biosens. Bioelectron.*, **14**, pp. 683–688.

62. Kao, K.-C., Lee, C.-H., Lin, T.-S., Mou, C.-Y. (2010). Cytochrome c covalently immobilized on mesoporous silicas as a peroxidase: orientation effect, *J. Mater. Chem.*, **20**, pp. 4653–4662.

63. Marets, N., Kanno, S., Ogata, S., Ishii, A., Kawaguchi, S., Hasegawa, M. (2019). Lanthanide-oligomeric brush films: from luminescence properties to structure resolution, *ACS Omega*, **4**, pp. 15512–15520.

64. Massart, R. (1981). Preparation of aqueous magnetic liquids in alkalineand acidic media, *IEEE Trans. Magn.*, **17**, pp. 1247–1248.

65. Skwierczyńska, M., Runowski, M., Goderski, S., Szczytko, J., Rybusiński, J., Kulpiński, P., Lis, S. (2018). Luminescent-magnetic cellulose fibers, modified with lanthanide-doped core/shell nanostructures, *ACS Omega*, **3**, pp. 10383–10390.

66. Yi, D. K., Selvan, S. T., Lee, S. S., Papaefthymiou, G. C., Kundaliya, D., Ying, J. Y. (2005). Silica-coated nanocomposites of magnetic nanoparticles and quantum dots. *J. Am. Chem. Soc.*, **127**, pp. 4990–4991.

67. Patsula, V., Kosinová, L., Lovrić, M., Ferhatovic Hamzić, L., Rabyk, M., Konefal, R., Paruzel, A., Šlouf, M., Herynek, V., Gajović, S., Horák, D. (2016). Superparamagnetic Fe_3O_4 nanoparticles: synthesis by thermal decomposition of iron(III) glucuronate and application in magnetic resonance imaging, *ACS Appl. Mater. Interfaces*, **8**, pp. 7238–7247.

68. Salgueiriño-Maceira, V., Correa-Duarte, M. A., Spasova, M., Liz-Marzán, L. M., Farle, M. (2006). Composite silica spheres with magnetic and luminescent functionalities, *Adv. Funct. Mater.*, **16**, pp. 509–514.

69. Goderski, S., Kanno, S., Yoshihara, K., Komiya, H., Goto, K., Tanaka, T., Kawaguchi, S., Ishii, A., Shimoyama, J., Hasegawa, M., Lis, S. (2020). Lanthanide luminescence enhancement of core-shell magnetite-SiO_2 nanoparticles covered with chain-structured helical Eu/Tb complexes, *ACS Omega*, **5**, pp. 32930–32938.

70. Sato, S., Ishii, A., Yamada, C., Kim, J., Song, C.H., Fujiwara, A., Takata, M., Hasegawa, M. (2015). Luminescence of fusion materials of polymeric chain-structured lanthanide complexes, *Polymer J.*, **47**, 195-200.

71. Hasegawa, M., Ohmagari, H. (2020) Helicate lanthanide complexes: the luminescent elements, *Chem. Lett.*, **49**, pp. 845–854.

72. Hasegawa, M., Sakurai, S., Yamaguchi, M.A., Iwasawa, D., Yajima, N., Ogata, S., Inazuka, Y., Ishii, A., Suzuki, K. (2020). Aspects of lanthanide

complexes for selectivity, intensity and sharpness in luminescence bands from twenty-four praseodymium, europium and gadolinium complexes with differently distorted-hexadentate ligands, *Photochem. Photobiol. Sci.*, **19**, pp. 1054–1062.

73. Piguet, C., Bünzli, J.-C.G., (2010). Chapter 247 - Self-assembled lanthanide helicates: from basic thermodynamics to applications. In *Handbook on the Physics and Chemistry of Rare Earths*, Gschneidner, K.A., Bünzli, J.-C.G., Pecharski, V.K., eds., **40**, pp. 301–553 (Elsevier).
74. Riehl, J.P., Richardson, F.S. (1986). Circularly polarized luminescence spectroscopy, *Chem. Rev.*, **86**, pp. 1–16.
75. Carr, R., Evans, N.H., Parker, D. (2012). Lanthanide complexes as chiral probes exploiting circularly polarized luminescence, *Chem. Soc. Rev.*, **41**, pp. 7673–7686.
76. Sánchez-Carnerero, R.M., Agarrabeitia, A.R., Moreno, F., Maroto, B.L., Muller, G., Ortiz, M.J., de la Moya, S. (2015). Circularly polarized luminescence from simple organic molecules, *Chem. Euro. J.*, **21**, pp. 13488–13500.
77. Mori, T., ed. (2020). *Circularly Polarized Luminescence of Isolated Small Organic Molecules* (Springer).
78. Sato, S., Yoshii, A., Takahashi, S., Furumi, S., Takeuchi, M., Isobe, H. (2017). Chiral intertwisted spirals and magnetic transition dipole moments dictated by cylinder helicity, *PNAS*, **114**, pp. 13097–13101.
79. Shen, Z., Wang, T., Shi. L., Tang, Z., Liu, M. (2015). Strong circularly polarized luminescence from the supramolecular gels of an achiral factor: tunable intensity and handedness, *Chem. Sci.*, **6**, pp. 4267–4272.
80. Hasegawa, Y., Miura, Y., Kitagawa, Y., Wada, S., Nakanishi, T., Fushimi, K., Seki, T., Ito, H., Iwasa, T., Taketsugu, T., Con, M., Tanaka, K., Chujo, Y., Hattori, S., Karasawa, M., Ishii, K. (2018). Spiral Eu(III) coordination polymers with circularly polarized luminescence, *Chem., Commn.*, **54**, pp. 10695–10697.
81. Lunkley, J. L., Shirotani, D., Yamanari, K., Kaizaki, S., Muller, G. (2008). Extraordinary circularly polarized luminescence activity exhibited by cesium tetrakis(3-heptafluoro-butylryl-(+)-camphorato) Eu(III) complexes in EtOH and $CHCl_3$ solutions, *J. Am. Chem. Soc.*, **130**, pp. 13814–13815.
82. Shirotani, D., Suzuki, T., Kaizaki, S. (2006). Novel configurational structures of sodium tetrakis(3-heptafluorobutylryl-(+)-camphorato) Ln(III) complexes with a trapped Na^+ by $Na^+\cdots FC$ interactions in the solid state and in solution, *Inorg. Chem.*, **45**, pp. 6111–6113.

83. Lunkley, J.L., Shirotani, D., Yamanari, K., Kaizaki, S., Muller, G. (2011). Chiroptical spectra of a series of tetrakis((+)-3-heptafluorobutylyrylcamphorato)lanthanide(III) with an encapsulated alkali metal ion: circularly polarized luminescence and absolute chiral structures for the Eu(III) and Sm(III) complexes, *Inorg. Chem.*, **50**, pp. 12724–12732.

84. Kumar, J., Marydasan, B., Nakashima, T., Kawai, T., Yuasa, J. (2016). Chiral supramolecular polymerization leading to eye differentiable circular polarization in luminescence, *Chem. Commun.*, **52**, pp. 9885–9888.

85. Sato, K., Haseagwa, M., Nojima, Y., Hara, N., Nishiuchi, T., Imai, Y., Yasuhiro, M. (2021) Circularly polarized luminescence of a stereogenic curved paraphenylene anchorning a chiral binaphthyl in solution and solid state, *Chem. Eur. J.*, **27**, pp. 1323–1329.

86. Koike, H., Nozaki, K., Iwamura, M. (2020). Microscopic imaging of chiral amino acid agar gel through circularly polarized luminescence of Eu^{III} complex, *Chem. Asian J.*, **15**, pp. 85–90.

87. Ishii, A., Miyasaka, T. (2020). Direct detection of circular polarized light in helical 1D perovskite-based photodiode, *Sci. Adv.*, **6**, p. eabd3274.

88. Kawamura, Y., Sasabe, H., Adachi, C. (2004). Simple accurate system for measuring absolute photoluminescence quantum efficiency in organic solid-state thin films, *Jpn. J. Appl. Phys.*, **43**, pp. 7729–7730.

89. Porrè, L., Holland, A., Pålsson, L.-O., Monkman, A.P., Kemp, C., Beeby, A. (2006). Absolute measurements of photoluminescence quantum yields of solutions using an integrating sphere, *J. Fluoresc.*, **16**, pp. 267–272.

90. Suzuki, K., Kobayashi, A., Kaneko, S., Takehira, K., Yoshihara, T., Ishida, H., Shiina, Y., Oishi, S., Tobita, S. (2009). Reevaluation of absolute luminescence quantum yields of standard solutions using a spectrometer with an integrating sphere and a back-thinned CCD detector, *PhysChemChemPhys*, **14**, pp. 9850–9860.

91. Hasobe, N., Suzuki K., Horiuchi, H., Suzuki, H., Yoshihara, T., Okusu, T., Tobita, S. (2015). Absolute phosphorescence quantum yields of singlet molecular oxygen in solution determined using an integrating sphere instrument, *Anal. Chem.*, **87**, pp. 2360–2366.

92. Blodgett, K.B. (1934). Monomolecular films of fatty acids on glass, *J. Am. Chem. Soc.*, **56**, p. 495.

93. Blodgett, K.B. (1935). Films built by depositing successive monomolecular layers on a solid surface, *J. Am. Chem. Soc.*, **57**, pp. 1007–1022.

94. Blodgett, K.B., Langmuir, I. (1937). Built-up films of barium stearate and their optical properties, *Phys. Rev.*, **51**, pp. 964–982.

95. Kitchen, J.A., Barry, D.E., Mercs, L., Albrecht, M., Peacock, R.D. Gunnlaugsson, T. (2012). Circularly polarized lanthanide luminescence from Langmuir–Blodgett films formed from optically active and amphiphilic Eu^{III}-based self-assembly complexes, *Chem. Int. Ed.*, **51**, pp. 704–708.

96. Barry, D.E., Gunnlaugsson, T., Kitchen, J.A., Mercs, L., Albrecht, M., Peacock, R.D. (2019). Chiral luminescent lanthanide complexes possessing strong (samarium, Sm^{III}) circularly polarised luminescence (CPL), and their self-assembly into Langmuir–Blodgett films, *Dalton Trans.*, **29**, pp. 11506–11515.

97. Yoshihara, K., Yamanaka, M., Kanno, S., Mizushima, S., Tsuchiyagaito, J., Kondo, K., Kondo., T., Iwasawa, D., Komiya, H., Saso, A., Kawaguchi, S., Goto, K., Ogata, S., Takahashi, H., Ishii, A., Hasegawa, M. (2019). Europium amphiphilic naphthalene based complex for the enhancement of linearly polarized luminescence in Langmuir–Blodgett films, *New J. Chem.*, **43**, pp. 6472–6479.

98. Hasegawa, M., Kunisaki, S., Ohtsu, H., Werner, F. (2009). Ultra-thin emissive molecular devices: polarized emission of Ln(III) complex films, *Monatsh. Chem.*, **140**, pp. 751–763.

99. Yang, C.Y., Srdanov, V., Dobinson, M.R., Bazan, G.C., Heeger, A.J. (2002). Orienting $Eu(dnm)_3phen$ by tensile drawing in polyethylene: polarized Eu^{3+} emission, *Adv. Mater.*, **14**, pp. 980–983.

100. Kato, M., Ito, H., Haseagwa, M., Ishii, K. (2019). Soft crystals: flexible response systems with high structural order, *Chem. Eur. J.*, **25**, pp. 2105–5112.

101. Dong, H., Sun, L.-D., Yan, C.-H. (2015). Energy transfer in lanthanide upconversion studies for extended optical applications, *Chem. Soc. Rev.*, **44**, pp. 1608–1634.

102. Fan, Y., Liu, L., Zhang, F. (2019). Exploiting lanthanide-doped upconversion nanoparticles with core/shell structures, *Nano Today*, **25**, pp. 68–84.

103. Vetrone, F., Naccache, R., Mahalingam, V., Morgan, C.G., Capobianco, J.A. (2009). The active-core/active-shell approach: a strategy to enhance the upconversion luminescence in lanthanide-doped nanoparticles, *Adv. Func. Mater.*, **19**, pp. 2924–2929.

104. Wang, F., Deng, R., Wang, J., Wang, Q., Han, Y., Zhu, H., Chen, X., Liu, X. (2011). Tuning upconversion through energy migration in core–shell nanoparticles, *Nat. Mater.*, **10**, pp. 968–973.

105. Sun, L.-N., Zhang, H.-J., Yu, J.-B., Meng, Q.-G., Liu, F.-Y., Peng, C.-Y. (2008). Performance of near-IR luminescent xerogel materials covalently bonded with ternary lanthanide (Er^{III}, Nd^{III}, Yb^{III}) complexes, *J. Photochem. Photobiol. A*, **193**, pp. 153–160.

106. Clough, T.J., Jiang, L., Wong, K.L., Long, N.J. (2019). Ligand design strategies to increase stability of gadolinium-based magnetic resonance imaging contrast agents, *Nat. Commun.*, **10**, p. 1420.

107. Law, G.-L., Wong, W.-T. (2015). An introduction to molecular imaging. In *The Chemistry of Molecular Imaging* (eds. Long, N. J. & Wong, W.-T.) pp. 1–24 (John Wiley & Sons, Hoboken, New Jersey, USA).

108. Mi, P., Cabral, H., Kokuryo, D., Rafi, M., Terada, Y., Aoki, I., Saga, T., Ishii, T., Nishiyama, N., Kataoka, K. (2013). Gd-DTPA-loaded polymer-metal complex micelles with high relaxivity for MR cancer imaging, *Biomaterials*, **34**, pp. 492–500.

109. Chen Z, Yu D, Liu C, Yang X, Zhang N, Ma C, Song J, Lu Z. (2011). Gadolinium-conjugated PLA-PEG nanoparticles as liver targeted molecular MRI contrast agent, *J. Drug Target.*, **19**, pp. 657–665.

110. Morales Ramírez, Á.-J., Sarabia Domínguez, F., Medina Velazquez, D.Y., Jaramillo Vigueras, D., García Hernández, M., Dorantes Rosales, H.J. (2014). Synthesis and photoluminescent properties of Y_2O_3:Eu^{3+} thin films prepared from F127-containing solution, *J. Ceramic Soc. Jpn.*, **112**, pp. 701–707.

111. Yazawa, M., Takahashi, K., Uchida, H., Hasegawa, Y. (2019). Development of LB film taste sensor for beer degradation evaluation and its physical properties, *Electron. Comm. Jpn.*, **102**, pp. 47–54.

Chapter 4

Correlating Luminescence Characteristics to Gas-Sensing Performance of Metal Oxides Based Heterostructures

Katekani Shingange[a] and Gugu H. Mhlongo[a,b]
[a]*Centre for Nanostructures and Advanced Materials, Council for Scientific and Industrial Research, Pretoria, 0001, South Africa*
[b]*Department of Physics, University of Free State, Bloemfontein 9300, South Africa*
kshingange@csir.co.za

Gas sensors based on metal oxides (MOX) play a very crucial role in everyday life for the application in environmental pollution monitoring, disease diagnosis, and food quality monitoring. The gas-sensing performance of MOX based sensors heavily relies on the type of material used and the condition of the material surface, thus a lot of factors affect the sensing performance. This chapter is focused on briefly reviewing gas-sensing performance attributed to surface defects factors, with the focus mainly on heterostructures

Luminescent Nanomaterials
Edited by Odireleng Martin Ntwaeaborwa

ISBN 978-981-4968-11-9 (Hardcover), 978-1-003-27738-5 (eBook)
www.jennystanford.com

MOX. Defects on MOX based sensors are usually characterized through their luminescence properties which are then correlated to the gas-sensing performance.

4.1 Introduction

Nanostructured metal oxide (MOX) based gas sensors have been comprehensively studied for the use in different applications including environmental air quality monitoring, food industry, and medical diagnosis [1–5]. The use of nano-MOX based materials for gas sensing comes with many advantages such as low cost, ease of maintenance, and the small size. In these MOX based sensors, electrical resistance measurements are used to identify the presence of the gas which can be affected by different variables such as choice of material, material shape, and size. A lot of literature on gas sensing has demonstrated that defects in MOX play a crucial role in gas sensing [6–8]. Defects include different types, anion vacancies (oxygen vacancies), cation vacancies, distortions, and vacancy associates [9]. Through engineering, one of the defects could tune the properties of the material such as electronic, microstructural, and electrical conductivity properties. Defects in gas sensing have been confirmed to have positive impact on the gas-sensing performance of MOX based sensors. Such can be demonstrated through a study done by Gu et al. [10] where the effect of defects engineering in ZnO was demonstrated. Also Wang et al. [11] demonstrated the oxygen vacancy defects engineering on Ce doped α-Fe_2O_3 in gas sensors.

4.2 Heterostructures Application as Gas Sensor Materials

Most of the MOX used have wide band gaps and require high operating temperatures in the range of 300–500°C or UV-light illumination to excite their valence band electrons, thus making it difficult to apply them into practical applications. Further, MOX are generally sensitive to a variety of gases hence they are unable to discriminate a specific gas in the presence of mixture gases. MOX based gas sensors also endure material poisoning interactions which make MOX surfaces to be less accessible and less sensitive to reactive gases over time.

Due to these flaws, interest in obtaining novel gas-sensing materials with higher response and selectivity at low temperatures have seen researchers trying different approaches including developing different dimensional microstructures from 0 to 3D and combining of dissimilar materials to form heterostructures [12–18].

The notion of coupling different materials to form heterostructures is an interesting modern approach which was motivated by the need to improve gas-sensing performance of MOX based sensors. It has been demonstrated through experiments and subsequent theoretical simulations that all types of heterostructures generally provide potential for the development of functional gas-sensing materials. Improved surface-related properties important for gas sensor applications such as electronic, luminescence, catalytic, mechanical, and chemical properties can be obtained through advanced heterostructures fabricated from different materials. These include mixed-metal oxides, polymers mixed with metals or metal oxides, decorated metal oxides as well as core-shells [19–22]. Heterostructures show enhanced sensor performances in comparison to single MOX materials due to the strong interactions between closely packed interfaces [23]. In general, when two different materials are in contact at the interface, the charge carriers from individual material migrate to the other and vice versa until Fermi levels equilibrate to the same energy, resulting in the creation of depletion layer and high resistance [23]. Even though the formed depletion layer is one of the most evident results from the combination of dissimilar materials, other influences such as the ratio of the combined contents may be responsible for the enhanced gas-sensing performance in heterostructures.

This chapter focuses on nanostructured MOX heterostructures for gas-sensing applications, in which the focus is on methods used to determine the existence of defects in heterostructures. Methods of interest are those that characterize the luminescence properties of sensing materials to explain the gas-sensing processes.

4.3 Defects and Gas-Sensing Properties

Gas-sensing performance is affected by a lot of factors, and in heterostructures; the key players usually considered for the gas-sensing processes and enhancement are the role of the interface between the dissimilar materials and the activated adsorption sites. With high activated adsorption sites, the response and overall sensing performance of the material would increase. High adsorption sites are usually obtained through increasing the surface area of the sensor material. With increasing surface area, the structural defects in the material are also increased, thus influencing the sensing performance and processes. Defects are known to manifest themselves in the band structure of a material as an energy level within the band gap that depend on the properties of the defect [24]. With that said; in the case of MOX materials, electrical properties are usually linked with the defects in the crystal lattice, particularly oxygen vacancies (Vo) [25, 26]. In gas sensing, the surface oxygen vacancies play a crucial role in determining the adsorption characteristics of the sensor.

4.4 Luminescence Properties and Gas-Sensing Correlation

Various factors have been demonstrated to influence the gas-sensing performance of metal oxide-based gas sensors including morphology, surface chemistry, surface defects, and crystal structure [24, 27]. The importance of these factors is normally explained using the combination of different characterization techniques to understand the sensing behavior and mechanism of the gas sensors. Since this chapter is on the role of defects in gas sensing, the focus will be on the correlation of luminescence properties to gas-sensing performance.

4.4.1 Correlating Luminescence Properties to Gas-Sensing Performance

4.4.1.1 UV-visible spectroscopy (UV-Vis) correlation to gas-sensing performance

In the classical band theory, the defects introduce discrete energy levels in the band of the material. These levels can be donor or acceptor depending on the nature of the defects that are present.

Therefore, the introduction of defects provokes the creation of charge carriers, free electrons, or holes. These introduced carriers condition most electric and optical properties of the material. For this reason, to track the changes in the band structure of metal oxides, which in turn gives an indication to the concentration of defects, UV-vis is often used. In a study by Kohli et al. [28] on In_2O_3/MWCNT heterostructures as acetone gas sensor; UV-Vis was used to determine the optical band gap (*Eg*) by extrapolation of $(\alpha h\nu)^2$ vs $h\nu$ plot to $h\nu = 0$ as depicted in Fig. 4.1a. The determined *Eg* values were found to be 4.11, 3.97, and 4.21 for samples labelled 0c, 0.5C and 1C, respectively. The band gap reduction observed for 0.5C was attributed to generation of a new energy level and enrichment of chemical defects present in intergranular regions resulting from addition of carbon nanotubes into In_2O_3. As presented in Fig. 4.1b, the 0.5C sample revealed enhanced sensing response in comparison to the rest of the other samples and this has been assigned to high concentration of defects. Other contribution factors mentioned include the high surface area and p-n heterojunctions. This clearly indicates that not only one factor can be pointed out as the contributor towards improved sensing behaviour and hence the importance of exploring different surface science characterization techniques to get a clear understanding of the synergistic effects that can be brought upon by heterostructures. In another study by Huyen et al. [29], it was discovered that the combination of TiO_2 and PANi leads to structural modification due to their interaction, which yielded high sensitivity towards O_2 and NH_3 adsorption due to variety of shallow donors and acceptor levels in the PANi band gap which enhanced the physical adsorption of gas molecules [29].

4.4.1.2 Photoluminescence (PL) correlation to gas-sensing performance

The PL has been normally used to characterize transition semiconductor MOX. The MOX can capture photons from the excitation source with energy equal or larger than their band energy. The photons are then absorbed by the MOX nanocrystal and causes the production of photo-excited electrons and holes in the conduction and valence bands, respectively. The excited electron-hole pairs, known as excitons, are then subject to recombination, on the defect centers to give trap-state luminescence [30, 31].

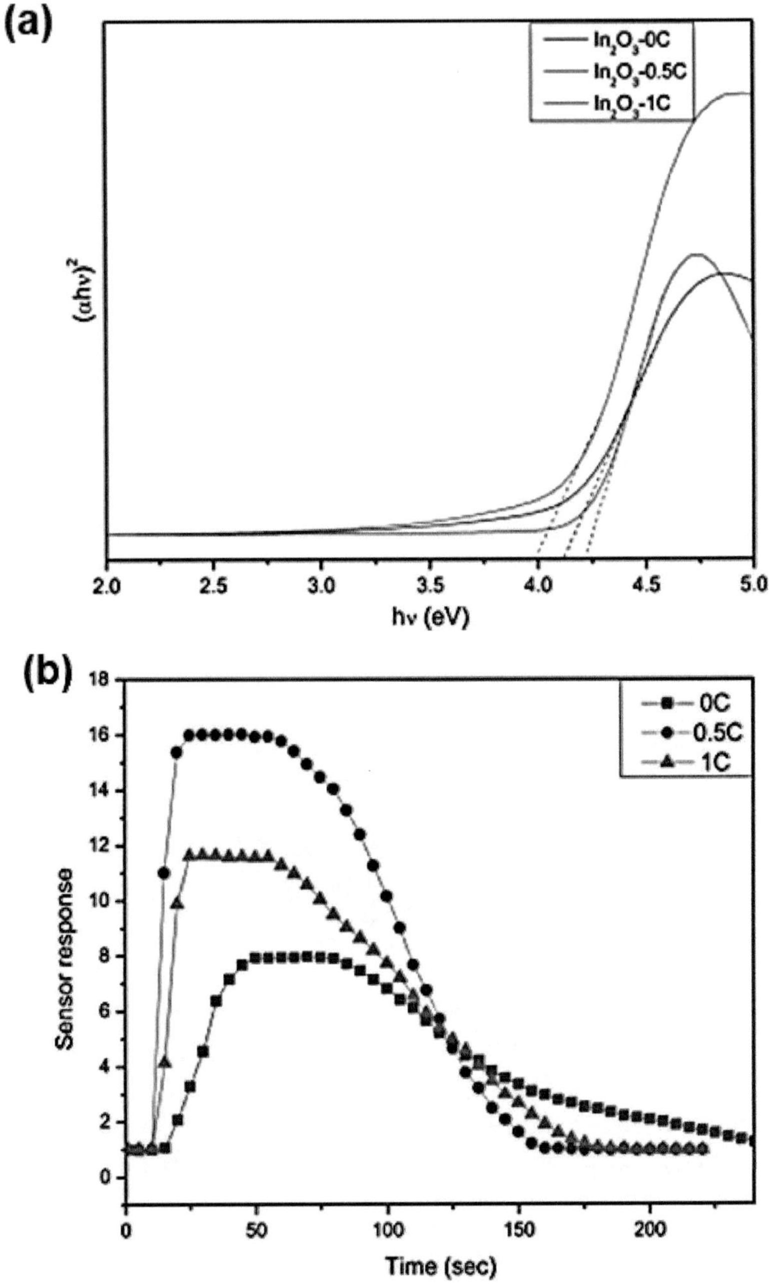

Figure 4.1 (a) Optical band gap determination of In_2O_3 and In_2O_3/CNT nanocomposites using Tauc plots and sensor response versus time variation of samples 0C, 0.5C, and 1C towards 250 ppm acetone vapour at 300°C, re-used with permission from Ref. [28].

PL is a common technique in the gas-sensing field that is used to study the defect structures of MOX based gas sensors. For instance, Rehman et al. [32] reported improved sensor response towards CO at an operating temperature of 300°C for SnO_2/Au multilayered heterostructured based gas sensor as compared to pristine SnO_2 as shown in Fig. 4.2a. Rehman et al. argued that the enhancement in response could be correlated with the oxygen deficiencies in the room temperature PL spectra excited at 266 nm shown in Fig. 4.2b. From their obtained PL spectra, a small peak in the UV region at 325 nm together with a visible emission band (except for pristine SnO_2 sample) centred around 545 nm was observed. The small peak in the UV region could be due to the near band emission (NBE) and is in close agreement with the optical band gap of SnO_2. The visible light emission of SnO_2 was ascribed to defect levels within the band gap of SnO_2, related to Sn interstitials. The visible emission band was further Gaussian de-convoluted into three peaks to determine the origin of the defects. The bands I and II were assigned to the bridging oxygen vacancies while the band III was associated with in-plane oxygen vacancies [33]. The origin of band I emission located at 433.6, 441.9, 453.7, and 469.9 nm for samples SA1, SA2, SA3, and SA4 were attributed to the structural defects that formed during the deposition of thin films [34]. Furthermore, an increase in the intensity of the in-plane oxygen vacancies with an increase in the number of layers was observed, and these vacancies were held responsible for improvement of gas-sensing properties [32].

PL measurements were also performed by Tian et al. [35] on La doped ZnO films at room temperature utilizing a He-Cd laser source at an excitation wavelength of 325 nm. It was discovered that addition of La as a dopant decreased the formation of Vo and gave rise to the formation of $La_{\dot{Z}n}$. Both defects belong to a donor and $La_{\dot{Z}n}$ had a longer wavelength (~570 nm) as compared to Vo (~520 nm), thus implying that $La_{\dot{Z}n}$ is a shallower donor [36]. Due to the presence of this type of defect ($La_{\dot{Z}n}$), the gas-sensing operating temperature was lowered from 350 to 250°C. This improvement was attributed to the shallow donor defects ($La_{\dot{Z}n}$) as they are reported to be able to supply electrons to the adsorbed oxygen molecules for the creation of oxygen species at lower temperatures. Further, the level of $La_{\dot{Z}n}$ is located closer to the conduction band of ZnO as compared to Vo, thus $La_{\dot{Z}n}$ only requires lower energy to activate its electrons to the

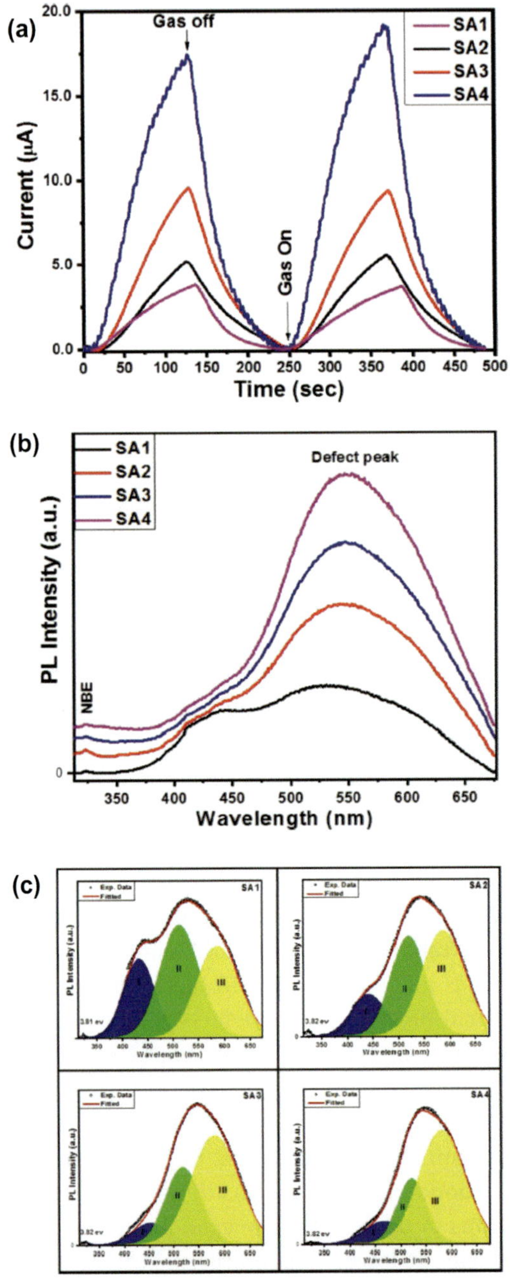

Figure 4.2 (a) CO-sensing (100 ppm) of, (b) Room temperature PL and (c) de-convoluted Gaussian peaks of the of SnO_2/Au heterostructures, figure reprinted with permission from Ref. [32].

conduction band for the adsorption of oxygen species. Other studies such as XPS and DFT calculations were performed to support this argument and detailed discussion can be obtained from Ref. [36]. Further reading correlating PL studies to gas-sensing performance can be found from Refs. [37–40].

4.4.1.3 Cathodoluminescence (CL)

CL is used for the analysis of defect states in MOX materials. The operating process entails an electron beam bombarding the sample to excite electrons to a higher energy state; usually the conduction band, and then the electrons recombine with holes and emit photons. These photons are then collected and converted into a spectrum where each peak signifies a radiative recombination whereby the center of the peak energy is the difference between the initial and final energy levels [41, 42]. Usually the CL is equipped in a scanning electron microscopy. Korotcenkov et al. [43] analysed the CL and gas-sensing properties of SnO_2 doped by Pd, Pt, Cu, Co, and Fe, and the conclusion was that there was a correlation between the CL intensity and the magnitude of the sensor response. As much as the luminescence properties of MOX can assist in relating the role of defects to the gas-sensing performance, additional characterization of gas-sensing materials should be performed.

4.4.2 Other Techniques for Defects Analysis in Gas Sensing

Usually the luminescence characterization techniques are used in combination with other techniques to identify defects in MOX heterostructured based sensing materials. Additional characterization techniques include:

(i) *Electron paramagnetic resonance (EPR)*: EPR provides information of unpaired electrons on the surface and bulk of materials, which is an indication of oxygen vacancies [44]. Studies that combined EPR as part of defects identification can be found in Refs. [38, 40, 45, 46].

(ii) *Raman spectroscopy*: Raman spectroscopy studies the molecular vibration and rotation of the molecular structure [47, 48]. The change of molecular vibrational level gives rise to Raman shift. Defects in materials can influence the vibration modes, thus leading to Raman shift or the appearance of new peaks [47]. More reading on correlation of Raman spectroscopy to gas-sensing performance can be obtained from Refs. [45, 49, 50].

(iii) *X-ray photoelectron spectroscopy (XPS)*: XPS is one of the mostly used surface analysis technique for the identifying of materials surface chemical state. The defects in materials will change the bonding energies and this could be noticed through peak shifting or appearance of new peak. More studies on correlating XPS studies with gas-sensing performance of MOX heterostructures can be obtained from Refs. [51–53].

4.5 Summary and Outlook

Studies on the utilization of metal oxide semiconductors as gas-sensing materials still attracts massive interest in the research and innovation space. A wide range of structures based on metal oxide semiconductors coupled together to form unique heterostructures give better chances in the achievement of potential applicable materials in the gas sensor technology space. A variety of factors contribute to much needed enhanced gas-sensing capabilities obtained from heterostructures, and mostly are reported as synergistic effects since they all act towards the enhancement of the gas sensor performance. This brief review focuses on the luminescence characteristic of these materials for better understanding of the gas-sensing behavior. Luminescence properties give an indication of the defects state of the material. Defects such as oxygen vacancies, metal interstitials, and surface defects, these defects are very important in gas sensing as they lead to a larger concentration of gas adsorption sites for the oxygen and the target gas, ultimately resulting in enhanced gas-sensing performance. Also, the characteristics of the defects can vary in different ways for different heterostructures of interest as the ratio of the materials being combined play a huge role in the concentration and type of defects. Since gas-sensing performance involves several factors, it would be interesting in future to look at the sensing performance of these heterostructures from a broader view. Gas sensors form part of our everyday life and it would be interesting to get a broader understanding of the different influences from different factors into the behavior of these sensor materials.

Acknowledgments

The authors would like to acknowledge the Department of Science and Innovation (DSI) and the Council for Scientific and Industrial Research (CSIR) Project no (CGER85X, HCARD03, and CGER77S) for the financial support.

References

1. Atanasova, G., Dikovska, A. O., Dilova, T., Georgieva, B., Avdeev, G. V., Stefanov, P., Nedyalkov, N. N., Metal-oxide nanostructures produced by PLD in open air for gas sensor applications. *Applied Surface Science* **2019,** *470,* 861–869.
2. Dral, A. P., ten Elshof, J. E., 2D metal oxide nanoflakes for sensing applications: Review and perspective. *Sensors and Actuators B: Chemical* **2018,** *272,* 369–392.
3. Lerma-García, M. J., Simó-Alfonso, E. F., Bendini, A., Cerretani, L., Metal oxide semiconductor sensors for monitoring of oxidative status evolution and sensory analysis of virgin olive oils with different phenolic content. *Food Chemistry* **2009,** *117* (4), 608–614.
4. Mahmoudi-Moghaddam, H., Tajik, S., Beitollahi, H., Highly sensitive electrochemical sensor based on La^{3+}-doped Co_3O_4 nanocubes for determination of sudan I content in food samples. *Food Chemistry* **2019,** *286,* 191–196.
5. Zakharova, G. S., Liu, Y., Enyashin, A. N., Yang, X., Zhou, J., Jin, W., Chen, W., Metal cations doped vanadium oxide nanotubes: Synthesis, electronic structure, and gas sensing properties. *Sensors and Actuators B: Chemical* **2018,** *256,* 1021–1029.
6. Ge, Y., Wei, Z., Li, Y., Qu, J., Zu, B., Dou, X., Highly sensitive and rapid chemiresistive sensor towards trace nitro-explosive vapors based on oxygen vacancy-rich and defective crystallized in-doped ZnO. *Sensors and Actuators B: Chemical* **2017,** *244,* 983–991.
7. Bonu, V., Das, A., Prasad, A. K., Krishna, N. G., Dhara, S., Tyagi, A., Influence of in-plane and bridging oxygen vacancies of SnO_2 nanostructures on CH_4 sensing at low operating temperatures. *Applied Physics Letters* **2014,** *105* (24), 243102.

8. Wang, Z., Zhang, T., Han, T., Fei, T., Liu, S., Lu, G., Oxygen vacancy engineering for enhanced sensing performances: A case of SnO_2 nanoparticles-reduced graphene oxide hybrids for ultrasensitive ppb-level room-temperature NO_2 sensing. *Sensors and Actuators B: Chemical* **2018,** *266,* 812–822.

9. Ye, K., Li, K., Lu, Y., Guo, Z., Ni, N., Liu, H., Huang, Y., Ji, H., Wang, P., An overview of advanced methods for the characterization of oxygen vacancies in materials. *TrAC Trends in Analytical Chemistry* **2019,** *116,* 102–108.

10. Gu, F., You, D., Wang, Z., Han, D., Guo, G., Improvement of gas-sensing property by defect engineering in microwave-assisted synthesized 3D ZnO nanostructures. *Sensors and Actuators B: Chemical* **2014,** *204,* 342–350.

11. Wang, X., Wang, T., Si, G., Li, Y., Zhang, S., Deng, X., Xu, X., Oxygen vacancy defects engineering on Ce-doped α-Fe_2O_3 gas sensor for reducing gases. *Sensors and Actuators B: Chemical* **2020,** *302,* 127165.

12. Bag, A., Lee, N.-E., Gas sensing with heterostructures based on two-dimensional nanostructured materials: a review. *Journal of Materials Chemistry C* **2019,** *7* (43), 13367–13383.

13. Lyson-Sypien, B., Kusior, A., Rekas, M., Zukrowski, J., Gajewska, M., Michalow-Mauke, K., Graule, T., Radecka, M., Zakrzewska, K., Nanocrystalline TiO_2/SnO_2 heterostructures for gas sensing. *Beilstein Journal of Nanotechnology* **2017,** *8* (1), 108–122.

14. Zappa, D., Galstyan, V., Kaur, N., Arachchige, H. M. M., Sisman, O., Comini, E., Metal oxide-based heterostructures for gas sensors: A review. *Analytica Chimica Acta* **2018,** *1039,* 1–23.

15. Li, T., Zeng, W., Wang, Z., Quasi-one-dimensional metal-oxide-based heterostructural gas-sensing materials: A review. *Sensors and Actuators B: Chemical* **2015,** *221,* 1570–1585.

16. Cheng, X., Xu, Y., Gao, S., Zhao, H., Huo, L., Ag nanoparticles modified TiO_2 spherical heterostructures with enhanced gas-sensing performance. *Sensors and Actuators B: Chemical* **2011,** *155* (2), 716–721.

17. Yu, Y., Zeng, W., Xu, M., Peng, X., Hydrothermal synthesis of $WO_3\cdot H_2O$ with different nanostructures from 0D to 3D and their gas sensing properties. *Physica E: Low-dimensional Systems and Nanostructures* **2016,** *79,* 127–132.

18. Shingange, K., Tshabalala, Z., Dhonge, B., Ntwaeaborwa, O., Motaung, D., Mhlongo, G., 0D to 3D ZnO nanostructures and their luminescence, magnetic and sensing properties: Influence of pH and annealing. *Materials Research Bulletin* **2017,** *85,* 52–63.

19. Sun, P., Zhou, X., Wang, C., Wang, B., Xu, X., Lu, G., One-step synthesis and gas sensing properties of hierarchical Cd-doped SnO_2 nanostructures. *Sensors and Actuators B: Chemical* **2014,** *190*, 32–39.

20. Tian, H., Fan, H., Ma, J., Liu, Z., Ma, L., Lei, S., Fang, J., Long, C., Pt-decorated zinc oxide nanorod arrays with graphitic carbon nitride nanosheets for highly efficient dual-functional gas sensing. *Journal of Hazardous Materials* **2018,** *341*, 102–111.

21. Khoang, N. D., Van Duy, N., Hoa, N. D., Van Hieu, N., Design of SnO_2/ZnO hierarchical nanostructures for enhanced ethanol gas-sensing performance. *Sensors and Actuators B: Chemical* **2012,** *174*, 594–601.

22. Karnati, P., Akbar, S., Morris, P. A., Conduction mechanisms in one dimensional core-shell nanostructures for gas sensing: A review. *Sensors and Actuators B: Chemical* **2019,** *295*, 127–143.

23. Walker, J. M., Akbar, S. A., Morris, P. A., Synergistic effects in gas sensing semiconducting oxide nano-heterostructures: A review. *Sensors and Actuators B: Chemical* **2019,** *286*, 624–640.

24. Al-Hashem, M., Akbar, S., Morris, P., Role of oxygen vacancies in nanostructured metal-oxide gas sensors: A review. *Sensors and Actuators B: Chemical* **2019,** *301*, 126845.

25. Zhao, M., Pan, W., Effect of lattice defects on thermal conductivity of Ti-doped, Y_2O_3-stabilized ZrO_2. *Acta Materialia* **2013,** *61* (14), 5496–5503.

26. Zhang, C., Liu, G., Geng, X., Wu, K., Debliquy, M., Metal oxide semiconductors with highly concentrated oxygen vacancies for gas sensing materials: A review. *Sensors and Actuators A: Physical* **2020,** 112026.

27. Miller, D. R., Akbar, S. A., Morris, P. A., Nanoscale metal oxide-based heterojunctions for gas sensing: A review. *Sensors and Actuators B: Chemical* **2014,** *204*, 250–272.

28. Kohli, N., Hastir, A., Kumari, M., Singh, R. C., Hydrothermally synthesized heterostructures of In_2O_3/MWCNT as acetone gas sensor. *Sensors and Actuators A: Physical* **2020,** *314*, 112240.

29. Huyen, D. N., Tung, N. T., Thien, N. D., Thanh, L. H., Effect of TiO_2 on the Gas Sensing Features of TiO_2/PANi Nanocomposites. *Sensors* **2011,** *11* (2), 1924–1931.

30. Seth, S., Ahmed, T., Samanta, A., Photoluminescence flickering and blinking of single $CsPbBr_3$ perovskite nanocrystals: Revealing explicit carrier recombination dynamics. *The Journal of Physical Chemistry Letters* **2018,** *9* (24), 7007–7014.

31. Brus, L., Luminescence of silicon materials: Chains, sheets, nanocrystals, nanowires, microcrystals, and porous silicon. *The Journal of Physical Chemistry* **1994,** *98* (14), 3575–3581.

32. Rehman, B., Bhalla, N. K., Vihari, S., Jain, S. K., Vashishtha, P., Gupta, G., SnO_2/Au multilayer heterostructure for efficient CO sensing. *Materials Chemistry and Physics* **2020,** *244*, 122741.

33. Prades, J. D., Arbiol, J., Cirera, A., Morante, J. R., Avella, M., Zanotti, L., Comini, E., Faglia, G., Sberveglieri, G., Defect study of SnO_2 nanostructures by cathodoluminescence analysis: Application to nanowires. *Sensors and Actuators B: Chemical* **2007,** *126* (1), 6–12.

34. Gu, F., Wang, S. F., Song, C. F., Lü, M. K., Qi, Y. X., Zhou, G. J., Xu, D., Yuan, D. R., Synthesis and luminescence properties of SnO_2 nanoparticles. *Chemical Physics Letters* **2003,** *372* (3–4), 451–454.

35. Tian, S., Zhang, Y., Zeng, D., Wang, H., Li, N., Xie, C., Pan, C., Zhao, X., Surface doping of La ions into ZnO nanocrystals to lower the optimal working temperature for HCHO sensing properties. *Physical Chemistry Chemical Physics* **2015,** *17* (41), 27437–27445.

36. Chen, M., Wang, Z., Han, D., Gu, F., Guo, G., Porous ZnO polygonal nanoflakes: Synthesis, use in high-sensitivity NO_2 gas sensor, and proposed mechanism of gas sensing. *The Journal of Physical Chemistry C* **2011,** *115* (26), 12763–12773.

37. Mhlongo, G. H., Motaung, D. E., Swart, H. C., Pd^{2+} doped ZnO nanostructures: Structural, luminescence and gas sensing properties. *Materials Letters* **2015,** *160*, 200–205.

38. Mhlongo, G. H., Shingange, K., Tshabalala, Z. P., Dhonge, B. P., Mahmoud, F. A., Mwakikunga, B. W., Motaung, D. E., Room temperature ferromagnetism and gas sensing in ZnO nanostructures: Influence of intrinsic defects and Mn, Co, Cu doping. *Applied Surface Science* **2016,** *390*, 804–815.

39. Shingange, K., Swart, H. C., Mhlongo, G. H., $LaBO_3$ (B = Fe, Co) nanofibers and their structural, luminescence and gas sensing characteristics. *Physica B: Condensed Matter* **2020,** *578*, 411883.

40. Mhlongo, G. H., Motaung, D. E., Nkosi, S. S., Swart, H. C., Malgas, G. F., Hillie, K. T., Mwakikunga, B. W., Temperature-dependence on the structural, optical, and paramagnetic properties of ZnO nanostructures. *Applied Surface Science* **2014,** *293*, 62–70.

41. Prades, J. D., Arbiol, J., Cirera, A., Morante, J., Avella, M., Zanotti, L., Comini, E., Faglia, G., Sberveglieri, G., Defect study of SnO_2 nanostructures by cathodoluminescence analysis: Application to nanowires. *Sensors and Actuators B: Chemical* **2007,** *126* (1), 6–12.

42. Wu, J.-M., Characterizing and comparing the cathodoluminesence and field emission properties of Sb doped SnO_2 and SnO_2 nanowires. *Thin Solid Films* **2008,** *517* (3), 1289–1293.

43. Korotcenkov, G., Cho, B., Bulk doping influence on the response of conductometric SnO_2 gas sensors: Understanding through cathodoluminescence study. *Sensors and Actuators B: Chemical* **2014,** *196*, 80–98.

44. Stefaniuk, I., Electron paramagnetic resonance study of impurities and point defects in oxide crystals. *Opto-Electronics Review* **2018,** *26* (2), 81–91.

45. Shingange, K., Mhlongo, G., Motaung, D., Ntwaeaborwa, O., Tailoring the sensing properties of microwave-assisted grown ZnO nanorods: Effect of irradiation time on luminescence and magnetic behaviour. *Journal of Alloys and Compounds* **2016,** *657*, 917–926.

46. Kortidis, I., Swart, H. C., Ray, S. S., Motaung, D. E., Characteristics of point defects on the room temperature ferromagnetic and highly NO_2 selectivity gas sensing of p-type Mn_3O_4 nanorods. *Sensors and Actuators B: Chemical* **2019,** *285*, 92–107.

47. Moskovits, M., Surface-enhanced Raman spectroscopy: A brief retrospective. *Journal of Raman Spectroscopy: An International Journal for Original Work in all Aspects of Raman Spectroscopy, Including Higher Order Processes, and also Brillouin and Rayleigh Scattering* **2005,** *36* (6–7), 485–496.

48. Stiles, P. L., Dieringer, J. A., Shah, N. C., Van Duyne, R. P., Surface-enhanced Raman spectroscopy. *Annual Review of Analytical Chemistry* **2008,** *1*, 601–626.

49. Gautam, M., Jayatissa, A. H., Gas sensing properties of graphene synthesized by chemical vapor deposition. *Materials Science and Engineering: C* **2011,** *31* (7), 1405–1411.

50. Sänze, S., Hess, C., Ethanol gas sensing by indium oxide: An operando spectroscopic Raman-FTIR Study. *The Journal of Physical Chemistry C* **2014,** *118* (44), 25603–25613.

51. Shingange, K., Tshabalala, Z., Ntwaeaborwa, O., Motaung, D., Mhlongo, G., Highly selective NH_3 gas sensor based on Au loaded ZnO nanostructures prepared using microwave-assisted method. *Journal of Colloid and Interface Science* **2016,** *479*, 127–138.

52. Shingange, K., Swart, H., Mhlongo, G., H_2S detection capabilities with fibrous-like La-doped ZnO nanostructures: A comparative study on the combined effects of La-doping and post-annealing. *Journal of Alloys and Compounds* **2019,** *797*, 284–301.

53. Shingange, K., Swart, H., Mhlongo, G. H., Design of porous p-type $LaCoO_3$ nanofibers with remarkable response and selectivity to ethanol at low operating temperature. *Sensors and Actuators B: Chemical* **2020,** *308,* 127670.

Chapter 5

Metal Nanoparticle Enhanced Upconversion of Lanthanide Ions

R. E. Kroon
Department of Physics, University of the Free State,
Bloemfontein 9300, South Africa
kroonre@ufs.ac.za

Upconversion phosphors, which absorb infrared light and emit visible light, have a number of potential applications in the energy, sensor, biomedical, and security fields which may bring disruptive change to applied technologies. However, they remain inefficient and continued research is required to improve their conversion quantum yield. Metal enhanced fluorescence has been applied for some decades to increase the emission of downshifting phosphors and similar principles have more recently also been explored for upconversion phosphors. This chapter considers the motivation behind plasmonic enhancement and gives background on upconversion processes and the resonance characteristics of metal nanoparticles, after which the key aspects forming the basis for plasmon enhanced upconversion are discussed. Models for interpreting plasmon

Luminescent Nanomaterials
Edited by Odireleng Martin Ntwaeaborwa

ISBN 978-981-4968-11-9 (Hardcover), 978-1-003-27738-5 (eBook)
www.jennystanford.com

enhanced upconversion via the excited state absorption and energy transfer upconversion mechanisms are compared to that for Raman scattering and downshifted photoluminescence, with distinction between excitation and emission enhancements. Care when interpreting a decrease in the measured lifetime and the principle of matching plasmon resonance(s) with excitation and/or emission bands are stressed. Energy transfer between the ions and metal nanoparticles is examined and some references to recent work on plasmonic enhancement of energy transfer are provided.

5.1 Introduction

Upconversion phosphors absorb more than one photon of some excitation wavelength and consequently emit a single photon of shorter wavelength (i.e. higher energy) [1]. Such phosphors have potential applications for solar spectrum manipulation to increase the efficiency of photovoltaic cells [2, 3], as biomedical theranostic agents (integrating imaging and therapy) that can be excited by deeply penetrating infrared light which itself does not cause fluorescence or damage of biological tissue [4, 5], for nanoscale temperature sensors [6] and for forensics and security applications [7]. Generally, research efforts on the class of "shorter than excitation wavelength" emitting materials has been predicted to increase and bring disruptive changes to science and engineering [8]. While applications support a drive towards more efficient but complex upconversion material systems, this can make it challenging to identify and model the physical processes involved. Although the quantum efficiency of standard (downshifting) phosphors can approach 100% (or even exceed this in the case of downconversion/quantum-cutting phosphors), the theoretical maximum quantum yield for upconversion phosphors is limited to $\frac{100\%}{n}$ where n is the number of photons absorbed to allow emission of a photon [9] and the maximum reported values are limited to a few percent [10]. This provides continuing scope and impetus for research into ways to enhance the upconversion yield. The emitted intensity of upconversion phosphors, involving a non-linear process, generally follows a power law $I = k(I_0)^n$, where k is a constant, while the absorption is proportional to the excitation intensity I_0. Therefore the quantum yield is dependent on the excitation intensity and scales with $(I_0)^{n-1}$, so unless measurements are done using the same excitation power, comparable efficiencies are only obtained when the quantum yield is divided by this factor [1]. Experimental values of n derived from the power dependence often show deviations from the expected integer values [11] and the emitted power saturates for high excitation power [12], while experimentally maintaining a spatially uniform excitation intensity is a further challenge that makes obtaining reliable and comparable upconversion efficiencies difficult. It is therefore vital to make direct comparisons with reference samples to determine improved performance, which is generally done when evaluating potential plasmonic enhancement of upconverting phosphors.

Despite upconversion phosphors having a history of over 50 years [13] of development, they remain an active research field in which some mechanisms are still being clarified [14]. A large number of studies have reported on different hosts, activators, and sensitizers for upconversion phosphors, as well as synthesis parameters affecting the crystal quality, defects, and optically active centre concentrations. A significant aspect of engineering upconversion phosphors has been the study of the crystallite size of nanophosphors on their emission properties. A review of advances in lanthanide-doped upconversion nanomaterials [15] highlighted the importance of this class of materials, particularly for biomedical applications, and discussed the synthesis methods, morphology and surface modifications. Looking beyond the upconversion phosphor material itself, its environment can strongly influence its interaction with the excitation radiation and change its upconversion efficiency. Taking advantage of the influence of the environment has been immensely successful for the enhancement of Raman scattering using SERS (surface enhanced Raman scattering) and is therefore of great interest to apply in other areas, e.g. surface-enhanced infrared absorption and surface-enhanced fluorescence [16], as well as being a research area for upconversion phosphors. Surface-enhanced fluorescence requires an optimal spacing between the surface and luminescent centres and decays quickly for large separations, therefore Chen et al. [17] have proposed a cylindrical gold groove structure in which enhancement is possible in the far field region.

They predicted an 8.25 times increase in the emission intensity for an upconverting nanoparticle positioned at a distance 4.4 μm from the surface. Despite being challenging to control, the distribution of metal nanoparticles throughout a phosphor (in the bulk or on the small particle surfaces) extends their (near field) plasmonic enhancement effects to a large volume of the material, so this approach has good potential for powder samples and thick films.

Enhancement of downshifting photoluminescence of Eu^{3+} ions due to silver nanoparticles was reported by Malta et al. in 1985 [18]. Some earlier pioneering work is discussed by Moskovits [19], Barnes [20] and Lacowicz et al. [21, 22]. Plasmon-enhanced luminescence has subsequently been reported in a large number of lanthanide-based phosphors, e.g. downconversion enhancement of Er^{3+} ions [23] or Tb^{3+} ions [24] due to plasmonic silver nanoparticles. Recent reviews of plasmon-enhanced fluorescence are provided by Knoblauch and Geddes [25], Li et al. [26] and Dong et al. [27], while Kühn et al. [28] give a useful summary of early work in the field. In comparison, investigations of plasmonic enhancement for upconversion phosphors based on lanthanide ions are more recent, e.g. the effects of silver nanoparticles on Er^{3+} ions in heavy metal oxide glass by da Silva et al. [29] or on Pr^{3+} ions in tellurite glass by Kassab et al. [30], both published in 2007. Studies of plasmon enhanced upconversion encompass not only the use of metallic nanoparticles, but a wide variety of plasmonic systems including surfaces and core-shell structures. This field has generated strong and growing interest over the last decade (Fig. 5.1), both because of the increasing envisaged applications of upconversion phosphors and also due to the realisation that upconversion, as a non-linear process, has a power law dependence on the intensity which makes more sensitive to changes in the local electric field than fluorescence, hence giving it a greater potential for larger enhancements compared to downshifting phosphors (e.g. compare potential enhancement factors of ~20,000 for a linear process, but 3.8×10^8 for the non-linear two-photon absorption excitation [31]). In this review the focus will be towards some basic principles and models involved in considering the effects of plasmonic metal nanoparticles on the emission properties of upconversion phosphors doped with lanthanide ions.

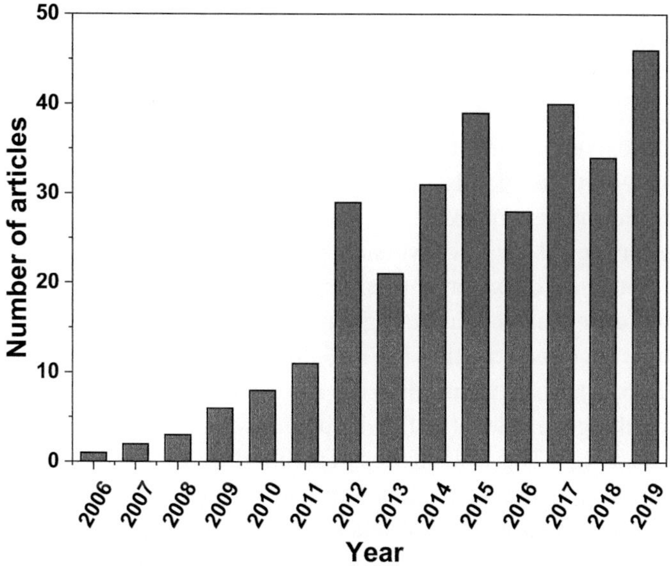

Figure 5.1 Number of articles per year in the Scopus database relating to plasmonic enhancement of upconversion, matching the query TITLE-ABS-KEY (plasmon or plasmonic, enhancement, and up-conversion or upconversion).

5.2 Upconversion Phosphors Incorporating Lanthanide Ions

Although over a dozen distinct physical processes can lead to the emission of light having a wavelength shorter than that of the excitation radiation [8], most are very inefficient or do not result in a large shift in the wavelength (e.g. anti-Stokes Raman scattering) and are therefore not applied in practical upconversion phosphors. Selected processes are illustrated in Fig. 5.2. The numerous and isolated energy levels of various lanthanide ions, e.g. Er^{3+}, Tm^{3+}, Ho^{3+}, Dy^{3+}, Pr^{3+} [32], make them good candidates for supporting the processes involving real rather than virtual energy levels, which are of importance for practical upconversion phosphor materials.

The feasibility of upconversion phosphors was heralded by the discovery of sequential energy transfer upconversion by Auzel in 1966 [33] which he called APTE (addition de photon par transferts d'energie) but which has become more generally known as ETU (energy transfer upconversion). This may be represented by Fig. 5.2f or Fig. 5.2g depending on whether the energy transfers originate from two separate excited neighbours or from a single neighbour that was re-excited after the first energy transfer (e.g. compare representations of this process in Figs. 3 and 4 of [1]). Prior to this, the process of excited state absorption (ESA, Fig. 5.2e) had been proposed for upconversion (infrared quantum counting) but had not been considered feasible [13], even though its efficiency is an order of magnitude greater than cooperative processes (cooperative sensitization in Fig. 5.2d, or cooperative luminescence in Fig. 5.2c), which themselves are much more probable than second harmonic generation (in Fig. 5.2b) or two photon absorption (in Fig. 5.2a) [1]. The ETU process exhibited an efficiency about two orders of magnitude greater than ESA [1] and boosted interest in upconversion phosphors. The key innovation was the non-resonant energy transfer occurring to an ion that was already in an excited state (after the initial non-radiative (Dexter) energy transfer to the ion in its ground state). The ETU process also occurs in Fig. 5.2h, although not with a sequential energy transfer but rather after GSA of the acceptor. The process illustrated in Fig. 5.2i has ESA as the final step and does not involve energy transfer *to an already excited state*, so it involves well-known energy transfer rather than ETU and on this basis should be distinguished from the group of ETU-related processes in Figs. 5.2f–h. The photon avalanche effect (Fig. 5.2j) involves ESA following which cross-relaxation may return the ion to its originally excited state while also exciting a neighbouring ion to such as excited state, leaving both ions in the excited state. This process can repeat for both ions, driving a positive feedback loop to populate the excited state. Above a threshold pump intensity the excited state therefore becomes highly populated. If upconverted luminescence also occurs from the state reached after ESA to the ground state (i.e. as a competing process to the cross-relaxation, not shown in Fig. 5.2j) then this luminescence also increases, i.e. a photon 'avalanche' occurs. The principle of photon avalanche therefore involve ESA and a positive feedback loop resulting from cross-relaxation populating the excited state, which can also occur in more complex energy level schemes like Pr^{3+} for which the effect was first observed in 1979 [34].

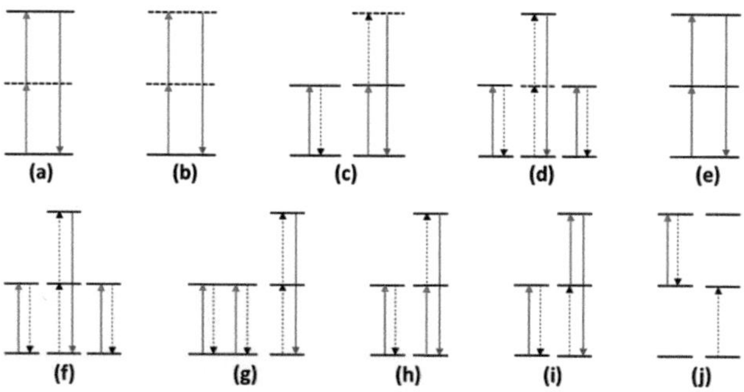

Figure 5.2 Selected upconversion processes. Vertical upward arrows indicate absorption and vertical down arrows indicate emission, while pairs of dashed arrows represent non-radiative energy transfer in the form of cross-relaxation. Horizontal solid and dashed lines represent real and virtual energy levels, respectively. (a) Two photon absorption; (b) second harmonic generation; (c) cooperative luminescence; (d) cooperative sensitization; (e) excited state absorption (ESA) following ground state absorption (GSA); (f) energy transfer (ET) from one donor followed by energy transfer upconversion (ETU) from a different donor; (g) ET from one donor followed by ETU from the same re-excited donor; (h) ETU following GSA; (i) ESA following ET (not ETU); (j) photon avalanche mechanism of ESA and positive feedback cross-relaxation populating the excited state.

The originally discovered and still generally most viable upconversion systems entail co-doping hosts with Er^{3+} and Yb^{3+} ions [13]. Similarly to materials co-doped with Tm^{3+}/Yb^{3+} and Ho^{3+}/Yb^{3+}, these systems produce effective visible luminescence when pumped near 980 nm where Yb^{3+} ions are strongly absorbing, since the favourable energy level distributions allows efficient energy transfer upconversion as illustrated in Fig. 5.3 [35]. Most often the doping of Yb^{3+} is high (~10–20%), while the concentration of the luminescent ions in which upconversion occurs is much lower (~1–2%). In systems where the energy scheme is less favourable for energy transfer upconversion, such as Eu^{3+}/Yb^{3+} and Tb^{3+}/Yb^{3+}, the co-doping with Yb^{3+} still enables excitation via cooperative sensitization (Fig. 5.2d), although this process is less efficient and visible emission intensity is less. In single Yb^{3+} doped materials, cooperative luminescence (Fig. 5.2c) can occur. In other single doped systems where the activator ion has a series of spread out energy levels, e.g. Er^{3+}, Ho^{3+}, Dy^{3+}, and Pr^{3+}, upconversion can still occur due to the ESA mechanism (Fig. 5.2e), while numerous possible cross-

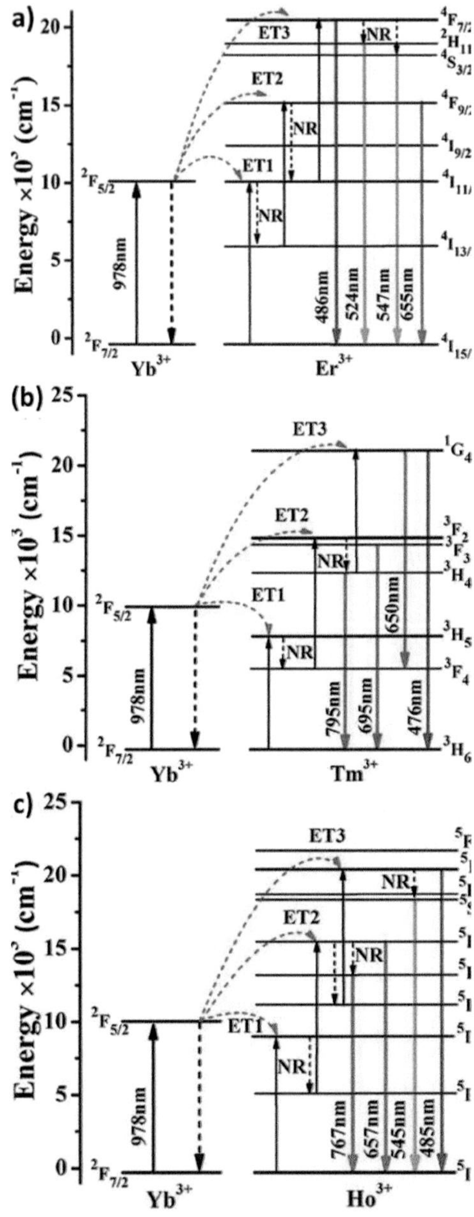

Figure 5.3 Energy level schemes for energy transfer upconversion systems sensitized with Yb^{3+} and activated by (a) Er^{3+}, (b) Tm^{3+}, and (c) Ho^{3+}. Adapted under the terms of the Creative Commons Attribution 4.0 licence (http://creativecommons.org/licenses/by/4.0/) by combining parts of Figs. 4, 5, and 6 from Ref. [35]. Copyright 2019 The Author(s). Published by National Institute for Materials Science in partnership with Taylor & Francis Group.

relaxation effects may complicate the modelling of such systems. A large amount of research has been done on upconversion phosphors to study different hosts, synthesis methods, morphology and doping as well as to discuss applications. For further information the reader is referred to the literature already discussed in the introduction, while the plasmonic enhancement of such upconversion phosphors will be discussed later in this review.

5.3 Plasmonic Effects of Metal Nanoparticles

Consider an electromagnetic plane wave of vacuum wavelength λ travelling in a non-absorbing medium of (real) refractive index n so that its wavenumber is $k = \dfrac{2\pi}{\lambda/n}$. If it is incident with intensity I_0 on a uniform spherical particle (e.g. a metal nanoparticle) of radius R and (complex) refractive index n_p then according to Mie theory the wavelength dependent scattering and absorption cross sections (σ), or efficiencies (Q), are given by [36, 37]

$$\sigma_{\text{sca}} = (\pi R^2)Q_{\text{sca}} = \frac{P_{\text{sca}}}{I_0} = \frac{2\pi}{k^2}\sum_{j=1}^{\infty}(2j+1)\left[\left|a_j\right|^2 + \left|b_j\right|^2\right] \tag{5.1a}$$

$$\sigma_{\text{abs}} = (\pi R^2)Q_{\text{abs}} = \frac{P_{\text{abs}}}{I_0} = \frac{2\pi}{k^2}\sum_{j=1}^{\infty}(2j+1)\,\text{Re}\,[a_j + b_j] - \sigma_{\text{sca}} \tag{5.1b}$$

where P_{sca} and P_{abs} are the scattered and absorbed powers and

$$a_j = \frac{\dfrac{n_p}{n}\psi_j\left(\dfrac{n_p}{n}kR\right)\psi_j'(kR) - \psi_j(kR)\psi_j'\left(\dfrac{n_p}{n}kR\right)}{\dfrac{n_p}{n}\psi_j\left(\dfrac{n_p}{n}kR\right)\xi_j'(kR) - \xi_j(kR)\psi_j'\left(\dfrac{n_p}{n}kR\right)}$$

and
$$b_j = \frac{\psi_j\left(\dfrac{n_p}{n}kR\right)\psi_j'(kR) - \dfrac{n_p}{n}\psi_j(kR)\psi_j'\left(\dfrac{n_p}{n}kR\right)}{\psi_j\left(\dfrac{n_p}{n}kR\right)\xi_j'(kR) - \dfrac{n_p}{n}\xi_j(kR)\psi_j'\left(\dfrac{n_p}{n}kR\right)}$$

where ψ_j and ξ_j are the Riccati-Bessel functions and the primes indicate differentiation with respect to the argument in brackets. The MiePlot software (www.philiplaven.com/mieplot.htm) is freely available to implement such calculations, although wavelength dependent refractive index data (or equivalent relative permittivity data) for bulk noble metals, e.g. silver, gold, etc. [38, 39], should be corrected for small nanoparticles [40].

For nanoparticles which are small compared to the wavelength of light (i.e. $kR << 1$), simplification is possible since the higher terms in the expansion of Eq. 5.1 can be neglected. Then the cross sections can be written as [36, 41]

$$\sigma_{\text{sca}} = (\pi R^2)Q_{\text{sca}} = \frac{8\pi}{3}k^4R^6\left|\frac{n_p^2 - n^2}{n_p^2 + 2n^2}\right|^2$$

$$= \frac{8\pi}{3}k^4R^6\left|\frac{\varepsilon_{rp} - \varepsilon_r}{\varepsilon_{rp} + 2\varepsilon_r}\right|^2 = \frac{1}{6\pi}k^4|\alpha|^2 \quad (5.2a)$$

and

$$\sigma_{\text{abs}} = (\pi R^2)Q_{\text{abs}} = 4\pi kR^3 \ \text{Im}\left[\frac{n_p^2 - n^2}{n_p^2 + 2n^2}\right]$$

$$= 4\pi kR^3 \ \text{Im}\left[\frac{\varepsilon_{rp} - \varepsilon_r}{\varepsilon_{rp} + 2\varepsilon_r}\right] = k \ \text{Im}[\alpha] \quad (5.2b)$$

where $\varepsilon_{rp} = n_p^2$ is the (complex) relative permittivity of the nanoparticle material, $\varepsilon_r = n^2$ is the (real) relative permittivity of the non-absorbing medium. The quantity $\alpha = 4\pi R^3\left(\frac{\varepsilon_{rp} - \varepsilon_r}{\varepsilon_{rp} + 2\varepsilon_r}\right)$ is the polarizability of the spherical nanoparticle which exhibits a dipole moment $p = \varepsilon_0\varepsilon_r\alpha E_{\text{ext}}$ in an external electric field E_{ext}. The polarized nanoparticle then produces the electric field of a dipole in the region around it. The scattering, absorption, and dipole electric field produced by the nanoparticle all depend on the polarizability α, which grows very large when the denominator $\varepsilon_{rp} + 2\varepsilon_r \approx 0$, so this is the criterion (often referred to as the Fröhlich condition) for when the nanoparticle will couple most strongly with the incident radiation and have the most profound effect on the electric field in the region surrounding it. Although tabulated data for specific metals [38, 39] corrected for size effects [40] is preferable for computer modelling, the Drude model for the relative permittivity of a metal can be used to obtain a general understanding of this criterion. According to the Drude model [39] the relative permittivity can be expressed as $\varepsilon_{rp} = 1 - \frac{\omega_p^2}{\omega^2 + i\omega\gamma_p}$ where $\omega_p = \sqrt{\frac{n'e^2}{\varepsilon_0 m}}$ is the plasmon frequency determined by the free electron density n', while e and m are the charge and mass of an electron, ε_0 is the permittivity of free space and γ_p represents a damping term. This model is poor for some metals (e.g. Ag [36], Au [42]) and may therefore often adjusted to $\varepsilon_{rp} = \varepsilon_\infty - \frac{\omega_p^2}{\omega^2 + i\omega\gamma_p}$ in order to better account for the susceptibility of the bound electrons [43]. Neglecting the damping associated with γ_p that imparts the relative permittivity with an imaginary part, the Fröhlich condition $\varepsilon_{rp} + 2\varepsilon_r = 0$ yields the localized surface plasmon resonance (LSPR) frequency

$$\omega_{\text{LSPR}} = \frac{\omega_p}{\sqrt{\varepsilon_\infty + 2\varepsilon_r}}, \tag{5.3}$$

close to which scattering, absorption, and electric field enhancement by the metal nanoparticles is maximized. The corresponding wavelength $\lambda_{\text{LSPR}} = \frac{2\pi c}{\omega_{\text{LSPR}}} = \frac{2\pi c}{\omega_p}\sqrt{\varepsilon_\infty + 2\varepsilon_r}$ may be correlated to experimental data found using absorption or reflectance measurements and increases as the relative permittivity ε_r of the surrounding medium increases, resulting in a red shift and allowing the tuning of λ_{LSPR} in different host materials [24]. Equation 5.3 is often applied with $\varepsilon_\infty = 1$ (i.e. the unmodified Drude model) for nanoparticles in air ($\varepsilon_r \cong 1$) to obtain $\omega_{\text{LSPR}} = \frac{\omega_p}{\sqrt{3}}$, showing that in general the LSPR frequency is lowered relative to the bulk plasmon frequency. It is also interesting to compare this to the case for a surface plasmon polariton (SPP) which can propagate along a flat interface, for which the corresponding formula is $\omega_{\text{SPP}} = \frac{\omega_p}{\sqrt{1+\varepsilon_r}}$ [20, 44]. For an air-metal interface this reduces to $\omega_{\text{SPP}} = \frac{\omega_p}{\sqrt{2}}$ which is intermediate between the bulk plasmon frequency and the LSPR frequency.

Around a nanoparticle much smaller than the wavelength of the incident light, the total electric field is the sum of the dipole field of the nanoparticle and the applied field. Taking the applied field at some moment to be $E_{\text{ext}}\hat{z}$ (considered uniform over the nanoparticle due to its small size relative to the wavelength) the total field is $\vec{E} = E_{\text{ext}}\hat{z} + \frac{p}{4\pi\varepsilon_0\varepsilon_r r^3}(2\cos\theta\hat{r} + \sin\theta\hat{\theta})$, from which it can be shown that $\frac{\vec{E}}{E_{\text{ext}}} = \hat{z}\left[1 + \frac{\alpha}{4\pi r^3}(3\cos^2\theta - 1)\right] + \hat{s}\left[\frac{3\alpha}{4\pi r^3}\cos\theta\sin\theta\right]$. This corresponds to a position dependent intensity enhancement of

$$\left|\frac{\vec{E}}{E_{\text{ext}}}\right|^2 = \left(1 - \frac{\alpha}{4\pi r^3}\right)^2 + 3\left(\frac{\alpha}{4\pi r^3}\right)\left(2 + \frac{\alpha}{4\pi r^3}\right)\cos^2\theta. \tag{5.4}$$

This expression is equivalent to a result previously derived in Cartesian coordinates [45]. Under the conditions for LSPR the polarizability $\alpha = 4\pi R^3\left(\frac{\varepsilon_{rp} - \varepsilon_r}{\varepsilon_{rp} + 2\varepsilon_r}\right)$ becomes large and when $\frac{\alpha}{4\pi r^3} \gg 1$ the enhancement factor becomes $\left|\frac{\vec{E}}{E_{\text{ext}}}\right|^2 = \left(\frac{\alpha}{4\pi r^3}\right)^2(1 + 3\cos^2\theta)$. This enhancement pattern is strongest near the 'poles' (i.e. $\theta = 0$ or π) and decreases towards the 'equator' (i.e. $\theta = \frac{\pi}{2}$) as well as quickly with distance (r^{-6}) from the nanoparticle surface. Such an enhancement pattern (Fig. 5.4) has been physically verified [46] by experimentally surrounding nanoparticles by a photopolymerizable formulation and then exciting them with light below the normal threshold intensity required for polymerization. In this way only the regions near the nanoparticles which experience an enhancement of intensity to above the threshold are polymerized and thereby capture this enhanced intensity pattern.

The polarizability of an ellipsoid with semi-axes $R_1 \geq R_2 \geq R_3$ along the ith axis is $\alpha_i = 4\pi R_1 R_2 R_3\left(\frac{\varepsilon_{rp} - \varepsilon_r}{3\varepsilon_r + 3f_i(\varepsilon_{rp} - \varepsilon_r)}\right)$ where $f_1 \leq f_2 \leq f_3$ are geometrical factors that decrease as the length of the ellipsoid along the respective axes increase and $\Sigma f_i = 1$ [36]. For a sphere all these factors are one third, while a prolate spheroid has $R_1 > R_2 = R_3$ with $f_2 = f_3 = \frac{1-f_1}{2}$ and $f_1 < \frac{1}{3}$. In general, the polarizability along the ith axis becomes large when the denominator $3\varepsilon_r + 3f_i(\varepsilon_{rp} - \varepsilon_r)$ tends to zero, i.e. when $\varepsilon_{rp} = \frac{f_i - 1}{f_i}\varepsilon_r$. Applying the Drude model for relative permittivity and neglecting the damping then leads to

$$\omega_{\text{LSPR},i} = \frac{\omega_p}{\sqrt{\varepsilon_\infty + \frac{1-f_i}{f_i}\varepsilon_r}} \qquad (5.5)$$

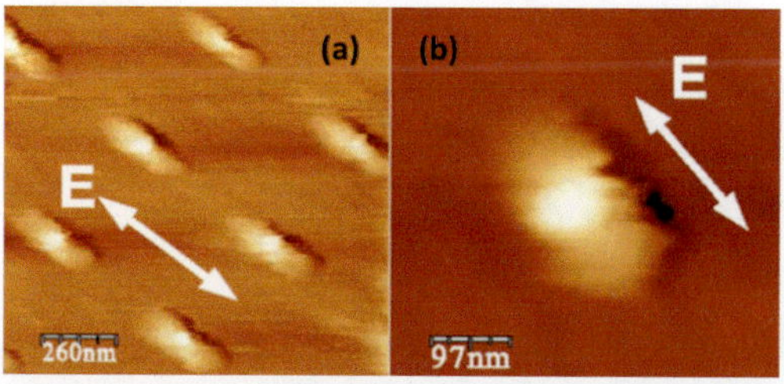

Figure 5.4 (a) AFM image recorded after irradiation and developing of a silver nanoparticle array covered with photopolymerizable formulation. (b) Enlarged region around a single nanoparticle showing polymer material around the silver nanoparticle in the region of the plasmon enhanced dipole electric field. Figure reprinted with permission from Ref. [46].

for an ellipsoid, so $\lambda_{\text{LSPR},i} = \frac{2\pi c}{\omega_p}\sqrt{\varepsilon_\infty + \frac{1-f_i}{f_i}\varepsilon_r}$. Since $f_1 < \frac{1}{3}$ along the extended axis, the value of $\lambda_{\text{LSPR},i}$ is increased compared to a sphere because, while along the two shorter axes it is expected to increase. Therefore if randomly oriented, small rod-like nanoparticles (with maximum dimension much smaller than the wavelength of light) have two resonance frequencies which can be tuned by changing their aspect ratio as illustrated in Fig. 5.5 [47]. Figure 5.6a illustrates a system used to enhance the luminescence of Er^{3+} ions excited at 532 nm and emitting in the infrared near 1540 nm. This was done by overlaying Er-implanted SiO_2 with silver nanorods. By increasing the length of the nanorods (Fig. 5.6b), Mertens and Polman [23] were able to change their plasmon frequencies. The measured enhancement factor (Fig. 5.6c) changed as a function of the nanorod length and achieved a maximum for a length of 400 nm, where the plasmon resonance frequency correlated with the emission wavelength from the Er^{3+} ions.

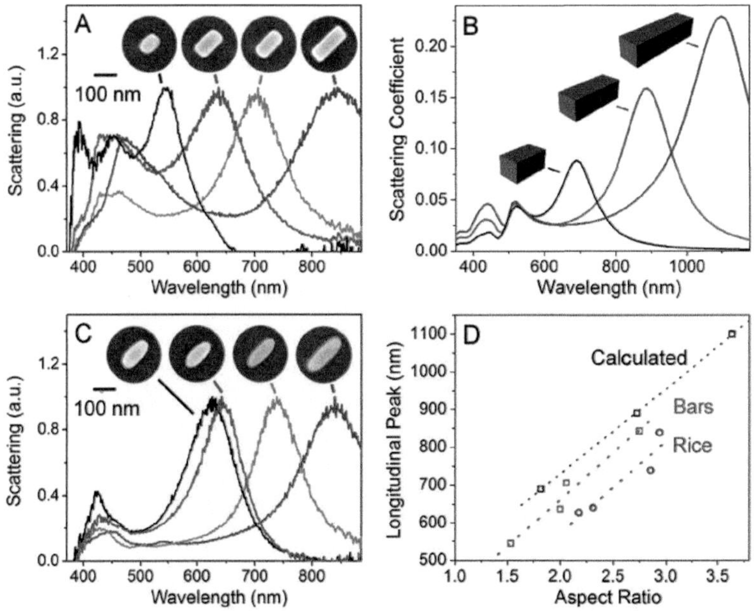

Figure 5.5 (A) SEM images of individual nanobars with their corresponding normalized scattering spectra. (B) Calculated scattering spectra of nanobars 100, 150, and 200 nm in length, keeping width 55 nm and height 50 nm. (C) SEM images of individual nanorice with their corresponding normalized scattering spectra. (D) Plot of longitudinal plasmon peak location versus aspect ratio. Reprinted with permission from Ref. [47]. Copyright 2007 American Chemical Society.

For larger spherical particles higher order terms from the Mie theory, like the quadrupole term [48], must be added to the dipole term. If the particles are not spherical, the fields and plasmon resonance become more complex and must generally be computed numerically. The reader is referred to the literature for further information on these topics [48–51]. Finally, it should be noted that if the density of nanoparticles is high, interactions between them become likely and results obtained for modelling for isolated nanoparticles may be inaccurate.

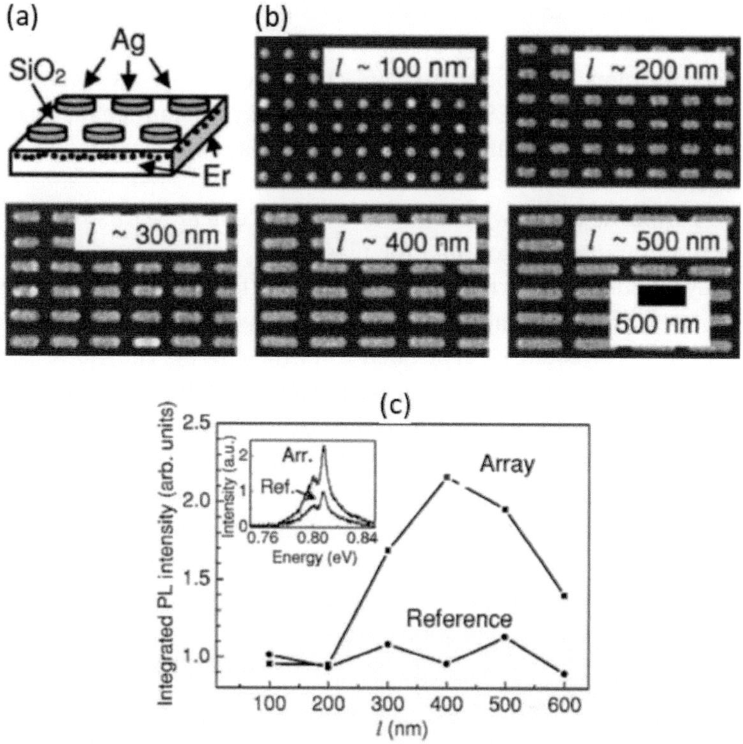

Figure 5.6 (a) Schematic view of system having Er implanted into SiO_2 and overlaid with silver nanoparticles using electron beam lithography. (b) Scanning electron microscope images of silver nanorods with constant width but different lengths. (a) Integrated luminescence from the Er^{3+} ions at 1540 µm as a function of the length of the silver nanorods. Reprinted from Ref. [23] with the permission of AIP Publishing.

5.4 Enhancement of Upconversion Phosphors Incorporating Metallic Nanoparticles

When a rhodamine B solution sandwiched between quartz plates was excited at 852 nm, Gryczynski et al. [31] reported an increase of about 4 times in the emission at 580 nm if the plates were coated with silver islands, hence demonstrating plasmonic enhanced upconversion for the two-photon absorption mechanism (Fig. 5.2a). However, most practical upconversion phosphors are co-doped with

Yb^{3+} and either Er^{3+}, Tm^{3+} or Ho^{3+} and modelled using single donor sequential ET/ETU (Fig. 5.2g), while alternatively single doped upconversion phosphors (e.g. Er^{3+} [52], Nd^{3+} [53]) are generally modelled using GSA/ESA (Fig. 5.2e), therefore plasmon enhancement for two photon upconversion based on these mechanisms will be discussed.

Park et al. [54] have summarized enhancement factors reported in the literature up to 2005 for plasmonic upconversion systems. They found a large variation in enhancement factors reported mostly between 1 and 100 and that no clear advantage for the use of either gold or silver could be concluded from the survey. They have also recently provided updated data for the period 2015 to 2019 [55], as reproduced in Fig. 5.7, for which they note a trend for more studies where the plasmon resonance is tuned to near to the common excitation wavelength of 980 nm. The fact that enhancement values reported vary over three orders of magnitude indicate the great potential for these materials, but also that further development is still necessary. Dong et al. [56] have also reviewed provided a summary of recent applications in the field.

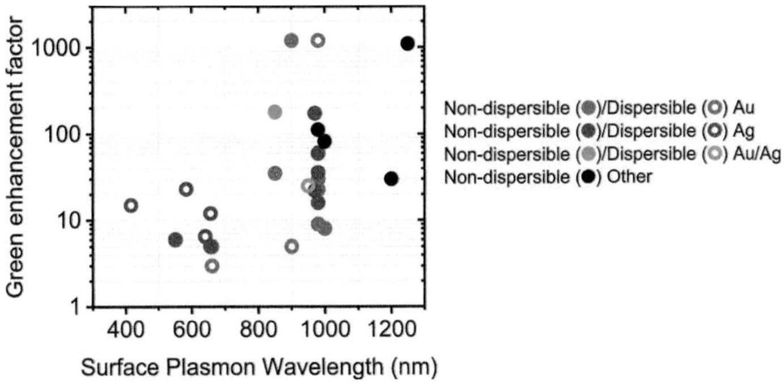

Figure 5.7 Survey of plasmonic enhancement factors for upconversion reported between 2015–2019. Reprinted from Ref. [55]. Copyright 2020 Ananda Das et al., published by De Gruyter. This work is licensed under the Creative Commons Attribution 4.0 Public License.

Before discussing plasmon enhancement of upconversion systems, first consider the simpler process of plasmon enhanced luminescence, since many of the principles also apply to upconversion. In 2006 Kühn et al. [57] reported shifted a 100 nm diameter gold nanoparticle near a single terrylene molecule and found this to cause up to 20 times enhancement of its fluorescence intensity when excited at 532 nm, which was the wavelength where the nanoparticle showed plasmonic resonance and also where the molecule exhibited fluorescence. The measured luminescence signal *I* was modelled by $I = k\left|e|\vec{E}\bullet\vec{p}|g\right|^2 \eta$ where $|g\rangle$ and $|e\rangle$ are the ground and excited states of the molecule, $\vec{E}$ is the electric field and $\vec{p}$ is the molecular dipole moment operator, η is the quantum yield of the molecule and k is a proportionality constant. Plasmonic enhancement of the electric field will therefore increase the transition matrix $e|\vec{E}\bullet\vec{p}|g$, so that the excitation rate increases with the square of the electric field enhancement. However, the luminescence intensity is also affected by the quantum yield, given by $\eta = \frac{\gamma_r}{\gamma_r + \gamma_{nr}}$ where γ_r and γ_{nr} are the radiative and non-radiative decay rates which are also altered in the vicinity of the gold nanoparticle. Since for isolated terrylene molecules the original quantum yield was close to 100%, this could not be significantly improved in the vicinity of the plasmonic nanoparticle and was actually found to decrease to about 50% under the experimental conditions of maximum luminescent intensity. Therefore the fluorescence was enhanced by a considerable increase in the excitation rate through plasmonic enhancement of the electric field, although it was lessened to a degree by a decrease in the quantum yield. The plasmonic effect on the total decay rate can be experimentally monitored since the measured lifetime corresponds to $\frac{1}{\gamma_r + \gamma_{nr}}$. This decreased, i.e. the total decay rate increased, as the separation between the luminescent molecule and the gold nanoparticle decreased. However, the quantum yield and therefore luminescence intensity depends not on the absolute decay rates, but instead their ratio $\left(\eta = \frac{1}{1 + \frac{\gamma_{nr}}{\gamma_r}}\right)$ so from this it is apparent that an increase in the luminescence intensity can only be caused if the increase of the radiative decay rate is more than the increase of the non-radiative decay rate. Additionally, the enhancement could only increase the quantum yield up to a maximum of 100% and so this enhancement can only be appreciable in cases where the quantum yield is originally low. Figure 5.8 illustrates calculated changes in radiative and non-radiative decays rates for a dipole as a function of distance from a silver nanoparticle simulated using exact electrodynamics theory [58]. Although both radiative and non-radiative rates increase as the separation decreases, the non-radiative decay rate only begins to increase very quickly for smaller separations and therefore an optimum separation exists for maximising the quantum yield, which for this simulation had a low value of 1% in the absence of the sphere but was enhanced to over 10% at the optimal separation of ~6 nm.

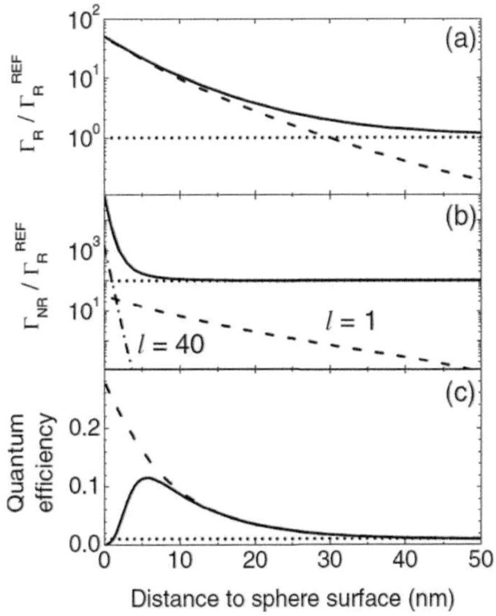

Figure 5.8 Exact electrodynamics simulation of radiative and non-radiative rates for a dipole at varying distances from a 60 nm diameter silver sphere. The quantum efficiency of the emitter in the absence of the sphere is taken as 0.01 (i.e. 1%), the refractive index of the embedding medium is 1.3 and the emission wavelength is 433 nm. The dipole emitter orientation is averaged over all solid angles. (a) Radiative rates—dotted: in the absence of the sphere; dashed: in the presence of the sphere, and solid: total radiative rate. (b) Non-radiative decay rate—dotted: in the absence of the sphere; dashed: contribution of lowest (dipole) mode $l = 1$; dot-dash: contribution of mode $l = 40$, and solid: total non-radiative rate up to mode $l = 60$. The decay rates in both (a) and (b) are normalized to the radiative decay rate in the absence of the sphere. (c) Luminescence quantum efficiency—dotted: in the absence of the sphere; dashed: considering only lowest (dipole) mode $l = 1$ and solid: considering all modes up to $l = 60$. Figure reprinted with permission from Ref. [58].

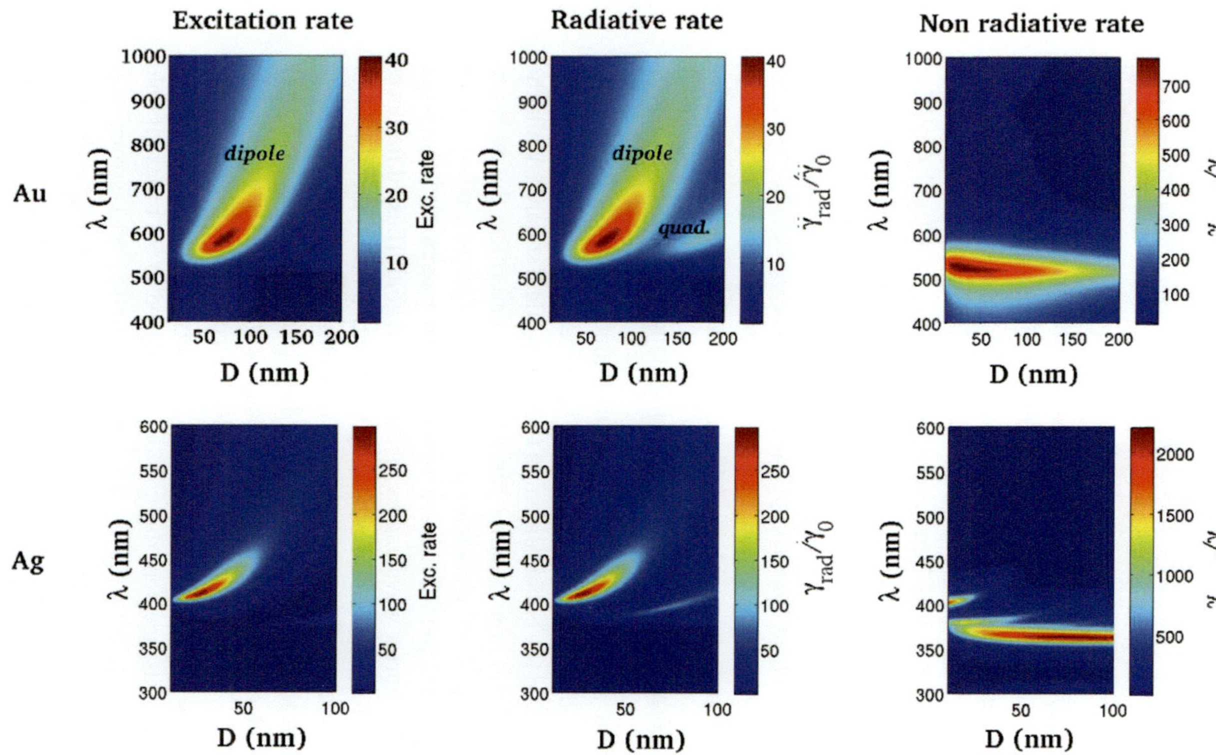

Figure 5.9 Enhancement as a function of wavelength and metal nanoparticle diameter of the excitation rate (left), radiative decay rate (middle) and non-radiative rate (right) for a dipole placed perpendicular 5 nm from the metal nanoparticle surface (gold on the top and silver at the bottom). Reprinted from Ref. [59]. Published by IOP Publishing.

Figure 5.9 illustrates the effect of different metals (gold above and silver below) on the factors required to calculate the plasmon enhancement factor near a spherical metal particle, i.e. [59]

$$EF(\vec{r},\omega_{\text{exc}},\omega_{\text{em}})=\frac{\left|\vec{p}\bullet\vec{E}\left(\vec{r},\omega_{\text{exc}}\right)\right|^{2}\eta(\vec{r},\omega_{\text{em}})}{p_0^2\left|E_0\right|^2\eta_0}. \tag{5.6}$$

In this formula original values are subscripted with '0' and although for generality a change in the dipole moment $\vec{p}$ of the luminescent centre is allowed, this is not normally considered in simulations. The angle between the dipole and the electric field is important and if luminescent centres have no fixed direction then the enhancement factor must be averaged over all possible orientations. Note that the electric field enhancement is associated with excitation/absorption at the angular frequency ω_{exc}, while the quantum yield factor is computed for the emission angular frequency ω_{em}. It is important to note that the emission enhancement depends on the change in quantum yield and not on the change of radiative rates, i.e. not on the so-called Purcell enhancement factor $\frac{\gamma_r}{\gamma_{r0}}$ [60], and that a decrease in the measured lifetime cannot be directly correlated to enhancement [61]. The increase in excitation rate is represented the enhancement of the electric field squared, while the radiative and non-radiative rates together determine the quantum yield. In Fig. 5.9 the scale for all quantities for silver are larger than for gold (i.e. the effects are stronger) and plasmonic effects can occur at shorter wavelengths for silver than gold. These differences are due to the different relative permittivities of the metals. Silver nanoparticles tend to produced more electric field enhancement because they have lower intrinsic loss (damping) than gold [54], although often gold is favoured because it offers less risk of toxicity [62, 63]. Under similar experimental conditions $NaYF_4$:Yb, Er separated from metal nanoparticles by an Al_2O_3 spacer showed about 45 times enhancement due to silver compared to 5.2 times for gold [64]. For both metals the increase in non-radiative rate is strongest at a nearly constant wavelength near the lower range, suggesting that the use of larger metal nanoparticles where the maximum excitation rate and radiative rate enhancements occur at longer wavelengths

(~70 nm for gold and ~30 nm for silver) may be preferable to avoid a strong increase in the non-radiative rate. If luminescent centres are distributed randomly in the space around the metal nanoparticle, the resulting enhancement factor must also be averaged over all possible positions to obtain an average enhancement factor for the system in order to compare with experimental data and this averaging may result in more modest enhancement predictions.

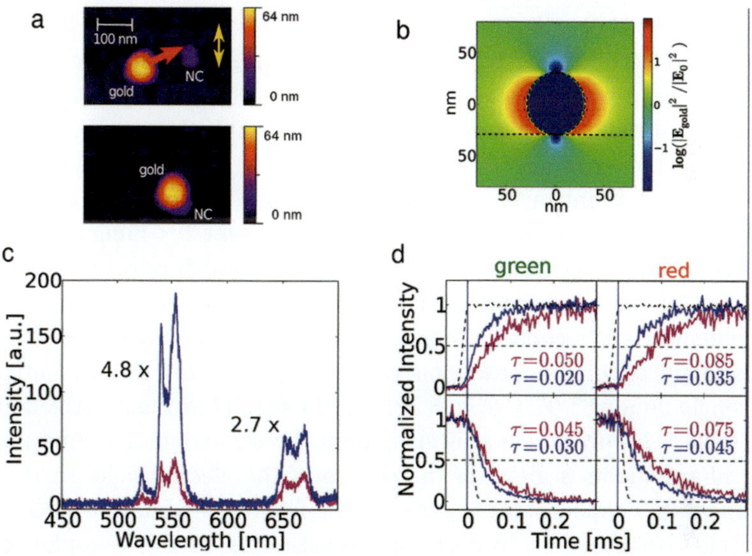

Figure 5.10 (a) AFM images showing a 60 nm gold nanosphere being attached to a $NaYF_4$:Yb^{3+}/Er^{3+} nanoparticle with the help of the AFM tip. The yellow arrow indicates the polarization axis of the excitation light. (b) Simulated intensity enhancement of the excitation light around a 60 nm gold sphere. (c) UC spectra without (violet) and with (blue) the gold nanosphere in close vicinity of the nanoparticle. The emission is strongly enhanced. (d) Rise (upper) and decay times (lower) of the green (left) and red (right) UC emission with the colour code as in part (c). Reprinted with permission from Ref. [65]. Copyright 2010 American Chemical Society.

Figure 5.10 illustrates plasmon enhanced upconversion of a $NaYF_4$:Yb^{3+}/Er^{3+} nanoparticle close to a gold nanoparticle [65]. The experimental setup is shown in Fig. 5.10a, while the simulation in Fig. 5.10b maps the enhancement in the electric field. For luminescence the excitation rate enhancement scales with the square of the electric field (shown), although for two photon upconversion the excitation rate is more sensitive to the electric field enhancement and scales with the fourth power (as will be discussed shortly), allowing higher enhancements but also putting more constraint on the region in which high enhancement occurs. Figure 5.10c shows the upconversion intensity enhancement, which is more pronounced for the green than the red emission and shows that the plasmonic enhancement is not uniform but wavelength dependent. Finally, Fig. 5.10d shows that both the decay lifetime and rise times for this phosphor are decreased, indicative of the enhanced transition rates. However, since the measured lifetime depends on both the radiative and non-radiative rates it cannot be directly linked to luminescence enhancement.

Plasmon enhancement for different physical processes are not the same. Based on Eq. 5.6 the enhancement for luminescence scales with the square of the electric field enhancement as well as with the quantum yield factor. Esteban et al. [66] derived the enhancement factors for other processes, namely Raman scattering and upconversion based on GSA/ESA (Fig. 5.2e) or sequential single donor ET/ETU (Fig. 5.2g) and the results are summarized in Table 5.1. For Raman scattering both the excitation and emission enhancements scale with the square of the electric field, so that the total enhancement scales with the fourth power of the electric field enhancement. In contrast the excitation rate of two-photon upconversion processes (ESA and ETU) depend on the fourth power of the electric field enhancement. However, the plasmon enhancement of the recombination rate has a detrimental effect of allowing faster decay from the first excited state to the ground state before upconversion can occur to the desired second excited state, which introduces a relaxation factor of $\left(\frac{\gamma_{\text{tot}}^{\text{ex}}}{\gamma_{\text{tot},0}^{\text{ex}}}\right)^{-1}$ for the ESA process. The enhancement factor given in Table 5.1 for ESA has been used to model this process near a gold nanorod and predict an enhancement of 160 times for green upconversion at 540 nm [60]. The rate equations for ETU show that it is affected more severely [66] by the relaxation factor, i.e. by $\left(\frac{\gamma_{\text{tot}}^{\text{ex}}}{\gamma_{\text{tot},0}^{\text{ex}}}\right)^{-2}$.

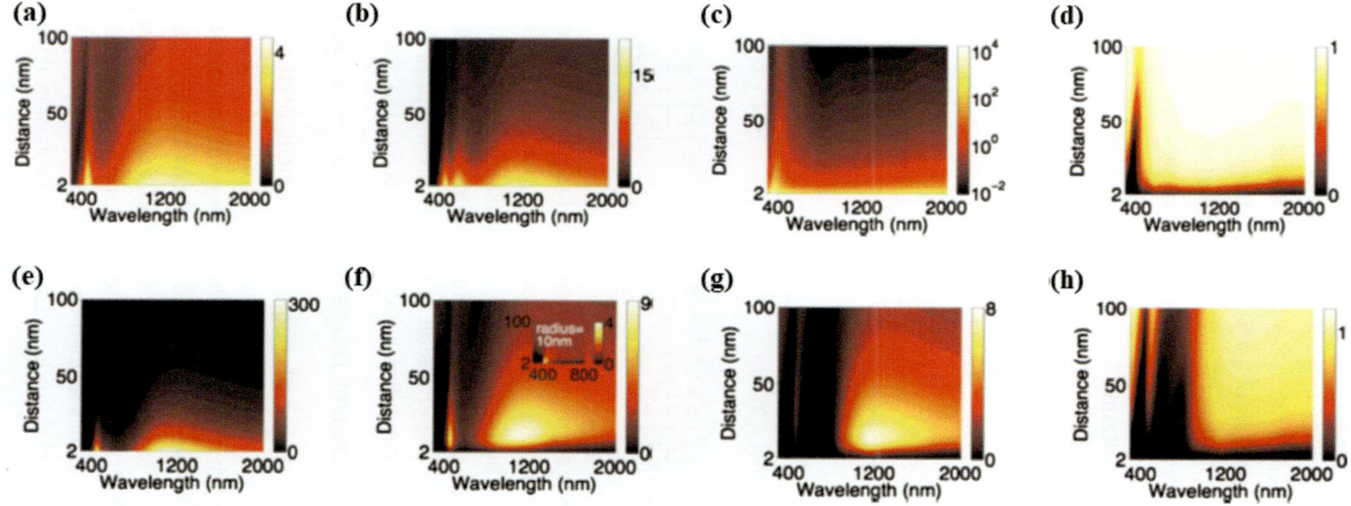

Figure 5.11 Simulations of a quantum emitter oriented radially near a silver nanoparticle in glass matrix. The top row shows (a) the electric field enhancement, (b) the radiative rate enhancement, (c) the non-radiative rate enhancement, and (d) the ratio of quantum yield (log scale, assuming the total non-radiative rate stems from the plasmonic interaction) as a function of wavelength and the separation. The bottom row shows the predicted enhancement in emission intensity for various mechanisms: (e) Raman scattering, (f) Luminescence, (g) ESA and (h) ETU. The inset in (f) shows a similar result for a much smaller 10 nm silver particle. Reprinted from Ref. [66], with the permission of AIP Publishing.

Table 5.1 Excitation, emission and total plasmon enhancement for various physical processes. E represents the electric field, γ_r the radiative recombination rate, η the quantum efficiency, $\gamma_{\text{tot}}^{\text{ex}}$ the total recombination rate of the first excited state of the ion undergoing ESA and $\gamma_{\text{tot}}^{\text{ex*}}$ the effective total recombination rate of the first excited states of the two ions in the system undergoing ETU

Process	Excitation enhancement	Emission enhancement	Total intensity enhancement
Raman scattering	$\left\|\frac{E}{E_0}\right\|^2$	$\frac{\gamma_r}{\gamma_{r0}} \cong \left\|\frac{E}{E_0}\right\|^2$	$\sim \left\|\frac{E}{E_0}\right\|^4$
Luminescence	$\left\|\frac{E}{E_0}\right\|^2$	$\frac{\eta}{\eta_0}$	$\left\|\frac{E}{E_0}\right\|^2 \frac{\eta}{\eta_0}$
ESA upconversion	$\left\|\frac{E}{E_0}\right\|^4 \left(\frac{\gamma_{\text{tot}}^{\text{ex}}}{\gamma_{\text{tot},0}^{\text{ex}}}\right)^{-1}$	$\frac{\eta}{\eta_0}$	$\left\|\frac{E}{E_0}\right\|^4 \left(\frac{\gamma_{\text{tot}}^{\text{ex}}}{\gamma_{\text{tot},0}^{\text{ex}}}\right)^{-1} \frac{\eta}{\eta_0}$
ETU upconversion	$\left\|\frac{E}{E_0}\right\|^4 \left(\frac{\gamma_{\text{tot}}^{\text{ex*}}}{\gamma_{\text{tot},0}^{\text{ex*}}}\right)^{-2}$	$\frac{\eta}{\eta_0}$	$\left\|\frac{E}{E_0}\right\|^4 \left(\frac{\gamma_{\text{tot}}^{\text{ex}}}{\gamma_{\text{tot},0}^{\text{ex}}}\right)^{-2} \frac{\eta}{\eta_0}$

The analysis of plasmon enhancement for ETU upconversion was also done by Park et al. using in particular the Yb^{3+}, Er^{3+} energy level scheme [54]. The results are not identical, but compatible, to those given in Table 5.1 based on three points. Firstly, Park et al. [54] considered that the emission enhancement would make only a modest contribution and dismissed it compared to the absorption (i.e. excitation) enhancement, i.e. they considered $\frac{\eta}{\eta_0} \sim 1$. Secondly, they included factors for plasmon enhancement of the energy transfer processes which had been not been done by Esteban et al. [66] and concluded that a factor for the relative enhancement of the ETU transfer rate between the excited Yb^{3+} donor and Er^{3+} acceptor initially in the $^4I_{11/2}$ state should be included, reporting a value of 1.7 [67]. Thirdly, Park et al. [54] considered saturation effects which would affect the dynamics at high excitation powers, and provided the additional result that under such strong excitation conditions the enhancement of $\sim\left|\frac{E}{E_0}\right|^2$ is predicted, similar as for luminescence enhancement. Wu et al. [61] have noted that upconversion plasmon enhancement is power dependent, similarly to the case of quantum yields, so enhancement should increase as excitation intensity increases, so that it is difficult to compare results in the literature that may have been performed using different power densities. On the other hand, the changing dynamics of upconversion in the weak and strong excitation regime also leads to different models for the enhancement and to decreasing enhancement with increasing excitation intensity when saturation occurs [67].

Based on the work surveyed, it is clear that a strong framework on which to model plasmon enhancement of ESA and ETU upconversion has been established and that this will improve the ability to plan and assess experimental work. However, the question of saturation needs to be carefully addressed during ongoing experimentation and therefore the increased reporting of power dependence of plasmon enhanced upconversion systems would be of immense value. The influence of plasmons on energy transfer enhancement has received much recent interest [68–74] and enhancements of 9 orders of magnitude in the transfer rate have been calculated [75]. This may lead to significant improvements in upconversion efficiency, while work on plasmon enhanced ETU upconversion systems may also provide experimental data in this field. Therefore the upconversion phosphors serve the dual purpose of their direct applications as well as serving as sensors for plasmonic research, much as upconversion luminescence has been used to map electric field 'hotspots' (plasmonic modes) [76] or local density of photonic states [77].

5.5 Conclusion

The mechanisms underlying plasmon enhancement of luminescence, including upconversion, have been outlined in this review. It should be noted that energy transfer from the plasmons to the luminescent centres was not included, and although it is sometimes depicted in energy level diagrams (e.g. Fig. 5 of [52], Fig. 3 of [56]; Fig. 32 of

[78]; Fig. 12 of [79], Fig. 6 of [80], Fig. 7 of [81], Fig. 6 of [82], Fig. 7 of [83]) this should be considered carefully (as done for Fig. 7 in [84]). Energy transfer to the luminescent ions from the metal nanoparticles is unexpected due to the very short lifetime of plasmons (~10^{-14} s) [85], whereas the opposite, namely non-radiative de-excitation of the luminescent ion via interaction with the plasmonic metal, is likely to occur and is manifested through the increase in the non-radiative transition rate. However, plasmonic noble metal nanoparticles are known to exhibit some weak luminescence when excited [86, 87]. Therefore some energy transfer from the metal nanoparticles to the luminescent ions is certainly theoretically possible, but this is unlikely to have a significant influence. The possible importance of such energy transfer might be considered by comparing the intensity of the luminescence from the metal nanoparticles in the absence of the phosphor to the luminescence intensity enhancement from the ions that such energy transfer from the nanoparticles to the luminescent ions is suggested to cause. It is more likely that the factors contributing to the enhancement of the luminescence from the ions are instead their much stronger excitation (absorption) and possibly their increased quantum yield if the radiative transitions are enhanced much more than the non-radiative transitions and the original quantum yield was relatively low. Since the most likely scenario for enhancement is increased excitation, it is generally preferable that the plasmonic resonance frequency is matched to the excitation wavelength, i.e. to the infrared for upconversion phosphors [54] if it is not broad enough or it does not have multiple resonances that allow matching to both the excitation and emission wavelengths. A drawback to this for upconversion phosphors is that it increases the transition rate from the first excited state to the ground state, depopulating the first excited state and thereby reducing the likelihood of electrons reaching the second excited state from which luminescence is desired. This is more severe for the ETU than the ESA mechanism, but the energy transfer for ETU may also potentially be increased substantially to offset this. Although one often considers the plasmonic nanoparticles as affecting the ions, mutual interaction occurs as recently demonstrated by a demonstration of mutually increased luminescence from both a fluorescence nanodiamond and a gold nanoparticle when brought into proximity of one another. It is clear that vast potential exists for plasmonically engineered phosphors.

References

1. F. Auzel. Upconversion and anti-Stokes processes with f and d ions in solids. *Chemical Reviews* 104 (2004) 139–173. doi: 10.1021/cr020357g
2. B.M. van der Ende, L. Aarts and A. Meijerink. Lanthanide ions as spectral converters for solar cells. *Physical Chemistry Chemical Physics* 11 (2009) 11081–11095. doi: 10.1039/B913877C

   A. Khare. A critical review on the efficiency improvement of upconversion assisted solar cells. *Journal of Alloys and Compounds* 821 (2020) 153214 (31 pages). doi: 10.1016/j.jallcom.2019.153214
3. G. Chen, H. Qiu, P.N. Prasad, X. Chen. Upconversion nanoparticles: design, nanochemistry, and applications in theranostics. *Chemical Reviews* 114 (2014) 5161–5214. doi: 10.1021/cr400425h

   J. Yao, C. Huang, C. Liu, M. Yang. Upconversion luminescence nanomaterials: A versatile platform for imaging, sensing, and therapy. *Talanta* 208 (2020) 120157 (23 pages). doi: 10.1016/j.talanta.2019.120157
4. B.P. Kore, A. Kumar, L. Erasmus, R.E. Kroon, J.J. Terblans, S.J. Dhoble, H.C. Swart. Energy transfer mechanisms and optical thermometry of $BaMgF_4$:Yb^{3+}, Er^{3+} phosphor. *Inorganic Chemistry* 57 (2018) 288–299. doi: 10.1021/acs.inorgchem.7b02436
5. A. Kumar, S.P. Tiwari, H.C. Swart, J.C.G.E. da Silva. Infrared interceded YF_3: Er^{3+}/Yb^{3+} upconversion phosphor for crime scene and anti-counterfeiting applications. *Optical Materials* 92 (2019) 347–351. doi: 10.1016/j.optmat.2019.04.050
6. J.E. Moffatt, G. Tsiminis, E. Klantsataya, T.J. de Prinse, D. Ottaway, N.A. Spooner. A practical review of shorter than excitation wavelength light emission processes. *Applied Spectroscopy Reviews* 55:4 (2020) 327–349. doi: 10.1080/05704928.2019.1672712
7. X. Liu, R. Deng, Y. Zhang, Y. Wang, H. Chang, L. Huang, X. Liu. Probing the nature of upconversion nanocrystals: Instrumentation matters. *Chemical Society Reviews* 44 (2015) 1479–1508. doi: 10.1039/c4cs00356j
8. A. Nadort, J. Zhao, E.M. Goldys. Lanthanide upconversion luminescence at the nanoscale: fundamentals and optical properties. *Nanoscale* 8 (2016) 13099–13130. doi: 10.1039/C5NR08477F
9. B.P. Kore, A. Kumar, A. Pandey, R.E. Kroon, J.J. Terblans, S.J. Dhoble, H.C. Swart. Spectroscopic investigation of up-conversion properties

in green emitting $BaMgF_4$:Yb^{3+},Tb^{3+} phosphor. *Inorganic Chemistry* 56 (2017) 4996–5005. doi: 10.1021/acs.inorgchem.7b00044

10. N.J. Greybush, M. Saboktakin X. Ye, C.D. Giovampaola, S.J. Oh, N.E. Berry, N. Engheta, C.B. Murray, C.R. Kagan. Plasmon-enhanced upconversion luminescence in single nanophosphor–nanorod heterodimers formed through template-assisted self-assembly. *ACS Nano* 8 (2014) 9482–9491. doi:10.1021/nn503675a

    F. Auzel. History of upconversion discovery and its evolution. *Journal of Luminescence* 223 (2020) 116900 (7 pages). doi: 10.1016/j.jlumin.2019.116900

11. P. Villanueva-Delgado, K.W. Krämer, R. Valiente, M. de Jong, A. Meijerink. Modeling blue to UV upconversion in β-$NaYF_4$:Tm^{3+}. *Physical Chemistry Chemical Physics* 18 (2016) 27396–27404. doi: 10.1039/c6cp04347j

12. P. Qiu, N. Zhou, H. Chen, C. Zhang, G. Gao, D. Cui. Recent advances in lanthanide-doped upconversion nanomaterials: Synthesis, nanostructures and surface modification. *Nanoscale* 5 (2013) 11512–11525. doi: 10.1039/c3nr03642a

13. J. Yi, E. You, S. Ding, Z. Tian. Unveiling the molecule–plasmon interactions in surface-enhanced infrared absorption spectroscopy. *National Science Review* 7 (2020) 1228–1238. doi: 10.1093/nsr/nwaa054

14. Z. Chen, L. Liang, Y. Wang, Y. Yang. Spatial remote luminescence enhancement by a half-cylindrical Au groove. *Journal of Materials Chemistry C* 4 (2016) 11321–11327. doi: 10.1039/c6tc04074

15. O.L. Malta, P.A. Santa-Cruz, G.F. De Sá, F. Auzel. Fluorescence enhancement induced by the presence of small silver particles in Eu^{3+} doped materials. *Journal of Luminescence* 33 (1985) 261–272. doi: 10.1016/0022-2313(85)90003-1

16. M. Moskovits. Surface-enhanced spectroscopy. *Reviews of Modern Physics* 57 (1985) 783–826. doi: 10.1103/RevModPhys.57.783

17. W.L. Barnes. Fluorescence near interfaces: The role of photonic mode density. *Journal of Modern Optics* 45 (1998) 661–699. doi: 10.1080/09500349808230614

18. J.R. Lakowicz, C.D. Geddes, I. Gryczynski, J. Malicka, Z. Gryczynski, K. Aslan, J Lukomska, E. Matveeva, J. Zhang, R. Badugu, J. Huang. Advances in surface-enhanced fluorescence. *Journal of Fluorescence* 14 (2004) 425–441. doi: 10.1023/B:JOFL.0000031824.48401.5c

19. J.R. Lakowicz. Radiative decay engineering 5: Metal-enhanced fluorescence and plasmon emission. *Analytical Biochemistry* 337 (2005) 171–194. doi: 10.1016/j.ab.2004.11.026

H. Mertens, A. Polman. Plasmon-enhanced erbium luminescence. *Applied Physics Letters* 89 (2006) 211107 (3 pages). doi: 10.1063/1.2392827

20. A.E. Abbass, H.C. Swart, E. Coetsee, R.E. Kroon. Enhanced terbium emission due to plasmonic silver nanoparticles in bismuth silicate. *Journal of the American Ceramic Society* 99 (2016) 876–880. doi: 10.1111/jace.14011
21. R. Knoblauch, C.D. Geddes. Review of advances in metal-enhanced fluorescence. *Reviews in Plasmonics* 2017 (2019) 253–283. doi: 10.1007/978-3-030-18834-4_10
22. J. Li, C. Li, R.F. Aroca. Plasmon-enhanced fluorescence spectroscopy. *Chemical Society Reviews* 46 (2017) 3962–1508. doi: 10.1039/c4cs00356j
23. J. Dong, Z. Zhang, H. Zheng, M. Sun. Recent progress on plasmon-enhanced fluorescence. *Nanophotonics* 4 (2015) 472–490. doi: 10.1515/nanoph-2015-0028
24. S. Kühn , G. Mori, M. Agio, V. Sandoghdar. Modification of single molecule fluorescence close to a nanostructure: radiation pattern, spontaneous emission and quenching. *Molecular Physics* 106 (2008) 893–908. doi: 10.1080/00268970802002510

    D.M. da Silva, L.R.P. Kassab, S.R. Lüthi, C.B. de Araújo, A.S.L. Gomes, M.J.V. Bell. Frequency upconversion in Er^{3+} doped PbO-GeO_2 glasses containing metallic nanoparticles. *Applied Physics Letters* 90 (2007) 081913 (3 pages). doi: 10.1063/1.2679798

    L.R. Kassab, C.B. de Araújo, R.A. Kobayashi, R. de A. Pinto, D.M. da Silva. Influence of silver nanoparticles in the luminescence efficiency of Pr^{3+}-doped tellurite glasses. *Journal of Applied Physics* 102 (2007) 103515 (4 pages). doi: 10.1063/1.2817980
25. I. Gryczynski, J. Malicka, Y. Shen, Z. Gryczynski, J.R. Lakowicz. Multiphoton excitation of fluorescence near metallic particles: enhanced and localized excitation. *Journal of Physical Chemistry B* 106 (2002) 2191–2195. doi: 10.1021/jp013013n
26. G. Liu. Advances in the theoretical understanding of photon upconversion in rare-earth activated nanophosphors. *Chemical Society Reviews* 44 (2015) 1635–1652. doi: 10.1039/c4cs00187g
27. F. Auzel, D. Pecile. Comparison and efficiency of materials for summation of photons assisted by energy transfer. *Journal of Luminescence* 8 (1973) 32–43. doi: 10.1016/0022-2313(73)90033-1
28. J.S. Chivian, W.E. Case, D.D. Eden. The photon avalanche: A new phenomenon in Pr^{3+}-based infrared quantum counters. *Applied Physics Letters* 35 (1979) 124–125. doi: 10.1063/1.91044

29. W. Liu, X. Wang, Q. Zhu, X. Li, X. Sun, J. Li. Upconversion luminescence and favorable temperature sensing performance of eulytite type $Sr_3Y(PO_4)_3$:Yb^{3+}/Ln^{3+} phosphors (Ln = Ho, Er, Tm). *Science and Technology of Advanced Materials* 20 (2019) 949–963. doi: 10.1080/14686996.2019.1659090

30. C.F. Bohren, D.R. Huffman. *Absorption and Scattering of Light by Small Particles*. Wiley, New York (1983). ISBN: 047105772X

31. W.T. Doyle. Optical properties of a suspension of metal spheres. *Physical Review B* 39 (1989) 9852–9858. doi: 10.1103/physrevb.39.9852

    D. Barchiesi, T. Grosges. Fitting the optical constants of gold, silver, chromium, titanium, and aluminum in the visible bandwidth. *Journal of Nanophotonics* 8 (2014) 083097 (16 pages). doi: 10.1117/1.JNP.8.083097

32. A. Derkachova, K. Kolwas, I. Demchenko. Dielectric function for gold in plasmonics applications: size dependence of plasmon resonance frequencies and damping rates for nanospheres. *Plasmonics* 11 (2016) 941–951. doi: 10.1007/s11468-015-0128-7

33. V. Amendola, O.M. Bakr, F. Stellacci. A study of the surface plasmon resonance of silver nanoparticles by the discrete dipole approximation method: Effect of shape, size, structure, and assembly. *Plasmonics* 5 (2010) 85–97. doi: 10.1007/s11468-009-9120-4

34. M. A. van Dijk, A.L. Tchebotareva, M. Orrit, M. Lippitz, S. Berciaud, D. Lasne, L. Cognet, B. Lounis. Absorption and scattering microscopy of single metal nanoparticles. *Physical Chemistry Chemical Physics* 8 (2006) 3486–3495. doi: 10.1039/b606090k

    A. Vial, A. Grimault, D. Macías, D. Barchiesi, M.L. de la Chapelle. Improved analytical fit of gold dispersion: Application to the modeling of extinction spectra with a finite-difference time-domain method. *Physical review B* 71 (2005) 085416 (7 pages). doi: 10.1103/PhysRevB.71.085416

35. R.E. Kroon. The classical oscillator model and dielectric constants extracted from infrared reflectivity measurements. *Infrared Physics & Technology* 51 (2007) 31–43. doi: 10.1016/j.infrared.2007.02.002

    J.M. Nápoles-Duarte, M.A. Chavez-Rojo, M.E. Fuentes-Montero, L.M. Rodríguez-Valdez, R. García-Llamas, J.A. Gaspar-Armenta. Surface plasmon resonances in Drude metal cylinders: radius dependence and quality factor. *Journal of Optics* 17 (2015) 065003 (6 pages). doi: 10.1088/2040-8978/17/6/065003

36. H. Zhao, F. Song, F. Wang, J. Liu, Y. Liu, J. Zhang, G. Chen, S. Liu. The role of silver and gold nanoparticles in enhancing luminescence of

europium complexes. *Journal of Nanoscience and Nanotechnology* 14 (2014) 4357–4362. doi: 10.1166/jnn.2014.8262

H. Ibn El Ahrach, R. Bachelot, A. Vial, G. Lérondel, J. Plain, P. Royer, O. Soppera. Spectral degeneracy breaking of the plasmon resonance of single metal nanoparticles by nanoscale near-field photopolymerization. *Physical Review Letters* 98 (2007) 107402 (4 pages). doi: 10.1103/PhysRevLett.98.107402

37. B.J. Wiley, Y. Chen, J.M. McLellan, Y. Xiong, Z. Li, D. Ginger, Y. Xia. Synthesis and optical properties of silver nanobars and nanorice. *Nano Letters* 7 (2007) 1032–1036. doi: 10.1021/nl070214f

38. K.L. Kelly, E. Coronado, L.L. Zhao, G.C. Schatz. The optical properties of metal nanoparticles: the influence of size, shape, and dielectric environment. *Journal of Physical Chemistry B* 107 (2003) 668–677. doi: 10.1021/jp026731y

39. C. Noguez. Surface plasmons on metal nanoparticles: The influence of shape and physical environment. *Journal of Physical Chemistry C* 111 (2007) 3806–3819. doi: 10.1021/jp066539m

40. X. Lu, M. Rycenga, S.E. Skrabalak, B. Wiley, Y. Xia. Chemical synthesis of novel plasmonic nanoparticles. *Annual Review of Physical Chemistry* 60 (2009) 167–192. doi: 10.1146/annurev.physchem.040808.090434

41. Y. Jin. Engineering plasmonic gold nanostructures and metamaterials for biosensing and nanomedicine. *Advanced Materials* 24 (2012) 5153–5165. doi: 10.1002/adma.201200622

M.R. Dousti, M.R. Sahar, R.J. Amjad, S.K. Ghoshal, A. Khorramnazari, A.D. Basirabad, A. Samavati. Enhanced frequency upconversion in Er^{3+}-doped sodium lead tellurite glass containing silver nanoparticles. *European Physical Journal D* 66 (2012) 237 (6 pages). doi: 10.1140/epjd/e2012-30089-1

42. D.S. da Silva, T.A.A. de Assumpção, L.R.P. Kassab, C.B. de Araújo. Frequency upconversion in Nd^{3+} doped $PbO–GeO_2$ glasses containing silver nanoparticles. *Journal of Alloys and Compounds* 586 (2014) S516–S519. doi: 10.1016/j.jallcom.2012.12.070

43. W. Park, D. Lu, S. Ahn. Plasmon enhancement of luminescence upconversion. *Chemical Society Reviews* 44 (2015) 2940–2962. doi: 10.1039/c5cs00050e

44. A. Das, K. Bae, W. Park. Enhancement of upconversion luminescence using photonic nanostructures. *Nanophotonics* 9 (2020) 1359–1371. doi: 10.1515/nanoph-2020-0159

J. Dong, W. Gao, Q. Han, Y. Wang, J. Qi, X. Yan, M. Sun. Plasmon-enhanced upconversion photoluminescence: Mechanism and application.

Reviews in Physics 4 (2019) 100026 (20 pages). doi: 10.1016/j.revip.2018.100026

S. Kühn, U. Håkanson, L. Rogobete, V, Sandoghdar. Enhancement of single-molecule fluorescence using a gold nanoparticle as an optical nanoantenna. *Physical Review Letters* 97 (2006) 017402 (4 pages). doi: 10.1103/PhysRevLett.97.017402

H. Mertens, A.F. Koenderink, A. Polman. Plasmon-enhanced luminescence near noble-metal nanospheres: Comparison of exact theory and an improved Gersten and Nitzan model. *Physical Review B* 76 (2007) 115123 (12 pages). doi: 10.1103/PhysRevB.76.115123

S. Derom, A. Berthelot, A. Pillonnet, O. Benamara, A.M. Jurdyc, C. Girard, G.C. des Francs. Metal enhanced fluorescence in rare earth doped plasmonic core–shell nanoparticles. *Nanotechnology* 24 (2013) 495704 (8 pages) doi:10.1088/0957-4484/24/49/495704

X. Liu, D.Y. Lei. Simultaneous excitation and emission enhancements in upconversion luminescence using plasmonic double-resonant gold nanorods. *Scientific Reports* 5 (2015) 15235 (12 pages). doi: 10.1038/srep15235

45. D.M. Wu, A. García-Etxarri, A. Salleo, J.A. Dionne. Plasmon-enhanced upconversion. *Journal of Physical Chemistry Letters* 5 (2014) 4020–4031. doi: 10.1021/jz5019042

46. P.V. Asharani , Y. Lianwu, Z. Gong, S. Valiyaveettil. Comparison of the toxicity of silver, gold and platinum nanoparticles in developing zebrafish embryos. *Nanotoxicology* 5 (2010) 43–54. doi: 10.3109/17435390.2010.489207

J. Asselin, M.L. Viger, D. Boudreau. Metal-enhanced fluorescence and FRET in multilayer core-shell nanoparticles. *Advances in Chemistry* 2014 (2014) 812313 (16 pages). doi: 10.1155/2014/812313

47. M. Saboktakin, X. Ye, S.J. Oh, S. Hong, A.T. Fafarman, U.K. Chettiar, N. Engheta, C.B. Murray, C.R. Kagan. Metal-enhanced upconversion luminescence tunable through metal nanoparticle-nanophosphor separation. *ACS Nano* 6 (2012) 8758–8766. doi: 10.1021/nn302466r

48. S. Schietinger, T. Aichele, H. Wang, T. Nann, O. Benson. Plasmon-enhanced upconversion in single $NaYF_4$:Yb^{3+}/Er^{3+} codoped nanocrystals. *Nano Letters* 10 (2010) 134–138. doi: 10.1021/nl903046r

R. Esteban, M. Laroche, J.-J. Greffet. Influence of metallic nanoparticles on upconversion processes. *Journal of Applied Physics* 105 (2009) 033107 (10 pages). doi: 10.1063/1.3075848

49. D. Lu, S.K. Cho, S. Ahn, L. Brun, C.J. Summers, W. Park. Plasmon enhancement mechanism for the upconversion processes in $NaYF_4$:Yb^{3+},Er^{3+} nanoparticles: Maxwell versus Förster. *ACS Nano* 8 (2014) 7780–7792. doi: 10.1021/nn5011254

50. P. Ghenuche, J. de Torres, S. B. Moparthi, V. Grigoriev, J. Wenger. Nanophotonic enhancement of the Förster resonance energy-transfer rate with single nanoapertures. *Nano Letters* 14 (2014) 4707–4714. doi: 10.1021/nl5018145

    F.T. Rabouw, S.A. den Hartog, T. Senden, A. Meijerink. Photonic effects on the Förster resonance energy transfer efficiency. *Nature Communications* 5 (2014) 3610 (6 pages). doi: 10.1038/ncomms4610

    A.N. Poddubny. Collective Förster energy transfer modified by a planar metallic mirror. *Physical Review B* 92 (2015) 155418 (9 pages). doi: 10.1103/PhysRevB.92.155418

51. T.U. Tumkur, J.K. Kitur, C.E. Bonner, A.N. Poddubny, E.E. Narimanov, M.A. Noginov. Control of Förster energy transfer in the vicinity of metallic surfaces and hyperbolic metamaterials. *Faraday Discussions* 178 (2015) 395–412. doi: 10.1039/c4fd00184b

52. J. de Torres, M. Mivelle, S.B. Moparthi, H. Rigneault, N.F. van Hulst, M.F. Garcia-Parajo, E. Margeat, J. Wenger. Plasmonic nanoantennas enable forbidden Förster dipole-dipole energy transfer and enhance the FRET efficiency. *Nano Letters* 16 (2016) 6222–6230. doi: 10.1021/acs.nanolett.6b02470

    T.V. Shahbazyan. Local density of states for nanoplasmonics. *Physical Review Letters* 117 (2016) 207401 (6 pages). doi: 10.1103/PhysRevLett.117.207401

    K. Rustomji, M. Dubois, B. Kuhlmey, C.M. de Sterke, S. Enoch, R. Abdeddaim, J. Wenger. Direct imaging of the energy-transfer enhancement between two dipoles in a photonic cavity. *Physical Review X* 9 (2019) 011041 (11 pages). doi: 10.1103/PhysRevX.9.011041

    J. Ren, T. Wu, B. Yang, X. Zhang. Simultaneously giant enhancement of Förster resonance energy transfer rate and efficiency based on plasmonic excitations. *Physical Review B* 94 (2016) 125416 (8 pages). doi: 10.1103/PhysRevB.94.125416

53. R. Vogelgesang, A. Dmitriev. Real-space imaging of nanoplasmonic resonances. *Analyst* 135 (2010) 1175–1181. doi: 10.1039/c000887g

54. F.T. Rabouw, P.T. Prins, D.J. Norris. Europium-doped $NaYF_4$ nanocrystals as probes for the electric and magnetic local density of optical states

throughout the visible spectral range. *Nano Letters* 16 (2016) 7254–7260. doi: 10.1021/acs.nanolett.6b03730

55. Q. Jianbei, J. Qing, Z. Dacheng, Y. Zhengwen. Recent progress on upconversion luminescence enhancement in rare-earth doped transparent glass-ceramics. *Journal of Rare Earths* 34 (2016) 341–367. doi: 10.1016/S1002-0721(16)60034-0

56. S.P. Tiwari, K. Kumar, V.K. Rai. Plasmonic enhancement in upconversion emission of La_2O_3:Er^{3+}/Yb^{3+} phosphor via introducing silver metal nanoparticles. *Applied Physics B* 121 (2015) 221–228. doi: 10.1007/s00340-015-6223-9

57. Y. Qi, Y. Zhou, L. Wu, F. Yang, S. Peng, S. Zheng, D. Yin. Enhanced upconversion emissions in Ho^{3+}/Yb^{3+} codoped tellurite glasses containing silver NPs. *Journal of Non-Crystalline Solids* 402 (2014) 21–27. doi: 10.1016/j.jnoncrysol.2014.05.014

58. A. Awang, S.K. Ghoshal, M.R. Sahar, R. Arifin, F. Nawaz. Non-spherical gold nanoparticles mediated surface plasmon resonance in Er^{3+} doped zinc–sodium tellurite glasses: Role of heat treatment. *Journal of Luminescence* 149 (2014) 138–143. doi: 10.1016/j.jlumin.2014.01.027

59. M.R. Dousti, M.R. Sahar, S.K. Ghoshal, R.J. Amjad, R. Arifin. Plasmonic enhanced luminescence in Er^{3+}: Ag co-doped tellurite glass. *Journal of Molecular Structure* 1033 (2013) 79–83. doi: 10.1016/j.molstruc.2012.08.022

60. Z. Pan, A. Ueda, R. Aga Jr., A. Burger, R. Mu, S.H. Morgan. Spectroscopic studies of Er^{3+} doped Ge-Ga-S glass containing silver nanoparticles. *Journal of Non-Crystalline Solids* 356 (2010) 1097–1101. doi: 10.1016/j.jnoncrysol.2010.04.014

    T. Som, B. Karmakar. Nanosilver enhanced upconversion fluorescence of erbium ions in Er^{3+}:Ag-antimony glass nanocomposites. *Journal of Applied Physics* 105 (2009) 013102 (8 pages). doi: 10.1063/1.3054918

61. O.L. Malta, M.A. Couto dos Santos. Theoretical analysis of the fluorescence yield of rare earth ions in glasses containing small metallic particles. *Chemical Physics Letters* 174 (1990) 13–18. doi: 10.1016/0009-2614(90)85319-8

62. R. Das, S. Sarkar. Optical properties of silver nano-cubes. *Optical Materials* 48 (2015) 203–208. doi: 10.1016/j.optmat.2015.07.038

63. T.V. Shahbazyan. Theory of plasmon-enhanced metal photoluminescence. *Nano Letters* 13 (2013) 194–198. doi: 10.1021/nl303851z

Chapter 6

Luminescence Properties of Zinc Silicate Nanophosphors for Lighting Display and Dosimetry Applications

P. S. Mbule
Department of Physics, College of Science, Engineering and Technology, University of South Africa, Johannesburg 1710, South Africa
mbuleps1@unisa.ac.za

Zinc silicate (Zn_2SiO_2) has been identified as one of the promising host matrix for many rare-earth elements and transition metal dopant ions with excellent luminescence properties in the blue, green, and red spectral range. Many devices are composed of lighting components made with light emitting diodes (LEDs) which are built up from one or more phosphor materials. The basic principle of those phosphors is associated with luminescence properties of the transition metal and/or rare-earth ions doped in the selected host matrix. In this chapter, transition metal and rare-earth ions doped-Zn_2SiO_2 nanophosphors were investigated. Manganese (Mn^{2+}), dysprosium (Dy^{3+}), and europium (Eu^{3+}) ions were incorporated in Zn_2SiO_2

Luminescent Nanomaterials
Edited by Odireleng Martin Ntwaeaborwa

ISBN 978-981-4968-11-9 (Hardcover), 978-1-003-27738-5 (eBook)
www.jennystanford.com

host matrix via sol-gel method to investigate their luminescence properties and subsequently evaluate their viability for lighting display and dosimetry application. The synthesized nanophosphors have a variation of doping concentration from 0.015 to 0.75 mol% for Mn^{2+} ions while the concentration for Dy^{3+} and Eu^{3+} ions was optimized to 2 mol%. Ultimately, a composition of a Mn^{2+}-Eu^{3+}-Dy^{3+} doped and co-doped Zn_2SiO_4 nanophosphors (Zn_2SiO_4:Mn^{2+}, Eu^{3+}, Dy^{3+}) was prepared. The X-ray diffraction revealed the crystallite phase corresponding to α-Zn_2SiO_4. The crystallinity was improved at an annealing temperature of 1000°C. Particle morphology showed by scanning electron microscopy (SEM) appeared to be non-uniform with spherical shape. Upon UV-excitation by a xenon lamp, photoluminescence (PL) spectra of Mn^{2+}-doped Zn_2SiO_4 revealed a green emission at ~520 nm, corresponding to the $^4T_1(^4G) \rightarrow {}^6A_1(^6S)$ electronic transition. Zn_2SiO_4:Eu^{3+} showed a prominent emission band at 615 nm, corresponding to $^5D_0 \rightarrow {}^7F_2$ transition of Eu^{3+} ions and Zn_2SiO_4:Dy^{3+} showed two prominent emission bands located at 481 and 578 nm corresponding respectively to the $^4F_{9/2} \rightarrow {}^6H_{15/2}$ and $^4F_{9/2} \rightarrow {}^6H_{13/2}$ transitions of Dy^{3+}. The CIE calculated from the emission spectra showed coordinates which varied from green, orange to red through white with the change of dopant in Zn_2SiO_4 host lattice and the thermoluminescence (TL) glow curves showed adequate glow for dosimetry, in terms of maxima location and intensity.

6.1 Introduction

The efficient synthesis of metal oxide based nanophosphors is needed for their luminecsence in advanced photonic applications [1]. Zinc silicate matrix is one of the promising materials to host rare-earth ions and transition metal dopants for excellent luminescence properties. Depending on the dopant, it has the ability to emit light in the blue, green, and red spectral range [2]. For example, zinc silicate doped with manganese ions (Zn_2SiO_4:Mn^{2+}) is a green phosphor used particularly in field-emission displays [3]. Similarly, Zn_2SiO_4:Eu^{3+} and Zn_2SiO_4:Dy^{3+} have strong red and white emissions, respectively. Therefore, co-doping these rare-earth ions into Zn_2SiO_4 improves the phosphor's luminescence and also give desirable features for lighting displays and dosimetry applications [4]. In addition, singly doping Zn_2SiO_4 with Dy^{3+} ions gives white light emission by balancing the ratio of yellow to blue emission bands [5, 6]. Thus making this phosphor a potential material for new generation of light source due to its long lifetime, low power consumption, energy efficiency and environmental friendliness. In numerous reports, emission spectra ranges from bands corresponding to wavelengths of the blue, yellow, and a weak red, demonstrating $^4F_{9/2} \rightarrow {}^6H_{15/2}$, $^4F_{9/2} \rightarrow {}^6H_{13/2}$ and $^4F_{9/2} \rightarrow {}^6H_{11/2}$ transitions of Dy^{3+} ions, respectively. It is noteworthy to mention that studies for a white light emission are carried through suitable combinations of emission peak positions, photoluminescence (PL) intensities and, controlled chemical compositions in the host matrix. Shasmal et al. [7], demonstrated Dy^{3+}-doped chloroborosilicate host matrix showing a white light emission. They observed a visible luminescence when the phosphor was excited with 447 nm wavelength source [7] and the emitted light was from the sample compositions of $35.7SiO_2$–$25.5B_2O_3$–$17BaO$–$3.4K_2O$–$3.4Al_2O_3$–$15BaCl_2$ (in mol%). They further showed that the sample with 0.5 wt% of Dy_2O_3 was found to be emitting white light suggesting such sample to be extensively useful for white light LEDs and/or various photonic applications.

Thermoluminescence (TL) spectroscopy is an extraordinary technique for investigating electron trap centers in materials evaluated for dosimetry applications. To evaluate the characteristics that govern the functionality of dosimeters, a few parameters are considered. These include dose response, dose rate dependence, repeatability, and signal fading. Furthermore, it is of interest to investigate electron trap centers in the material, thus getting information about activation energies/trap depths. For instance, when phosphors are subjected to thermal irradiation, electron and holes traps are created and they need external stimulation to get de-trapped [8]. De-trapped electron–hole pairs cause thermally stimulated luminescence by recombination process, and therefore thermoluminescence spectroscopy is used to detect the recombination emissions from surface states. In addition, to de-trap electron–hole pairs, amounts of activation energies needs to be determined and this is processed through the analysis of the glow curve provided that the glow peak position, shape, and symmetry are known [8]. In the literature, synthesis methods for nanophosphors are widely reported, particularly the sol-gel method proved to have advantages over conventional methods due to its particle size, morphology and processing conditions control. Additionally, low temperature processing by sol-gel produces highly pure nanophosphors [9]. In this work photoluminescence and thermoluminescence properties of $Zn_2SiO_4:Mn^{2+}$, Eu^{3+}, Dy^{3+} nanophosphors are demonstrated for possible application in lighting display and dosimetry. In addition, the energy transfer mechanism in co-doped Zn_2SiO_4 nanophosphor is proposed and explained.

6.2 Experimental Section

$Zn_{2-x}SiO_4:Mn_x$ (x = 0.015–0.75 mol%), $Zn_{2-x}SiO_4:Eu_x$ (x = 2 mol%), $Zn_{2-x}SiO_4:Dy_x$ (x = 2 mol%) and $Zn_2SiO_4:Mn^{2+}$, Eu^{3+}, Dy^{3+} nanophosphors were prepared using the sol-gel method. Fixed amount of 10 ml tetraethylorthosilicate (TEOS) (Aldrich, 98%), 3 g zinc nitrate hexahydrate (Aldrich, 99%) and 0.002 g manganese (ii) nitrate pentahydrate (Aldrich, 99%), 0.087 g europium (iii) nitrate pentahydrate, and 0.071 g dysprosium (iii) nitrate hydrate were used as starting materials while ethanol was used as a solvent. TEOS was mixed with ethanol and water (1:1:0.5, volume ratio), the mixture was stirred for 10–15 min. Zinc nitrate hexahydrate was dissolved with 10 ml ethanol and stirred for 10–15 min. Then with a constant stirring, TEOS solution was slowly added to the zinc solution. The solution was stirred for about 30 min until a sol was formed. Furthermore, 0.5 ml of HCl was added and the solution was stirred until a gel formed. Finally, $Zn_2SiO_4:Mn^{2+}$, $Zn_2SiO_4:Eu^{3+}$, $Zn_2SiO_4:Dy^{3+}$, and $Zn_2SiO_4:Mn^{2+}$, Eu^{3+}, Dy^{3+} nanophosphor powders were obtained after drying in air. The thermal treatment was carried out in a preheated muffle furnace at 1000°C for 2 h.

XRD measurements were carried out to evaluate crystalline phase and for these measurements Rigaku smartLab X-ray diffractometer (Cu Kα, 45 kV and 200 mA) was used. The morphology was studied using JEOL JSM-7800F thermal field emission SEM coupled with Oxford Aztec 350 X-Max80 EDS, that was used to analyse the chemical composition of the phosphors under typical vacuum of 9.634×10^{-5} Pa. Photoluminescence (PL) measurements were recorded at room temperature using a 450W Xenon lamp as an excitation source and

HR320 detector. The PL spectra were recorded using Horiba Jobin-Yvon fluorolog-3. RISO TL/OSL reader supplied by DTU Nutech was used to collect thermoluminescence glow curves at room temperature.

6.3 Results and Discussion

6.3.1 XRD Analysis

The XRD pattern of Zn_2SiO_4:Mn^{2+}, Eu^{3+}, Dy^{3+} nanophosphor is shown in Fig. 6.1. The as-prepared nanophosphors showed an amorphous phase, however, upon annealing at 1000°C crystallized α-Zn_2SiO_4 phase was observed. The diffraction peaks matched well with the standard pattern JCPDS no. 9014832 of zinc orthosilicate (willemite). The concentration of Mn^{2+}, Eu^{3+}, and Dy^{3+} dopants had no significant effect on the XRD pattern and the crystallinity. The XRD results are in correlation with the results of Babu et al. [10] were they indicated that upon doping Zn_2SiO_4 with Mn^{2+} concentration variation, no impurity peaks were detected. Upon thermal annealing at 1000°C, the nanophosphors had an average crystallite size of $\sim$30 – 35 ± 0.5 nm, as estimated using Scherer's equation. It was previously demonstrated [11] that Zn_2SiO_4 nanophosphors obtained by sol-gel method are not crystalline but amorphous-like precursors that become crystalline when calcined at high temperatures (up to 1100°C) to form α-phase Zn_2SiO_4. This transition was shown by annealing from temperatures of 600°C to 1100°C [11].

6.3.2 SEM and EDS Analysis

The SEM images and EDS spectrum of Zn_2SiO_4:Mn^{2+}, Eu^{3+}, Dy^{3+} nanophosphor are shown in Figs. 6.2a and b. As shown in Fig. 6.2a, the particle morphology of the nanophosphors appeared to be non-uniform and spherical in shape. It is reported that the use of spherical nanophoshors increases the screen brightness and improves the resolution because of its lower scattering of evolved light, better slurry properties, smoother light-intensity distribution and higher packing density than irregularly shaped nanophosphors. The spherical shaped nanoparticles can also optimize the optical and geometrical structure of a phosphor layer [12]. As it appears in Fig. 6.2b, EDS spectrum indicates that all elements making a composition of a Zn_2SiO_4:Mn^{2+}, Eu^{3+}, Dy^{3+} nanophosphor are present. In addition, it can be inferred in the spectrum that oxygen (O), silicon (Si), and zinc (Zn) had the relatively higher number of counts. The TEM image shown in Fig. 6.2c confirms the spherical particles within the nano-range.

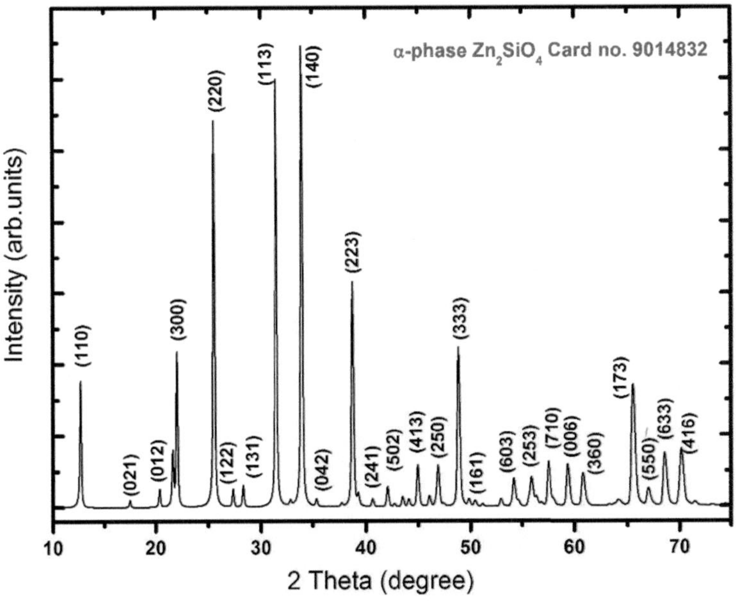

Figure 6.1 XRD pattern of Zn_2SiO_4:Mn^{2+}, Eu^{3+}, Dy^{3+} calcined at 1000°C corresponding to α-phase Zn_2SiO_4.

6.3.3 Photoluminescence Analysis

Figure 6.3a–e shows the photoluminescence spectra of single doped and co-doped-Zn_2SiO_4 nanophosphors. In Fig. 6.3a, the PL excitation and emission spectra of $Zn_{2-x}SiO_4$:Mn_x (x = 0.015 – 0.75 mol%) are shown. The optimized PL emission from Zn_2SiO_4:Mn^{2+} nanophosphor was from the doping concentration of 0.09 mol% and the excitation spectra peaking at ~240–243 nm are assigned to the excitation of the Mn^{2+} charge transfer transition from the ground state $^6A_1(S)$ to the

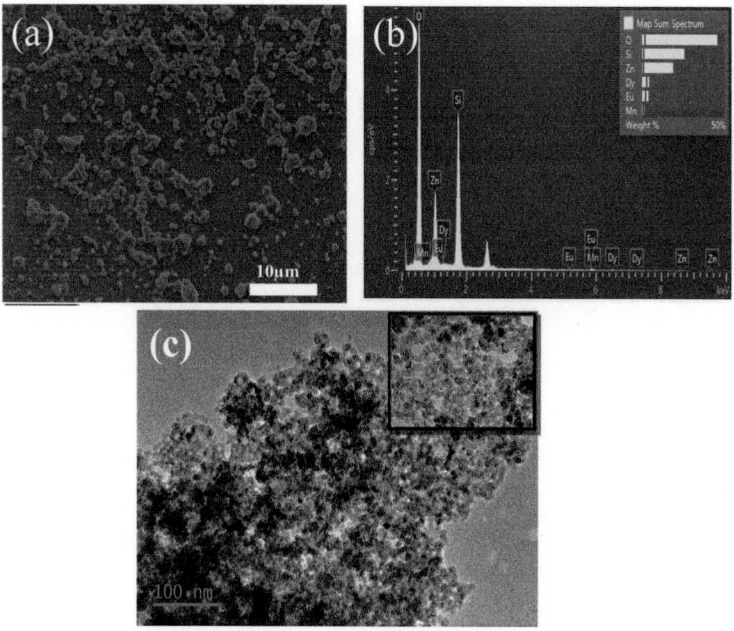

Figure 6.2 (a) SEM image, (b) EDS spectrum, and (c) TEM image of Zn_2SiO_4:Mn^{2+}, Eu^{3+}, Dy^{3+} nanophosphor.

conduction band [13]. The broad-band PL emission spectra extends over a range of wavelengths from 500–700 nm with a peak maximum at ~520 nm. In the rhombohedral-crystal-structured Zn_2SiO_4 host, the Mn^{2+} ion is incorporated substitutionally at the Zn^{2+} lattice site with a tetrahedral site symmetry forming the Mn^{2+} luminescent center. The 3d orbital electrons in the Mn^{2+} ion relax radiatively from the $^4T_1(^4G)$ excited state to the $^6A_1(^6S)$ ground state, leading to the green PL emission [14]. The emission spectrum of Zn_2SiO_4:Eu^{3+} (2 mol%) nanophosphor is shown in Fig. 6.3b. The insert shows the excitation spectrum with peaks located at ~307, 360, 379, and 394 nm attributed to the f-f transitions of Eu^{3+}. As shown in the figure, the emission spectrum exhibits five typical emission peaks in the range of 550–750 nm, which result from $^5D_0 \rightarrow {}^7F_J$ (J = 0, 1, 2, 3, and 4) transitions of Eu^{3+}. The emission peak at 615 nm is ascribed to the electric dipole transition from $^5D_0 \rightarrow {}^7F_2$, while the emission at ~590 nm is assigned to the magnetic dipole transition from the

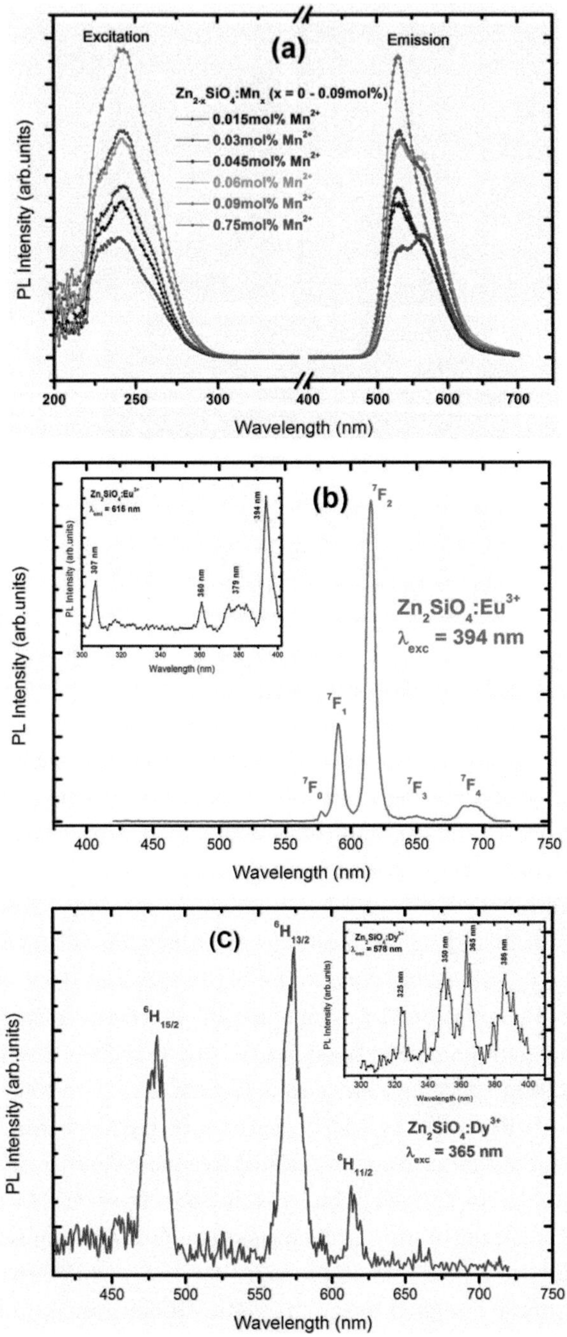
(a)
Excitation
Emission
Zn2-xSiO4:Mnx (x = 0 - 0.09mol%)
0.015mol% Mn2+
0.03mol% Mn2+
0.045mol% Mn2+
0.06mol% Mn2+
0.09mol% Mn2+
0.75mol% Mn2+
PL Intensity (arb.units)
Wavelength (nm)
200 250 300 400 500 600 700
(b)
7F2
7F1
7F0
7F3
7F4
Zn2SiO4:Eu3+
λexc = 394 nm
PL Intensity (arb.units)
Wavelength (nm)
400 450 500 550 600 650 700 750
307 nm
360 nm
379 nm
394 nm
(C)
6H13/2
6H15/2
6H11/2
Zn2SiO4:Dy3+
λexc = 365 nm
325 nm
350 nm
365 nm
386 nm
PL Intensity (arb.units)
Wavelength (nm)
450 500 550 600 650 700 750

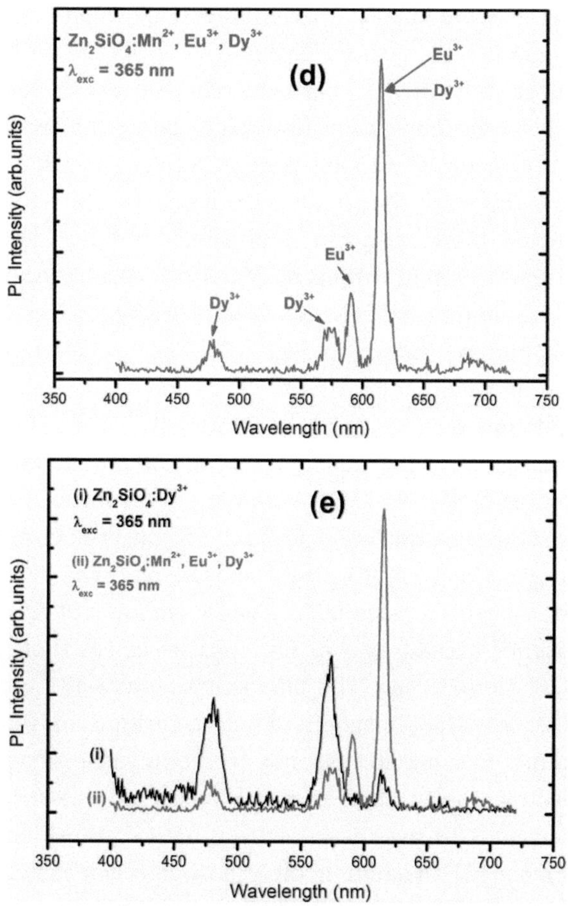

Figure 6.3 PL Excitation and emission spectra of (a) Zn_2SiO_4:Mn^{2+}, (b) Zn_2SiO_4:Eu^{3+}, (c) Zn_2SiO_4:Dy^{3+}, (d) Zn_2SiO_4:Mn^{2+}, Eu^{3+}, Dy^{3+} nanophosphors, and (e) comparison emission intensity of Dy^{3+} and Eu^{3+} ions.

$^5D_0 \rightarrow {}^7F_1$ of Eu^{3+} [15]. Moreover, there are weak emission peaks located at ~576, 650, and 690 nm, which are assigned to $^5D_0 \rightarrow {}^7F_0$, $^5D_0 \rightarrow {}^7F_3$ and $^5D_0 \rightarrow {}^7F_4$ transitions of Eu^{3+}, respectively. According to the parity selection rule, when Eu^{3+} ions are located at the site with an inversion symmetric center, the $^5D_0 \rightarrow {}^7F_1$ magnetic dipole transition is permitted, which result in orange-red emission at ~590 nm. But if located at the site without an inversion symmetric center, because the opposite parity 5d-configuration is mixed into $4f^n$ configuration, the parity selection rule is able to be lifted, and f-f forbidden transition is partially released, the hypersensitive $^5D_0 \rightarrow {}^7F_2$ electric dipole transition will be permitted, which results in the red emission at ~615 nm. For the Zn_2SiO_4:Eu^{3+} nanophosphor prepared in this work, the prominent emission peak at ~615 nm will be dominated, therefore it can be presumed that Eu^{3+} ions mainly occupies non-inversion symmetric center in a host lattice [16].

The PL excitation and emission spectra of Zn_2SiO_4:Dy^{3+} (2 mol%) nanophosphor are shown in Fig. 6.3c. The excitation spectrum was recorded when monitoring the emission at the wavelength of 578 nm. It consist of sharp peaks located at ~325, 350, 365, and 386 nm assigned to the transitions from the ground $^6H_{15/2}$ state to the excited $^4L_{19/2}$, $^6P_{7/2}$, $^6P_{5/2}$, and $^4I_{13/2}$, respectively [17]. The intensity of the main excitation peak at ~365 nm is much prominent, which suggests that Zn_2SiO_4:Dy^{3+} nanophosphor can be efficiently excited by near-UV or blue LEDs. The PL emission of Zn_2SiO_4:Dy^{3+} nanophosphor excited by 365 nm consists of emission peaks at ~481, 578, and 613 nm. The blue emission at ~481 correspond to the $^4F_{9/2} \rightarrow {}^6H_{15/2}$ magnetic dipole transition and the yellow emission at ~578 nm corresponds to the hypersensitive $^4F_{9/2} \rightarrow {}^6H_{13/2}$ electric dipole transition and the weak red emission at ~613 nm corresponds to the $^4F_{9/2} \rightarrow {}^6H_{11/2}$ optical transition state of the Dy^{3+} ions [18]. In addition, the yellow emission corresponds to the hypersensitive with $\Delta J = 2$ which is strongly influenced by the crystal field environment and the intensity of this peak can increase or decrease depending on the local environment. The blue emission is related to the magnetic dipole transition and is insensitive to the site symmetry and they are parity allowed [19]. PL emission of Zn_2SiO_4:Mn^{2+}, Eu^{3+}, Dy^{3+} nanophosphor is shown in Fig. 6.3d. The emission was recorded under the excitation of 365 nm wavelength. We observed the emission peaks at ~475, 573, 590, and a prominent peak at ~615 nm. These emission peaks are due to transitions of Dy^{3+} and/or Eu^{3+}. The emission from Mn^{2+} ions was not observed whereas the emission intensity of Dy^{3+} ions at ~481 nm and ~578 nm was considerably decreased as compared to the single-doped sample. This suggests that there was possible energy transfer in the Zn_2SiO_4:Mn^{2+}, Eu^{3+}, Dy^{3+} nanophosphor from the Dy^{3+} to the Eu^{3+} ions. It is also important to note that there exists an overlapping radiative transitions, $^5D_0 \rightarrow {}^7F_2$ of Eu^{3+} and $^4F_{9/2} \rightarrow {}^6H_{11/2}$ of Dy^{3+} and this may have caused the quenching of emission intensity at 481 nm and 578 nm resulting the relative increase of the red emission (see Fig. 6.3e). In addition, it is important to note that the $^4F_{9/2} \rightarrow {}^6H_{11/2}$ transition (613 nm) of Dy^{3+} ions has fully overlapped with the $^5D_0 \rightarrow {}^7F_2$ transition (615 nm) of Eu^{3+} ions. Therefore, this may also suggests that the intense emission at ~615 nm was enforced by Dy^{3+} and we attribute the energy transfer from Dy^{3+} to Eu^{3+}. The emission intensity Dy^{3+} ion is considerably reduced as seen in Fig. 6.3e.

In general, the color of any phosphor material is represented by means of color coordinates. The Commission International have L'Eclairage (CIE) chromaticity coordinates for Zn_2SiO_4:Mn^{2+}, Zn_2SiO_4:Eu^{3+}, Zn_2SiO_4:Dy^{3+}, and Zn_2SiO_4:Mn^{2+}, Eu^{3+}, Dy^{3+} nanophosphors were calculated by using the color calculator software [20] under the excitation of 243, 394, 365, and 365 nm, respectively. The CIE color coordinates of all the samples are presented in Fig. 6.4. The emission spectrum of nanophosphors was converted to the CIE chromaticity using the photoluminescence data and interactive CIE software as shown in Fig. 6.4. Every color of the nanophosphors can be identified by (*x*, *y*) coordinates that are disposed inside the 'chromatic shoe' representing the saturated colors. The CIE chromaticity coordinates (0.32, 0.48) of Zn_2SiO_4:Mn^{2+} nanophosphor falls in the green region, Zn_2SiO_4:Eu^{3+} nanophosphor (0.59, 0.34) in the red region of the spectrum, Zn_2SiO_4:Dy^{3+} nanophosphor (0.33, 0.37) showed white color and Zn_2SiO_4:Mn^{2+}, Eu^{3+}, Dy^{3+} nanophosphor (0.55, 0.36) showed orange-red color. The CIE coordinated of Zn_2SiO_4:Dy^{3+} nanophosphor showed coordinates which are close to the standard coordinates (0.3333, 0.3333) for a white photoluminescence [21]. Therefore, this observation correlates with the PL results as expected.

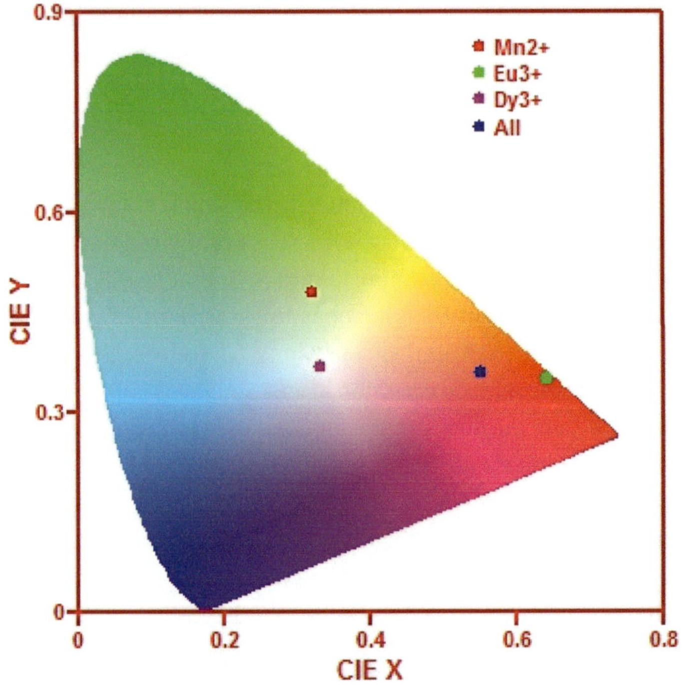

Figure 6.4 The CIE chromaticity diagram for Zn_2SiO_4:Mn^{2+}, Zn_2SiO_4:Eu^{3+}, Zn_2SiO_4:Dy^{3+}, and Zn_2SiO_4:Mn^{2+}, Eu^{3+}, Dy^{3+} nanophosphors.

The energy level diagram of Mn^{2+} ion doped Zn_2SiO_4 host lattice with all the possible excitation and emission transitions is shown in Fig. 6.5. When Zn_2SiO_4:Mn^{2+} nanophosphor is excited with various excitation wavelength in the range of 240–243 nm, the Mn^{2+} ions are excited from the ground state $^6A_1(S)$ to these excited states $^4T_1(P)$, $^4E(D)$, $^4T_2(D)$, $^4A_1$, $^4E(G)$, $^4T_2(G)$, $^4T_1(G)$. Due to the small energy difference between the energy level higher than the$^4T_1(G)$ energy state, the initial electron population at higher energy levels relaxes to the $^4T_1(G)$ energy state through non-radiative relaxations (indicated by red arrows) processes. Therefore, from the $^4T_1(G)$ energy level of Mn^{2+}, the radiative transitions give rise to two intense emission peaks located at ~520 and ~550 nm.

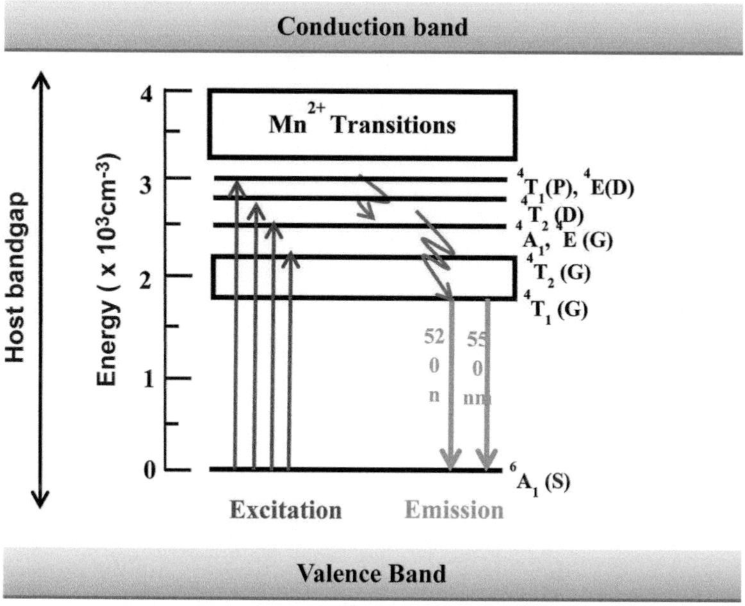

Figure 6.5 The excitation and emission transitions of Mn^{2+} ions.

The pathways for excitation, energy transfer and subsequent emission are presented in schematic diagram of Fig. 6.6 for Zn_2SiO_4:Mn^{2+}, Eu^{3+}, Dy^{3+} nanophosphor. According to our observation, the process of energy transfer in the Zn_2SiO_4:Mn^{2+}, Eu^{3+}, Dy^{3+} system may be from Dy^{3+} ions to Eu^{3+} ions or from Eu^{3+} ions to Dy^{3+} ions. However, we propose that the energy is transferred from Dy^{3+} ions energy states to Eu^{3+} energy states. Upon UV excitation, the absorbed photon energy by Dy^{3+} ions is likely transferred to excited states of Eu^{3+} ions ($^5D_3$ and $^5D_0$) through resonant energy transfer from Dy^{3+} non-radiative transitions. The possibility of resonant absorption of 365 nm photons directly by the Eu^{3+} ion itself cannot be ruled out as we have observed the PL emission band at ~615 nm that could be emission overlap of Eu^{3+} and Dy^{3+}. Table 6.1 summarizes examples of the representative single-phased white emitting phosphors with multiple energy transfer and the comparison is made with the Zn_2SiO_4:Mn^{2+}, Eu^{3+}, Dy^{3+} nanophosphor.

Table 6.1 Comparison with single-phased white-emitting phosphors with multiple energy transfer

Phosphor	$\mathbf{Zn_2SiO_4{:}Mn^{2+}, Eu^{3+}, Dy^{3+}}$	$\mathbf{Ca_3Y(GaO)_3(BO_3)_4{:} Ce^{3+}, Tb^{3+}, Mn^{2+}}$	$\mathbf{Mg_2Y_8(SiO_4)_6O_2{:} Ce^{3+}, Tb^{3+}, Mn^{2+}}$	$\mathbf{NaCaBO_3{:} Ce^{3+}, Tb^{3+}, Mn^{2+}}$	$\mathbf{BaMg_2Al_6Si_9O_{30}{:} Eu^{2+}, Tb^{3+}, Mn^{2+}}$
Excitation Wavelength (nm)	365 nm (UV)	365 nm (UV)	324 nm (UV)	347 nm (UV)	330 (UV)
Emission Wavelength (nm)	475 (G), 573(G), 590 (O), 615 (R)	409(V-B), 544 (G), 589 (O)	456 (B), 544 (G), 613 (R)	427 (B), 545 (G), 600 (R)	450 (B), 542 (G), 610 (R)
Energy transfer mechanism	Radiative transfer (Dy^{3+}–Eu^{3+})	Dipole–quadrupole (Ce^{3+}–Mn^{2+}) Dipole–quadrupole (Ce^{3+}–Tb^{3+})	Dipole–quadrupole (Ce^{3+}–Mn^{2+}) Dipole–quadrupole (Ce^{3+}–Tb^{3+})	Dipole–quadrupole (Ce^{3+}–Mn^{2+}) Dipole–quadrupole (Ce^{3+}–Tb^{3+})	Dipole–quadrupole (Eu^{2+}–Mn^{2+}) Dipole–quadrupole (Eu^{2+}–Tb^{3+})
References	28	29	30	31	32

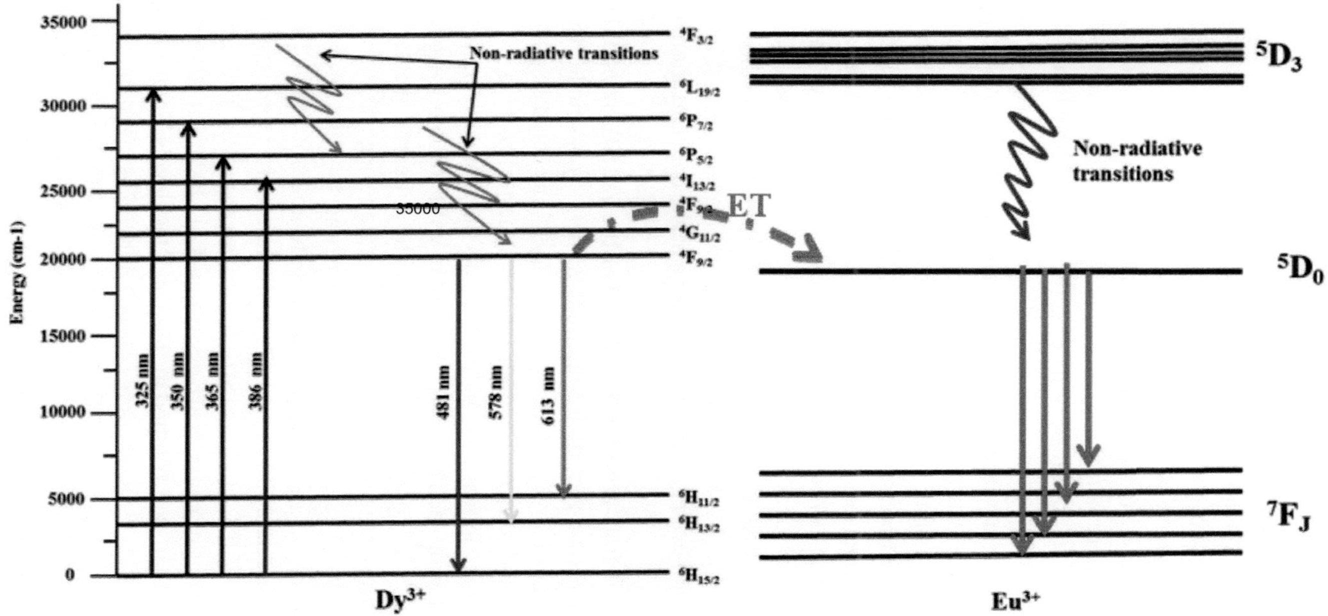

Figure 6.6 The proposed $Dy^{3+} \rightarrow Eu^{3+}$ energy transfer mechanism in $Zn_2SiO_4:Mn^{2+}$, Eu^{3+}, Dy^{3+} nanophosphor.

6.3.4 Thermoluminescence Analysis

Figure 6.7 shows the TL glow curves of undoped and Dy^{3+} doped Zn_2SiO_4 nanophosphors. In Fig. 6.7a the TL glow curves for different concentrations of Dy^{3+} doped Zn_2SiO_4 nanophosphors after exposure to a beta dose are observed. The glow curves from all samples have adequate glow for dosimetry, in terms of maxima location and intensity. There is a significant peak position shift (from ~120°C) when different concentrations of Dy^{3+} ions are introduced and the 2 mol% concentration showed a relatively high intensity at ~130°C. The peak shift may be due to alterations in the population of trapping/luminescence centers as a result of addition of Dy^{3+} impurities [22] and the significant increase of intensity may possibly be due to charge compensation effect caused by introducing Dy^{3+} ions. It has been reported [23] that in bulk phosphor materials, a strong luminescence can be obtained with activators such as Dy^{3+} ions, even in small quantities, owing to the facilitated energy transfer from the host to the doped rare-earth ions [23]. Therefore, we can suggest that the allocations of traps observed after irradiation of beta-rays can be influenced greatly by the change in the concentration of Dy^{3+} ion doping.

As Dy^{3+} ions are introduced within the Zn_2SiO_4 crystal lattice, it is well known that their ionic radius (0.91 Å) is smaller than the ionic radius of Si^{2-} (1.18 Å). Therefore, Dy^{3+} ions could be positioned in interstitial sites, promoting the formation of oxygen vacancies, causing the appearance of new peaks in comparison with the undoped sample. This effect is similar with the one recorded by Gupta et al. [22] and Molefe et al. [24]. The appearance of new glow peaks towards the higher temperature region suggest the formation of different types of trapping levels such as shallow, intermediate, and deep traps. The TL intensity of lower temperature peak (~120°C, for undoped sample) is superior to that at higher temperatures of ~130°C, ~230°C, and ~415°C, for doped samples. This indicates that the sum of charge carriers trapped at shallower defect levels are more than that trapped at deeper defect levels. Figure 6.7b and c illustrate the TL glow curves of Dy^{3+} doped Zn_2SiO_4 nanophosphor (2 mol%), exposed to different doses of beta irradiation in the range of 100–700 Gy and heating rates in the range of 2–5°C·s^{-1}, respectively. In Fig. 6.7b, the sample with the most prominent glow

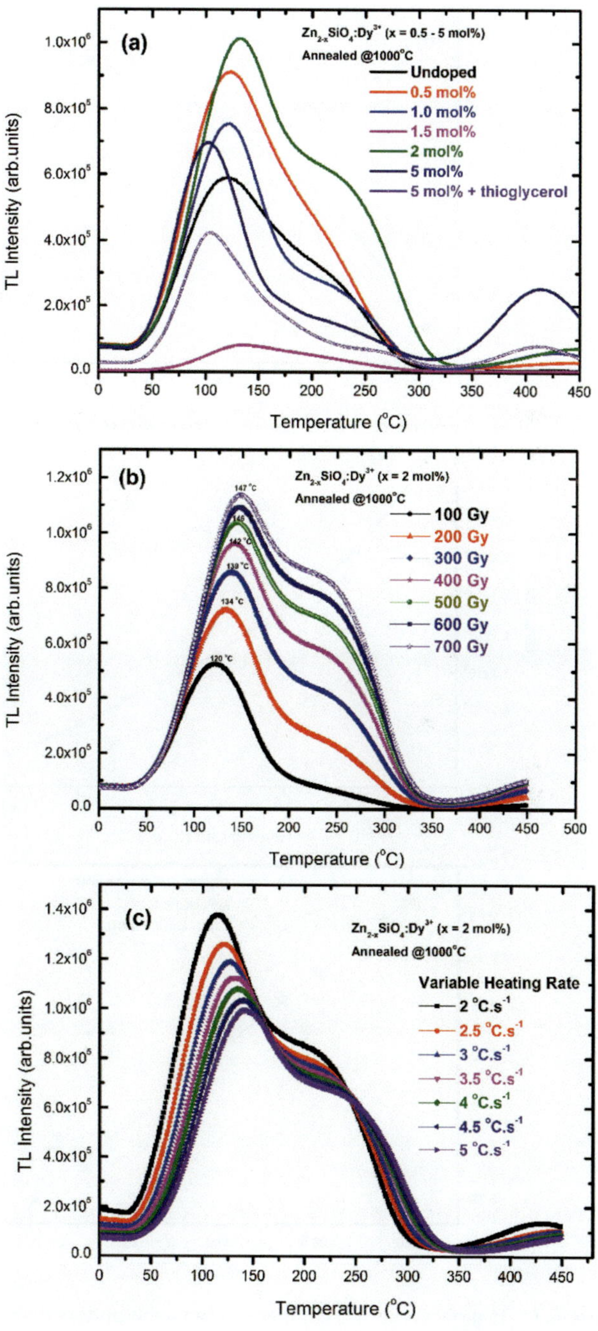
(a)
Zn$_{2-x}$SiO$_4$:Dy^{3+} (x = 0.5 - 5 mol%)
Annealed @1000°C
Undoped
0.5 mol%
1.0 mol%
1.5 mol%
2 mol%
5 mol%
5 mol% + thioglycerol
TL Intensity (arb.units)
Temperature (°C)
(b)
Zn$_{2-x}$SiO$_4$:Dy^{3+} (x = 2 mol%)
Annealed @1000°C
147 °C
142 °C
134 °C
120 °C
100 Gy
200 Gy
300 Gy
400 Gy
500 Gy
600 Gy
700 Gy
TL Intensity (arb.units)
Temperature (°C)
(c)
Zn$_{2-x}$SiO$_4$:Dy^{3+} (x = 2 mol%)
Annealed @1000°C
Variable Heating Rate
2 °C.s^{-1}
2.5 °C.s^{-1}
3 °C.s^{-1}
3.5 °C.s^{-1}
4 °C.s^{-1}
4.5 °C.s^{-1}
5 °C.s^{-1}
TL Intensity (arb.units)
Temperature (°C)

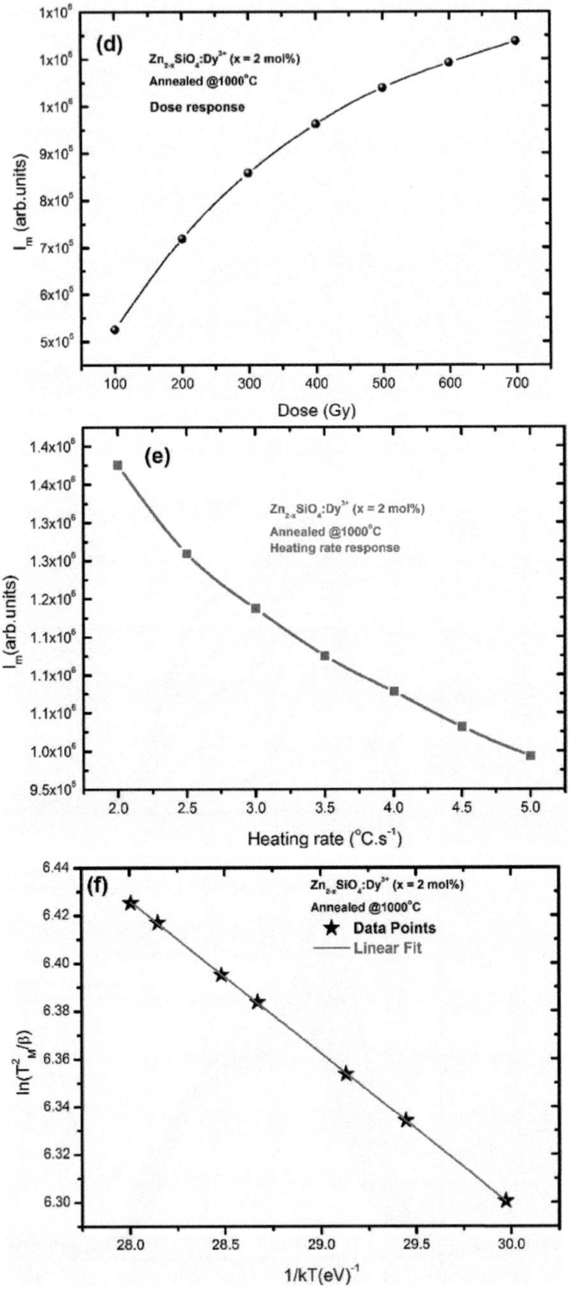

Figure 6.7 TL glow curves (a) comparison between undoped and Dy^{3+} doped samples, (b) acquired at different doses of beta radiation, (c) acquired at variable heating rates, (d) dose response curve, (e) heating rate response curve and (f) $\ln(T_M^2/\beta)$ as a function of $1/kT_M$ plot used to estimate the activation energy for variable heating rate method.

peak was considered for variation of beta dose and it was observed that the glow curve patterns did not change with an increase of beta dose but only the TL intensity was increased with the slight shift to the higher temperatures (from 120 to 147°C). The TL response increased with the increase of beta dose, as shown in Fig. 6.7d. The intensities of the glow curves acquired at different doses and heating rates are presented in Fig. 6.7c and Fig. 6.7e, respectively. Figure 6.7e demonstrates the increase of TL intensity as a function of beta dose. The decrease in TL intensity as heating rate is increased was observed, with a slight shift of peak position to the higher temperatures. This effect may be due to thermal quenching [25] mainly due to recombinations which slow down as the heating rate increases [26]. By applying the variable heating rate method for the TL glow curves of the Dy^{3+} (2 mol%) ion activated sample, the activation energy can be estimated using Eq. (6.1) [27]:

$$E = \ln(T_M^2/\beta)kT_M \tag{6.1}$$

where E is the activation energy or trap depth, in the case of electrons, T_M is the temperature at the maximum intensity of the glow curve, β is the heating rate, and k is the Boltzmann constant. By linearizing Eq. (6.1), the activation energy can be evaluated from the slope of $\ln(T_M^2/\beta)$ as a function of $1/kT_M$ as shown by a plot in Fig. 6.7f. The value of the activation energy was estimated to be ~0.06 ±0.02 eV.

6.4 Conclusion

Zn_2SiO_4 nanophosphors, doped and co-doped with Mn^{2+}, Eu^{3+}, and Dy^{3+} ions were synthesized by sol-gel method. The XRD results revealed that nanophosphors crystallized to form α-Zn_2SiO_4 phase after annealing at 1000°C. Upon thermal annealing, the nanophosphors had an average crystallite size of ~30 – 35 ± 0.5 nm and showed non-uniform spherical shape. PL spectra showed that Zn_2SiO_4:Mn^{2+}, Eu^{3+}, Dy^{3+} nanophosphor can be excited by UV light to emit blue, green, and red and according to this observation the energy transfer mechanism was proposed. The energy is transferred from Dy^{3+} ions energy states to Eu^{3+} energy states. The absorbed photon energy by Dy^{3+} ions is likely transferred to excited states of Eu^{3+} ions through resonant energy transfer from Dy^{3+} non-radiative transitions. The CIE chromaticity coordinates of Zn_2SiO_4:Mn^{2+}

nanophosphor falls in the green region, Zn_2SiO_4:Eu^{3+} nanophosphor in the red region of the spectrum, Zn_2SiO_4:Dy^{3+} nanophosphor showed white color and Zn_2SiO_4:Mn^{2+}, Eu^{3+}, Dy^{3+} nanophosphor showed orange-red color. The CIE calculated from the emission spectra showed coordinates which varied from green, orange to red through white with the change of dopant in Zn_2SiO_4 host lattice. The TL glow curves from all samples showed adequate glow for dosimetry, in terms of maxima location and intensity for Zn_2SiO_4:Dy^{3+} nanophosphor. Therefore, it can be concluded that Zn_2SiO_4:Mn^{2+}, Eu^{3+}, Dy^{3+} nanophosphor is a potential material for application in UV-pumped multicolor and white light emitting diodes as well as medium and high dosimetry applications.

Acknowledgments

The author would like to acknowledge the College of Science, Engineering and Technology at the University of South Africa (UNISA) for financial support (Grant no. 88028).

References

1. M-T. Tsai, F-H. Lu, J-M. Wu, Y-K. Wang and J-Y. Lin. *Mater. Sci. Eng.* 18 (2011) 032012-1–032012-4.
2. Q.Y. Zhang, K. Pita and H. Kam. *J. Phys. Chem. Solids.* 64 (2003) 333–336.
3. K.H. Yoon and J.H. Kim. *J. Korean Phys. Soc.* 58 (2011) 1668–1671.
4. V. Dubey, R. Tiwari, M.K. Pradhan, G.S. Rathore, C. Sharma and R.K Tamrakar. *J. Lumin. Appl.* 1 (2014) 30–39.
5. L.H. Cheng, X.P. Li, J.S. Sun, H.Y. Zhong, Y. Tian and J. Wan. *Physica B.* 405 (2010) 4457–4461.
6. Q. Su, Z.W. Pei, L.S. Chi, H.J. Zhang, Z.Y. Zhang and F. Zou. *J. Alloys Compd.* 192 (1993) 25–27.
7. N. Shasmal and B. Karmakar. White light-emitting Dy^{3+}-doped transparent chloroborosilicate glass: Synthesis and optical properties, *J. Asian Ceram. Soc.* 7 (2019) 42–52.
8. C. Mnjunath, M.S. Rudresha, K.R. Nagabhushana, R. Hari Krishna, B.M. Nagabhushana, and B.M. Walsh. Thermoluminescence glow curve analysis of gamma irradiated Sr2SiO4:Dy^{3+} nanophosphor, *Physica B Condens. Matter.* 585 (2020) 412113.

9. P.S. Mbule, B.M. Mothudi and M.S. Dhlamini. Mn^{2+}-Eu^{3+}-Dy^{3+} doped and co-doped Zn_2SiO_4 phosphors: Study of the structure, photoluminescence and surface properties, *J. Lumin.*192 (2017) 853–859.
10. B.C. Babu, B.V. Rao, M. Ravi and S. Babu. *J. Mol. Struct.* 1127 (2017) 6–14.
11. P.S. Mbule, O.M. Ntwaeaborwa, B.M. Mothudi and M.S. Dhlamini. *J. Lumin.* 179 (2016) 74–82.
12. S. Nakajima and M. Tamatani. Chapter 18 in: *Phosphor Handbook*, S. Shionoya and W. M. Yen (Eds.), CRC Press, Boca Raton, FL, 1999.
13. Y. Jiang, J. Chen, Z. Xie and L. Zheng. *J. Mater. Chem. Phys.* 120 (2010) 313–318.
14. G. Blasse and B.C. Grabmaier, *Luminescent Materials* (Springer-Verlag, Berlin, Heidelberg, 1994).
15. H. Wang, P. He, H. Yan and M. Gong. *Sens. Actuators, B.* 156 (2011) 6–11.
16. L.D. Sun, C. Qian, C.S. Liao, X.L. Wang and C.H. Yan. *Solid State Commun.* 119 (2001) 393.
17. Y.C. Li, Y.H. Chang, Y.F. Lin, Y.S. Chang and Y.J. Lin. *J. Alloys Compd.* 439 (2007) 367–375.
18. Q. Liu, Y. Liu, Z.Yang, Y. Han, X. Li and G. Fu. *J. Alloys Compd.* 515 (2012) 16–19.
19. M. Manhas, V. Kumar, O.M. Ntwaeaborwa and H.C. Swart. *Ceram. Int.* 42 (2016) 5743–5753.
20. ⟨http://in.mathworks.com/matlabcentral/fileexchange/29620-ciecoordinate-calculator⟩
21. M.A. Tshabalala, H.C. Swart, F.B. Dejene, E. Coetsee and O.M. Ntwaeaborwa. *Appl. Surf. Sci.* 360 (2016) 409–418.
22. K.K. Gupta, S.N. Dhoble, K.D. Burghate and J.S. Dhoble. Luminescence properties of nanocrystalline $Mg_2P_2O_7$:Eu phosphor, *Luminescence.* 33 (2018) 947–953.
23. F. Gu, C. Li, H. Cao, W. Shao, Y. Hu, J. Chen and A. Chen. Crystallinity of Li-doped MgO:Dy^{3+} nanocrystals via combustion process and their photoluminescence properties, *J. Alloy. Compd.* 453 (2008) 361–365.
24. F.V. Molefe, L.L. Noto, M.S. Dhlamini, B.M. Mothudi and V.R. Orante-Barrón. Structural, photoluminescence and thermoluminescence study of novel Li^+ co-activated lanthanum oxide activated with Dy^{3+} and Eu^{3+} obtained by microwave-assisted solution combustion synthesis, *Opt. Mater.* 88 (2019) 540–550.

25. P.R. González, C. Furetta, E. Cruz-Zaragoza and J. Azorín. Heating rate effects on thermoluminescence of $BaSO_4$:Eu+PTFE prepared at ININ-Mexico, *Mod. Phys. Lett. B.* 24 (2010) 717–726.

26. I. Veronese. The thermoluminescence peaks of quartz at intermediate temperatures and their use in dating and dose reconstruction, Ph.D. Thesis Universita Degli Studi Milano (2004).

27. L.L. Noto, M.L. Chithambo, O.M. Ntwaeaborwa and H.C. Swart. The greenish-blue emission and thermoluminescent properties of $CaTa_2O_6$:Pr^{3+}, *J. Alloy. Compd.* 589 (2014) 88–93.

Chapter 7

Luminescent Materials as Spectral Converters for Enhancing Solar Cells' Efficiency

M. Y. A. Yagoub, H. C. Swart, and E. Coetsee
Department of Physics, University of the Free State, Bloemfontein, Free State 9300, South Africa
yagoubm@ufs.ac.za, coetseee@ufs.ac.za, swarthc@ufs.ac.za

7.1 Introduction

This chapter presents the reader with a brief overview on the development of luminescent materials for solar cells' application, instead of presenting an extensive review on literature. The focus will therefore be on the processes that have developed for photon conversion. The requirements of luminescent materials for solar cells' application and then the luminescent materials and the practical application of these materials will be discussed.

Figure 7.1 shows the different losses of a single-junction solar cell that limits the performance of the solar cell. Different approaches

Luminescent Nanomaterials
Edited by Odireleng Martin Ntwaeaborwa

ISBN 978-981-4968-11-9 (Hardcover), 978-1-003-27738-5 (eBook)
www.jennystanford.com

have been suggested to increase the single junction's performance beyond the Shockley-Queisser limit [1–6]. The thermalization and transparency losses due to photon mismatches were considered as the major mechanism that limit the performance of the solar cells. Approaches such as space-separated quantum cutting, intermediate band solar cells and multiple exciton generation have been developed to modify the structure of semiconductor junctions to better match the solar spectrum [1–4]. Luminescent materials have also been suggested as one of the solutions to overcome the Shockley-Queisser limit of the single-junction solar cells [5–7]. The luminescent materials can also externally be added to the solar cells without affecting the highly optimized solar cell's design [7]. The solution was based on the photon conversion process to deal with the thermalization and transparency losses. Thermalization loss can be eliminated with the help of downshifting and downconversion luminescent materials. Upconversion luminescent materials could take care of the energy loss due to photons with lesser energy than the bandgap of the solar cell's junction.

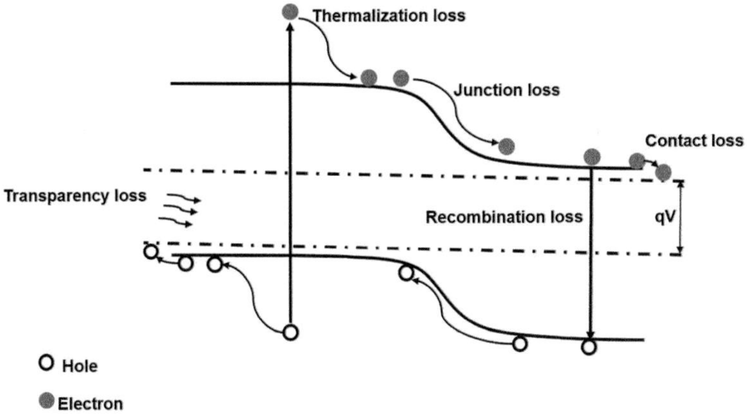

Figure 7.1 Different losses in a single-junction solar cell.

Photon conversion processes for solar cell application was first studied by Dexter in 1953 [8]. Energy conversion efficiencies of 47.6% [9] and 38.6% [10] were theoretically predicted by using upconversion and downconversion methods, respectively, in conjunction with a Si solar cell in non-concentrated sunlight. Since then, luminescent materials based on lanthanide ions were the most common investigated materials for up- and downconversion processes for spectral conversion due to their unique and rich energy level structure. Choosing the ideal phosphor material for a specific solar cell has the potential to enhance the solar cell's efficiency [5–7]. The luminescent materials therefore used for solar cells should therefore exhibit the desired properties. The properties of the luminescent materials and the problems to overcome in such luminescent materials will therefore also be outlined in this chapter.

7.2 Requirements for Luminescent Materials as Spectral Converters

The optical up- and downconversion approaches modify the solar spectrum to better match the bandgap of absorption of single-junction solar cells. Up- and downconversion are photon conversion processes that aim to increase the solar cells' efficiencies while keeping the cells' design unchanged.

Specific materials that can convert the photons that causes thermalization and transparency losses have to be designed. The materials that luminesce are called luminescent materials (phosphors). This is therefore a great task with various challenges for the luminescent materials' community. The following are specific required properties for luminescent materials for solar cell applications:

- The materials must be transparent in the region where the solar cells absorb efficiently.
- The materials must have strong and broad absorption bands in the region where the absorption of the solar cell's semiconductor is weak.
- The materials must have sharp emission bands in the solar cell's desired absorption part.
- The materials must show large Stoke shifts to avoid reabsorption of the emitted photons.
- The materials should possess high quantum efficiency of the desired wavelength.
- The materials should have desired reflective indexes to reduce/eliminate reflection and scattering.

- The materials should be optically and chemically stable over the lifespan of the solar cell.
- The materials must show low cost-effective production.
- The materials should have low phonon energy. This would minimize the non-radiative emission and may enhance any possible energy transfers that happen during the conversion processes for better energy conversion.
- The materials should have the ability to accommodate enough percentages of the activators without changing their structural properties.
- The materials should be easily fabricated.

To achieve photon conversion an inorganic material (host) activated with lanthanide ions that consist of possible energy levels (to emit at the desired wavelength) must be used. Studies showed that there are commonly used activators for solar cells' application, and these will be highlighted in the next section.

7.3 Activators for Solar Cell Application

7.3.1 Commonly Used Activators for Solar Cell Application

Lanthanide ions have been extensively utilized for photon conversion processes as luminescent materials for a wide range of applications [11–13]. The lanthanide ions are normally doped into insulators or semiconductor inorganic materials. The inorganic materials are known as hosts that form the bulk of the phosphors. The host is of vital importance in the photon conversion processes. The host should therefore be able to optimize the distribution of the activators and prevent the occurrence of non-radiative processes [14]. Most of the studies on luminescent materials for solar cells' converting processes have focused on phosphors with co-doped lanthanide ions, Ln^{3+}-Yb^{3+}. Ln^{3+} act as a donor and Yb^{3+} as an acceptor. The 4f-4f transitions of Ln^{3+} ions are mainly responsible of the photon conversion processes. Therefore, they are known as the narrowband conversion processes ions. The reason of using the Ln^{3+}-Yb^{3+} activated phosphors for solar cells application include:

- Ln^{3+}-Yb^{3+} couples are suitable for all photon conversion processes like downconversion and upconversion.
- Yb^{3+} consists of only two energy levels that emits photons in the NIR, which is suitable for c-Si solar cell.
- Ln^{3+} ion mainly absorb photons in the UV and visible regions and therefore can feed Yb^{3+} ions through different energy transfer routes. Ln^{3+} ions therefore consist of energy levels that are either the same or that are twice the energy spacing of the Yb^{3+} excited level. This results in the interaction with Yb^{3+} ions through both first order and cooperative second order energy transfer, which can emit two photons in the NIR region.

On the other hand, ions with broad transition bands (such as Ce^{3+}, Eu^{2+}, Yb^{2+}, and Bi^{3+}) are also investigated for photon conversion processes [7]. The broad band ions have specifically been investigated as sensitizers for Yb^{3+} ions. The couples were mostly studied in downconversion processes for emitting NIR emission for solar cells' application. The advantages of broad band transition ions may include the following:

- Emission from these ions undergo d-f transitions instead of f-f transition.
- The absorption strength of the d-f electronic transitions is electrically allowed transitions that are stronger than 4f-4f transitions.
- The d orbital of these ions is sensitive to the environment. This result in broad and tuneable emission. The tune-ability of the wavelength can be achieved by changing the host lattice of the phosphor.

Broad band ions have relatively strong absorption strength and tuneable emission that can be utilized for harvesting the UV and visible photons. The 5d orbital is strongly affected by the crystal filed of the host materials. The crystal field interactions would degenerate and shift the whole 5d level towards lower/higher energy depending on the nature of the interaction, which depend on the site symmetry of the crystal [15]. The emission of such ions can therefore be tuned all the way from blue to red photons.

7.3.2 Energy Transfer between Activator Ions

The energy transfer between activators is a vital physical phenomenon for understanding the photon conversion processes that occur in luminescent materials. In the process of energy transfer, the excitation energy transfers from luminescent centres (donor ion) to the accepter ions which in turn release the emission photons.

$$D + h\nu \rightarrow D^*$$

$$D^* + A \rightarrow D + A^*$$

$$A^* \rightarrow A + h\nu'$$

Where D and A stand for donor and acceptor, respectively. $h\nu$ is the excited energy and the asterisk (*) represents the excited ion. The energy transfer processes occur through different mechanisms [16]. The resonant energy transfer process (radiative and non-radiative) requires spectral overlap between the emission of the donor ion and the absorption of the accepter ion, Fig. 7.2. A large spectral overlap results in strong interaction. The significant difference between radiative and non-radiative resonant energy transfer is the decay time of the donor's emission. In radiative energy transfer the decay time of the donor remains unchanged with acceptor concentration. The resonant energy transfer requires a resonance condition to occur between a donor and an acceptor. Thus, the interaction between the activators occur through either exchange or multipolar interaction [17, 18]. The exchange interaction is known as Dexter's energy transfer, where the wavefunctions of the activators overlap. Therefore, the exchange interaction occurs only over very short distances between the activators. The multipolar interaction/Förster energy transfer relies on the strength of the optical transitions which occurs over relatively large distances. The resonant energy transfer probability is therefore approximated by the following equation [16].

$$W_{\mathrm{DA}} = \frac{2\pi}{\hbar}\left|< DA^* \middle| H_{\mathrm{DA}} \,|\, D^* A >\right|^2 \int g_{\mathrm{D}}(E) g_{\mathrm{A}}(E) dE$$

The symbols hold the same meaning as described in Ref. [19]. The integral represents the spectral overlap ($J(\lambda)$). The distance dependent of the energy transfer between donor and acceptor is expressed in the square matrix element term. The distance dependent is exponential for exchange interaction, whereas for electric multipolar it is given by R^n (where n = 6, 8, 10) for different multipolar interactions. In general, the resonant energy transfer rate depends on various factors. These factors may include the extent of spectral overlap between the activator pair (Fig. 7.2), the dipole moments' relative orientation of the donor–acceptor transition, the donor's quantum yield and the distance between the activator [19].

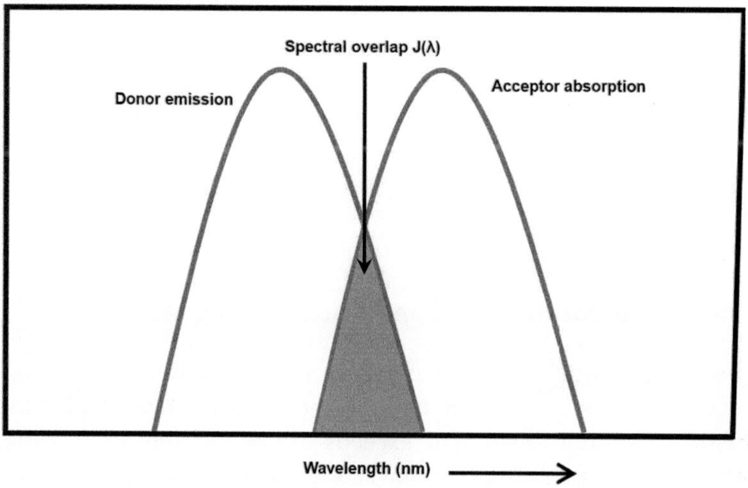

Figure 7.2 The spectral overlap between donor emission and acceptor absorption.

In some cases, the resonance condition is not fulfilled therefore energy transfer can occur through a phonon assistant process, Fig. 7.3c. The probability of phonon-assistant energy transfer can be approximated by Miyakawa-Dexter theory. However, the energy can also occur through cross-relaxation, Fig. 7.3d. The cross-relaxation energy transfer therefore takes place when the donor and the acceptor possess similar energy levels. The cross-relaxation can lead to emission quenching where the excited energy diffuses between the activators.

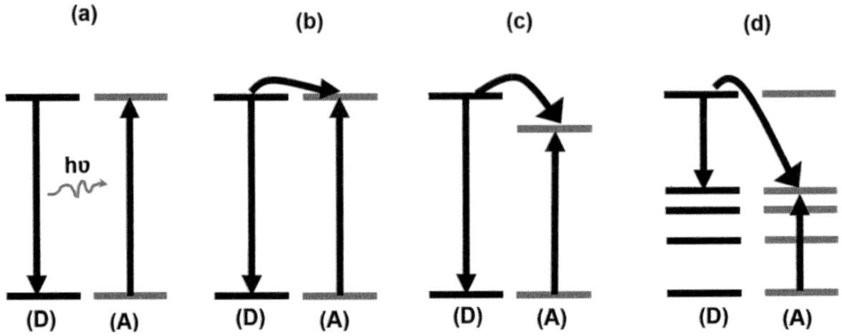

Figure 7.3 Schematic diagram of different basic energy transfer processes between donor and acceptor ions. (a) Resonant radiative transfer, (b) non-radiative transfer, (c) phonon assisted transfer and (d) cross-relaxation between two identical ions. D and A stand for donor and acceptor, respectively.

7.4 Converter Materials for Solar Cell Application

7.4.1 Downconversion and Downshifting Materials

Downconversion is a process where the absorption of one high energy photon results in two or more lower energy photons with a quantum yield that exceeds 100%. On the other hand, the downshifting process converts a high energy photon into lower energy photons. In the downshifting process, the quantum efficiency does not exceed unity. The downconversion process is therefore known as the quantum cutting process. The difference between downconversion and downshifting is summarised in Figs. 7.4a and b. Both processes require an intermediate level. In downconversion there are two steps of emission of two lower energy photons. Whereas, in downshifting the excited electron relaxes non-radiatively to the intermediate level from where it can relax to the ground state resulting in only one emitted photon. Only one photon is required for downconversion and downshifting and the intensity of the emitted photons is therefore linearly dependent on the intensity of the incident light [7].

Both downconversion and downshifting materials can be used as layers on top of all type of existing solar cells, see Fig. 7.4. Different materials can be used as conversion layers such as lanthanide ion complexes, organic dyes and quantum dots. However, downconversion or downshifting layers require certain properties for solar cells' application. Some of these properties are:

- The quantum yield must be high (near 200% or more) for downconversion and near unity for downshifting.
- There must be a relatively broad absorption spectrum in the UV-vis region where the external quantum efficiency of the solar cell is low.
- The layer must have a high absorption coefficient in the high energy region, and it must be nearly transparent in the low energy region.
- The layer must possess a narrow emission band at the solar cell's desired absorption part.
- There must be extended spectral overlap and the Stokes shift must be large enough to prevent energy losses due to re-absorption of the emitted photons.

For c-Si solar cells, the downconversion luminescent materials should convert UV and blue photon energies to the NIR region. The emission profile of the luminescent materials should therefore be in the same range than the solar cell's junction. The best emission profile for c-Si solar cells that are currently reported is a profile that extends in the ~980 nm wavelength [7].

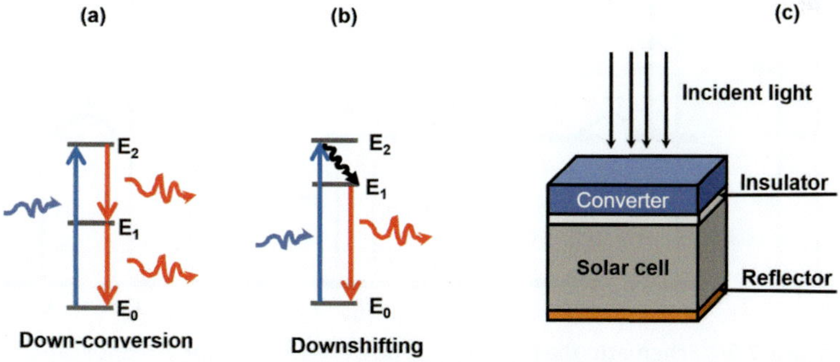

Figure 7.4 Schematic energy diagrams of different photon conversion processes. (a) Downconversion where one high energy photon (blue) cuts into two lower energy photons (red), (b) downshifting where a high energy photon (blue) results in one lower energy photon (red) after non-radiative relaxation, and (c) a schematic representation of a solar cell with a conversion layer (downconversion/downshifting).

It is also very crucial to optimize the reflectivity of the surface on the top side of the solar cell since the conversion layer should be located on top of the solar cell. A conversion layer with a refractive index close to 1.0 is therefore desirable in order to minimize the reflections from the top side of the solar cell. On the other hand, the re-emitted photons from the conversion layers need to be directed to the solar cells by total internal reflection. In this case a conversion layer with the same refractivity as the solar cell is ideal. However, a conversion layer with a high refractive index will maximize the surface reflections. The optimum refractive index for a conversion layer can be calculated from the following equation.

$$n_c = \sqrt{n_0 \times n_{sub}}$$

n_0 and n_{sub} are the refractive index for the surrounding medium (air) and the substrate/solar cell, respectively. The optimum value for the conversion layer n_c therefore spans in the range of 1.5–2.0. Fortunately, most of the materials that are used in solar cells as anti-reflection coatings possess a refractive index in this range. The thickness of the layer is also very important to be optimized. Attention must also be given to the decoupling effect when studying the downconversion and downshifting processes.

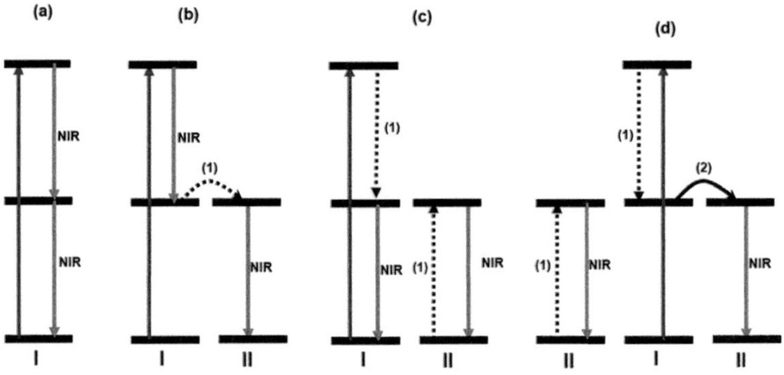

Figure 7.5 Schematic diagram of NIR downconversion for PV applications. (a) Single ion NIR downconversion by sequential emission of two NIR photons. (b–d) NIR quantum cutting due to resonant energy transfer from ion I to ion II.

Figure 7.5 schematically demonstrates NIR quantum cutting for solar cell application. Quantum cutting can occur with only one activator ion or with a combination of different activator ions. The donor ion absorbs the UV/Vis photons and then feeds two sensitizer ions through energy transfer. The resonance condition must be fulfilled for the processes to occur. In one activator ion, the non-radiative emission competes with the NIR photons and therefore unwanted UV/Vis photons can be emitted [18].

Quantum cutting may also occur via second order energy transfer. There are two types of second order energy transfer, cooperative and accretive [20, 21]. Both energy transfer mechanisms require polarization induced through a virtual state of opposite parity. In cooperative energy transfer the virtual state locates in the donor ion, while in accretive energy transfer the virtual state locates in one of the acceptor ions. The second order energy transfer is well known to occur in the Tb^{3+}-Yb^{3+} couple [21]. The mechanisms are explained schematically in Fig. 7.6. After excitation of the donor ion (Tb^{3+}), the two acceptor ions (Yb^{3+}) receive equal amounts of energy and then emit two NIR photons for solar cells' application.

Downconversion has been demonstrated in most of the Ln^{3+}-Yb^{3+} couples in various inorganic phosphors [7, 22]. The downconversion mechanism through resonant energy transfer and cooperative energy transfer dominated all studies of Ln^{3+}-Yb^{3+} couples. The most efficient downconversion has been reported for the Pr^{3+}-Yb^{3+} and Tm^{3+}-Yb^{3+} lanthanide couples [7]. Pr^{3+}-Yb^{3+} is the most studied downconversion couple in different inorganic hosts [23–25]. The Pr^{3+} ion acts as a donor ion that can feed the Yb^{3+} acceptor ion through energy transfer. Pr^{3+} transfers the absorbed photons to Yb^{3+} with two energy transfer routes. Pr^{3+} absorbs blue photons (around 440 nm excitation wavelength) which populate the $^1I_6$ and ^3P_j (j = 0, 1, 2) levels, from where a depopulation occurs to the metastable $^3P_0$ level. The energy transfer can therefore occur through second-order cooperative energy transfer and first-order resonant energy transfer. In cooperative energy transfer the Pr^{3+}:$^3P_0$ level act as a virtual level as it is situated at about twice of the energy of the $^2F_{5/2}$ excited level of Yb^{3+}. The Pr^{3+}:$^3P_0$-$^3H_4$ transition can therefore excite two Yb^{3+} ions where two NIR photons are released. In resonant energy transfer the $^1G_4$ acts as an intermediate level where a two-step energy transfer occurs from Pr^{3+} to Yb^{3+}, Pr^{3+}: [$^3P_0$-$^1G_4$,$^1G_4$-$^3H_4$] $\rightarrow 2 \times$ Yb^{3+}: [$^2F_{5/2}$-$^2F_{7/2}$]. Yb^{3+} ion therefore releases two NIR photons that are suitable for c-Si solar cells' absorption.

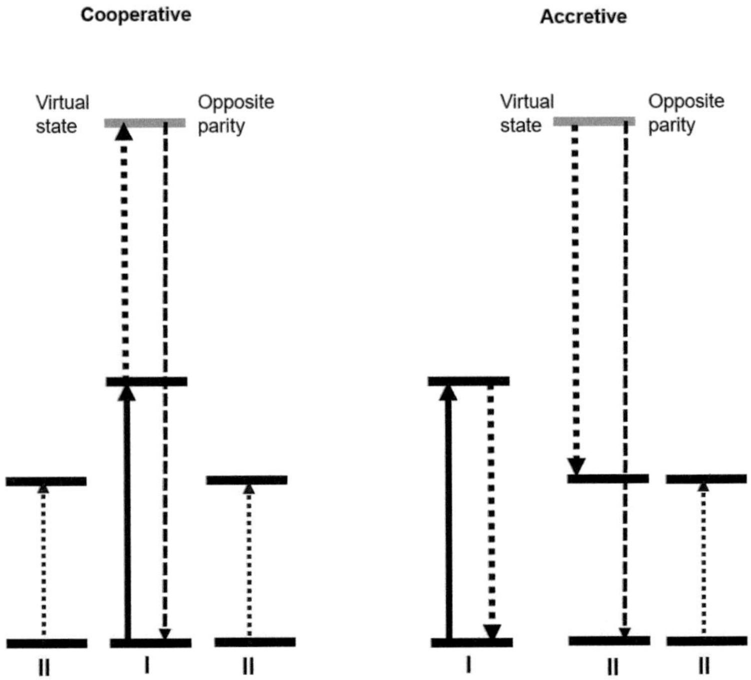

Figure 7.6 Schematic diagram of cooperative and accretive energy transfer routes from ion I to ion II.

The promising result of the Pr^{3+}-Yb^{3+} couple has been reported for fluoride materials [7, 23]. Alkaline earth fluoride materials possess high transparency, low refractive index as well as mechanical durability that are suitable for a luminescent layer [26]. The structure of these materials consists of cation ions that are surrounded by empty cubes that can accommodate high concentration of dopant ions without influencing their structure [26]. The SrF_2:Pr^{3+}-Yb^{3+} phosphor has exhibited NIR photons with efficient energy transfer and quantum efficiency close to 200% [23]. NIR emission was quenched at high Yb^{3+} concentration due to concentration quenching between the Yb^{3+} ions.

The energy transfer process can be investigated by both steady state and time-decay curve luminescence spectroscopy [27]. Equations (7.1) and (7.2) are usually used to calculate the energy transfer and quantum efficiency, respectively.

$$\eta_{ET} = \eta_{x\%Acc} = 1 - \frac{\int I_{x\%Acc} dt}{\int I_{0\%Acc} dt} \tag{7.1}$$

$$\eta_{QE} = \eta_{Don}(1 - \eta_{ET}) + 2\eta_{ET} \tag{7.2}$$

I is the emission intensity, τ is the lifetime, x%Acc stands for the acceptor concentration and η_{Don} represents the quantum efficiency for the donor and is set to 1 [18].

The sub-unity efficiency of the downshifting process can also be useful for solar cells application. Downshifting has been investigated for luminescent solar concentrators as well as to minimize the thermalization loss to overcome the limitation of solar cells. Trivalent Eu^{3+} ions are the most use ions as downshifting converters [28]. Eu^{3+} exhibits efficient red emission with long Stoke shifts, see Fig. 7.7. The ion has therefore been utilized in different materials as photon convertors [28–30]. For solar cell application, the downshifter converter are mostly used for cells with larger bandgap than c-Si solar cells such as amorphous Si and dye sensitized solar cells to minimize the thermalization loss.

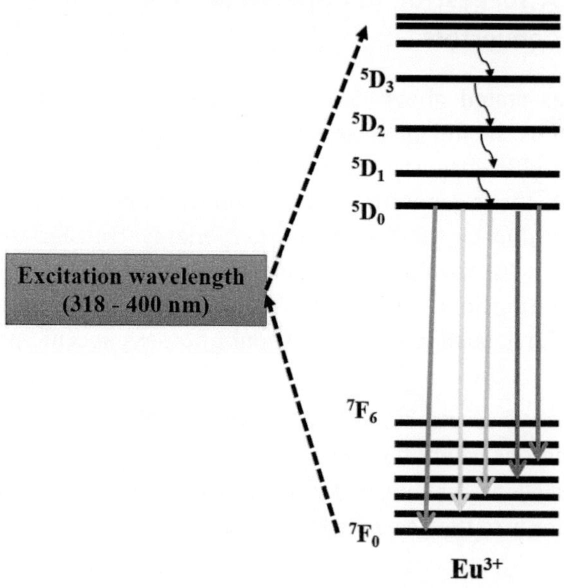

Figure 7.7 Energy diagram of Eu^{3+} ion with the possible excitation and downshifting emissions.

Trivalent lanthanide ions are also known as the 4f-4f transition ions [18]. There are two more issues regarding downconversion and downshifting of 4f-4f transition ions to be considered for better energy harvesting. The absorption of 4f-4f transition of Ln^{3+} ions are relatively weak in terms of the harvesting the energy that needs to be increased. The second issue is the concentration quenching that occurs between Yb^{3+} ions at high concentration. The high Yb^{3+} concentrations are needed for better quantum efficiency as every two Yb^{3+} ions need only one Ln^{3+} ion.

On the other hand, the broadband transition ions have also been investigated for possible use as downconversion and downshifting layers [27, 31]. Investigations were reported that these ions coupled with Yb^{3+} ion interacted through second-order energy transfer [7, 18]. The most studied couples are Ce^{3+}-Yb^{3+}, Eu^{2+}-Yb^{3+}, and Bi^{3+}-Yb^{3+} in different host materials [7, 18, 27, 32]. Although the absorption of these couples is optimal, they exhibit relatively weak NIR emissions. The reason of such behaviour was attributed to the interaction mechanism between such ions and Yb^{3+}.

7.4.2 A Suggestion to Improve the 4f-4f Luminescent Converters

As it was stated above, the conversion materials for solar cell applications should possess strong absorption as well as a high quantum efficiency. On the other hand, the broadband ions that couple Yb^{3+} ions generate NIR emission with low external quantum efficiency due to their second-order energy transfer to Yb^{3+}. A sensitizer with strong absorption has been suggested to increase the harvesting ability of Ln^{3+}-Yb^{3+} couples for downconversion and downshifting. Such a sensitizer should possess specific properties like:

- High absorption ability. It can efficiently absorb the UV/blue photons where the solar cells weakly absorb.
- Excellent sensitizer. It can efficiently transfer its absorbed energy to the acceptor Ln^{3+} ions.

In general, the idea of using broadband ions as sensitizers for 4f-4f transition ions has been investigated for various ions [32–38]. Preliminary results suggested that the concentration of the sensitizer need to be optimized. The optimization of the concentration of both the sensitizer and the activator would result in efficient NIR emission. The solution for concentration quenching can be obtained through optimizing the dopant concentration and synthesis conditions. The synthesis of small particles (nanocrystals) may be useful as the energy migration is limited. Nanocrystal defects are therefore limited, and it would result in high quantum yields. The idea is therefore to increase the quantum yield of quantum dots versus the quantum yield in bulk semiconductors.

The quantum dots or nanorods may also be used in downshifting processes as they possess broad absorption spectra. The NIR emission and high quantum efficiency (larger than the present ~70%) are required when using such materials. The materials also should exhibit large Stokes shift, which would be possible as Stokes shift is influenced by the quantum dots' size [39].

The use of sensitizer ions for Ln^{3+}-Yb^{3+} in luminescent materials could be practically harmful for some of the inorganic materials which might result in structural deformations. The Eu^{2+} sensitized Pr^{3+}-Yb^{3+} couple in SrF_2 resulted in a mixed structural phase with the compensation of a Na^+ ion [33]. The system exhibited weak NIR emission of Pr^{3+} ions and the emission from Yb^{3+} ions was unobserved. In general, the suggestion has been carefully investigated and some further studies are needed.

7.4.3 Upconversion Materials

The upconversion process combines two or more low energy photons into one high energy (visible) photon [40]. Upconversion is a process that violates Stoke's low (anti-Stoke law). The upconversion phenomenon was discovered by Bloembergen in 1959 and was connected to the development of infrared detectors [16]. Upconversion has then investigated in several different fields, such as solar cells, temperature sensors, biomedical imaging as well as lightings and displays [41–44]. In solar cells' application, the upconversion layer modifies the photons' spectral range in order to be absorbed. Therefore, the upconversion materials should be coated beneath a bifacial solar cell to absorb and convert the transmitted photons (those photons that have lesser energies than the bandgap of the solar cell) to higher energy photons back to the solar cell where they can be efficiently absorbed, see Fig. 7.8. The materials can also be coated on both side of the bifacial Si solar cells for better solar cell performance [45].

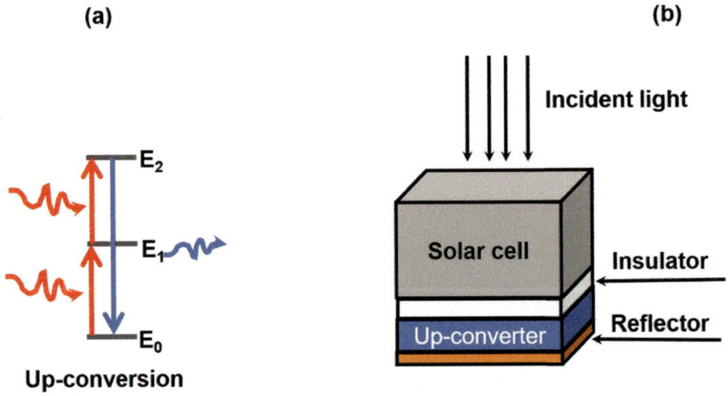

Figure 7.8 Schematic representation of (a) upconversion where two low energy photons (red) combine to give one high energy photon (blue), and (b) a solar cell with an upconverter.

There are different upconversion mechanisms and it all involve a subsequent energy transfer from a sensitizer to an activator. The mechanisms are excited state absorption (ESA), energy transfer upconversion (ETU) and photon avalanche (PA) [16]. In the ESA mechanism, a successive absorption of one or two photons occur in the activator ion. The ETU process is like ESA, but ETU is much more efficient and independent of the pump power. Upconversion is therefore a two-step absorption process where photons are absorbed by the ground-state followed by an excited state. Upconversion is a non-linear optical process which requires high intensity of incident light for generating small intensities of upconversion emission.

Lanthanide ions have extensively been studied for upconverters to be attached underneath the bifacial solar cells. The ideal Ln^{3+} ion for converting NIR light is trivalent Er ion which consists of equally spaced energy levels. $NaYF_3$:Er^{3+}-Yb^{3+} luminescent material is one of the most efficient upconverters [43, 46]. The upconversion process involves two-steps and Yb^{3+} act as absorption centre. Er^{3+}-Yb^{3+} efficiently absorbs in the 920 to 980 nm range and emits in the visible wavelength range. The Er-Yb couple is therefore considered to be unsuitable for c-Si solar cells as they absorbed NIR emission where the cell is efficiently active. Such phosphors can be utilized in solar cells with large bandgaps such as the dye-sensitized solar cell (DSSC) or CdTe.

The emission intensity of upconversion emission depends on a higher order of the incoming light's intensity [44].

$$N_n \propto N_{n-1}N_s \propto (N_n)^2 \propto (P_{in})^2,$$

where N_n is the higher excited state of the acceptor, like Er^{3+}, n is the number of photons that excite the state of the upconversion ions and N_s is the excited state of the sensitizer ion, mostly Yb^{3+} and P_{in} is the intensity of the incoming light. The power law that is dependent on the emission intensity can be deviated if the higher levels are saturated which might cause other processes like non-radiative relaxations to occur [47, 48]. Upconversion emission intensity is proportional to N_n, so when the upconversion layer is placed underneath a solar cell a quadratic power dependent is expected. The increased photocurrent due to the upconversion emission is therefore given by [44]

$$I_{SC}^{UC} \propto P_{in}^n$$

I_{SC}^{UP} is the short-circuit photocurrent due to due to upconversion in the solar cell.

The refractive index of the upconversion layer is important for improving the solar cells photogenerated current [49]. An upconverting material with a refractive index the same as the solar cell could improve the cell's efficiency up to 10%.

7.5 Practical Applications

The synthesis parameters of luminescent materials for solar cells can easily be controlled to provide the desired properties [50]. Different techniques can be used to obtain certain specifications of samples such as combustion, sol-gel and hydrothermal techniques. The particle size of the materials mainly influences the photon conversion performance. This was observed by Ho et al. [51] who applied a downshifting layer on c-Si solar cell. They have found that the short-current circuit depended on the particle size of the material. The smaller the size the higher the short-current circuit. In general, different methods were used for downconversion and downshifting layers in order to optimize the solar cells performance. The downconversion/downshifting layer can be applied directly/encapsulated on top of the solar cells. The encapsulated downconversion layers within anti-reflection silica glass or organic materials (such as PMMP) have demonstrated to highly improve the quantum efficiency and short-current circuit the counterpart [52, 53]. Different studies have been reported on applying the downshifting and downconversion layers on different solar cells [51–55]. The reports showed significant increase in the solar cells' performance with the converter layers, which proved that the luminescent materials are useful in solar cell application.

The upconversion materials have also been applied in different forms of solar cells such as conventional solar cells and advanced solar cells like dye-sensitized, organic, and hybrid solar cells. The upconversion materials improved the photo-current for such devices which implies that these materials can have great potential in enhancing the solar cells efficiency irrespective of the solar cell's type [56]. $NaYF_4:Er^{3+}$-Yb^{3+} upconverter has been proved to be widely used in such application, but other upconverters can also still be used in solar cells application [45, 57, 58].

7.6 Conclusion and Outlook

Fundamental approaches of using luminescent materials as spectral converters for solar cells were discussed. The approaches were designed to minimize the energy losses associated with the spectral mismatch between the incident solar spectrum and the solar cell. The approaches therefore utilized the photon conversion processes of downshifting, downconversion and upconversion for shifting the high and low energy photons to better match the solar cells' absorption strength. These approaches were invented to be applied externally to existing solar cells. Solutions for boosting the solar cells' efficiency is well-known as the third-generation technology. Such innovative approaches have shown promising results as spectral converters irrespective of the solar cell's technology.

Downconversion collects the high energy photons above the bandgap of the materials to efficiently minimize the thermalization loss and increase the number of electron–hole generation. Yb^{3+} ion is an essential element in the downconversion materials as it can efficiently be sensitized by various Ln^{3+} ions and emit at the desired wavelength for solar cell applications. The Ln^{3+}-Yb^{3+} couples extensively demonstrated to efficiently produce narrow band NIR downconversion with quantum efficiency close to 200%. Downshifting layers can successfully minimize the thermalization loss due to the absorption of high energy photons (twice of the bandgap). Whereas, upconversion can utilize the sub-band photons with energies less than the bandgap and then reduces transparency loss.

A noticeable improvement of the solar cells' output has been observed when combined with luminescent material but solving the issue of the absorption strength and concentration quenching can still lead to enhancement. Optimization of the absorption strength of downshifting and downconversion materials will increase the efficiency of the solar cells beyond the theoretical limit. This can also be done for upconversion materials. Further investigations and exploration will certainly be fruitful in order to produce affordable luminescent materials as spectral converters in order to reduce or maintain the production costs and enhance the energy efficiency.

Acknowledgments

This work is based on the research supported by the South African Research Chairs Initiative of the Department of Science and Technology and National Research Foundation of South Africa (grand 84415). The financial assistance of the National Research Foundation (NRF) and the Cluster program of the University of the Free State towards this research is hereby acknowledged.

References

1. Yamaguchi, M., Lee, K. H., Araki, K. and Kojima N. (2018). A review of recent progress in heterogeneous silicon tandem solar cells, *J. Phys. D: Appl. Phys.*, 51, pp. 133002–133015.
2. Dimroth, F. (2006). High-efficiency solar cells from III-V compound semiconductors, *Phys. Status Solidi C*, **3**, pp. 373–379.
3. Schaller, R. D., Agranovich, V. M. and Klimov, V. I. (2005). High-efficiency carrier multiplication through direct photogeneration of multi-excitons via virtual single-exciton states, *Nat. Phys.*, **1**, pp. 189–194.

4. Timmerman, D., Izeddin, I., Stallinga, P., Yassievich, I. N. and Gregorkiewicz, T. (2008). Spaces-separated quantum cutting with silicon nanocrystals for photovoltaic applications, *Nat. Photonics*, **2**, pp. 105–109.
5. Shalav, A., Richardsc, B. S. and Green M. A. (2007). Luminescent layers for enhanced silicon solar cell performance: Up-conversion, *Sol. Ener. Mater. and Sol. Cells*, **91**, pp. 829–842.
6. Abrams, Z. R., Niv, A. and Zhang, X. (2011). Solar energy enhancement using down-converting particles: A rigorous approach, *J. Appl. Phys.*, **109**, pp. 114905–114914.
7. Van der Ende, B. M., Aarts, L. and Meijernk, A. (2009). Lanthanide ions as spectral converters for solar cells, *Phys. Chem. Chem. Phys.*, **11**, pp. 11081–11095.
8. Dexter, D. (1953). A theory of sensitized luminescence in solids, *J. Chem. Phys.*, **21**, pp. 836–850.
9. Trupke, T., Green, M. A. and Worfel, P. (2002). Improving solar cell efficiencies by up-conversion of sub-bandgap light, *J. Appl. Phys.,* **92**, pp. 4117–4122.
10. Trupke, T., Green, M. A. and Worfel, P. (2002). Improving solar cell efficiencies by down-conversion of high energy photons, *J Appl. Phys.*, **92**, pp. 1668–1674.
11. Bünzli, J.-C. G. and Piguet, C. (2005). Taking advantage of luminescent lanthanide ions, *Chem. Soc. Rev.*, **34**, pp. 1048–1077.
12. Urquhart, P. (1988). Review of rare earth doped fibre lasers and amplifiers, *IEE Proc. J Optoelectron*, **135**, pp. 385–407.
13. Kenyon, A. J. (2002). Recent developments in rare-earth doped materials for optoelectronics, *Prog. Quantum Electron.*, **26**, pp. 225–284.
14. Feldmann, C., Justel, T., Ronda, C. R. and Schmidt, P. J. (2003). Inorganic luminescent materials: 100 years of research and application, *Adv. Funct. Mater.*, **13**, pp. 511–516.
15. Solé, J. G. L., Bausa, E. and Jaque, D. (2025) *An Introduction to the Optical Spectroscopy of Inorganic Solids*, (John Wiley and Sons Ltd, England).
16. Auzel, F. (2004). Upconversion and anti-Stokes processes with f and d ions in solids, *Chem. Rev.,* **104**, pp. 139–173.
17. Liu, G. and Jacquier, B. (2005) *Spectroscopic Properties of Rare Earth in Optical Materials* (Springer, China).
18. Zhang, Q. Y. and Huang, X. Y. (2010). Recent progress in quantum cutting phosphors, *Prog. Mater. Sci.*, **55**, pp. 353–427.

19. Huang, X., Han, S., Huang, W. and Liu., X. (2013). Enhancing solar cell efficiency: The search for luminescent materials as spectral converters, *Chem. Soc. Rev.,* **42**, pp. 173–201.

20. Auzel, F. (1990). Upconversion processes in coupled ion systems, *J. Lumin.*, **45**, pp. 341–345.

21. Andrews, D. L. and Jenkins, R. D. (2001). A quantum electrodynamical theory of three-center energy transfer for upconversion and downconversion in rare earth doped materials, *J. Chem. Phys.,* **114**, pp. 1089–1100.

22. Miritello, M., Savio, L. R., Cardile, P. and Priolo, F. (2010). Enhanced down conversion of photons emitted by photoexcited $Er_xY_{2-x}Si_2O_7$ films grown on silicon, *Phys. Rev. B*, **81**, pp. 041411–041414.

23. Van der Ende, B. M., Aarts, L. and Meijerink, A. (2009). Near-infrared quantum cutting for photovoltaics, *Adv. Mater.*, **21**, pp. 3073–3077.

24. Zhanga, Q. Y., Yang, G. F. and Jiang, Z. H. (2007). Cooperative downconversion in $GdAl_3(BO_3)_4$:RE^{3+}, Yb^{3+} (RE = Pr, Tb, and Tm), *Appl. Phys. Lett.*, **91**, pp. 051903–051905.

25. Van Wijngaarden, J. T., Scheidelaar, S., Vlugt, T. J. H., Reid, M. F. and Meijerink, A. (2010). Energy transfer mechanism for downconversion in the (Pr^{3+}, Yb^{3+}) couple, *Phys. Rev. B*, **81**, pp. 155112–155117.

26. Bendall, P. J., Catlow, C. R. A., Corish, J. and Jacobs, P. W. M. (1984). Defect aggregation in anion-excess fluorites II. Clusters containing more than two impurity atoms, *J. Solid State Chem.,* **51**, pp. 159–169.

27. Zhou, J., Teng, Y., Lin, G. and J. Qiu, J. (2011). Ultraviolet to near-infrared spectral modification in Ce^{3+} and Yb^{3+} codoped phosphate glasses, *J. Non-Cryst. Solids*, **357**, pp. 2336–2339.

28. Dejneka, M., Snitzer, E. and Riman, R. E. (1995). Blue, green and red fluorescence and energy transfer of Eu^{3+} in fluoride glasses, *J. Lumin.,* **65**, pp. 227–245.

29. Hu, Y., Zhuang, W., Ye, H., Wang, D., Zhang, S. and Huang, X. (2005). A novel red phosphor for white light emitting diodes, *J. Alloys Compd.,* **390**, pp. 226–229.

30. Li, Y., Chang, Y., Lin, Y., Chang, Y. and Lin, Y. (2007). Synthesis and luminescent properties of Ln^{3+} (Eu^{3+}, Sm^{3+}, Dy^{3+})-doped lanthanum aluminium germinate $LaAlGe_2O_7$ phosphors, *J. Alloys Compd.,* **439**, pp. 367–375.

31. Yagoub, M. Y. A., Swart, H. C., Noto, L. L., O'Connell, J. H., Lee, M. E. and Coetsee, E. (2014). The effects of Eu-concentrations on the luminescent properties of SrF_2:Eu nanophosphor, *J. Lumin.*, **156**, pp. 150–156.

32. Yagoub, M. Y. A., Swart, H. C., Dhlamini, M. S. and Coetsee, E. (2016). Near infrared quantum cutting of Na^{+} and Eu^{2+}-Yb^{3+} couple activated SrF_2 crystal, *Opt. Mater.*, **60**, pp. 521–525.
33. Yagoub, M. Y., Swart, H. C. and Coetsse, E. (2017). Effect of Yb^{3+} ions on structural and NIR emission of SrF_2:Eu^{2+}/Pr^{3+} down-conversion containing Na^{+} ions, *Mater. Res. Bull.*, **93**, pp. 170–176.
34. Guille, A., Pereira, A., Breton, G., Bensalah-ledoux, A. and Moine, B, (2012). Energy transfer in $CaYAlO_4$:Ce^{3+}, Pr^{3+} for sensitization of quantum cutting with the Pr^{3+}-Yb^{3+} couple, *J. Appl. Phys.*, **111**, pp. 043104–043108.
35. Yagoub, M. Y., Swart, H. C. and Coetsse, E. (2019). Energy transfer study between Ce^{3+} and Tb^{3+} ions in a calcium fluoride crystal for solar cell applications, *J. Lumin.*, **187**, pp. 96–101.
36. Wanjun, T. and Donghua, C. (2009). Photoluminescence properties Pr^{3+} and Bi^{3+}-codoped $CaTiO_3$ phosphor prepared by a peroxide-based route, *Mater. Res. Bull.*, **44**, pp. 836–839.
37. Zorenko, Y., Gorbenko, V., Savchyn, V., Zorenko, T., Nikl, M., Mares, J. A., Beitlerova, A. and Jary, V. (2013). Bi^{3+}–Pr^{3+} energy transfer processes and luminescent properties of LuAG:Bi, Pr and YAG:Bi, Pr single crystalline films, *J. Lumin.*, **141**, pp. 137–143.
38. Chen, Y., Wang, J., Liu, C., Tang, J. Kuang, X., Wu, M. and Su, Q. (2013). UV-Vis-NIR luminescence properties and energy transfer mechanism of $LiSrPO_4$:Eu^{2+}, Pr^{3+} suitable for solar spectral convertor, *Opt. Express*, **21**, pp. 3161–3169.
39. Barnham K., Marques J. L., Hassard J. and O'Brien P. (2000). Quantum-dot concentrator and thermodynamic model for the global redshift, *Appl. Phys. Lett.*, **76**, pp. 1197–1199.
40. Wang, F., Banerjee, D., Liu, Y. S., Chen, X. Y. and Liu., X. G. (2010). Upconversion nanoparticles in biological labeling, imaging, and therapy, *Analyst*, **135**, pp. 1839–1854.
41. Wang, F., Deng, R. R., Wang, J., Wang, Q. X., Han, Y., Zhu, H. M., Chen, X. Y. and Liu, X. G. (2011). Tuning upconversion through energy migration in core–shell nanoparticles, *Nat. Mater.*, **10**, pp. 968–973.
42. Wang, F., Han, Y., Lim, C. S., Lu, Y. H., Wang, J., Xu, J., Chen, H. Y., Zhang, C., Hong, M. H. and Liu, X. G. (2010). Simultaneous phase and size control of upconversion nanocrystals through lanthanide doping, *Nature*, **463**, pp. 1061–1065.
43. Wang, F. and Liu, X. G. (2008). upconversion multicolor fine-tuning: Visible to near-infrared emission from lanthanide-doped $NaYF_4$ nanoparticles, *J. Am. Chem. Soc.*, **130**, pp. 5642–5643.

44. Van Sark, M. G., De Wild, J., Rath, J. K., Meijerink, A. and El Schropp, R. (2013). Upconversion in solar cell, *Nanoscale Res. Lett.*, **8**, pp. 81–91.

45. Pan, A. C., Canizo, C. D. and Luque, A. (2009). Characterization of upconverter layers on bifacial silicon solar cells, *Mater. Sci. Eng. B*, **159–160**, pp. 212–215.

46. Shalav, A., Richards, B. S. and Green, M. A. (2007). Luminescent layers for enhanced silicon solar cell performance: Up-conversion, *Sol. Energy Mater. Sol. Cells*, **91**, pp. 829–842.

47. Pollnau, M., Gamelin, D. R., Lüthi S. R., Güdel H. U. and Hehlen M. P. (2000). Power dependence of upconversion luminescence in lanthanide and transition metal-ion systems, *Phys. Rev. B*, **61**, pp. 3337–3346.

48. Suyver, J. F., Aebischer, A., García-Revilla, S., Gerner, P. and Güdel H. U. (2005). Anomalous power dependence of sensitized upconversion luminescence, *Phys. Rev. B*, **71**, pp. 125123–125132.

49. Atre, A. C. and Dionne, J. A. (2011). Realistic upconverter-enhanced solar cells with nonideal absorption and recombination efficiencies, *J. Appl. Phys.*, **110**, pp. 034505–034514.

50. Tikhomirov, V. K., Rodriguez, V. D., Ramos, J. M., Castillo, J. D., Kirilenko, D., Van Tendeloo, G. and Moshchalkov, V. V. (2012). Optimizing Er/Yb ratio and content in Er-Yb co-doped glass-ceramics for enhancement of the up-and down-conversion luminescence, *Sol. Energy Mater. Sol. Cells*, **100**, pp. 209–215.

51. Ho, W. J., Deng, Y. J., Liu, J. J, Feng, S. K. and Lin, J. C. (2017). Photovoltaic performance characterization of textured silicon solar cells using luminescent down-shifting Eu-doped phosphor particles of various dimensions, *Materials*, **10**, pp. 21–31.

52. Monzón-Hierro, T., Sanchiz, J., González-Pérez, S., González-Díaz, B., Holinski, S., Borchert, D., Hernández-Rodríguez, C and Guerrero-Lemus, R. (2015). A new cost-effective polymeric film containing an Eu(III) complex acting as UV protector and down-converter for Si-based solar cells and modules, *Sol. Energy Mater. Sol. Cells*, **136**, pp. 187–192.

53. Cheng, Y., Wang, Y., Teng, F., Dong, H., Chen, L., Mu, J., Sun, Q., Fan, J., Hu, X. and Miao, H. (2018). Down-conversion emission of Ce^{3+}-Tb^{3+} co-doped CaF_2 hollow spheres and application for solar cells, *Mater. Res. Express*, **5**, pp. 036206–036216.

54. Xu, H., Zhang, Q., Xiong, G., Liang, L., Wang, X., Shen, L. and Xu, W. (2011). Large enhancement in short-circuit current densities of dye-sensitized solar cells via photo conversion, *J. Phys. Conf. Ser.*, **276**, pp. 012195–012203.

55. Cheng, C. L., and Yang, Y. (2012). Hydrothermal synthesis of Eu^{3+} doped $Y(OH)_3$ nanotubes as down conversion materials for efficiency enhancement of screen-printed monocrystalline silicon solar cells, *IEEE Electron Device Lett.*, **33**, pp. 697–699.

56. Verma, D., Saetre, T. O. and Midtgard, O. M. (2012). Review on up/down conversion materials for solar cell application, *IEEE*, **978**, pp. 02608–02613.

57. Shan, G. B. and Demopoulos, G. P. (2010). Near infrared sunlight harvesting in dye-sensitized solar cells via the insertion of an upconverter-TiO_2 nanocomposite layer, *Adv. Mater.*, **22**, pp. 4373–4377.

58. Goldschmidt, J. E., Fischer, S., Loper, P., Kramer, K. W., Biner, D., Hermie, M. and Glunz S. W. (2011). Experimental analysis of upconversion with both coherent monochromatic irradiation and broad-spectrum illumination, *Sol. Energy Mater. Sol. Cells,* **95**, pp. 1960–1963.

Chapter 8

Enhanced UVB Photoluminescence Intensity of Calcium Phosphate Co-doped with Gadolinium and Praseodymium for Phototherapy Applications

P. P. Mokoena
Department of Physical and Earth Sciences, Sol Plaatje University, Kimberley ZA8300, South Africa
puse8814@gmail.com

This chapter reviews the synthesis and characterization of gadolinium (Gd^{3+}) and praseodymium (Pr^{3+}) co-doped calcium phosphate phosphors which were evaluated for a possible application in phototherapy lamps for treatment of various skin diseases. Pr^{3+}/Gd^{3+} co-doped $Ca_3(PO_4)_2$ and $Ca_5(PO_4)_3OH$ phosphors were prepared by various synthesis methods, namely co-precipitation, urea combustion and microwave-assisted methods. The phosphors were characterized by X-ray diffraction, X-ray photoelectron

Luminescent Nanomaterials
Edited by Odireleng Martin Ntwaeaborwa

ISBN 978-981-4968-11-9 (Hardcover), 978-1-003-27738-5 (eBook)
www.jennystanford.com

spectroscopy, field emission scanning electron microscopy, high resolution transmission electron microscopy, energy dispersive X-ray spectroscopy and photoluminescence spectroscopy. The phase formation of both $Ca_3(PO_4)_2$ and $Ca_5(PO_4)_3OH$ powders was confirmed by X-ray diffraction. The photoluminescence data exhibited a narrowband emission located at 313 nm, which is associated with the $^6P_{7/2} \rightarrow {}^8S_{7/2}$ transition of the Gd^{3+} ion. The emission intensity was enhanced considerably when co-doping with Pr^{3+}. This emission is classified under ultraviolet B (UVB) light and it is suitable for phototherapeutic treatment for skin disease such as hyperbilirubin, vitiligo, and psoriasis.

8.1 Introduction

Hydroxyapatite ($Ca_5(PO_4)_3OH$, HAP) and tricalcium phosphate ($Ca_3(PO_4)_2$, TCP) are recognized as calcium phosphate based bioceramics that can be widely used in biomedical applications because of their excellent properties such as non-toxicity, biocompatibility, osteoconductivity, bioactivity, non-immunogenicity and they are also noninflammatory [1–4]. As important members of inorganic biomaterials, they are widely used in tumor diagnosis, drug delivery, gene drug transfection, and biological imaging or treatment [5–10]. They are also used as host materials for rare-earth ions to prepare light emitting materials that can be used in different light emitting devices such as phototherapy lamps [11]. Phototherapy uses radiation from specific wavelengths in the natural spectrum of the sun for treatment of various skin diseases such as psoriasis, vitiligo, and hyperbilirubin [12, 13]. Initially, only natural sunlight was used for the treatment of skin diseases but due to insufficiency at the latitude, natural therapy has since changed over to the use of the artificial light source [14]. The skin diseases were treated by using three different types of radiation therapy, namely (1) a narrowband UVB-phototherapy, (2) psoralen UVA (PUVA) therapy and (3) phototherapy with near UV–visible light source. UVB narrowband emission consists only of a wavelength corresponding to the therapeutic range and it therefore has extremely minimal sunburn effects, explaining why it has preference over the UVB broadband in phototherapy lamps. Although the radiation emitted by Gd^{3+} at

~313 nm is narrowband UVB ideally suited to phototherapy lamps, its intensity is very low and this can affect the lifespan and the general performance of the phototherapy lamps. One way of improving the narrowband UVB emission at ~313 nm is by sensitization, that can be realized by co-doping with other rare-earths such as Ce^{3+}, Pr^{3+}, Tb^{3+}, Tm^{3+} that can absorb primary excitation energy and transfer it non-radiatively to Gd^{3+}. There are many synthesis methods that can be used to prepare HAP and β-TCP powders doped with the above mentioned rare earth ions. These include low and high temperature methods. Low temperature methods can be divided into 2 categories: dry and wet. Wet methods are precipitation [15], hydrolysis [16], hydrothermal [17], sol-gel [18], and emulsion [19]. These methods are capable of producing nanosized particles with narrow particle size distribution and uniform morphology [20, 21]. Dry techniques include solid-state and mechanochemical methods [22]. The particle size distribution of these methods are usually wide [23]. In this study, several wet methods such as precipitation, urea combustion and microwave were used for the preparation of $Ca_5(PO_4)_3OH$ and $Ca_3(PO_4)_2$ doped with Gd^{3+} and Pr^{3+}. When compared to dry methods, these methods are cost-effective, time-saving, have higher reaction rate; and they also have added advantage of ease of incorporation of dopant ions. This chapter gives a review of Gd^{3+}/Pr^{3+} co-doped $Ca_5(PO_4)_3OH$ and $Ca_3(PO_4)_2$ phosphors synthesized by using different wet synthesis methods. The fundamental properties of these phosphors such as the crystalline structure, particle morphology, optical, and photoluminescent properties are reported. In addition, the mechanism of energy transfer from Pr^{3+} to Gd^{3+} in the co-doped system aimed at improving the UVB photoemission intensity of the narrow emission band located at 313 nm is reported.

8.2 Experimental

Various methods (co-precipitation, urea combustion, and microwave-assisted) were used to synthesize calcium phosphate co-doped with gadolinium (Gd^{3+}) and praseodymium (Pr^{3+}). The precursors used were calcium nitrate ($Ca(NO_3)_2{\cdot}NO_2O$), di-ammonium hydrogen phosphate ($(NH_4)_2HPO_4$), gadolinium nitrate ($Gd(NO_3)_3{\cdot}6H_2O$), and praseodymium nitrate ($Pr(NO_3)_3{\cdot}6H_2O$).

Calcium nitrate tetrahydrate ($Ca(NO_3)_2{\cdot}4H_2O$) and diammonium hydrogen phosphate ($(NH_4)_2HPO_4$) were used as starting materials. These materials were weighed stoichiometrically according to the reaction equation below:

$$45Ca(NO_3)_2{\cdot}4H_2O + 27(NH_4)_2HPO_4 \rightarrow$$
$$9Ca_5(PO_4)_3OH + 64NH_3 + 80NO_3 + 66H_2O$$

For co-precipitation method, the pH of the system was controlled by adding NaOH with further stirring a white precipitate was formed. The precipitate was separated by centrifugation and was dried in an oven. The dried powder was grounded gently using a pestle and mortar resulting in a fine powder and that was later calcined at 900°C for 2 h.

For the combustion method, urea or $CO(NH_4)_2$ was used as a fuel. The solution that was kept in furnace maintained at a temperature of 550°C, boiled, underwent dehydration and decomposed while generating combustible gases. The combustion ashes were cooled to room temperature and were grounded gently using a pestle and mortar resulting in a fine powder. The powder was calcined at 900°C for 2 h.

For the microwave-assisted method, the precursors were dissolved together under magnetic stirring at room temperature, and the pH value was controlled by slow addition of NaOH. The resulting solution was loaded into a glass reactor, sealed, and transferred to microwave oven preheated to a temperature of 150°C and maintained in that condition for 1 h before allowing it to cool down to room temperature. The powder was calcined at 900°C for 2 h.

8.3 Characterization

X-ray diffraction (XRD) analysis was carried out using a Bruker AXS D8 ADVANCE X-ray diffractometer with Cu $K_{\alpha 1}$ radiation (λ = 1.5406 Å) in the range of 10°–60°. X-ray photoelectron spectroscopy (XPS) analyses were performed using a PHI 5000 Versaprobe system to examine the chemical states of the elements in the $Ca_5(PO_4)_3OH:Gd^{3+}$, Pr^{3+} powder phosphor. The particle morphology was analysed using a JEOL JSM 7800F thermal field emission scanning electron microscope (FE-SEM) and a Jeol JEM-ARM200F high resolution transmission electron microscope (HRTEM). The chemical composition was carried out using an Oxford Instruments AzTEC energy dispersive spectrometer (EDS), with X-Max80 silicon drift detector (SDD) system, attached to the FESEM. The absorption, and subsequently the bandgap energy, was evaluated using a Perkin Elmer Lambda 950 UV-vis spectrometer. The photoluminescence (PL) data were recorded using a Varian Cary Eclipse fluorescence spectrophotometer coupled with a monochromatized 150 W xenon lamp used as an excitation source.

8.4 Results and Discussion

Figure 8.1a shows the XRD patterns of $Ca_3(PO_4)_2$, $Ca_3(PO_4)_2$, and $Ca_5(PO_4)_3OH$ powders synthesized by urea combustion, microwave, and co-precipitation methods, respectively. The powders were all annealed at 900°C in air. For urea and microwave methods, the patterns are consistent with the standard rhombohedral phase of $Ca_3(PO_4)_2$ referenced in JCPDS Card No. 70-2065 and for co-precipitation method, the patterns are consistent with the hexagonal structure of $Ca_5(PO_4)_3(OH)$ referenced in JCPDS Card Number 73-0293 [24]. The position of the peaks signify that the powders did not have secondary phases pointing to crystallization of pure rhombohedral and hexagonal phases of $Ca_3(PO_4)_2$ and $Ca_5(PO_4)_3(OH)$ respectively. There is only minor reflection marked with asterisk (*) on the XRD pattern of microwave method which is possibly the secondary phases associated with $Ca(OH)_2$ [25]. The sharp and narrow diffraction peaks suggest that the powders were highly crystalline. The average crystallite sizes were calculated using the well-known Scherrer equation.

$$D = \frac{k\lambda}{\beta \cos\theta} \tag{8.1}$$

where β is the full width at half maximum, k is a constant referred to as the crystalline shape factor, λ is the X-ray wavelength (1.542 Å for copper), θ is the Bragg angle at the peak maximum. The average crystallite size calculated were ~60, ~76, and ~53 nm in diameter for urea combustion, microwave, and co-precipitation methods respectively.

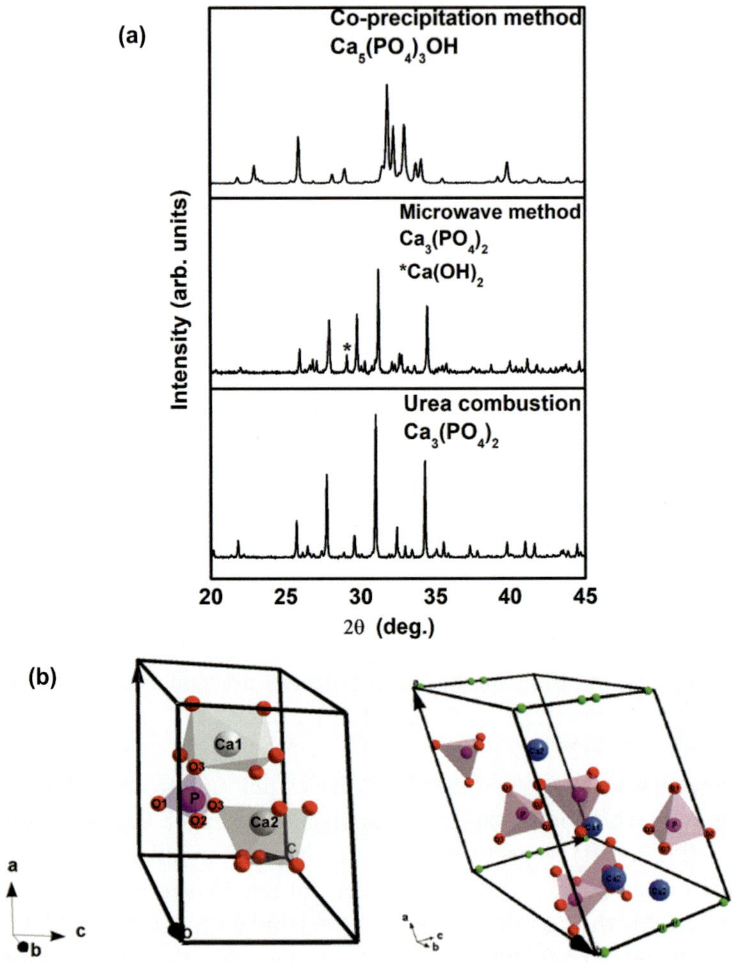

Figure 8.1 The (a) XRD pattern of $Ca_3(PO_4)_2$, $Ca_3(PO_4)_2$, and $Ca_5(PO_4)_3OH$ synthesized by urea combustion, microwave method, and co-precipitation method, respectively, and (b) crystal structure of $Ca_3(PO_4)_2$ and $Ca_5(PO_4)_3OH$ powders.

Shown in Fig. 8.1b are the crystal structures of $Ca_3(PO_4)_2$ and $Ca_5(PO_4)_3OH$ which were drawn by using Diamond-Visual Crystal Structure Information System program referenced in ICSD code No. 99358 and 151414. The two distinct sites are present in the unit cell: Ca (1) and Ca (2) in both $Ca_3(PO_4)_2$ and $Ca_5(PO_4)_3OH$. The Ca^{2+} are surrounded by O^{2-} in the hexagonal arrangement. The P ion is surrounded by O^{2-} ions in the tetrahedral arrangements. Generally, tri-calcium phosphate crystallizes in rhombohedral space group of *R3c* and it is usually described in the trigonal setting with unit cell dimensions varying from 10.35 to 10.53 Å for the "*a*" axis and 36.89 to 37.63 Å for the "*c*" axis [26–29]. The large variations are due to the presence of impurities in some analysed structures [30]. Tri-calcium phosphate has cell dimensions in the ranges: a = 10.435–10.438 Å and c = 37.39–37.43 Å [31, 32]. Hydroxyapatite compounds have hexagonal structures with a space group $P6_3/m$ and unit cell parameters of a = 9.43 Å and c = 6.88 Å. The structure consists of Ca, PO_4, and OH groups closely packed together.

Figures 8.2a shows the XPS spectrum of the $Ca_5(PO_4)_3OH:Gd^{3+}$, Pr^{3+} system. The peak positions match up to those expected for calcium phosphate found in the literature [33, 34]. The C1s peak observed at around 300 eV is attributed to C-H/C-C bonds present from the surface hydrocarbons [35]. Figures 8.2a–e show respectively the fitted Ca 2p, P 2p, O 1s, Gd 3p and Pr 3d high resolution XPS peaks of the $Ca_5(PO_4)_3OH:Gd^{3+}$, Pr^{3+} system [36]. The Ca 2p spectrum in Fig. 8.2b consists of two peaks ascribed to the crystal field split components of the $2p_{1/2}$ and $2p_{3/2}$ energy states. These peaks have been deconvoluted into four peaks due to Ca occupying two sites in the hexagonal $Ca_5(PO_4)_3OH$ host. All the peaks at binding energies of 346, 347, 349, and 350 eV are assigned to Ca bonded to O in the O-Ca-O system due to two different environments of Ca with different coordination. Figure 8.2b shows the P 2p XPS peak deconvulated into two peaks at 133 (P $2p_{3/2}$) and 134 (P $2p_{1/2}$) eV which are due to the crystal field splitting of the P 2p energy state. These peaks are assigned to P tetrahedrally coordinated to O atoms [37, 38]. As shown in Fig. 8.2c, the high resolution O 1s peak was deconvoluted into four peaks sites labelled I1, I2, I3, and I4 corresponding to 530, 531, 533, and 529 eV respectively. These peaks correspond to the O 1s electron binding energies in the various oxygen-containing layers in the sample structure [39]. Oxygen sites labelled I2 and I3 are assigned to P – O (oxygen singly bonded to phosphorus) and O (I1) is assigned to P = O (oxygen doubly bonded to phosphorus). The less intense peak (I4) at 529 eV is assigned to O–H bond. The high-resolution spectrum of the Gd 3d XPS peak is shown in Fig. 8.2d. The figure shows peaks of Gd at $3d_{5/2}$ (m = 1187 eV, s = 1196 eV, and t = 1198 eV) and $3d_{3/2}$ (m' = 1218 eV, s' = 1223 eV, and t' = 1230 eV) energy states respectively. The peaks are assigned to the $3d$-$4f$ transition and the separation between the peaks is due to the spin-orbit splitting of the $3d_{5/2}$ and $3d_{3/2}$ level. The reason for the multiplet feature in each of the peaks is the spin-orbit moment coupling of $3d$ and $4f$ states [40]. The Gd $3d$ peaks and their satellite structures are in agreement with the multiplet calculation. The data exhibit main and satellite structures as in other trivalent Gd compounds. The positions and the full width at half maximum (FWHM) of the peaks appear to have almost the same values for all rare–earth compounds [41]. Figure 8.2e shows the Pr $3d_{5/2}$ and $3d_{3/2}$ peaks of $Ca_5(PO_4)_3OH:Gd^{3+}$, Pr^{3+} powder phosphor [42]. The complex structure of $3d_{3/2}$ might be due to $4f$-induced multiplet. The data exhibit $3d_{5/2}$ (m = 929 eV and s = 933 eV) and $3d_{3/2}$ (m' = 950 eV and s' = 954 eV) structures. The extra peaks (t = 934 and t' = 958 eV) at higher binding energies for trivalent Pr compound are due to $3d4f^3$ final state multiplet structure [43, 44].

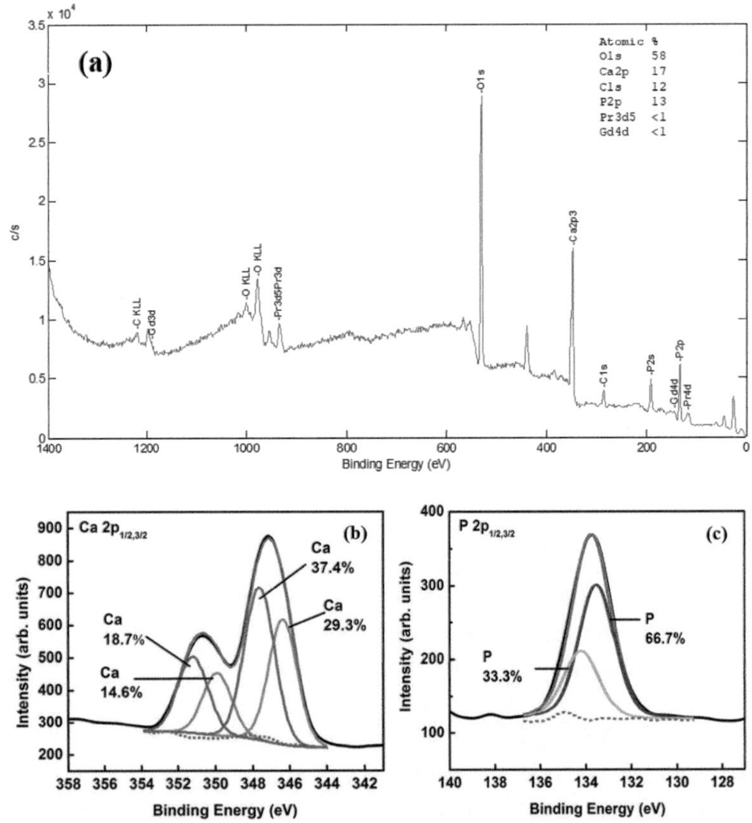

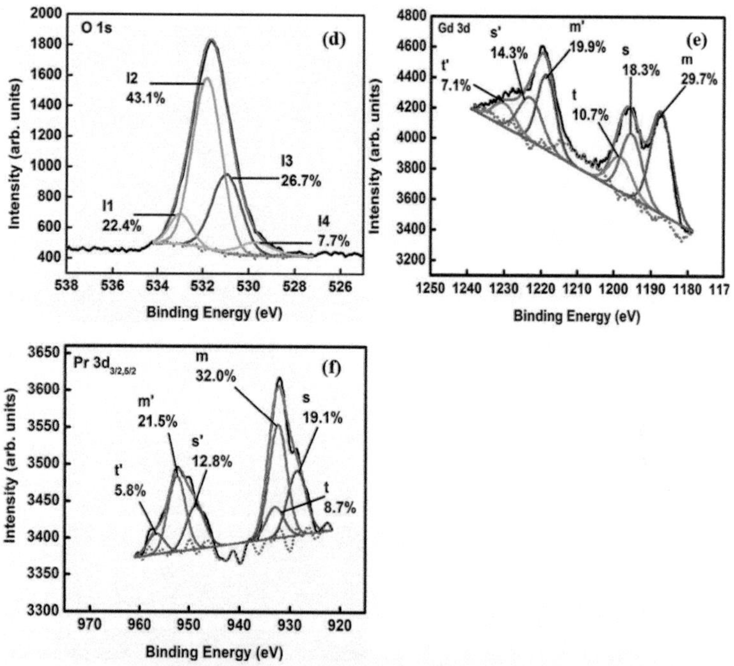

Figure 8.2 (a) XPS spectrum and (b–f) deconvulated Ca (2p), P (2p), O (1s), Gd (3d), and Pr (3d) peaks of $Ca_5(PO_4)_3OH:Gd^{3+}$, Pr^{3+} phosphor.

Figure 8.3a shows the SEM images of $Ca_3(PO_4)_2$ powder phosphor. The particle morphology composed of agglomerated nanosphere particles [45]. Figure 8.3b shows FE-SEM micrograph of $Ca_3(PO_4)_2:Gd^{3+}$, Pr^{3+} captured using extremely low landing energies achieved by deceleration of highly unstigmated electron beam leaving the objective lens. This imaging process is called a gentle beam landing [46] or specimen bias or retarding field. Landing energy is defined as the difference between the accelerating (column) voltage and stage-bias voltage. The gentle beam imaging is suitable for non-conductive specimens or insulators which experience a lot of surface charging. By using extremely low landing energies charging can be eliminated thereby enhancing the image quality without coating the surface [47]. The FE SEM images taken using the gentle beam confirm that the powder consisted of an agglomeration of hexagonal particles with some degrees of faceting. Figure 8.3c shows the HRTEM image of $Ca_3(PO_4)_2:Gd^{3+}$, Pr^{3+} showing faceted edges, forming a hexagonal

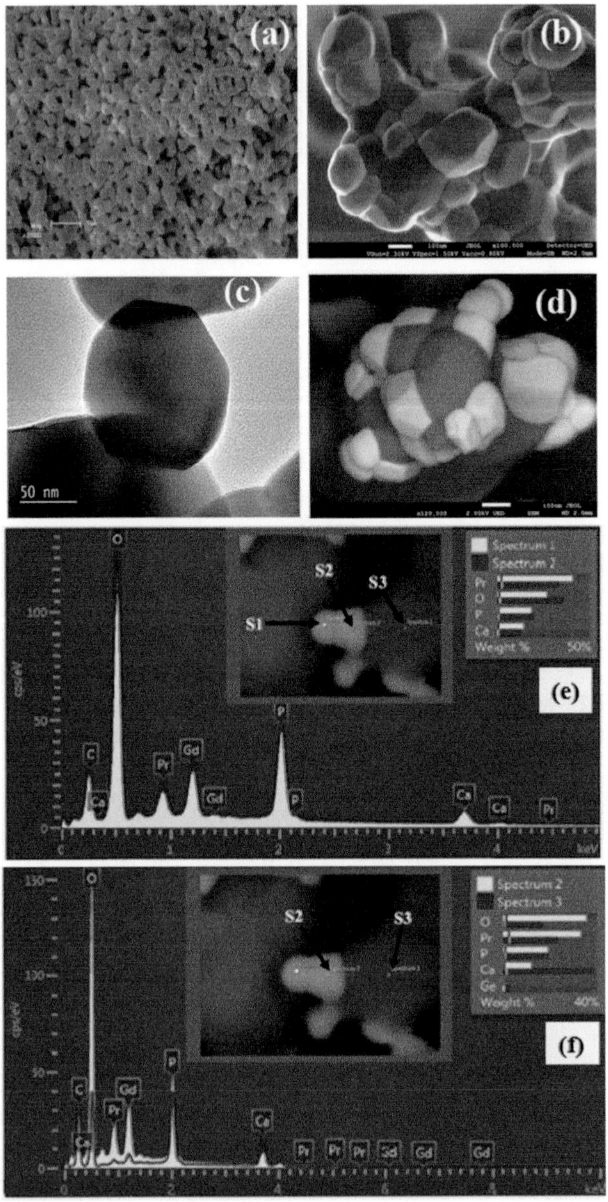

Figure 8.3 (a and b) SEM images of $Ca_3(PO_4)_2$ phosphors, (c) TEM images of $Ca_5(PO_4)_3(OH)$ phosphor, (d) FE-SEM image of $Ca_5(PO_4)_3(OH):Gd^{3+}$, Pr^{3+} phosphor captured using extremely low landing energies, and (e and f) EDS spectra of $Ca_5(PO_4)_3(OH):Gd^{3+}$, Pr^{3+} phosphor powders.

shape. Figure 8.3d shows high resolution FE-SEM images of the $Ca_5(PO_4)_3(OH)$:Gd^{3+}, Pr^{3+} phosphor powder. The images show that the powder consisted of agglomerated dark and bright particles. It is most likely that particles containing heavier atoms such as Gd^{3+} and Pr^{3+} backscattered electrons more strongly than particles containing lighter atoms and appeared to be much brighter. Figures 8.3e and f show the EDS spectra of $Ca_5(PO_4)_3(OH)$:Gd^{3+}, Pr^{3+} powder phosphor confirming the presence of all the elements in the material. Carbon (C) was probably coming from the carbon tape on which the powder was mounted. The inset in the top middle shows the FE-SEM image and the positions where EDS analysis was performed and these positions are marked S1, S2, and S3 while the one on the top right corner is the bar graph comparing the wt% of the elements at the chosen positions. Figure 8.3e compares the EDS spectra recorded at positions S1 and S2. The concentration of Pr^{3+} at S1 is 40% and it is approximately 33% at S2. Oxygen is about 25 and 35% at S1 and S2, respectively. Phosphorus (P) and calcium (Ca) are approximately 20 and 10% at and O were greater than that of Pr^{3+} at position S3. These results confirm that heavier atoms (Pr^{3+}) backscattered electrons more strongly than the lighter atoms resulting in the brighter contrast. In other words, particles with brighter contrast were dominated by Pr^{3+} while darker particles were dominated by relatively lighter O and Ca atoms. In Fig. 8.3f, the concentrations of Ca and O were greater than that of Pr^{3+} at position S3. Again, these results confirm that heavier atoms (Pr^{3+}) backscattered electrons more strongly than the lighter atoms resulting in the brighter contrast.

Figures 8.4a and b show the optical absorption spectra of $Ca_5(PO_4)_3(OH)$ and $Ca_3(PO_4)_2$ phosphor powders, respectively. Figures 8.4a and b consist of two absorption peaks at 207, 214, 285, and 295 nm. The peak at 207 nm (5.99 eV) and 214 nm are due to interband absorption of the host while the 285 nm and 295 nm (4.20 eV) absorption is assigned to intrinsic defects (e.g. vacancies or incidental impurities) in the host. Figure 8.4c shows a direct bandgap energy (E_g) of $Ca_5(PO_4)_3(OH)$ from a plot of $(\alpha h\upsilon)^2$ vs. $(h\upsilon)$. For crystalline materials with a direct band gap the dependence of the absorption coefficient (α) on the frequency υ can be approximated by:

$$\alpha h\upsilon = A(h\upsilon - E_g)^{\frac{1}{2}} \tag{8.2}$$

where A is a proportionality constant, h is Planck's constant and E_g is the bandgap energy and υ is the frequency of the radiation. The estimated E_g values were obtained by extrapolating the straight line to the axis where $(\alpha h\upsilon)^2 = 0$ and were found to be 5.00 ± 0.05 eV and 4.90 ± 0.05 eV for $Ca_5(PO_4)_3(OH)$ and $Ca_3(PO_4)_2$ phosphor powders, respectively. These values are similar to the values reported in the literature cited [48, 49].

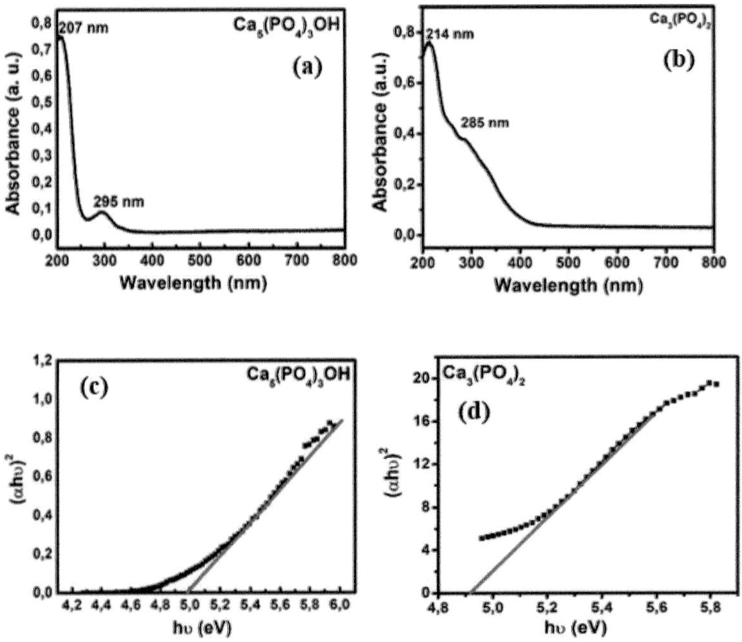

Figure 8.4 (a and b) Optical absorption spectra and (c and d) $(\alpha h\upsilon)^2$ vs. photon energy ($h\upsilon$) plot of $Ca_5(PO_4)_3(OH)$ and $Ca_3(PO_4)_2$ phosphor powders.

Figure 8.5a shows the PL (i) excitation and (ii) emission spectra of $Ca_3(PO_4)_2:Gd^{3+}$. The excitation was recorded when monitoring the 313 nm emission peak from Gd^{3+}. The excitation peaks at 245, 254, and 275 nm are assigned to $^8S_{7/2} \rightarrow {}^6D_j$ and $^8S_{7/2} \rightarrow {}^6I_j$ transitions of Gd^{3+}, respectively [50–52]. The excitation peak at 275 nm is more intense and sharper and it is assigned to $^8S_{7/2} \rightarrow {}^6I_j$ transitions of Gd^{3+} ion. The photoluminescence emission spectrum was recorded when monitoring the 275 nm excitation peak from Gd^{3+} ion. A sharp intense band was observed at 313 nm attributed to the $^6P_{7/2} \rightarrow {}^8S_{7/2}$ transition of the Gd^{3+} ion. The $^6P_{7/2}$ is more intense as it is the lowest lying level of the ^6P_j multiplet and gets populated by the fast multi-phonon de-excitation [53].

Figure 8.5b shows the PL (i) excitation and (ii) emission spectra of $Ca_3(PO_4)_2$:Pr^{3+}. The excitation spectra were recorded when monitoring the Pr^{3+} emission at 603 nm. The excitation peaks at 447 nm and 483 nm are due to Laporte forbidden 4f → 4f ($^3H_4 \rightarrow {}^3P_2$ and $^3H_4 \rightarrow {}^3P_0$) transitions of Pr^{3+} respectively [54, 55]. The emission spectra consist of a narrow-line emission peaking at 603 nm, assigned to the $^3P_0 \rightarrow {}^3H_4$ transition of Pr^{3+} ion [56].

Figure 8.5c shows the (i) excitation and (ii) emission spectra of $Ca_3(PO_4)_2$:Gd^{3+}, Pr^{3+} powder phosphor. The excitation spectra consist of minor peak at 275 nm from Gd^{3+} transition and a broad, intense peak centred at 222 nm. Although the broad band at 222 nm band is close to the 207 and 214 nm peak assigned to the band-to-band absorption of the host in the UV-vis spectrum shown in Fig. 8.4a and b, it cannot be assigned to the band-to-band excitation of the host because neither the undoped nor the Gd^{3+} single doped host could be effectively excited by the wavelength of 222 nm. Instead this is assigned to the $4f \rightarrow 4f5d$ transition of Pr^{3+} ion, consistent with the predicted location of the $4f5d$ band of Pr^{3+} in different inorganic solid hosts [57, 58]. The *f-f* excitation peaks from Pr^{3+} single doped $Ca_3(PO_4)_2$ observed in Fig. 8.5b do not occur in the excitation spectrum because of the emission of Gd^{3+} at 313 nm that was being monitored rather than the emission of Pr^{3+} at 603 nm. Although the Gd^{3+} excitation peaks are less intense than those in Fig. 8.5a, the UV emission intensity of Gd^{3+} at 313 nm improved considerably after incorporating Pr^{3+}. The maximum emission intensity was observed when the molar ratio of Gd^{3+} to Pr^{3+} was 10:2. The emission spectra of the co-doped samples show only the emission peak at 313 nm corresponding to the $^8S_{7/2} \rightarrow {}^6G_j$ transitions of Gd^{3+}, while the narrow line peak at 603 nm corresponding to $^3P_0 \rightarrow {}^3H_4$ transition of Pr^{3+} was suppressed. These data suggest that Pr^{3+} acted to harvest the primary excitation energy and transfer it non-radiatively to Gd^{3+}. The fact that the red emission from Pr^{3+} at 603 nm was suppressed implies that the rate of energy transfer from Pr^{3+} to Gd^{3+} was faster than the radiative emission by Pr^{3+} [59].

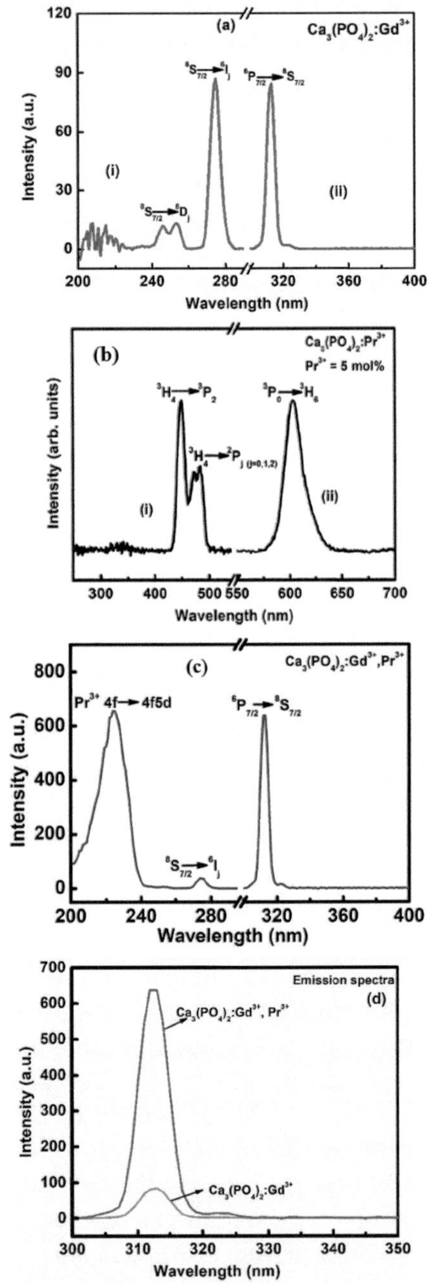

Figure 8.5 (a) PL spectra (i) excitation (λ_{emi} = 603 nm) and (ii) emission (λ_{exc} = 444 nm) of $Ca_3(PO_4)_2:Gd^{3+}$, (b) PL spectra (i) excitation (λ_{emi} = 603 nm) and (ii) emission (λ_{exc} = 444 nm) of $Ca_3(PO_4)_2:Gd^{3+}$, Pr^{3+}, (c) PL (i) excitation (λ_{emi} = 603 nm) and (ii) emission (λ_{exc} = 444 nm) spectra of $Ca_3(PO_4)_2:Pr^{3+}$, and (d) comparison of the maximum PL emission intensity of the $Ca_3(PO_4)_2:Gd^{3+}$ and $Ca_3(PO_4)_2:Gd^{3+}$, Pr^{3+} systems.

Figure 8.5d shows the emission spectra of $Ca_3(PO_4)_2$:Gd^{3+} and $Ca_3(PO_4)_2$:Gd^{3+}, Pr^{3+} phosphor powders. Comparing the emission spectra of the two systems, it was found that Pr^{3+} co-doping enhanced the Gd^{3+} emission at 313 nm [60]. The Gd^{3+} peak from $Ca_3(PO_4)_2$:Gd^{3+}, Pr^{3+} system is 5× more intense than the same peak observed from the Gd^{3+} single doped $Ca_3(PO_4)_2$.

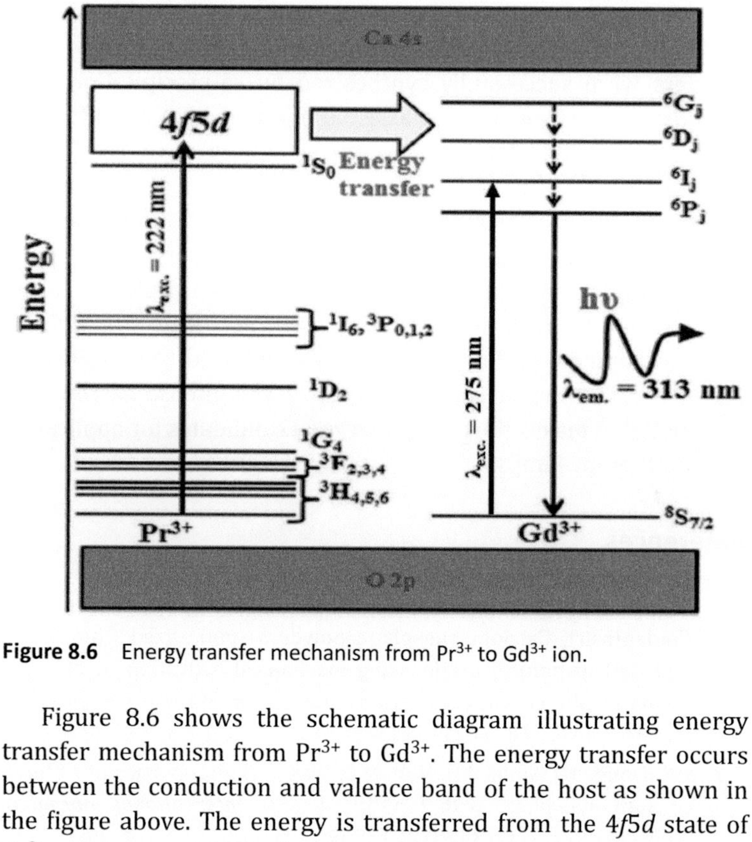

Figure 8.6 Energy transfer mechanism from Pr^{3+} to Gd^{3+} ion.

Figure 8.6 shows the schematic diagram illustrating energy transfer mechanism from Pr^{3+} to Gd^{3+}. The energy transfer occurs between the conduction and valence band of the host as shown in the figure above. The energy is transferred from the 4*f*5*d* state of Pr^{3+} to *f*-*f* states of Gd^{3+}. The energy difference between ground state and excited 4*f*5*d* state of Pr^{3+} ions matches well with the difference between ground state and excited state of *f*-*f* energy levels of Gd^{3+}. The electrons in Pr^{3+} ions get excited to 4*f*5*d* level by absorption of primary excitation energy, followed by non-radiative transfer to excited level of Gd^{3+} ions. Gd^{3+} relaxes to the ground state by emitting ultraviolet photons at 313 nm [60]. Energy transfer can occur in three ways, namely spectral overlap, resonance, and phonon mediated process. The energy transfer is attributed to phonon mediated process [61]. The rate of energy transfers from Pr^{3+} was faster than an electron capture by a hole that could result in radiative emission of Pr^{3+}. This results shows that Pr^{3+} is a good sensitizer of the 313 nm UVB emission of Gd^{3+}.

8.5 Conclusion

$Ca_3(PO_4)_2$:Gd^{3+}, Pr^{3+} and $Ca_5(PO_4)_3(OH)$:Gd^{3+}, Pr^3 phosphor powders were successfully synthesized by co-precipitation, urea combustion, and microwave-assisted methods. X-ray diffraction analyses confirmed formation and crystallinity of $Ca_5(PO_4)_3(OH)$ and $Ca_3(PO_4)_2$. The FE-SEM images showed an agglomeration of faceted spherical nanoparticles. The EDS data confirmed the presence of all the elements present in our phosphors. Singly doped $Ca_3(PO_4)_2$:Gd^{3+} exhibited a narrowband emission peak at 313 nm when excited through the $^8S_{7/2} \rightarrow {}^6G_j$ excitation of Gd^{3+} at 275 nm. Upon co-doping with Pr^{3+}, 313 nm emission of Gd^{3+} was enhanced due to energy transfer from Pr^{3+} to Gd^{3+}. The intense narrowband UVB emission makes the phosphors good candidates for application in phototherapy lamps.

References

1. T. Laonapakul, R. Sutthi, P. Chaaikool, S. Talangkun, A. Booma, P. Cindaprasirt, Calcium phosphate powders synthesized from $CaCO_3$ and CaO of natural origin using mechanical activation in different media combined with solid-state interaction, *Materials Science and Engineering C*, 118 (2021) 111333.
2. I.M. Hung, W.J. Shih, M.H. Hon, M.C. Wang, The properties of sintered calcium phosphate with [Ca]/[P] = 1.50, *International Journal of Molecular Sciences*, 13 (2012) 13569.
3. B. Mirhadi, B. Mehdikhani, N. Askari, Synthesis of nano-sized β-tricalcium phosphate via wet precipitation, *Processing and Application of Ceramics*, 5 (4) (2011) 193.
4. H. J. Kang, P. Makkar, A. R. Padalhin, G. H. Lee, S. B. Im, B. T. Lee, Comparative study on biodegradation and biocompatibility of multichannel calcium phosphate based bone substitutes, *Materials Science & Engineering C*, 110 (2020) 110694.

5. W. Feng, C. Feng, B. Wang, A. Jing, G. Li, X. Xia, G. Liang, An amorphous calcium phosphate for drug delivery: ATP provides a phosphorus source and microwave-assisted hydrothermal synthesis, *Materials Today Communications*, 25 (2020) 101455.
6. L.J. Yi, J.F. Li, M.G. Ma, Y.J. Zhu, Nanostructured calcium-based biomaterials and their application in drug delivery, *Current Medicinal Chemistry*, 27 (3) (2020) 5189.
7. D. Huang, B. He, P. Mi, Calcium phosphate nanocarriers for drug delivery to tumors: Imaging, therapy and theranostics, *Biomaterials Science*, 7 (10) (2019) 3942.
8. S. Bisso, S. Mura, B.P. Castagner, P. Couvreur, J.C. Leroux, Dual delivery of nucleic acids and PEGylated-bisphosphonates via calcium phosphate nanoparticles, *European Journal of Pharmaceutics and Biopharmaceutics*, 142 (2019) 142.
9. C.W. Lo, T.H. Lin, M. Ueno, M. Romero-Lopez, M. Maruyama, Y. Kohno, C. Rhee, Z.Y. Yao, M. Perez-Cruz, E. Meyer, S.B. Goodman, Optimization and characterization of calcium phosphate transfection in mesenchymal stem cells, *Tissue Engineering Part C Methods*, 9 (2019) 543.
10. C. Qi, S. Musetti, L.H. Fu, Y.J. Zhu, L. Huang, Biomolecule-assisted green synthesis of nanostructured calcium phosphates and their biomedical applications, *Chemical Society Review*, 10 (2019) 2698.
11. Y.R. Parauha, V. Chopra, S.J. Dhoble, Synthesis and luminescence properties of RE^{3+} (re = Eu^{3+}, Dy^{3+}) activated $CaSr_2(PO_4)_2$ phosphors for lighting and dosimetric applications, *Materials Research Bulletin*, 131 (2020) 110971.
12. V. Singh, G. Sivaramaiah, N. Singh, J.L. Rao, P.K. Singh, M.S. Pathak, D.A. Hakeem, EPR and PL studies on UVB-emitting gadolinium-doped $SrAl_{12}O_{19}$ phosphors, *Optik*, 158 (2018) 1227.
13. A. Hak, V.R. Shinde, A.K. Rengan, A review of advanced nanoformulations in phototherapy for cancer therapeutics, *Photodiagnosis and Photodynamic Therapy*, 33 (2021) 102205.
14. R.G. Kunghaktar, V.L. Barai, S.J. Dhoble, Synthesis route dependent characterization of $CaMgP_2O_7{:}Gd^{3+}$ phosphor, *Results in Physics*, 13 (2019) 102295.
15. L. Sinusaite, I. Grigoraviciute-Puroniene, A. Popov, K. Ishikawa, A. Kareiva, A. Zarkov, Controllable synthesis of tricalcium phosphate (TCP) polymorphs by wet precipitation: Effect of washing procedure, *Ceramics International*, 45 (2019) 12423.

16. Y.E. Greish, J.D. Bender, A. Singh, L.S. Nair, P.W. Brown, H.R. Allcock, C.T. Laurencin, Hydrolysis of Ca-deficient hydroxyapatite precursors in the presence of alanine-functionalized polyphosphazene nanofibers, *Ceramics International*, 39 (2013), 519.
17. A.F.M. Sunarso, S.R. Noor, R. Kasim, I.D. Othman, K. Ana, Ishikawa, synthesis of biphasic calcium phosphate by hydrothermal route and conversion to porous sintered scaffold, *Journal of Biomaterials and Nanobiotechnology*, 4 (2013) 273.
18. K.P. Sanosh, M.C. Chu, A. Balakrishnan, T.N. Kim, S.J. Cho, Preparation and characterization of nano-hydroxyapatite powder using sol–gel technique, *Bulletin of Materials Science*, 32 (2009) 465.
19. B.H. Chena, K.I. Chenb, M.L. Hoc, H.N. Chend, W.C. Chene, C.K. Wang, Synthesis of calcium phosphates and porous hydroxyapatite beads prepared by emulsion method, *Materials Chemistry and Physics*, 113 (2009) 365.
20. S.K. Ghosh, S.K. Roy, B. Kundu, S. Datta, D. Basu, Synthesis of nano-sized hydroxyapatite powders through solution combustion route under different reaction conditions, *Materials Science and Engineering B*, 176 (2011) 14.
21. F. Kermani, S. Kargozar, Z. Tayarani-Najaran, A. Yousefi, S.M. Beidokhti, M.H. Moayed, Synthesis of nano HA/βTCP mesoporous particles using a simple modification in granulation method, *Materials Science & Engineering C*, 96 (2019) 859.
22. G.D. Webler, M.J.M. Zapata, L.C. Agra, E. Barreto, A.O.S. Silva, J.M. Hickmann, E.J.S. Fonseca, Characterization and evaluation of cytotoxicity of biphasic calcium phosphate synthesized by a solid state reaction route, *Current Applied Physics*, 14 (2014) 876.
23. A.L. Daltin, S. Beaufils, T. Rouillon, P. Millet, J.P. Chopart, Calcium phosphate powder synthesis by out-of-phase- pulsed sonoelectrochemistry, *Ultrasonic – Sonochemistry*, 58 (2019) 104662.
24. J.G. Acheon, L. Robison, S. McKillop, S. Wilson, M.J. McIvor, B.J. Meenan, A.R. Boyd, TOFSIMS and XPS characterization of strontium in amorphous calcium phosphate sputter deposited coatings, *Material Characterization*, 171 (2021) 110739.
25. L. Reddy, T.J. Nkosi, P.L. Masiteng, A. Balakrishna, H.C. Swart, Violet-blue-shift of emission and enhanced luminescent properties of $Ca_3(PO_4)_2{:}Ce^{3+}$ phosphor induced by substitution of Gd^{3+} ions, *Current Applied Physics*, 20 (2020) 696.
26. I.R. Gibson, I. Rehman, S.M. Best, W. Bonfield, Characterization of the transformation from calcium-deficient apatite to beta-tricalcium

phosphate, *Journal of Materials Science: Materials in Medicine*, 11 (2000) 533.

27. K. Matsunaga T. Kubota, K. Toyoura, A. Nakamura, First-principles calculations of divalent substitution of Ca^{2+} in tricalcium phosphate, *Acta Biomaterialia*, 23 (2015) 329.
28. M. Gallo, B. Le Gars Santoni, T. Douillard, F. Zhang, L. Gremillard, S. Dolder, W. Hofstetter, S. Meille, M. Bohner, J. Chevalier, S. Tadier, Effect of grain orientation and magnesium doping on β-tricalcium phosphate resorption behaviour, *Acta Biomaterialia*, 89 (2019) 391.
29. M. Frasnelli, V.M. Sglavo, Effect of Mg^{2+} doping on beta-alpha phase transition in calcium phosphate (TCP) bioceramics, *Acta Biomaterialia*, 33 (2016) 283.
30. M. Bohner, B. Le Gars Santoni, N. Dobelin, β-tricalcium phosphate for bone substitution: Synthesis and properties, *Acta Biomaterialia*, 113 (2020) 23.
31. S. Hesaraki, S. Farhangdoust, K. Ahmadi, R. Nemati, M. Khorami, Phase evaluation and lattice parameters of strontium/magnesium incorporated beta-tricalcium phosphate bioceramics, *Journal of the Australian Ceramic Society*, 48 (2012) 166.
32. S. Kannan, F. Goetz-Neunhoeffer, J. Neubauer, S. Pina, P.M.C. Torres, J.M.F. Ferreira, Synthesis and structural characterization of strontium- and magnesium-co-substituted β-tricalcium phosphate, *Acta Biomaterilia*, 6 (2010) 571.
33. N. Lowry, M. Brolly, Y. Han, S. McKillop, B.J. Meenan, A.R. Boyd, Synthesis and characterisation of nanophase hydroxyapatite co-substituted with strontium and zinc, *Ceramics International*, 44 (2018) 7761–7770, https://doi.org/10.1016/j. ceramint.2018.01.206.
34. N. Lowry, Y. Han, B.J. Meenan, A.R. Boyd, Strontium and zinc co-substituted nanophase hydroxyapatite, *Ceramics International*, 43 (2017) 12070–12078, https://doi.org/ 10.1016/j.ceramint.2017.06.062.
35. J.G. Acheson, L. Robinson, S. McKillop, S. Wilson, M.J. McIvor, B.J. Meenan, A.R. Boyd, TOFSIMS and XPS characterisation of strontium in amorphous calcium phosphate sputter deposited coatings, *Materials Characterization*, 171 (2021) 110739.
36. V. Uskoković, X-ray photoelectron and ion scattering spectroscopic surface analyses of amorphous and crystalline calcium phosphate nanoparticles with different chemical histories, *Physical Chemistry Chemical Physics*, 22 (2020) 5531.
37. J.N. Hart, P.W. May, N.L. Allan, K.R. Hallam, F.Claeyssens, G.M. Fuge, M. Ruda, P.J. Heard, Towards new binary compounds: Synthesis of

amorphous phosphorus carbide by pulsed laser deposition, *Journal of Solid State Chemistry*, 198 (2013) 466.

38. W. Zhang, X. Li, R. Yang, Study on the change of silicon and phosphorus content in the condensed phase during the combustion of epoxy resin with OPS/DOPO, *Polymer Degradation and Stability*, 99 (2014) 289.
39. G.D. Khattak, A. Mekki, L.E. Wenger, X-ray photoelectron spectroscopy (XPS) and magnetic susceptibility studies of vanadium phosphate glasses, *Journal of Non-Crystalline Solids*, 355 (2009) 2148.
40. M. Ahren, L. Selegard, F. Soderlind, M. Linares, J. Kauczor, P. Norman, P.V. Kall, K. Uvdal, *Journal of Nanoparticle Research*, 14 (2012) 10061–10067.
41. J. Chung, J. Park, J.-G. Park, B.-H. Choi, S.-J. Oh, E.-J. Cho, H.-D. Kim, Y.S. Kwon, Photoemission study of rare-earth ditelluride compounds (ReTe2:Re = La, Pr, Sm, and Gd), *Journal of the Korean Physical Society*, 38 (6) (2001) 744.
42. A. Mekki, A. Ziq Kh, D. Holland, C.F. McConville, Magnetic properties of praseodymium ions in Na_2O–Pr_2O_3–SiO_2 glasse, *Journal of Magnetism and Magnetic Materials*, 260 (2003) 60.
43. H. Ogasawara, A. Kotani, R. Potze, G.A. Sawatzky, B.T. Thole, Praseodymium 3d- and 4d-core photoemission spectra of Pr_2O_3, *Physical Review B*, 44 (11) (1991) 5465.
44. S.W. Han, J.D. Lee, K.H Kim, H. Song, W.J. Kim, S.J Kwon, H.G Lee, C. Hwang, J.L. Jeong, J.S. Kang, Electronic structures of the CMR perovskites $R_{1-x}A_xMnO_3$ (R = La, Pr; A = Ca, Sr, Ce) using photoelectron spectroscopy, *Journal of the Korean Physical Society*, 40 (2002) 501.
45. N. Strutynska, O. Livitska, S. Prylutska, Y. Yumyna, P. Zelena, L. Skivka, A. Malyshenko, L. Vovchenko, V. Strelchuk, Y. Prylutskyy, N. Slobodyanik, U. Ritter, New nanostructured apatite-type (Na+, Zn^{2+}, CO_3^{2-})-doped calcium phosphates: Preparation, mechanical properties and antibacterial activity, *Journal of Molecular Structure*, 1222 (2020) 128932.
46. T. Liang, J. Qian, Y. Yuan, C. Liu, Synthesis of mesoporous hydroxyapatite nanoparticles using a template-free sonochemistry-assisted microwave method, *Journal of Material Science*, 48 (2013) 5334.
47. P.P. Mokena, M. Gohain, B.C.B Bezuidenhoudt, H.C. Swart, O.M. Ntwaeaborwa, Luminescent properties and particle morphology of $Ca_3(PO_4)_2$:Gd^{3+}, Pr^{3+} phosphor powder prepared by microwave assisted synthesis, *Journal of Luminescence*, 155 (2014) 288.

48. T. Mineharu, W. Masato, Y. Naoya, W. Toshiya, Band gap and photocatalytic properties of Ti-substituted hydroxyapatite: Comparison with anatase-TiO_2, *Journal of Molecular Catalysis A: Chemical*, 338 (2011) 18.

49. A. Mitsionis, T. Vaimakis, C. Trapalis, N. Todorova, D. Bahnemann, R. Dillert, Hydroxyapatite/titanium dioxide nanocomposites for controlled photocatalytic NO oxidation, *Applied Catalysis B*, 106 (2011) 398.

50. V. Singh, V.P. Singh, N. Singh, H. Jeong, V. Natarajan, Ultraviolet B radiation from Gd^{3+} doped $Sr_3P_4O_{13}$: Photoluminescence, electron paramagnetic resonance and evaluation of shielding parameters, *Radiation Physics and Chemistry*, 174 (2020) 108956.

51. S. Tamboli, G.B. Nair, S.J. Dhoble, D.K. Burghate, Energy transfer from Pr^{3+} to Gd^{3+} ions in BaB_8O_{13} phosphor for phototherapy lamps, *Physica B* 535 (2018) 232.

52. V. Singh, A. Prasad, A.S. Rao, S.W. Jung, N. Singh, M. Irfan, Luminescence and EPR studies of UVB emitting YPO_4 doped with Gd^{3+} ions, *Optik*, 225 (2021) 165804.

53. A. Ciric, S. Stojadinovic, Photoluminescence of ZrO_2:Gd^{3+} and ZrO2:Dy^{3+} coatings formed by the plasma electrolytic oxidation, *Journal of Alloys and Compounds*, 832 (2020) 154907.

54. S. Stojadinovic, A. Ciric, Highly-increased photoluminescence of Pr^{3+} due to the $Eu^{2+} \rightarrow Pr^{3+}$ energy transfer in Al_2O_3 coatings formed by plasma electrolytic oxidation, *Journal of Luminescence*, 226 (2020) 117407.

55. R. Demoulin, L. Khomenkova, C. Labbe, F. Gourbilleau, C. Castro, P. Pareige, Correlation of luminescence measurements to the structural characterization of Pr^{3+}-doped $HfSiO_x$, *Journal of Luminescence*, 235 (2021) 118004.

56. O.P. Golim, S. Huang, L. Yin, T. Yang, H. Zhou, W. Gao, T. Sohnel, P. Cao, Synthesis, neutron diffraction and photoluminescence properties of a whitlockite structured $Ca_9MgLi(PO_4)_7$:Pr^{3+} phosphor, *Ceramics International*, 46 (2020) 27476.

57. B. Kahouadji, L. Guerbous, D.J. Jovanovic, M.D. Dramicanin, A.H. Souici, Photoluminescence properties of nano-sized $(Lu_{1-x}Y_x)PO_4$:Pr^{3+} (x = 10, 20, 30, 40, 50 at. %) phosphor powders, *Optical Material*, 109 (2020) 110252.

58. Y. Sato, Y. Ohashi, Y. Iguchi, M. Kakihana, Synthesis and photoluminescence properties of $CaSnO_3$:Pr^{3+} prepared through

the amorphous metal complex method using a water-soluble Sn^{4+} complex, *Journal of Solid State Chemistry*, 277 (2019) 648.

59. P.P. Mokoena, I.M. Nagpure, Vinay Kumar, R.E. Kroon, E.J. Olivier, J.H. Neethling, H.C. Swart, O.M. Ntwaeaborwa, Enhanced UVB emission and analysis of chemical states of $Ca_5(PO_4)_3(OH):Gd^{3+}$, Pr^{3+} phosphor prepared by co-precipitation, *Journal of Physics and Chemistry of Solids*, 75 (2014) 998.
60. Y. Shimizu, Y. Takano, K. Ueda, UV emission from Gd^{3+} ions in Gd^{3+}–Pr^{3+} codoped $YAlO_3$ perovskite, *Journal of Luminescence*, 141 (2013) 44.
61. S. Tamboli, S.J. Dhoble, Influence of Li^+ charge compensator ion on the energy transfer from Pr^{3+} to Gd^{3+} ions in $Ca_9Mg(PO_4)_6F_2:Gd^{3+}$, Pr^{3+}, Li^+ phosphor, *Spectrochimica Acta Part A: Molecular and Biomolecular Spectroscopy*, 184 (2017) 119.

Chapter 9

The Luminescence Dynamics of Up- and Downconverting Rare Earth Ions Doped in $ZnTiO_3$ Nanoparticles

S. J. Mofokeng,[a] F. V. Molefe,[b] L. L. Noto,[c] and M. S. Dhlamini[c]

[a]*National Metrology Institute of South Africa (NMISA), CSIR Campus, Building 7, Meiring Naude Road, Brummeria Pretoria 0182, Private Bag X34, Lynnwood Ridge 0040, South Africa*

[b]*Department of Physics, Tshwane University of Technology, Private Bag X680, Pretoria 0001, South Africa*

[c]*Department of Physics, College of Science, Engineering and Technology, University of South Africa, Florida 1709, Johannesburg, South Africa*

sefakojmofokeng@gmail.com

Perovskite materials activated with rare-earth ions have been found potentially attractive for photoluminescence applications. The luminescence characteristics of rare earth ions activated perovskite $ZnTiO_3$ host lattice prepared by solid-state chemical reaction method is presented and discussed. The use of rare-earth ions in perovskite $ZnTiO_3$ phosphor as a luminescence centres is

Luminescent Nanomaterials
Edited by Odireleng Martin Ntwaeaborwa

ISBN 978-981-4968-11-9 (Hardcover), 978-1-003-27738-5 (eBook)
www.jennystanford.com

due to their interesting energy level structures that gives intense luminescence. The experimental results for the rare earth doped perovskite $ZnTiO_3$ phosphor materials and discussed their various potential application in the field of emission displays is presented. Particularly, the attention is focused on the principles of down- and upconversion in the prepared material to investigate their effects on the luminescence properties of rare earth ions. Finally, we discuss the advantages associated with advancing the luminescence properties for the prepared phosphor material.

9.1 Introduction

Luminescence is a phenomenon where light emission is produced from electron transitions when excited electrons de-excite to a ground state, radiatively [1]. There are various types of luminescence phenomena such as bioluminescence, phosphorescence, cathodoluminescence, and thermoluminescence, just to mention a few [2]. Bioluminescence is displayed by living organisms, like fireflies, glow-worms, and certain sea bacteria [3]. Phosphorescence and cathodoluminescence occur when a luminescent compound is irradiated with a photon and an electron beam (Fig. 9.1), respectively. Thermoluminescence occurs as a result of electron stimulation by thermal energy from electron trapping centres to a luminescent centre [1]. Figure 9.1 illustrates the luminescence phenomenon, where a luminescent compound is irradiated with an electron beam or a photon beam, which brings about electron excitation from the ground state to an excited state, from where they will de-excite radiatively back to the ground state.

It is common during the luminescence phenomenon for the energy of the excitation source, not to be the same as that of the emitted photons, due to internal energy losses. It is also common for lifetime of the emission not to be the same, because of different pathways of the electrons from the excited state to the ground state. The latter phenomena are explained in detail by Jablonski's diagram shown in Fig. 9.2. The first transition in Jablonski's diagram is the absorbance of a photon (a straight arrow pointing up). This transition is very fast, occurring at an order of 10^{-15} s. During this process, the energy of the photon is transferred to a particular electron, which is excited from the lower energy level to a higher energy level. Once the electrons are excited, they transit to a different state called eigenstate which correspond to the amount of energy transferred. The resulting electrons return to their initial states through several number of routes. The first transition includes a molecular vibrational relaxation and non-radiative process. The vibrational relaxation is also a very fast transition occurring at an order between 10^{-14} and 10^{-11} s, just immediately after the absorption process. In addition, some of the excited electrons lose energy and emit light radiation to an order between 10^{-9} and 10^{-7} s, and this process is called fluorescence [4, 5].

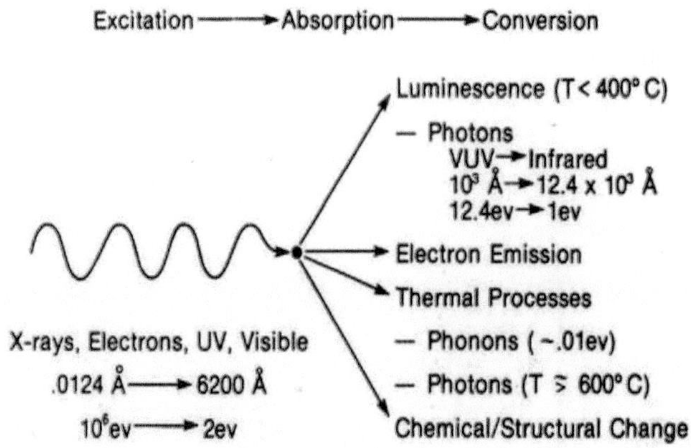

Figure 9.1 Electronic configuration of a generic luminescence system [3].

Yet another possible process that is shown by the Jablonski diagram is the intersystem crossing. This process involves slow forbidden transitions, where the excited electrons normally change spin multiplicity in an excited singlet state into an excited triplet state. This process allows the electrons to return to ground state through several routes. The direct transition in this case is a slow forbidden transition called phosphorescence which is in the order of 10^{-4}–10^{-1} s. This transition occurs when the radiative transition from the excited triplet state relaxes to a singlet ground state which emits prolonged light emission [6, 7].

Luminescent materials are mostly known as phosphors, which normally consist of a host matrix, rare earth (RE^{3+}) doped host matrix or combination of hosts matrix. Luminescent materials have many applications at scientifical and technological levels such as in optical communication and amplifiers, displays, solid-state lighting, solar cells, theft prevention, long afterglow, dosimetry, infrared detection, medical imaging, sensing, high density optical storage and many more [8, 9]. Research on applications of modern light technologies using individual luminescent host matrix at nanoscales is a hot topic. Several reports revealed that combining two or more materials on the nanoscales to form new material known as nanocomposites, doping and co-doping with suitable (RE^{3+}) ions can significantly improve the luminescent properties of the individual materials. The nanocomposites often consist of a hybrid of filler or a combination of nanomaterials in which one or both consist of different phases and dimensions (0-D, 1-D, 2-D, or 3-D) [10–13].

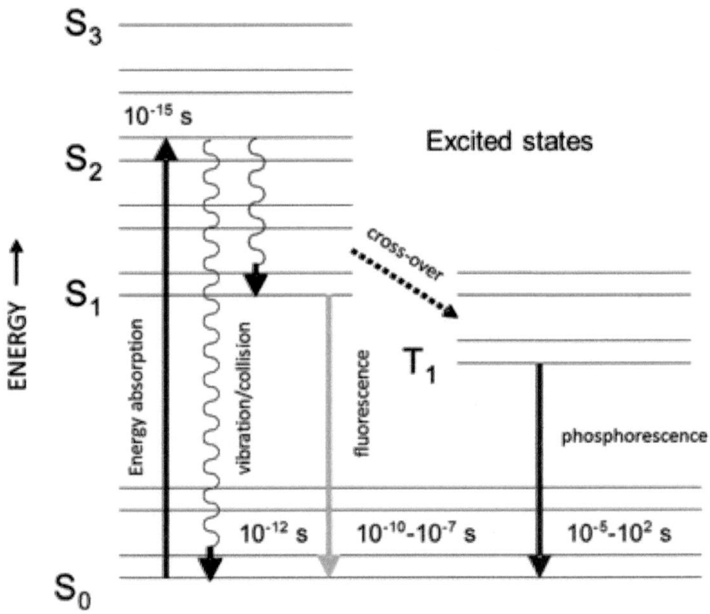

Figure 9.2 Jablonski diagram for luminescence system [4].

Among various host nanomaterials, $ZnTiO_3$ with a ABO_3 perovskite structure is one of the recent host materials at nanoscale used as microwave dielectric resonator, photocatalysis, gas sensors, detection of carbon monoxide and nitric acid, and efficient host matrix for luminescence of (RE^{3+}) ions because they can tune its catalytic and electronic properties [13]. In this chapter, the focus of our studies will be on $ZnTiO_3$ semiconductor as a host compound co-doped with Er^{3+}, Yb^{3+}, Eu^{3+} and Al^{3+} which can be used as luminescent materials in various applications. In addition, the investigation of energy transfer occurring through (RE^{3+}) ions is in line with the objective of the present chapter.

9.2 The Formation of the Crystal Structure of $ZnTiO_3$

Metal oxide semiconductors as luminescent materials are mostly designed and used as a host lattice that is doped with a (RE^{3+}) ion, which acts as the luminescence centers [11–14]. The role of a suitable host materials is to tune the luminescent lifetime for the luminescent materials that consist of intrinsic defects. These defects are known to trap and release the electrons to the luminescent centres, which result in extended duration of the emission that can last up to several hours or days. The host is the one that determines whether the f-f transition or the f-d transition of the dopant ion will be favoured, including the intensity of their signal [11]. Recently, researchers are working hard to develop different host materials that can produce high quantum yield. Some materials include mixed luminescent materials, doped luminescent materials and co-doped luminescent materials [12].

There are three crystallophic phases of zinc titanite compounds, originating from $ZnO–TiO_2$ composites. The two phases with a cubic structure are $Zn_2Ti_3O_8$ and Zn_2TiO_4, and the third one with both hexagonal and cubic structure is $ZnTiO_3$ [14]. As mentioned above, $ZnTiO_3$ is a perovskite structure which has important applications in the electronic industry as dielectric ceramic, bioimaging, photon catalysis, and light emitting diodes owing to its exceptional luminescence properties due to many oxygen vacancies [14, 15]. In our study, the compound was doped with up- and downconverting (RE^{3+}) ions, to convert unabsorbed solar spectra to absorbable spectra by dye sensitized solar cells. The polymorphs of zinc titanate are commonly produced from a combination of ZnO and TiO_2 materials at high temperatures. At lower synthetic temperatures, the cubic $Zn_2Ti_3O_8$ phase is formed. By further annealing the ZnO–TiO_2 compound at higher temperatures, the $ZnTiO_3$ perovskite is formed at temperatures $T < 900°C$. At annealing temperatures $T > 900°C$, the $ZnTiO_3$ decomposes into Zn_2TiO_4 phase [14].

$ZnTiO_3$ perovskites have unique properties which lend them as a promising luminescent material to host up- and downconverting RE^{3+} ions. These properties include wide energy band gap, high chemical and thermal stability, high electron mobility, cost-effective as well as being environmentally friendly [13]. The incorporation of up- and downconverting RE^{3+} gives the opportunity to tune the properties of $ZnTiO_3$ structures in a controlled manner to improve the luminescence properties. Some RE^{3+} ions act as activators and some as sensitizers in the host matrix [13–15]. By tri-doping the host matrix with activator, sensitizer and cations play a vital role to enhance the luminescence of the phosphor materials. The cations have small ionic radius which allows them to be easily substituted or introduced into interstitial sites of a lattice compound, to tailor the local crystal symmetry [13].

9.3 Rare Earth Ions Doped $ZnTiO_3$

Rare earth ions are mostly luminescent and are used to create luminescence centers of phosphor materials. Some RE^{3+} ions act as a wavelength converting system in a form of downconversion (DC) and some in a form of upconversion (UC) process to achieve a wealthy luminescent characteristic when incorporated in a suitable host lattices [14]. The well-known upconverting trivalent RE^{3+} ions from the infrared region to the visible region are Tm^{3+}, Er^{3+}, Pr^{3+}, Nd^{3+}, Ho^{3+} and Yb^{3+}. Ho^{3+}, Er^{3+}, Nd^{3+}, Pr^{3+}, and Tm^{3+} are mostly used as activators which serve as upconversion luminescence ions for 980 nm radiations. In addition, these RE^{3+} ions are employed as UC luminescence centres due to their intra-4f transitions that can easily be affected by external crystal field because the electrons at 4f orbital are shielded by the filled outer 5s and 5p orbits. Yb^{3+} ion is an interesting ion because it has proven to be an effective sensitizer in generating efficient upconversion luminescence when co-doped with UC activators. In addition, the selection of the host lattice is very important for RE^{3+} activating to achieve high UC luminescence since the fundamental luminescence properties for UC RE^{3+} ions depend on the phonon energy of the host, which result in the multi-phonon process leading to non-radiative decay of the excited energy [13–16]. Figure 9.3 depicts representative of upconversion mechanisms in RE^{3+} ions activated UC materials. The figure shows that the UC based lanthanides materials originate from various classes; namely, excited-state absorption (ESA), energy-transfer upconversion (ETU), cooperative sensitization upconversion (CSU), photon absorption (TPA) sometimes called photon avalanche and energy migration-mediated upconversion (EMU). An ESA process is a single emitting ion with energy levels absorbing two or more energy photons which promoted to higher excited state as shown in Fig. 9.3a. The main mechanisms for UC luminescence involve activator (GSA/ESA), activator–activator (GSA/ETU), sensitizer–activator (GSA/ETU) and energy migration upconversion (EMU) as illustrated in Fig. 9.3b–d [17, 18].

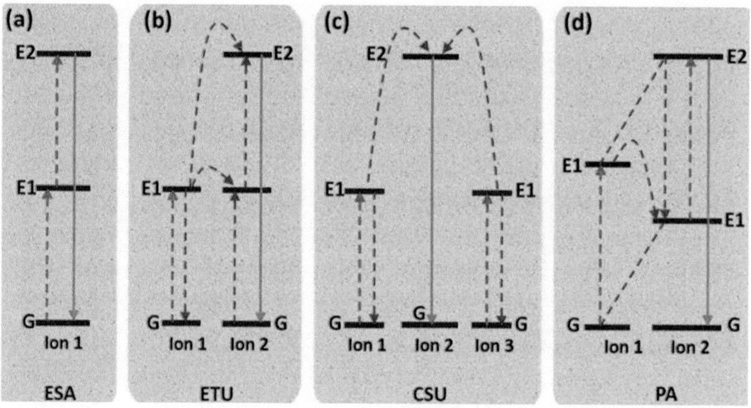

Figure 9.3 Upconversion mechanisms in RE^{3+} ions activated host materials [17].

Downconverting (DC) process involves RE^{3+} such as Dy^{3+}, Eu^{3+}, Tb^{3+}, Pr^{3+}, Ce^{3+}, Yb^{3+}, and Er^{3+}. To achieve efficient DC luminescent, these RE^{3+} ions are normally paired in this manner; Er^{3+}/Yb^{3+}, Pr^{3+}/Yb^{3+}, Dy^{3+}/Eu^{3+}, Tb^{3+}/Yb^{3+}, and Ce^{3+}/Tb^{3+} [19]. DC is an interesting technique to improve the quantum efficiency in luminescent nanomaterials. This phenomenon can be achieved in acquiring a high quantum yield through converting a single high photon energy into two or more lower photon energies. The advantage of incorporating up- and downconverting RE^{3+} ions in $ZnTiO_3$ include the production of a phosphor having several luminescence spectra, i.e., ultraviolet, and blue luminescence contributed by $ZnTiO_3$ [20] and various luminescence emission contributed by RE^{3+} starting from ultraviolet up until infrared emission bands [14–22].

Unlike UC, the DC mechanisms include processes such as quantum cutting (QC) and downshifting (DS) as shown in Fig. 9.4. The QC process is the simplest downconversion process in which the absorption of a single photon result in two-photon visible emission (Fig. 9.4a). The QC is always occurring in a single RE^{3+} ion doped luminescent materials. The DS also involves two or three RE^{3+} ions of which the excited photon from RE^{3+} ion (1) de-excite into the lower energy level. The latter transfer the energy to RE^{3+} ion 2 through path (2), where one or both ions subsequently emit a visible light (Fig. 9.4) [14].

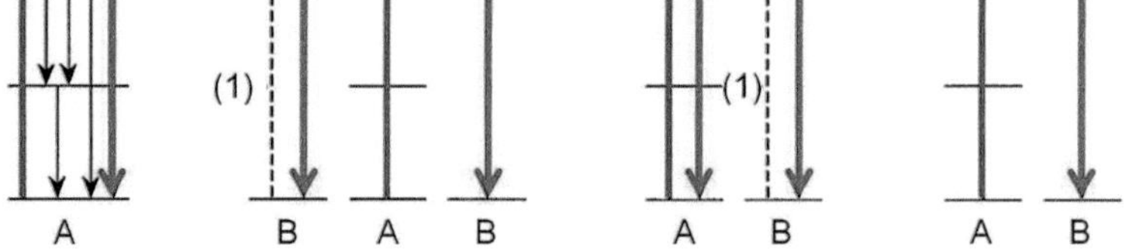

Figure 9.4 Downconversion mechanisms in RE^{3+} ions activated host materials [14].

9.4 Synthesis of $ZnTiO_3$ Nanoparticles

Technological advances have shown that new methods need to be developed to finetune materials in mesoscale and microscale towards attaining materials at smaller scale called nanomaterials. Downscaling materials to nanometer (1–100 nm) revealed many of their properties that were unknown. This introduced many materials applications in the textile, medicine, and electronics industries, just to mention a few. The attention of most researchers was attracted by an American physicist, Richard Feynman when he delivered a classical lecture where he mentioned that "There's Plenty of Room at the Bottom" [23]. It is without doubt that the innovative ways of producing nanomaterials have been at the heart of every scientist, globally. They have explored the immense possibilities afforded by miniaturization materials using different techniques, in pursuit of advancing technology. Numerous methods such as sol-gel [24], ceramic [25], solution combustion synthesis (SCS) [26] and conventional solid-state reaction [13] have been explored towards attaining $ZnTiO_3$ nanoparticles, irrespective of challenges to crystallize Zn_2TiO_4 and TiO_2 at ~1223 K [27]. These preparation methods undergo either bottom-up process (assemble using atoms/molecules) or top-down process (dissociate bulk into finer) to generate nanomaterials. Upon selecting the best method from afore mentioned, one needs to know the intended structure, size, shape, and dimension beforehand because the final product can be monocrystalline, polycrystalline, or amorphous [13, 24–27].

The current study focuses on the use of conventional solid-state reaction for preparation of $ZnTiO_3$ nanomaterials for luminescence applications due to outstanding results accrued [13, 15]. Conventional solid-state reaction is a top-down process frequently used to produce complex oxides using simple oxides, carbonates, nitrates, hydroxides, and various metal salts. An overview of conventional solid-state reaction is shown in Fig. 9.5 [28]. Preparation of complexes entails chemical decomposition reactions in which solid reactants are mixed by ball miller prior calcination at higher temperatures to promote interdiffusion of the cations. The final product and rate of solid-state reaction relies on many components such as reaction conditions, structural properties of the reactants, surface area of the solids, reactivity, and thermodynamic free energy [29]. This method is preferred because it is cost effective and large production can be achieved in relatively simple manner.

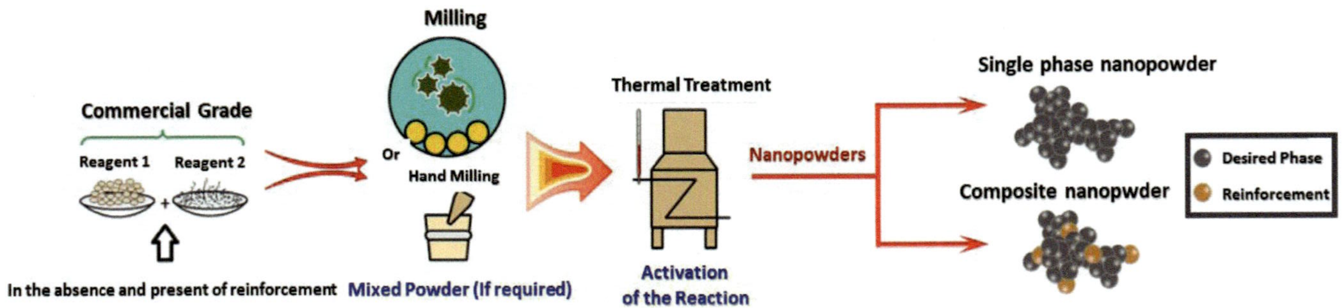

Figure 9.5 Schematic representation of steps followed during conventional solid-state reaction [28].

9.5 Er^{3+} Activated Crystal Structure of $ZnTiO_3$

The photoluminescence emission spectra of $ZnTiO_3$ doped with different concentrations of Er^{3+} ions that were measured under 980 nm, using a continuous laser diode is shown in Fig. 9.6. The figure shows that photoluminescence spectra of Er^{3+} doped $ZnTiO_3$ originate from $^2H_{11/2} \rightarrow {}^4I_{15/2}$, $^4S_{3/2} \rightarrow {}^4I_{15/2}$ and $^4F_{9/2} \rightarrow {}^4I_{15/2}$ intrinsic electronic transitions of Er^{3+} at room temperature. These transitions correspond to intense two green emissions centred at ~527 nm, ~545 nm, and a single red emission centred at ~665 nm. The results show that the green emissions were split in two bands, which is attributed to splitting sublevel manifolds induced by spin-orbit coupling and breakdown of the degeneracy of the electronic interaction. Figure 9.6b shows the corresponding photoluminescence intensity variation with Er^{3+} concentration. The intensity was first increasing from 0.2 to 0.8 mol% and then decreases from 1.0 to 1.2 mol% doping due to the concentration quenching. The increase in PL intensity indicated an increase in luminescent centres of Er^{3+} ions in $ZnTiO_3$ compound with increasing Er^{3+} doping concentration [13, 14]. However, the concentration quenching at higher doping concentration Er^{3+} can be explained in terms of distance between Er^{3+} luminescence centres. That is, the distance is reduced and the luminescent centres get close to each other and increase their trap carriers, which non-radiatively relaxes due to phonon interaction [11]. In addition, Fig. 9.6b shows that the integrated green to red intensity ratios varies from 1.76 to 2.67 for the doping concentrations. These emission intensity ratios and the variation in concentration of Er^{3+} ions did not influence the position of the luminescence emission bands [13, 14].

9.6 Yb^{3+} Co-activated Crystal Structure of $ZnTiO_3$:Er^{3+}

Figure 9.7 depicts the photoluminescence spectra of the optimized $ZnTiO_3$:Er^{3+}, Yb^{3+} phosphors. The green and red emission peaks from Er^{3+} transitions are centered at ~527 nm, ~545 nm, and ~665 nm, respectively.

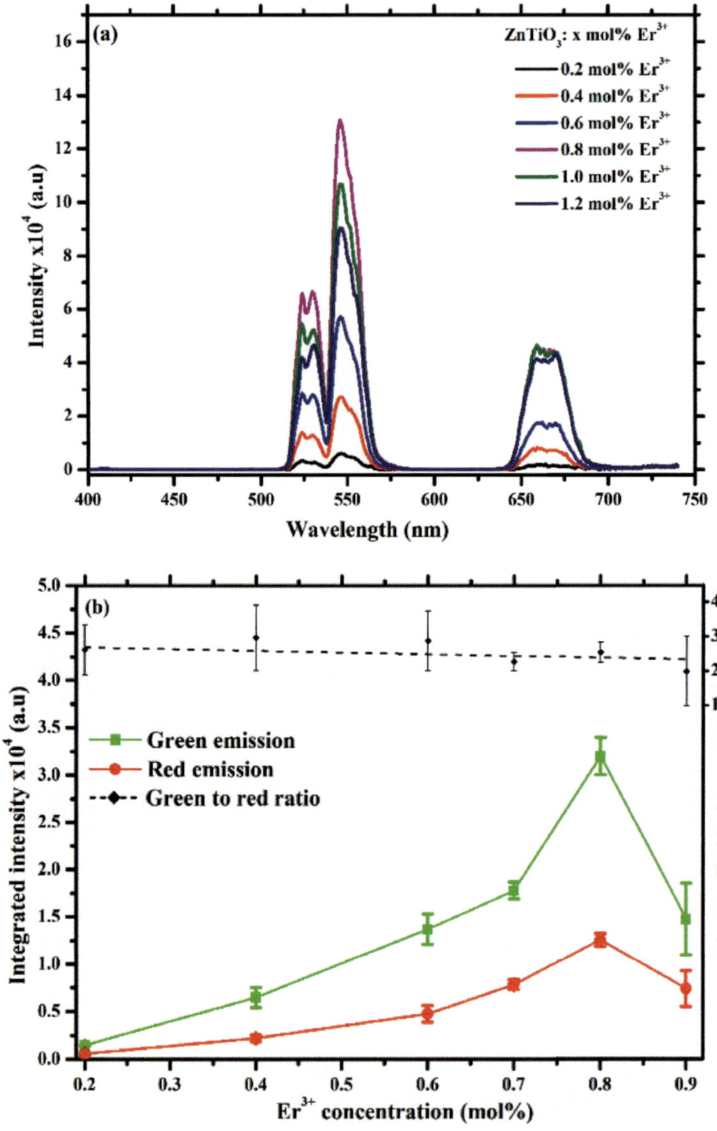

Figure 9.6 PL spectra of (a) $ZnTiO_3$:$x$$Er^{3+}$ phosphor and (b) comparison of integrated intensities of green and red emissions for the phosphors [13, 14].

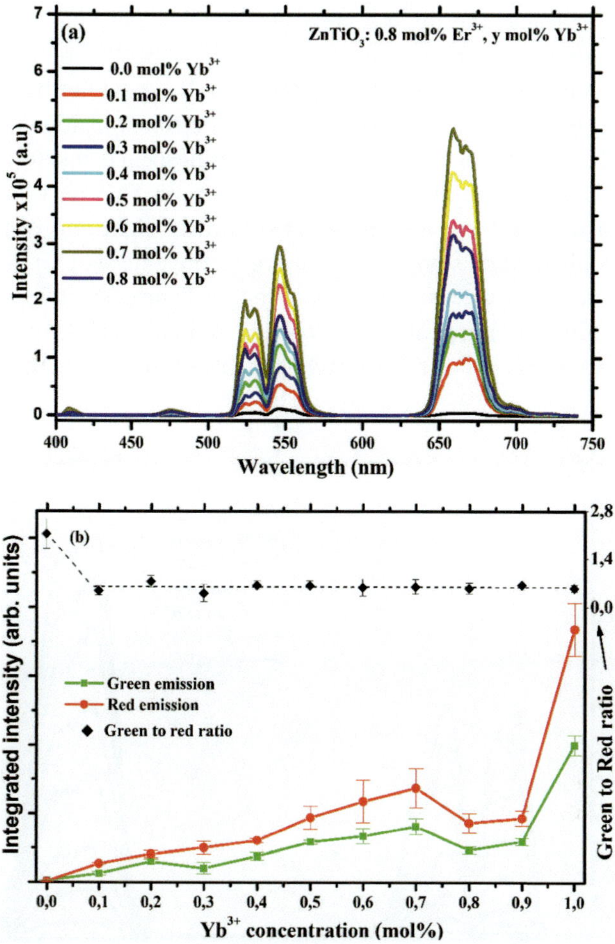

Figure 9.7 Upconversion emission spectra of (a) $ZnTiO_3$:08Er^{3+}, *y*Yb^{3+}(*y* = 0.1–0.8 mol%) and (b) comparison of integrated intensities of green and red emissions of $ZnTiO_3$:0.8Er^{3+}, *y*Yb^{3+} phosphors [13, 14].

As indicated earlier, the photoluminescence of Er^{3+} doped $ZnTiO_3$ shows the highest emission intensity at a concentration of 0.8 mol% of Er^{3+}. However, Fig. 9.7 shows that Yb^{3+}, which acts as a sensitizer, enhances the photoluminescence properties of Er^{3+} ions. The optimum emission intensity of the $ZnTiO_3$:Er^{3+}, Yb^{3+} phosphor was determined by fixing the optimum concentration of the Er^{3+} ion with variable concentration of Yb^{3+} ion, respectively. Upon variation of Yb^{3+} content, the distinct new weak violet and blue emissions due to effective energy transfer of $Yb^{3+} \rightarrow Er^{3+}$ were observed. These emissions are assigned to $^2H_{9/2} \rightarrow {}^4I_{15/2}$ (violet band) [14, 15, 30] and $^4F_{7/2} \rightarrow {}^4I_{15/2}$ (blue band) electron transitions of Er^{3+} [14, 15, 31]. To achieve high photoluminescence inside $ZnTiO_3$ matrix, the incorporation of Er^{3+} coupled with Yb^{3+} ions lead to a synergistic effect on absorbing more infrared-light and convert it into visible light by significantly enlarging the emission range of the electromagnetic waves in the visible region. Figure 9.8 indicates that the overall emission peaks shapes of Er^{3+} and Yb^{3+} remain unaffected by the variation in laser power when incorporated inside the matrix [13].

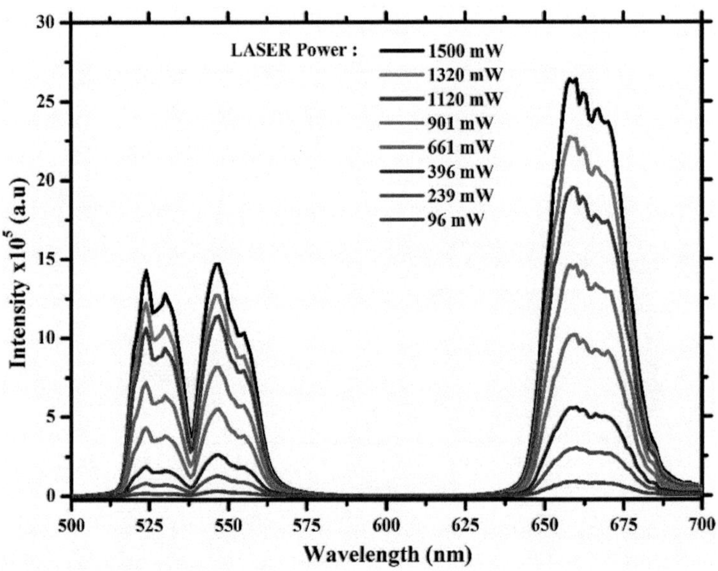

Figure 9.8 Power dependence of luminescence spectra of $ZnTiO_3$:Er^{3+}, Yb^{3+} within a laser power range of 96 to 1500 mW. The phosphor was excited at a wavelength of 980 nm [13, 14].

As mentioned earlier, Yb^{3+} tend to sensitize Er^{3+} by generation energy transfer in $ZnTiO_3$:Er^{3+}, Yb^{3+} phosphor (as illustrated in Fig. 9.9). There are absorption and emission pathways inside $ZnTiO_3$:Er^{3+}, Yb^{3+} phosphor for possible energy transfer between the RE^{3+} ions. In this case, the 980 nm excitation is absorbed by the ground state abortion (GSA) and the electrons moves to metastable levels where they de-excite to 4f–4f electron transitions of Er^{3+}. In the co-doped compound, it is commonly known that the absorption cross section of Yb^{3+} ions is relatively large compared to Er^{3+} ions at 980 nm. As shown in Fig. 9.9, the Yb^{3+} and Er^{3+}ions are excited from their fundamental ground state energy levels to the excited energy levels via GSA. The electrons from $^4I_{11/2}$ energy level of Er^{3+} which was populated through energy transfer from Yb^{3+} ions are further populating $^4I_{9/2}$, $^2F_{7/2}$, and $^2G_{11/2}$ energy levels of Er^{3+} through an excited state absorption (ESA) of Er^{3+}. During this process, some ions move to excited metastable $^2H_{9/2}$, $^4F_{7/2}$, $^4S_{3/2}$, and $^4F_{9/2}$ levels through non-radiative transition. The excited electrons from different metastable states de-excite to the $^4I_{15/2}$ state and emit violet, blue, green, and red light [13, 32, 33].

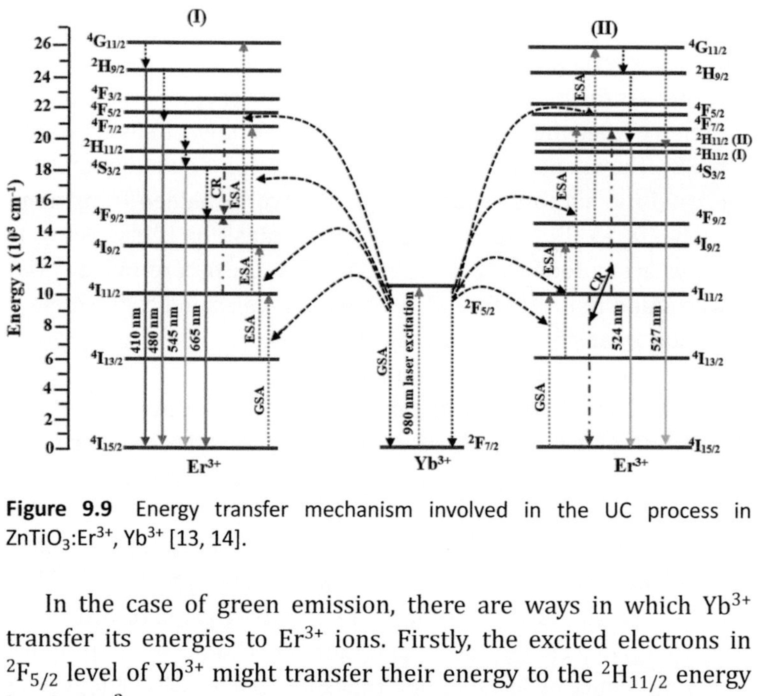

Figure 9.9 Energy transfer mechanism involved in the UC process in $ZnTiO_3$:Er^{3+}, Yb^{3+} [13, 14].

In the case of green emission, there are ways in which Yb^{3+} transfer its energies to Er^{3+} ions. Firstly, the excited electrons in $^2F_{5/2}$ level of Yb^{3+} might transfer their energy to the $^2H_{11/2}$ energy level of Er^{3+} via energy transfer (ET). Secondly, this phenomenon can happen through a direct absorption of a 980 nm NIR photon via GSA from $^4I_{15/2}$ energy level to $^4I_{13/2}$ energy level. However, the excited ions in $^2H_{11/2}$ energy level are influenced by a process called cross-relaxation (CR) energy transfer between Er–Er ions in the matrix.

The mechanism of this kind of energy transfer is given by $^4I_{11/2}$ + $^4I_{11/2} \rightarrow {}^4F_{7/2}$ + $^4I_{15/2}$. Lastly, the electrons in $^2H_{11/2}$ energy level of Er^{3+} decay radiatively to $^4I_{15/2}$ ground level in the form of two transitions, $^2H_{11/2}$ (I) and $^2H_{11/2}$ (II) of the UC green emission, which confirm the Stark splitting of the $^2H_{11/2}$ energy level of Er^{3+} ions in the compound [13, 32–34].

9.7 Al^{3+} Charge Compensated Crystal Structure of $ZnTiO_3$:Eu^{3+}

Figure 9.10 shows the emission spectra for $ZnTiO_3$:Eu^{3+} phosphor and charge compensated by Al^{3+}. The emission spectra consist of a series of peaks which are assigned to the intra-configurational 4f–4f transitions of Eu^{3+} ions. It is well known that Eu^{3+} ions occupy a site with inversion symmetry in the host lattice when the $^5D_0 \rightarrow {}^7F_1$ transition is the most prominent; however, it occupies a site without symmetry when the $^5D_0 \rightarrow {}^7F_2$ transition is the most intense emission [35–38]. Kumar et al. [35] also studied the photoluminescence of Eu^{3+} and the authors reported that the two forbidden transitions; namely, $^5D_0 \rightarrow {}^7F_1$ magnetic dipole (MD) transition is insensitive to the crystal environment and the $^5D_0 \rightarrow {}^7F_2$ to the nature. In Fig. 9.10a, it was found that the $^5D_0 \rightarrow {}^7F_2$ transition emission has the highest emission intensity. This is because the electric dipole transition is hypersensitive in crystal environment [35–38].

To enhance the photoluminescence and color purity of $ZnTiO_3$:Eu^{3+}, the concentration of Eu^{3+} was increased and the optimum concentration was found to be 2 mol% after which decreased due to concentration quenching (Fig. 9.10a). The optimized sample was further enhanced by introducing charge compensation as illustrated in Fig. 9.10b. The incorporation of Al^{3+} to improve the photoluminescence of Eu^{3+} in $ZnTiO_3$ can be attributed to the following effects: the compensation of charge, which promote a charge balance by forming electronegativity defects that neutralize the electron-positivity in the host lattice. This clearly states that the Al^{3+} cations occupy and substitute the adjacent lattice of Ti^{4+} ion in the form of Al^{3+} to maintain the charge balance in the $ZnTiO_3$:Eu^{3+} phosphor by suppressing the defects and leading to a higher and stronger emission of Eu^{3+}.

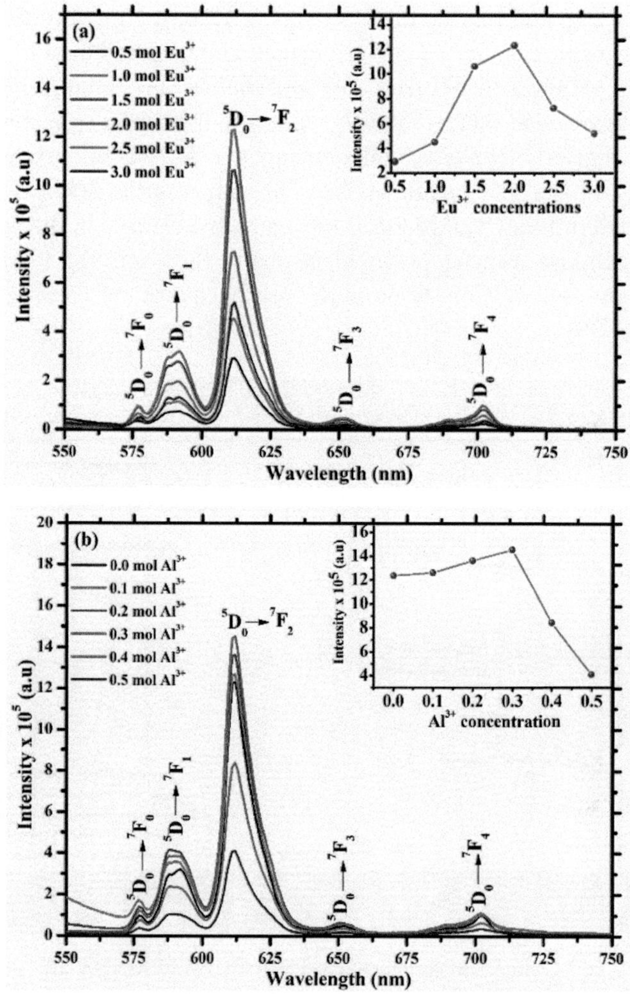

Figure 9.10 Photoluminescence spectra of (a) $ZnTiO_3:xEu^{3+}$ and (b) $ZnTiO_3$:2mol%Eu^{3+}, yAl^{3+} [15].

Figure 9.11 depicts the charge transfer mechanism in $ZnTiO_3:Eu^{3+}$, Al^{3+} phosphor. Upon the absorption of the excitation, the material tends to compensate for current charge imbalance by generating forbidden negative oxygen (O_i'') point defects which neutralizes the additional positive zinc (Zn_i'') point defect within the host. The O_i'' are temporary trapping and releasing the holes (h^+) to F_j (j = 0, 1, 2, 3, and 4) states, and also, positive zinc (Zn_i'') point defects trap and release the electrons (e^-). During this process, the negative zinc vacancies ($V_{Zn}^{\cdot\cdot}$) is formed and trap and hold h^+ while oxygen vacancies (Vo″) trap and hold the carriers. In addition, the carriers trapped by Zn_i'' and $V_{Zn}^{\cdot\cdot}$ point defects tend to recombine non-radiatively. However, introducing the Al^{3+} in the compound occupying the Ti^{4+} positions serve to suppress the effect of non-radiative process. The reduced non-radiative centres in the system ensures more transfer of the electrons to Eu^{3+}, and the improved emission of Eu^{3+} is observed in the presence of Al^{3+} as a co-dopant [15, 38, 39].

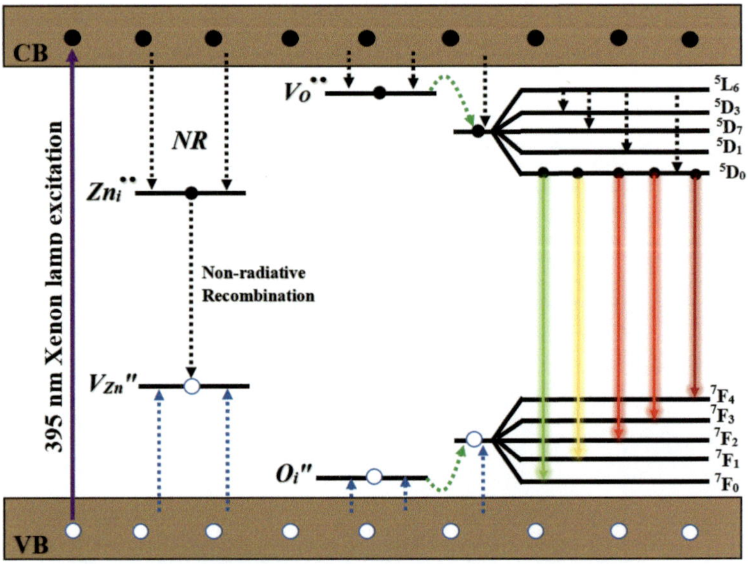

Figure 9.11 Charge transfer mechanism in $ZnTiO_3:Eu^{3+}$, Al^{3+} [14, 15].

9.8 Conclusion

Perovskite nanomaterials as up- and downconverting phosphors have been investigated and explored for different applications in different field of research due to tremendous physical and chemical properties. The synthesis method has an effect on the luminescence properties of perovskite luminescence materials. The solid-state chemical reaction, which involves direct firing of stoichiometric amounts of reagent oxide compounds, has proven to be a cost-effective method to prepare perovskite materials. The relatively low phonon energy of $ZnTiO_3$ makes it a suitable host lattice for upconverting RE^{3+} ions and achieve high luminescence characteristics due to efficient energy transfer. The cross-relaxation process of energy transfer mechanism was shown to be effective for the improvement of the luminescence properties of Er^{3+} and Yb^{3+} doped in $ZnTiO_3$. The optimum concentrations of activator and sensitizer in $ZnTiO_3$: Er^{3+}, Yb^{3+} were observed to be 0.8 and 0.7 mol%, respectively.

Acknowledgments

The academic work presented in this chapter was supported financially by NRF Freestanding, Innovation and Scarce Skills Masters and Doctoral Scholarships (Grant no. 102590).

References

1. Murthy, K.V.R. and Virk, H.S. (2014). Luminescence phenomena: An introduction, *Defect Diffus. Forum.*, **347**, pp. 1–34.
2. Peng, X., Xu, L., Liu, Y., Zhang, G., Cao, Y., Hong, F., and Yan, K. (2017). Experimental measurement of tip vortex flow field with/without cavitation in an elliptic hydrofoil, *J. Hydrodynam. B*, **29**, pp. 939–953.
3. DeLuca, J.A. (1980). An introduction to luminescence in inorganic solid, *J. Chem. Educ.*, **57**, pp. 541–545.
4. Garcia, J.A., Moreno, J.M., Perales, F.J., Romero, J., Sanchez, P., and Gomez-Robledo, L. (2012). Fluorescence: An interdisciplinary phenomenon for different education levels, *Eur. J. Phys.*, **3**, pp. 30–35.
5. Omar, M. and Patterson, H. (1999). *Encyclopedia of Spectroscopy and spectrometry*. (University of Maine, Orono, ME, USA) pp.1186–1207.
6. Jaffe, H.H. and Miller, A.L. (1966). The fates of electronic excitation energy, *J. Chem. Educ.*, **43**, pp. 469–478.
7. Priestley, E.B. and Haug, A. (1968). Phosphorescence spectrum of pure crystalline naphthalene, *J. Chem. Phys.*, **49**, pp. 622–631.
8. Naick, B.N., Damodaraiah, S., Prasad, V.R., Lakshmi, R.P.V., and Ratnakaram, Y.C. (2019). Judd-Ofelt analysis and luminescence studies on Dy^{3+}-doped different phosphate glasses for white light emitting material applications, *Optik*, **192**, pp. 162980–162989.

9. Zhou, A., Song, F., Han, Y., Song, F., Ju, D., Adnan, K., Liu, L., and Feng, M. (2018). Color-tunable emission by adjusting sensitizer (Yb^{3+}) and excitation power of 980 nm in NaGdTiO4:Yb3^{+}/Tm^{3+}/Er^{3+} phosphors for light emitting diodes, *J. Lumin.*, **194**, pp. 225–230.
10. Mofokeng, S.J., Kumar, V., Kroon, R.E., and Ntwaeaborwa, O.M. (2017). Structure and optical properties of Dy^{3+} activated sol-gel ZnO-TiO_2 nanocomposites, *J. Alloys Compd.*, **711**, pp. 121–131.
11. Noto, L.L., Mofokeng, S.J., Molefe, F.V., Swart, H.C., Tebele, A.S., and Dhlamini, M.S. (2020). Luminescence dynamics of rare earth-doped $CaTiO_3$ phosphors. In Dhoble, S.J., Pawade, V.B., Swart, H.C., and Chopra V. (eds.) *Spectroscopic of Lanthanide Doped Oxide Materials* (Elsevier) pp. 57–86.
12. Wang, D., Xu, B., Zou, k., Sun, M., Dong, G., and Liu, J. (2018). Effect of Er^{3+} concentration on the photoluminescence of Y_2O_3/ZnO upconversion films, *Opt. Mater.*, **83**, pp. 124–130.
13. Mofokeng, S.J., Noto, L.L., Kroon, R.E., Ntwaeaborwa, O.M., and Dhlamini, M.S. (2020). Up-conversion luminescence and energy transfer mechanism in $ZnTiO_3$: Er^{3+}, Yb^{3+} phosphor, *J. Lumin.*, **223**, pp. 117192–117200.
14. Mofokeng, S.J. (2020). Investigation of up and down-converting rare earth ions doped $ZnTiO_3$ for photovoltaic applications, PhD thesis (University of South Africa, South Africa) pp. 9–121.
15. Mofokeng, S.J., Noto, L.L., and Dhlamini, M.S. (2020). Photoluminescence properties of $ZnTiO_3$:Eu^{3+} phosphor with enhanced red emission by Al^{3+} charge compensation, *J. Lumin.*, **228**, pp. 117569–117575.
16. Luitel, H.N., Ikeue, K., Okuda, R., Chand, R., Torikai, T., Yada, M., and Watari, T. (2014). Bright orange upconversion in a ZnO–TiO_2 composite containing Er^{3+} and Yb^{3+}, *Opt. Mater.*, **36**, pp. 591–595.
17. Li, D., Agren, H., and Chen, G. (2018). Near infrared harvesting dye-sensitized solar cells enabled by rare-earth upconversion materials, *Chem. Soc. Rev.*, **47**, pp. 8526–8537.
18. Wang, Y., Zheng, K., Song, S., Fan, D., Zhang, H., and Liu, X. (2018). Remote manipulation of upconversion luminescence, *Chem. Soc. Rev.*, **47**, pp. 6473–6485.
19. Mofokeng, S.J. (2015). Synthesis and characterization of zinc oxide-titanium dioxide nanocomposites co-doped with dysprosium and europium, Master's thesis (University of the Free State, South Africa) pp. 8–109.

20. Ye, C., Wang, Y., Ye, Y., Zhang, J., and Li, G.H. (2009). Preparation and photoluminescence of undoped $ZnTiO_3$ thin films, *J. Appl. Phys.*, **106**, pp. 033520–033524.

21. Mendez-Ramos, J., Acosta-Mora, P., Ruiz-Morales, J.C., Sierra, M., Redondas, A., Ruggiero, E., Salassa, L., Borges, M.E., and Esparza, P. (2015). Rare-earth doped colour tuneable up-conversion ZBLAN phosphor for enhancing photocatalysis, *Opt. Mater.*, **41**, pp. 98–103.

22. Tamrakar, R.K., Bisen, D. P., Upadhyay, K., and Sahu, I.P. (2015). Comparative study and role of Er^{3+} and Yb^{3+} concentrations on upconversion process of Gd_2O_3:Er^{3+}, Yb^{3+} phosphors prepared by solid-state reaction and combustion method, *J. Phys. Chem. C*, **119**, pp. 21072–21086.

23. Feynman, R.P. (1959). Plenty of room at the bottom. *Am. Phys. Soc.*, pp. 1–7.

24. Mengting, L.I. and Baoxiang, J.I.A.O. (2015). Synthesis and photoluminescence properties of $ZnTiO_3$:Eu^{3+} red phosphors via sol-gel method, *J. Rare. Earths*, **33**, pp. 231–238.

25. Bernert, T., Ruiz-Fuertes, J., Bayarjargal, L., and Winker, B. (2015). Synthesis and high (pressure, temperature) stability of $ZnTiO_3$ polymorphs studied by Raman spectroscopy, *Solid State Sci.*, **43**, pp. 53–58.

26. Raveendra, R.S., Prashanth, P.A., Krishna, R.H., Bhagya, N.P., Nagabhushana, B.M., Naika, H.R., Lingaraju, K., Nagabhushana, H., and Prasad, B.D. (2014). Synthesis, structural characterization of nano $ZnTiO_3$ ceramic: An effective azo dye adsorbent and antibacterial agent, *J. Asian Ceram Soc.*, **2**, pp. 357–365.

27. Yamaguchi, O., Morimi, M., Kawabata, H., and Shimizu, K. (1978). Formation and transformation of $ZnTiO_3$, *J. Am. Ceram. Soc.*, **70** pp. C97–C98.

28. Nasiri-Tabrizi, B., Baradaran, S., Zalnezhad, E., and Basirun, W.J. (2014). Applications of electron microscopy in mechanochemistry. In A. Méndez-Vilas (ed.), *Microscopy: Advances in Scientific Research and Education* (Formatex) pp. 791–802.

29. Kumar, V., Tiwari, S.P., Ntwaeaborwa, O.M., and Swart, H.C. (2020). Luminescence properties of rare-earth doped oxide materials. In *Spectroscopy of Lanthanide-Doped Oxide Materials* (Elsevier) pp. 345–364.

30. Elleuch, R., Salhi, R., Deschanvresc, J.L., and Maalej, R. (2015). Highly efficient NIR to visible upconversion in a ZnO:Er, Yb thin film deposited

by a AACVD atmospheric pressure process, *RSC Adv.,* **5**, pp. 60246–60253.

31. Sales, A.J.M., Sousa, D.G., Rodrigues, H.O., Costa, M.M., Sombra, A.S.B., Freire, F.N.A., Soares, M.J., Graça, M.P.F., and Kumar, J.S. (2016). Power dependent upconversion in Er^{3+}/Yb^{3+} co-doped $BiNbO_4$ phosphors, *Ceram. Int.,* **42**, pp. 6899–6905.

32. Ma, Y., Xiang, G., Zhang, J., Liu, Z., Zhou, P., Liu, W., Tang, X., Jiang, S., Zhou, X., Li, L., Luo, Y., and Jin, Y. (2018). Upconversion properties and temperature sensing behaviors in visible and near-infrared region based on fluorescence intensity ratio in $LuVO_4$: Yb^{3+}/Er^{3+}, *J. Alloys Compd.,* **769,** pp. 325–331.

33. Cheng, X., Ma, X. Zhang, H., Ren, Y., and Zhu, K. (20117). Optical temperature sensing properties of Yb^{3+}/Er^{3+} codoped LaF_3 upconversion phosphor, *Physica B,* **521**, pp. 270–274.

34. Kim, D.H., Ryu, J.H., Chung, J.H., Shim, K.B., and Cho, S.Y. (2011). Green upconversion luminescence of $Li^{+}/Er^{3+}/Yb^{3+}$ tridoped $CaWO_4$ and effects of Er^{3+} concentration, *J. Electrochem. Soc.,* **158**, pp. J345–J348.

35. Kumar, A.N., Jnaneshwara, D.M., Kumar, M.R.A., Basavaraju, N., Ravikumar, C.R., Murthy, H.C.A., Nagabhushana, H., Girishe, K.M., Ashwini, S., and Naikg, R. (2020) Photoluminescence and electrochemical performances of Eu^{3+} doped $La_{10}Si_6O_{27}$ nanophosphor: Display and electrochemical sensor applications, *Appl. Surf. Sci. Adv.,* **1**, pp. 100026–100035.

36. Kolesnikov, I.E., Mamonova, D.V., Kurochkin, M.A., Kolesnikov, E.Y., and Lahderanta, E. (2021). Eu^{3+}-doped ratiometric optical thermometers: Experiment and Judd-Ofelt modelling, *Opt. Mater.,* **112**, pp. 110797–110802.

37. Padilla-Rosales, I., Falcony, C., Sosa, R., Aguilar-Frutis, M., Alarcon-Flores, G., and Gonzalez, F. (2020). Effect of crystalline ordering on the luminescent properties of Eu^{3+}-doped aluminum oxide nanophosphors, *J. Solid State Chem.,* **288** pp. 121427–121439.

38. Noto, L.L., Pitale, S.S., Gusowki, M.A., Terblans, J.J., Ntwaeaborwa, O.M., and Swart, H.C. (2013). Afterglow enhancement with In^{3+} codoping in $CaTiO_3$:Pr^{3+} red phosphor, *Powder Technol.,* **237**, pp. 141–146.

39. Baira, M., Bekhti-Siad, A., Hebali, K., Bouhani-Benziane, H., and Sahnoun, M. (2018). Charge compensation mechanisms in favor of the incorporation of the Eu^{3+} ion into the ZnO host lattice, *Physica B,* **537** pp. 296–300.

Chapter 10

Luminescent Nanomaterials: A Step Towards Solid-State Lighting and Display

Mohan Lal Meena,[a] Sudipta Som,[b] Chung-Hsin Lu,[a,b,c] Somrita Dutta,[d] Rajan Kumar Singh,[a] and Shawn D. Lin[a]

[a]*Department of Chemical Engineering, National Taiwan University of Science and Technology, Taipei 10617, Taiwan, R.O.C.*

[b]*Department of Chemical Engineering, National Taiwan University, Taipei 10617, Taiwan, R.O.C.*

[c]*Advanced Research Center of Green Materials Science & Technology, Taipei 10617, Taiwan, R.O.C.*

[d]*Department of material science and engineering, National Taiwan University, Taipei 10617, Taiwan, R.O.C.*

sudipta.som@gmail.com

Semiconductor nanocrystals have expanded excessive concern due to their unique characteristics and enhanced luminescence properties. A unique classification of semiconductor nanocrystals is quantum dots. In recent years, the quantum dots are being used in diverse regime including light-emitting diodes, display

Luminescent Nanomaterials
Edited by Odireleng Martin Ntwaeaborwa

ISBN 978-981-4968-11-9 (Hardcover), 978-1-003-27738-5 (eBook)
www.jennystanford.com

technology, photovoltaic industries, and biomedical applications. In the present chapter, ample info about altered features of QDs is delivered including the basics, structural and optical properties. The historical advancement of solid-state lighting from bulk phosphors to nanomaterials are discussed in the theoretical background of quantum confinement and compositional engineering. With a focus on the light emission and display aspects, the present chapter introduces the motive for the instability and the way out for the wide application of QDs in the light-emitting and display technologies.

10.1 Introduction and Background

Solid-state lighting technology in recent years is piercing profoundly into several emergent machinery sectors including indoor/outdoor and automotive lighting, medical applications and daily products. The modern encroachment of light-emitting diodes (LEDs) grabs the lighting industry as an alternative to the solution for a superior lighting system. LEDs are two-lead semiconductor light sources. As a sufficient voltage is applied to the device, the electron can recombine with holes and release energy in the form of light. Red and green LEDs were first developed in early years, and later on LEDs of low wavelength emanated into market. However, the recent trend clasps the production of white light from the LEDs. The white light emitting diodes (WLEDs) are now recognized as the next generation lighting devices and superior lighting technology compared to traditional incandescent and fluorescent lamps because of their inherent advantageous such as energy-saving, robust/tough devices, long-lifetime in use and environment-friendly features [1–4].

As the LED can only emit monochromatic light, several approaches have taken to obtain white light from LEDs. Until now, four methods are in the front row, such as mixing blue, green, and red LEDs, combining blue LED with a yellow phosphor or blue LED with green and red phosphors, and combing UV-LED with blue, green, and red phosphors [5]. Multichip LEDs uses direct combination of blue, green, and red LEDs to generate white light with high quality. However, this approach is seldom used owing to the drawbacks including high cost, different degradation trend and different operating voltage of each LED. Presently, a blue LED chip combined with $Y_3Al_5O_{12}:Ce^{3+}$ (YAG:Ce^{3+}) yellow phosphor is the most mature method for fabricating commercial WLEDs [6]. YAG:Ce^{3+} is an excellent phosphor for blue-chip (450–470 nm) excitation and displays high photoluminescence (PL) and quantum efficiency (QE). The method has produced the first WLEDs in the world. This approach is still the most efficient way to generate white light. However, because of deficiency of a red light in color quality of white emission is not upto the mark [7]. It is difficult to make a warm white LEDs with large color-rendering index (CRI), small correlated color temperature (CCT), and have great impact on general illumination via this approach. The third way is combing a blue LED with green and red phosphors. This method didn't attract attention due to the low performance of green and red phosphors. The fourth way is the combination of a UV LED with blue, green, and red phosphors. This method is beneficial to generate white light with high CRI, and phosphors suitable for UV excitation are much more than that suitable for blue excitation [8, 9]. As a result, many researchers have tried to develop phosphors for this kind of WLEDs.

Phosphors, generally known as luminescent materials, can be demarcated as the materials that can absorb energy convert it into radiation of light. Rare earth (RE) doped phosphor materials draws a vast attention among all kinds of phosphors. The doped inorganic materials with desirable luminescence properties can be obtained via a doping skill, which encompasses the substitution of a low concentration of atoms or ions into host lattices [10]. The influence of host crystal induce the rare earth elements to emit light upon excitation in the desirable wavelength from the ultraviolet (UV), through the visible region, to the infrared light area. The excitation radiation is generally absorbed by the activator ions in the host [11]. Absorbing the excitation energy, the activator ions propel to the excited state. Then the ions come back from excited to ground level via emission of radiation, which is luminescence [12]. However, in many other materials and ions non radiative return to the ground state generally prevents the luminescence and hence, suppressing non radiative process is the main strategy for the luminescent materials. Luminescence materials are generally characterized with the measurement of spectral energy distribution for the emission and excitation, known as emission and excitation spectra respectively and the conversion efficiency. The luminescent materials can be applied to street lighting, car lighting, automobile, indoor enlightenment, parking space display, digital display, and various display applications [13].

Recent development of displays devices including emissive and nonemissive displays draws significant requirement for the phosphors to produce visible light. The prerequisites for the phosphors in display devices should placate high quantum efficiency, reflectivity to visible light, longevity, and stability. Micro LED is deliberated to be the next generation display technology having extensive viewing angle, admirable contrast, low energy consumption, and nippy response time. For recent display technology morphology, particle size distribution (PSD), and zeta potential are the important physical requirements. However, the phosphors reported involve typically of micrometer-sized particles. The distended particles exhibit reduced packing density, which persuade high light-scattering effect. It results in inferior resolution. Moreover, comparatively large amount of materials are required [14]. To overcome this issue, nanoscopic phosphors with high packing density and low light-scattering ability comes into limelight [15]. LED surfaces with phosphors of large particle size can create large voids between crystals, triggering high light-scattering effects to diminute the resolution of LED. Luminescent Nanomaterials, a newly incipient arena at the edge of science, provide challenges to both fundamental research and breakthrough development of technologies in various areas such as electronics, photonics, nanotechnology, display, lasing, detection, optical amplification, fluorescent sensing to biomedical engineering, and environmental control.

10.1.1 Luminescent Nanomaterials

In the present time, available nanomaterials can be classified in two categories based on sources, artificial nanomaterials and naturally available nanomaterials.

For example, naturally available nanoparticles are lunar and volcanic ash along with mineral complexes, while artificial nanomaterials are man-made nanomaterials with diameter maximum diameter of 100 nm such as nanoparticles, nanotubes, nanophosphors, and quantum dots [16, 17].

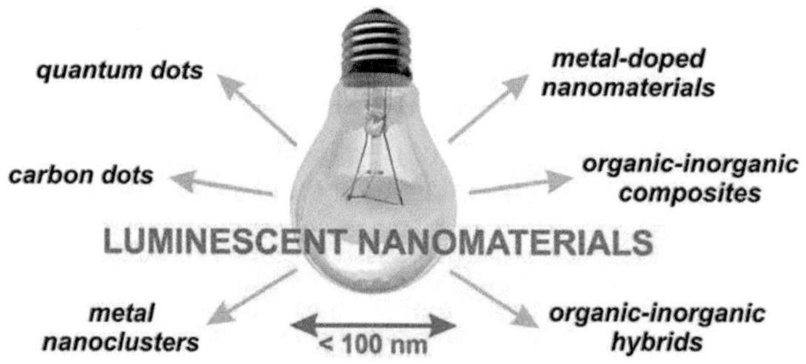

Figure 10.1 Broad range of luminescent nanomaterials. Reproduced by permission of The Royal Society of Chemistry [18].

Luminescent nanomaterials on a fundamental level can be applied to all the utilizations of traditional bulk-phosphors (i.e., X-ray detectors, fluorescent lamps, light-emitting diodes). Figure 10.1 shows the wide range of luminescent nanomaterials [18]. This chapter mainly focuses on quantum dots among various luminescence nanoparticles. Luminescent nanomaterials are broadly utilized due to unique optical properties [12].

10.1.2 Quantum Dots

Quantum dots are exceptional categories of semiconductor nanocrystals (NCs). The clear understanding of optical, electronic, and transport properties of quantum dots are established in recent years via the solid theoretical framework [19]. Among quantum dots, the cadmium and lead based colloidal quantum dots are extensively investigated in recent times as tunable light emitters with highly saturated, efficient, and pure emission color [20, 21]. Indium phosphide QDs are also capable substitute of the toxic heavy metals Pb and Cd and draws broad attention to consumer lighting and display market [22]. However, the basics of quantum dot is the prerequisite for further researches for light emitting applications.

10.1.3 Basics of Quantum Dots

One of the greatest assets of quantum dots in light emitting applications is their size-tunable color emission. Their tunable band gap is governed by the quantum size effect. For a bulk semiconductor, the energy dispersion curve of a direct band gap semiconductor is shown in Fig. 10.2. Quantum dots (QD) are widely used in lighting applications owing to size tunable color emission. The variable band gap is caused by quantum confinement effect [23]. Nanomaterials generally offers a bridge among molecular domain and bulk materials [24, 25], along with novel opportunities of electronic and optical properties based on the size of nanomaterials [26]. When nanomaterials show unique changes in optical properties depending on the confinement in any dimension is known as quantum confinement structure.

The dimension dependence of optical and electronic properties is major effect of quantum confinement effect. With decrease in the size of nanomaterials they behave like nano-clusters and results in evolution of discrete states [27]. These properties have great impact on absorption on nanostructures. Adequately nanoclusters act as semiconductors, as outlined in Fig. 10.3. Alternately, low-dimensional materials, which are originally semiconductor in bulk, do not possess efficient repetitive crystal lattice to converge onto dense bands and band gap energy and a bandgap energy comparing to the bulk bandgap (Fig. 10.3b). The above mentioned can be explained using a bottom up principle from M.O. theory and orbital energy splitting in conversion of M.O. to cluster orbitals from dense bands for bulk crystals. Top-down model can also be utilized to explain the above mentioned phenomenon via quantum confinement effect.

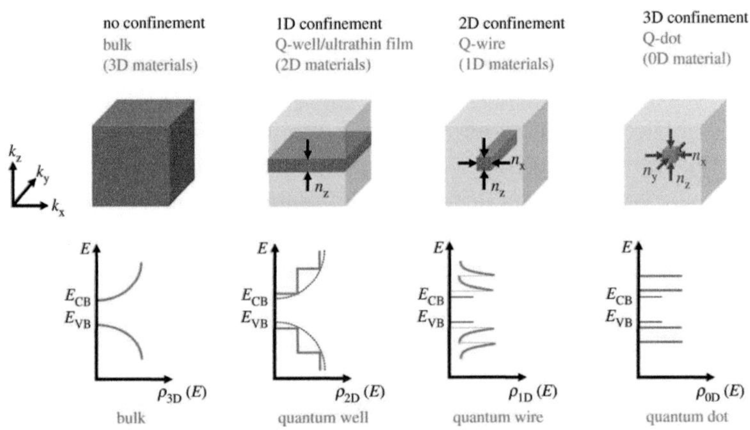

Figure 10.2 Energy dispersion curve of bulk to nanomaterials [23].

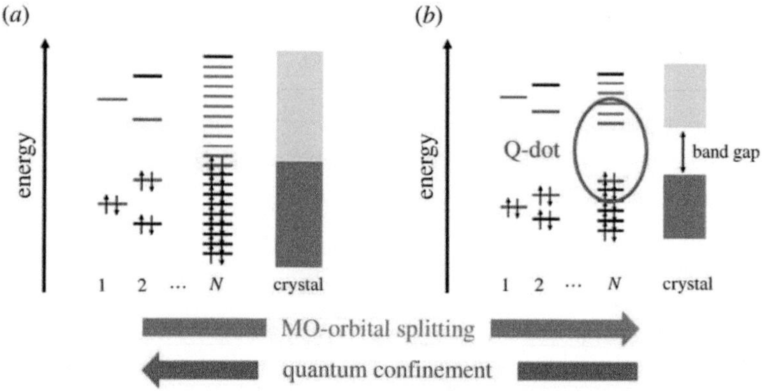

Figure 10.3 Size dependent quantum confinement in (a) metal and (b) semiconductor [23].

First way to explain the optical properties in quantum confinement semiconductors by Brus [24], the photo-generated electron–hole pair is approximated with a model Hamiltonian for a bound state inside effective mass approximation in an effective potential encompassing, is given by [24, 28]

$$H = \frac{-\hbar^2}{2m_e^*} + \nabla_e^2 \frac{-\hbar^2}{2m_h^*}\nabla_h^2 - \frac{e^2}{\varepsilon|r_e - r_h|} + \sum_{n=1}^{\infty} \alpha_n \frac{e^2}{2}\frac{(r_e^{2n} + r_h^{2n})}{R^{2n+1}} \qquad (10.1)$$

The effective masses of hole and electron are denoted by m^*_h and m^*_e. Polarizability and dielectric constants are shown via α and ε. r_e and r_h represents the radial position of electron and hole inside the spherical potential of radius R. According to Brus, electron hole wave function can be separated as $\psi = \Phi(r_e)\ \Phi(r_h)$. The effective polarization term is shown by summation in the above equation. In the absence of polarization term, above equation reduces to the screened Wannier Hamiltonian for a bound electron–hole pair.

The variation in emission band and absorption band is known as Stokes shift. The reabsorption effect is strongly related to the Stokes shift values. In general, traditional QDs shows average 40 meV Stokes shift for Cd based QDs. While Brennan et al. [29] showed Stokes shift in the range of 30–100 meV for PQDs with size variation in 12 nm to 4 mm. The PLQY of the QDs decreases with auger recombination of charged excitons. The average auger lifetime is found to be shorter than single excitons. The general principal of auger recombination suggests that non-radiative energy transfer to the nearby charge carriers rather than converting into a photon. The PLQY and decay dynamics can easily explain using various optical characterization techniques such as blinking, transient absorption, and time resolved PL spectroscopy [30, 31]. Hence, a detailed explanation of exciton behavior is needed for the application of PQDs in LEDs.

10.1.4 Perovskite Quantum Dots

The optical properties of PQDs are controlled via anion composition as well as particle size. In general quantum dots are classified into three major categories, namely, Cd-based, Cd-free, and PQDs [33]. PQDs generally have compositional formula of ABX_3, where A denotes Cs, FA, MA; B represents divalent metal cation (Pb) and X signifies halide anion (Br, Cl, I). Among various luminescent nanomaterials, PQDs attracted many researchers for optoelectronic utilizations based on distinctive nature such as variable emission, high absorption coefficient and excellent defect tolerance [34]. The band gap and size leads to the quantum confinement in PQDs and shown in Fig. 10.4.

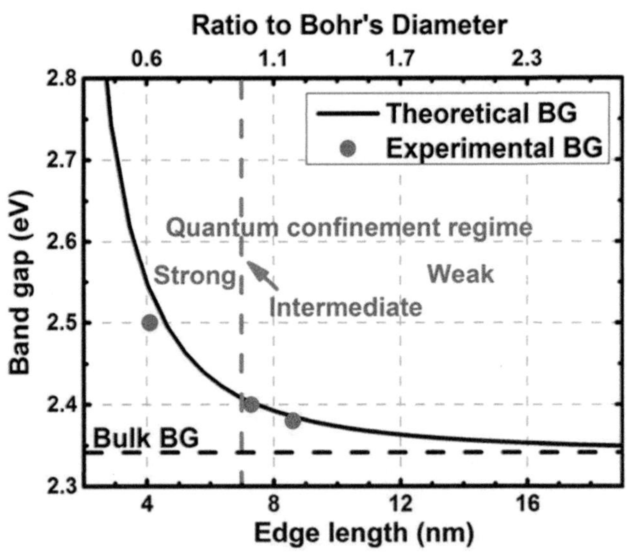

Figure 10.4 Band gap and size dependent quantum confinement with Bohr diameter of PQDs. Reproduced with permission from Ref. [32]. Copyright 2017, American Chemical Society.

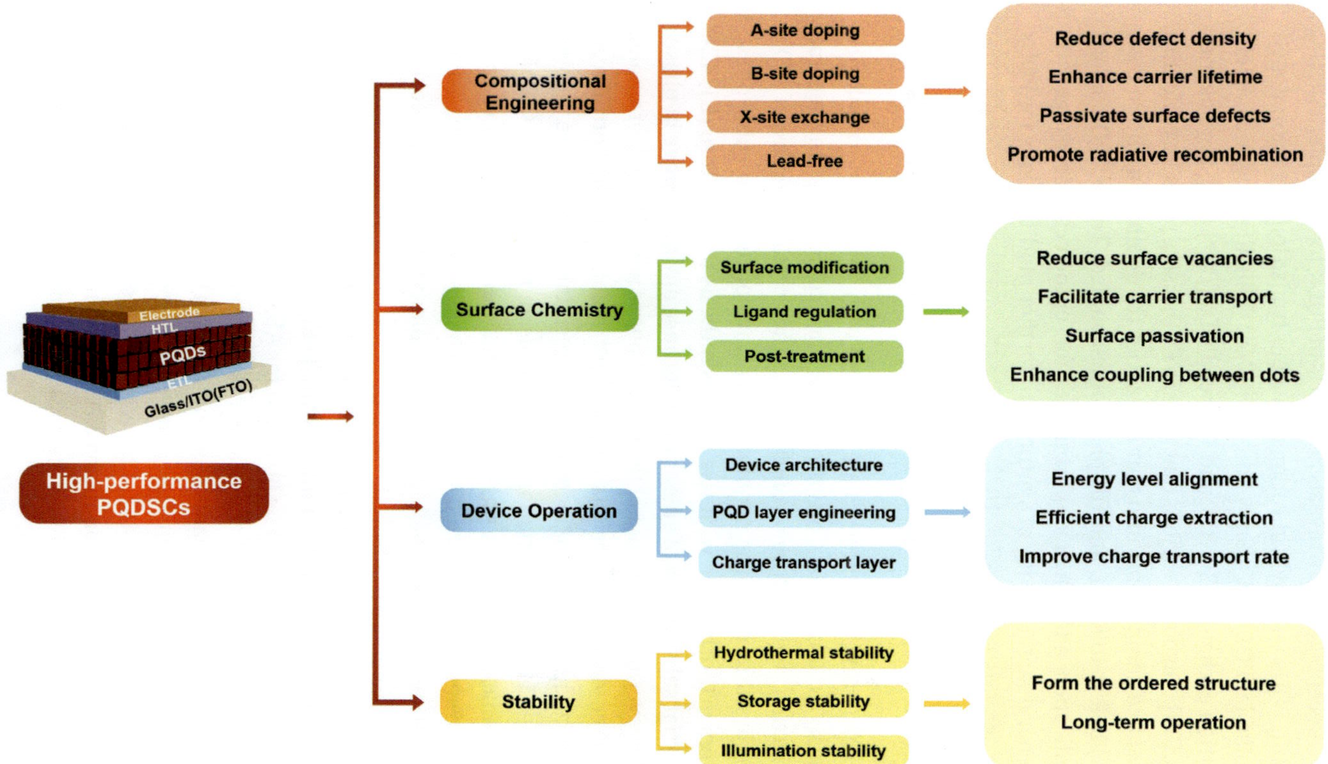

Figure 10.5 Schematic illustration of functionalization and application of PQDs. Reproduced from Ref. [34] with permission from The Royal Society of Chemistry.

Recent literature reveals new sorts of PQDs, for example, $FAPbBr_3$ [35], PQDs with hybrid cations, divalent transition metal ion doped colloidal PQDs (M = Sn^{2+}, Cd^{2+}, Zn^{2+}, Mn^{2+}) highlighting incomplete cation exchange [36, 37], and Pb free PQDs (Bi^{3+}) [38]. Swarnkar et al. [39] illustrated various functions in PQDs using periodic table, and isolated components into four categories (a) doping ion, (b) Pb-free, (c) defect-tolerant, and (d) plasmonic-coupled. These PQD analogs will open additional opportunities for optoelectronic gadgets. A general methodology of synthesis to further modifications for PQDs as demonstrated in Fig. 10.5.

10.1.5 Microstructures of ABX_3 Perovskite Materials

ABX_3 perovskite materials are reported with different morphologies such as tetragonal, orthorhombic and cubic [40]. In general, reaction temperature strongly affects the crystal structure of ABX_3 perovskites [41]. The nanomaterials transformed into cubic phase from orthorhombic with rise in the synthesis temperature [21, 42]. Partial substitution of anions did not cause any not change in phase. However, emission color of the PQDs changes with addition of anions. The co-doping of various cations and anions can be introduced via compositions engineering and resultant effects are shown in Fig. 10.6. The variation of synthesis temperature for $CsPbBr_3$ nanomaterials causes phase transformation, while emission color was remained unaffected with phase change [40]. However, $CsPbI_3$ materials changed to yellow from dark color with phase change from orthorhombic to cubic [40, 43].

The morphology of nanomaterials can be studied with the help of transmission electron microscopy (TEM). In general, cubical and spherical particles can be observed for ABX_3 PQDs [40].

Several published reports revealed other type of morphologies such as nanowires, nanorods, and nanoparticles synthesized with the variation of reaction time, temperature, precursor concentration, and surfactants [45, 46]. The various type of ligands affects the morphology of PQDs on the basis of their length. The particle size of PQDs is also dependent on the concentration of ligands utilized.

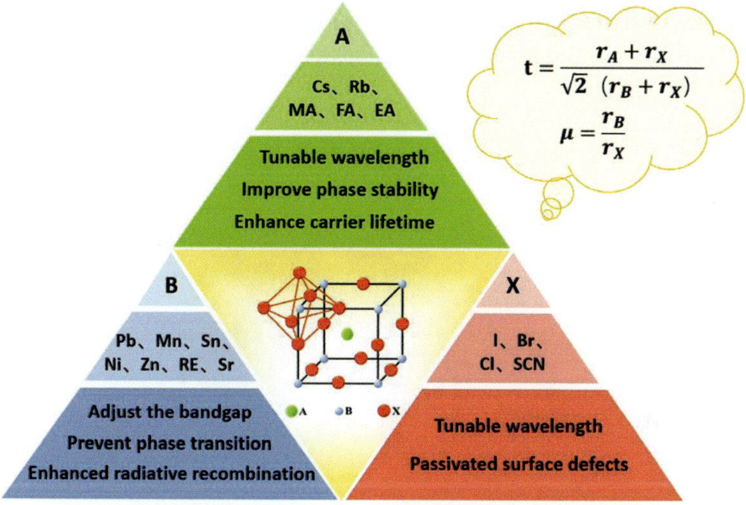

Figure 10.6 Possible compositional engineering and related effects on various properties of PQDs. Reproduced from Ref. [34] with permission from The Royal Society of Chemistry.

10.1.6 Optical Properties of ABX_3 PQDs

The emission peaks of PQDs can be obtained in broad visible spectral zone by varying the composition of halide ions and average size. A symbolic picture of different color emission is shown in Fig. 10.2a. Protesescu et al. [21] studied the emission properties of $CsPbX_3$ PQDs. The recorded PL spectra and absorption of $CsPbX_3$ PQDs is shown in Fig. 10.8b,c. The reported results showed emission peak position in the range 400–700 nm. When chlorine ion is used as halide ion, blue emission was observed for the synthesized PQDs. While bromine (Br) and iodide (I) ions showed emission in green and red region, respectively. However, the combination of halide ions showed different color. When Cl/Br mixture is utilized in the synthesis of PQDs, the emission peak indicated cyan color, and with combination of Br/I ions yellow/orange color was obtained depending on the concentration ratio of halide ions. The synthesized PQDs showed high PLQY in the range of 50–90%. The obtained lifetime of $CsPbX_3$ PQDs is shown in Fig. 10.2d. The reported results revealed short lifetime in the range of 1–29 ns. The narrow bandwidth of PQDs suggest high color purity and wide color gamut, thus PQDs are the potential candidates for backlight display devices [47, 48].

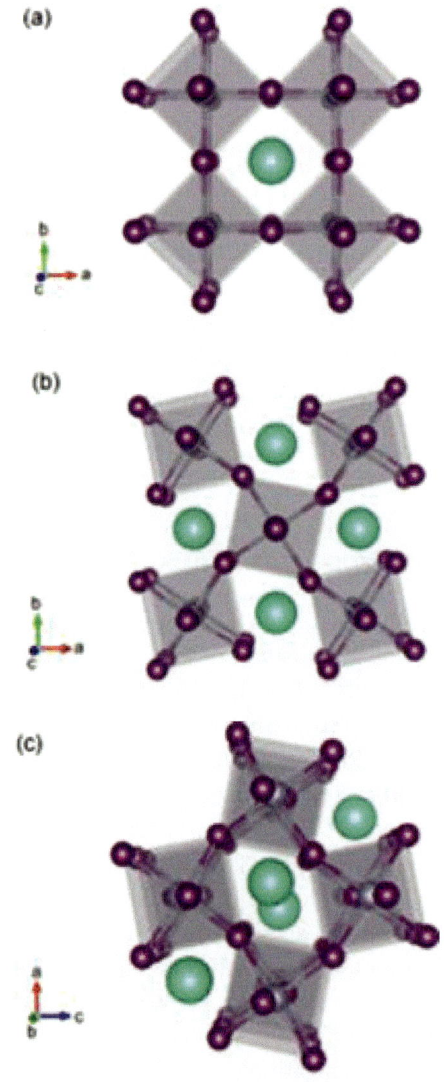

Figure 10.7 Various phases of perovskite materials (a) cubic, (b) tetragonal, and (c) orthorhombic. Reproduced from Ref. [44] with permission from American Physical Society. Copyright (2015) by the American Physical Society.

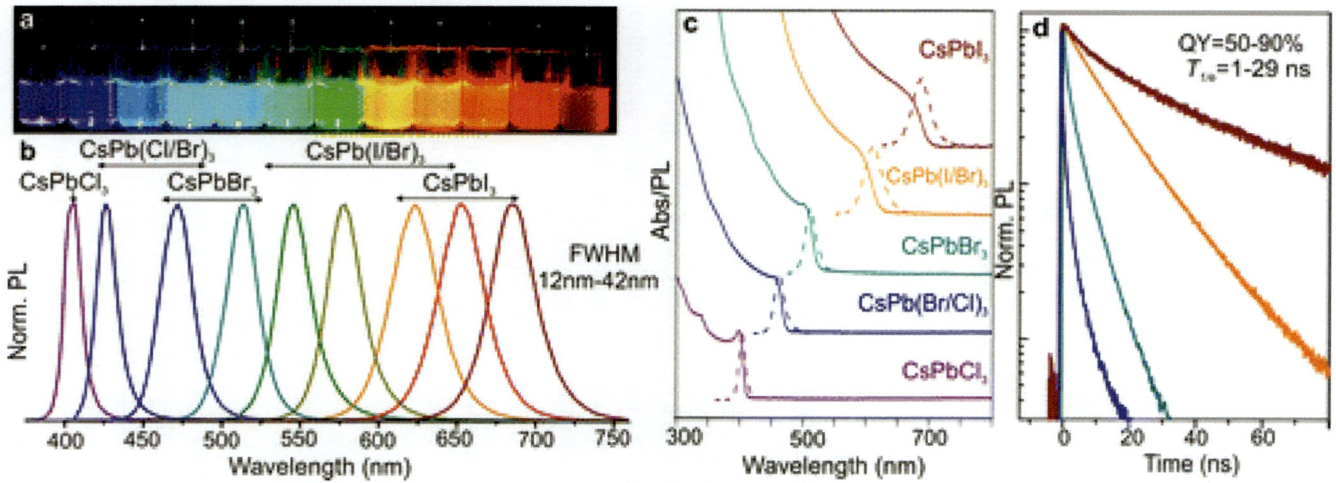

Figure 10.8 (a) Various colors of PQDs under UV light, (b) PL spectra of $CsPbX_3$ PQDs, (c) Absorption spectra of PQDs, (d) lifetime curves. Reproduced with permission from Ref. [21]. Copyright 2017, American Chemical Society.

10.1.7 Challenges with PQDs

Perovskites are $CaTiO_3$ mineral species named after L. A. Peroski in 1839. Compounds with chemical formula ABX_3 are described as perovskites [49]. The original structure of $CaTiO_3$ is reported as pseudo cubic. In general, ABX_3 PQDs have highly symmetric structure. Pb occupies the center of the cube, cation A (MA, FA, Cs) locates at vortex and the halogen atoms at the centre of faces. The central metals manage octahedral coordination with halides as PbX_6. The ideas phase (cubic) of the PQDs with Pm3m space group. However, several times PbX_6 octahedral is distorted (tilted) and found in less symmetrical orthorhombic phase [50]. The PbX_6 octahedral distortion occurs due to several factors such as (a) Jahn-Teller effect, (b) off center displacement, (c) filling of A cation (small radii) in the PbX_6 octahedral interstices, (d) cation deficiency, and (e) halide ion vacancy. The optical and electronic properties of PQDs changes with the distortion of octahedral [51]. However, the physical properties of PQDs can be fine-tuned via substituting ions in order to minimize the distortion effects [50].

The *A* cation which smaller in size than *B* cation in PQDs occupies 12-fold coordinated positions adjoining eight octahedral PbX_6. In order to have an ideal cubic shape of PQDs, the constituting elements should follow the relationship given below:

$$r_X + r_A = \sqrt{2}(r_X + r_B)$$

Also the distortion and structural stability of PQDs can be measured using Goldschmidt tolerance factor *t* as shown below [48, 50].

$$t = \frac{r_X + r_A}{\sqrt{2}(r_X + r_B)}$$

Where the radius of cations *A*, *B* and anion *X* is denoted by r_X, r_A, r_B, respectively.

The ideal range for the 3D structure of PQDs lies in $0.813 \leq t \leq 1.107$ [52]. If the value *t* goes beyond the above mentioned range, 3D structure will be collapsed. The octahedral stability of the PQDs can be evaluated using octahedral factor (μ). It can be calculated as $\mu = r_B/r_X$. Normally, the stability value of PbX_6 octahedral lies in between 0.442–0.895. The 3D structure formation possibility can be obtained using *t* and μ values [53].

PQDs suffers with the stability problem when exposed to atmospheric air. Even though PQDs are stabilized with various encapsulation matrices, the residual moisture and air affects the stability of PQDs [54]. A complete encapsulation of PQDs in inert atmosphere might increase the stability and mitigate the degradation. However, such a technique is costly and difficult to achieve in devices.

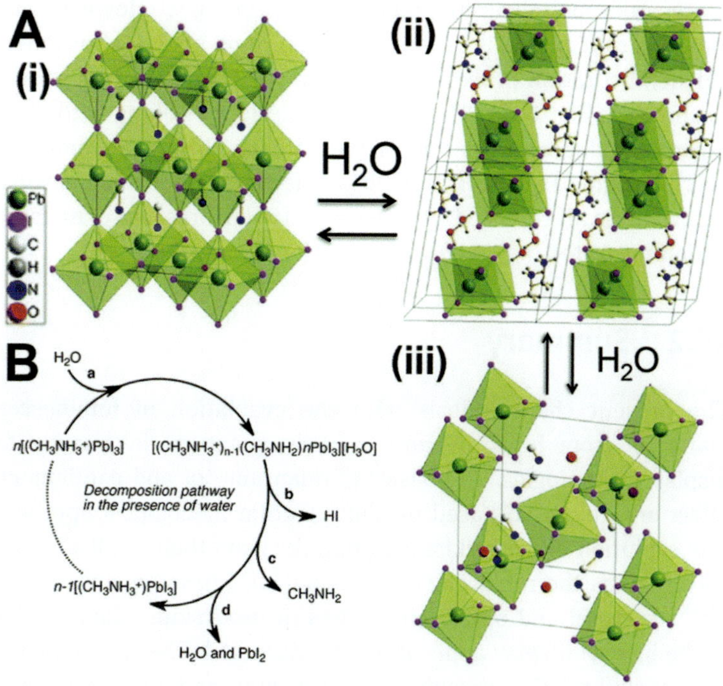

Figure 10.9 (a) Hydrated structures of PQDs and evaluation of structures. Reprinted with permission from Ref. [58]. Copyright 2014, American Chemical Society. (b) Degradation routes of PQDs in presence moisture. Reprinted with permission from Ref. [59]. Copyright 2014, American Chemical Society.

The stability and performance of PQDs also degrades under the continuous exposure to the light [55]. PQDs are utilized as the color conversion layer, PQDs absorbs the UV/blue light and emits light in visible region. Thus, PQDs used in a device are exposed to UV/blue for longer time when device is working. Hence, the QLEDs or pc-LEDs lifetime is crucial with the stability of PQDs. Perovskites also show the light soaking effect [56]. Due to the ionic nature of the PQDs, moisture/water easily degrades the PQDs [57]. In comparison to oxygen moisture can degrade even in dark conditions. The chemical reaction of PQDs degradation obtained from experimental and theoretical studies is given below [58].

$$4CH_3 3NH_3PbI_3 \underset{H_2O}{\Leftrightarrow} 4CH_3 3NH_3PbI_3 \cdot H_2O \underset{H_2O}{\Leftrightarrow} (CH_3 3NH_3)_4 PbI_6 \cdot 2H_2O + 3PbI_2$$

Traces of monohydrate phase form initially. However, the monohydrate phase is reversible when kept in dry atmosphere. Further, hydration process occurs with the formation of dihydrate as shown in Fig. 10.9a. The hydration accelerates along with phase separation due to formation of PbI_2. Hence, dihydrate compound degrades into CH_3NH_2, HI, and PbI_2. A decomposition mechanism of perovskites was reported by Walsh et al. [59] in presence of moisture as demonstrated in Fig. 10.9b.

10.2 Summary

The present chapter deals with the evaluation of luminescent materials from bulk to nano towards quantum dot mainly for display technologies. The basics of quantum dot and confinement effect are discussed based on the quantum mechanical approach. The evolution of perovskite quantum dots and their applicability is discussed as the replacement of colloidal Cd based quantum dots. The structural and optical properties of perovskite quantum dots in the light of display application are discussed. The reason behind the instability of perovskite quantum dots and their remedy is mentioned in the present chapter. Finally, the chapter provides an essence of luminescent nanomaterials for light-emitting and display applications.

References

1. A. Kitai, Luminescent materials and applications, *Lumin. Mater. Appl.* (2008) 1–278. https://doi.org/10.1002/9780470985687.
2. E.F. Schubert, J.K. Kim, Solid-state light sources getting smart, *Science (80-.).* 308 (2005) 1274–1278. https://doi.org/10.1126/science.1108712.

3. V. Kumar, V. Kumar, S. Som, L.P. Purohit, O.M. Ntwaeaborwa, H.C. Swart, Role of swift heavy ions irradiation on the emission of boron doped ZnO thin films for near white light application, *J. Alloys Compd.* 594 (2014) 32–38. https://doi.org/https://doi.org/10.1016/j.jallcom.2014.01.110.
4. S. Som, S.K. Sharma, Eu3+/Tb3+-codoped Y_2O_3 nanophosphors: Rietveld refinement, bandgap and photoluminescence optimization, *J. Phys. D. Appl. Phys.* 45 (2012) 415102. https://doi.org/10.1088/0022-3727/45/41/415102.
5. S.K. Sharma, S. Dutta, S. Som, P.S. Mandal, $CaMoO_4$:Dy^{3+}, K+ near white light emitting phosphor: Structural, optical and dielectric properties, *J. Mater. Sci. Technol.* 29 (2013) 633–638. https://doi.org/https://doi.org/10.1016/j.jmst.2013.03.014.
6. M.L. Meena, S. Som, R.K. Singh, C.H. Lu, Synthesis, spectroscopic characterization and estimation of Judd-Ofelt parameters for Dy^{3+} activated Li_2MgZrO_4 double perovskite materials, *Polyhedron.* 177 (2020) 114322. https://doi.org/10.1016/j.poly.2019.114322.
7. S. Dutta, S. Som, M. Lal Meena, R. Chaurasiya, T.-M. Chen, Multisite-occupancy-driven intense narrow-band blue emission from $Sr_5SiO_4Cl_6$:Eu^{2+} phosphor with excellent stability and color performance, *Inorg. Chem.* 59 (2020) 1928–1939. https://doi.org/10.1021/acs.inorgchem.9b03222.
8. S. Das, S. Som, C.Y. Yang, S. Chavhan, C.H. Lu, Structural evaluations and temperature dependent photoluminescence characterizations of Eu^{3+}-activated $SrZrO_3$ hollow spheres for luminescence thermometry applications, *Sci. Rep.* 6 (2016) 1–13. https://doi.org/10.1038/srep25787.
9. A. Yousif, V. Kumar, H.A.A. Seed Ahmed, S. Som, L.L. Noto, O.M. Ntwaeaborwa, H.C. Swart, Effect of Ga^{3+} doping on the photoluminescence properties of $Y_3Al_{5-x}Ga_xO_{12}$:Bi^{3+} phosphor, *ECS J. Solid State Sci. Technol.* 3 (2014) R222–R227. https://doi.org/10.1149/2.0021412jss.
10. S. Som, A.K. Kunti, V. Kumar, V. Kumar, S. Dutta, M. Chowdhury, S.K. Sharma, J.J. Terblans, H.C. Swart, Defect correlated fluorescent quenching and electron phonon coupling in the spectral transition of Eu^{3+} in $CaTiO_3$ for red emission in display application, *J. Appl. Phys.* 115 (2014). https://doi.org/10.1063/1.4876316.
11. S. Dutta, S. Som, J. Priya, S.K. Sharma, Band gap, CIE and trap depth parameters of rare earth molybdate phosphors for optoelectronic

applications, *Solid State Sci.* 18 (2013) 114–122. https://doi.org/https://doi.org/10.1016/j.solidstatesciences.2013.01.012.

12. C.R. Ronda, *Luminescence: From Theory to Applications*, Wiley (2007) 1–260. https://doi.org/10.1002/9783527621064.
13. S. Som, P. Mitra, V. Kumar, V. Kumar, J.J. Terblans, H.C. Swart, S.K. Sharma, The energy transfer phenomena and colour tunability in $Y_2O_2S:Eu^{3+}/Dy^{3+}$ micro-fibers for white emission in solid state lighting applications, *Dalton Trans.* 43 (2014) 9860–9871. https://doi.org/10.1039/C4DT00349G.
14. H. Cerjak, Nanomaterials: An introduction to synthesis, properties and applications, *Mater. Technol.* 24 (2009) 74–74. https://doi.org/10.1179/175355509x454827.
15. H. Terraschke, C. Wickleder, UV, blue, green, yellow, red, and small: Newest developments on Eu^{2+}-doped nanophosphors, *Chem. Rev.* 115 (2015) 11352–11378. https://doi.org/10.1021/acs.chemrev.5b00223.
16. R.C. Monica, R. Cremonini, Nanoparticles and higher plants, *Caryologia.* 62 (2009) 161–165. https://doi.org/10.1080/00087114.2004.10589681.
17. A. Dandia, S. Bansal, R. Sharma, D. Kumar Mahawar, K.S. Rathore, M. Lal Meena, V. Parewa, Nanoporous Cu doped ZnS nanoparticles an efficient photo catalyst for the chemoselective synthesis of 2-substituted azoles via C-N arylation/ CSp3– H oxidation/ cyclization/ dehydration sequence in visible light, *J. Photochem. Photobiol. A Chem.* 389 (2020) 112242. https://doi.org/https://doi.org/10.1016/j.jphotochem.2019.112242.
18. C. Feldmann, Luminescent nanomaterials, *Nanoscale.* 3 (2011) 1947–1948. https://doi.org/10.1039/c1nr90008k.
19. A.L. Efros, D.J. Nesbitt, Origin and control of blinking in quantum dots, *Nat. Nanotechnol.* 11 (2016) 661–671. https://doi.org/10.1038/nnano.2016.140.
20. X. Dai, Y. Deng, X. Peng, Y. Jin, Quantum-dot light-emitting diodes for large-area displays: Towards the dawn of commercialization, *Adv. Mater.* 29 (2017). https://doi.org/10.1002/adma.201607022.
21. L. Protesescu, S. Yakunin, M.I. Bodnarchuk, F. Krieg, R. Caputo, C.H. Hendon, R.X. Yang, A. Walsh, M. V. Kovalenko, Nanocrystals of cesium lead halide perovskites (CsPbX3, X = Cl, Br, and I): Novel optoelectronic materials showing bright emission with wide color gamut, *Nano Lett.* 15 (2015) 3692–3696. https://doi.org/10.1021/nl5048779.

22. E.M. Janke, N.E. Williams, C. She, D. Zherebetskyy, M.H. Hudson, L. Wang, D.J. Gosztola, R.D. Schaller, B. Lee, C. Sun, G.S. Engel, D.V. Talapin, Origin of broad emission spectra in InP quantum dots: Contributions from structural and electronic disorder, *J. Am. Chem. Soc.* 140 (2018) 15791–15803. https://doi.org/10.1021/jacs.8b08753.

23. T. Edvinsson, Optical quantum confinement and photocatalytic properties in two-, one- and zerodimensional nanostructures, *R. Soc. Open Sci.* 5 (2018). https://doi.org/10.1098/rsos.180387.

24. L.E. Brus, Electron-electron and electron-hole interactions in small semiconductor crystallites: The size dependence of the lowest excited electronic state, *J. Chem. Phys.* 80 (1984) 4403–4409. https://doi.org/10.1063/1.447218.

25. A.P. Alivisatos, Downloaded from www.sciencemag.org on March 4, 2011 Downloaded from www.sciencemag.org on March 4, 2011, *Science (80-.).* 271 (1996) 933–937.

26. T.J. Jacobsson, T. Edvinsson, Photoelectrochemical determination of the absolute band edge positions as a function of particle size for ZnO quantum dots, *J. Phys. Chem. C.* 116 (2012) 15692–15701. https://doi.org/10.1021/jp302220w.

27. A. Kawabata, Electronic properties of metallic fine particles, *Surf. Rev. Lett.* 3 (1996) 9–12. https://doi.org/10.1142/S0218625X96000048.

28. T.J. Jacobsson, T. Edvinsson, Absorption and fluorescence spectroscopy of growing ZnO quantum dots: Size and band gap correlation and evidence of mobile trap states, *Inorg. Chem.* 50 (2011) 9578–9586. https://doi.org/10.1021/ic201327n.

29. M.C. Brennan, J. Zinna, M. Kuno, Existence of a size-dependent Stokes shift in $CsPbBr_3$ perovskite nanocrystals, *ACS Energy Lett.* 2 (2017) 1487–1488. https://doi.org/10.1021/acsenergylett.7b00383.

30. N.S. Makarov, S. Guo, O. Isaienko, W. Liu, I. Robel, V.I. Klimov, Spectral and dynamical properties of single excitons, biexcitons, and trions in cesium-lead-halide perovskite quantum dots, *Nano Lett.* 16 (2016) 2349–2362. https://doi.org/10.1021/acs.nanolett.5b05077.

31. F. Hu, C. Yin, H. Zhang, C. Sun, W.W. Yu, C. Zhang, X. Wang, Y. Zhang, M. Xiao, Slow Auger recombination of charged excitons in nonblinking perovskite nanocrystals without spectral diffusion, *Nano Lett.* 16 (2016) 6425–6430. https://doi.org/10.1021/acs.nanolett.6b02874.

32. J. Butkus, P. Vashishtha, K. Chen, J.K. Gallaher, S.K.K. Prasad, D.Z. Metin, G. Laufersky, N. Gaston, J.E. Halpert, J.M. Hodgkiss, The evolution of quantum confinement in $CsPbBr_3$ perovskite nanocrystals,

Chem. Mater. 29 (2017) 3644–3652. https://doi.org/10.1021/acs.chemmater.7b00478.

33. H.C. Wang, Z. Bao, H.Y. Tsai, A.C. Tang, R.S. Liu, Perovskite quantum dots and their application in light-emitting diodes, *Small.* 14 (2018) 1–23. https://doi.org/10.1002/smll.201702433.
34. J. Chen, D. Jia, E.M.J. Johansson, A. Hagfeldt, X. Zhang, Emerging perovskite quantum dot solar cells: Feasible approaches to boost performance, *Energy Environ. Sci.* (2021) 224–261. https://doi.org/10.1039/d0ee02900a.
35. Y.H. Kim, G.H. Lee, Y.T. Kim, C. Wolf, H.J. Yun, W. Kwon, C.G. Park, T.W. Lee, High efficiency perovskite light-emitting diodes of ligand-engineered colloidal formamidinium lead bromide nanoparticles, *Nano Energy.* 38 (2017) 51–58. https://doi.org/10.1016/j.nanoen.2017.05.002.
36. W. Van der Stam, J.J. Geuchies, T. Altantzis, K.H.W. Van Den Bos, J.D. Meeldijk, S. Van Aert, S. Bals, D. Vanmaekelbergh, C. De Mello Donega, Highly emissive divalent-ion-doped colloidal $CsPb_{1-x}M_xBr_3$ perovskite nanocrystals through cation exchange, *J. Am. Chem. Soc.* 139 (2017) 4087–4097. https://doi.org/10.1021/jacs.6b13079.
37. A.K. Guria, S.K. Dutta, S. Das Adhikari, N. Pradhan, Doping Mn^{2+} in lead halide perovskite nanocrystals: Successes and challenges, *ACS Energy Lett.* 2 (2017) 1014–1021. https://doi.org/10.1021/acsenergylett.7b00177.
38. R. Begum, M.R. Parida, A.L. Abdelhady, B. Murali, N.M. Alyami, G.H. Ahmed, M.N. Hedhili, O.M. Bakr, O.F. Mohammed, Engineering interfacial charge transfer in $CsPbBr_3$ perovskite nanocrystals by heterovalent doping, *J. Am. Chem. Soc.* 139 (2017) 731–737. https://doi.org/10.1021/jacs.6b09575.
39. A. Swarnkar, V.K. Ravi, A. Nag, Beyond colloidal cesium lead halide perovskite nanocrystals: Analogous metal halides and doping, *ACS Energy Lett.* 2 (2017) 1089–1098. https://doi.org/10.1021/acsenergylett.7b00191.
40. D. Chen, X. Chen, Luminescent perovskite quantum dots: Synthesis, microstructures, optical properties and applications, *J. Mater. Chem. C.* 7 (2019) 1413–1446. https://doi.org/10.1039/C8TC05545A.
41. A. Swarnkar, R. Chulliyil, V.K. Ravi, M. Irfanullah, A. Chowdhury, A. Nag, Colloidal $CsPbBr_3$ perovskite nanocrystals: Luminescence beyond traditional quantum dots, *Angew. Chem. Int. Ed.* 54 (2015) 15424–15428. https://doi.org/10.1002/anie.201508276.
42. X. Li, Y. Wu, S. Zhang, B. Cai, Y. Gu, J. Song, H. Zeng, $CsPbX_3$ quantum dots for lighting and displays: Roomerature synthesis, photoluminescence

superiorities, underlying origins and white light-emitting diodes, *Adv. Funct. Mater.* 26 (2016) 2435–2445. https://doi.org/10.1002/adfm.201600109.

43. P.S. Whitfield, N. Herron, W.E. Guise, K. Page, Y.Q. Cheng, I. Milas, M.K. Crawford, Structures, phase transitions and tricritical behavior of the hybrid perovskite methyl ammonium lead iodide, *Sci. Rep.* 6 (2016) 1–16. https://doi.org/10.1038/srep35685.
44. E.L. Da Silva, J.M. Skelton, S.C. Parker, A. Walsh, Phase stability and transformations in the halide perovskite $CsSnI_3$, *Phys. Rev. B: Condens. Matter Mater. Phys.* 91 (2015) 1–12. https://doi.org/10.1103/PhysRevB.91.144107.
45. J.A. Sichert, Y. Tong, N. Mutz, M. Vollmer, S. Fischer, K.Z. Milowska, R. García Cortadella, B. Nickel, C. Cardenas-Daw, J.K. Stolarczyk, A.S. Urban, J. Feldmann, Quantum size effect in organometal halide perovskite nanoplatelets, *Nano Lett.* 15 (2015) 6521–6527. https://doi.org/10.1021/acs.nanolett.5b02985.
46. A. Pan, B. He, X. Fan, Z. Liu, J.J. Urban, A.P. Alivisatos, L. He, Y. Liu, Insight into the ligand-mediated synthesis of colloidal $CsPbBr_3$ perovskite nanocrystals: The role of organic acid, base, and cesium precursors, *ACS Nano.* 10 (2016) 7943–7954. https://doi.org/10.1021/acsnano.6b03863.
47. C.-H. Lu, Y.-H. Liu, M.L. Meena, S. Som, Synthesis and spectroscopic characterization of $CsPbBr_3/Cs_4PbBr_6$ perovskites synthesized via the microwave-assisted heating process for backlight display devices, *Org. Electron.* (2021) 106079. https://doi.org/https://doi.org/10.1016/j.orgel.2021.106079.
48. R.K. Singh, P. Sharma, R. Kumar, S. Som, S. Dutta, N. Jain, R. Chaurasiya, M.L. Meena, J.-S. Ho, S.-W. Dai, J. Singh, C.-H. Lu, H.-W. Lin, $CH_3NH_3Pb_{1-x}Co_xBr_{3-2x}Cl_{2x}$ perovskite quantum dots for wide-color backlighting, *ACS Appl. Nano Mater.* 4 (2021) 717–728. https://doi.org/10.1021/acsanm.0c03019.
49. Z. Cheng, J. Lin, Layered organic-inorganic hybrid perovskites: Structure, optical properties, film preparation, patterning and templating engineering, *CrystEngComm.* 12 (2010) 2646–2662. https://doi.org/10.1039/c001929a.
50. Y. Wei, Z. Cheng, J. Lin, An overview on enhancing the stability of lead halide perovskite quantum dots and their applications in phosphor-converted LEDs, *Chem. Soc. Rev.* 48 (2019) 310–350. https://doi.org/10.1039/c8cs00740c.

51. W.J. Yin, T. Shi, Y. Yan, Unique properties of halide perovskites as possible origins of the superior solar cell performance, *Adv. Mater.* 26 (2014) 4653–4658. https://doi.org/10.1002/adma.201306281.

52. G. Kieslich, S. Sun, A.K. Cheetham, Solid-state principles applied to organic–inorganic perovskites: New tricks for an old dog, *Chem. Sci.* 5 (2014) 4712–4715. https://doi.org/10.1039/c4sc02211d.

53. C. Li, X. Lu, W. Ding, L. Feng, Y. Gao, Z. Guo, Formability of ABX_3 (X = F, Cl, Br, I) halide perovskites, *Acta Crystallogr. Sect. B Struct. Sci.* 64 (2008) 702–707. https://doi.org/10.1107/S0108768108032734.

54. J. Ren, X. Dong, G. Zhang, T. Li, Y. Wang, Air-stable and water-resistant all-inorganic perovskite quantum dot films for white-light-emitting applications, *New J. Chem.* 41 (2017) 13961–13967. https://doi.org/10.1039/c7nj03017g.

55. F. Lang, O. Shargaieva, V.V. Brus, H.C. Neitzert, J. Rappich, N.H. Nickel, Influence of radiation on the properties and the stability of hybrid perovskites, *Adv. Mater.* 30 (2018) 1–22. https://doi.org/10.1002/adma.201702905.

56. W. Nie, J.C. Blancon, A.J. Neukirch, K. Appavoo, H. Tsai, M. Chhowalla, M.A. Alam, M.Y. Sfeir, C. Katan, J. Even, S. Tretiak, J.J. Crochet, G. Gupta, A.D. Mohite, Light-activated photocurrent degradation and self-healing in perovskite solar cells, *Nat. Commun.* 7 (2016) 1–9. https://doi.org/10.1038/ncomms11574.

57. A. Kojima, K. Teshima, Y. Shirai, T. Miyasaka, Organometal halide perovskites as visible-light sensitizers for photovoltaic cells, *J. Am. Chem. Soc.* 131 (2009) 6050–6051. https://doi.org/10.1021/ja809598r.

58. A.M.A. Leguy, Y. Hu, M. Campoy-Quiles, M.I. Alonso, O.J. Weber, P. Azarhoosh, M. Van Schilfgaarde, M.T. Weller, T. Bein, J. Nelson, P. Docampo, P.R.F. Barnes, Reversible hydration of $CH_3NH_3PbI_3$ in films, single crystals, and solar cells, *Chem. Mater.* 27 (2015) 3397–3407. https://doi.org/10.1021/acs.chemmater.5b00660.

59. J.M. Frost, K.T. Butler, F. Brivio, C.H. Hendon, M. Van Schilfgaarde, A. Walsh, Atomistic origins of high-performance in hybrid halide perovskite solar cells, *Nano Lett.* 14 (2014) 2584–2590. https://doi.org/10.1021/nl500390f.

Chapter 11

TL/OSL Properties of Calcium Oxide Nanophosphor

K. R. Nagabhushana,[a] Sonia Hatsue Tatumi,[a] and D. Prakash[b]

[a]*Universidade Federal de São Paulo, Campus Baixada Santista, Rua Carvalho de Mendonca, 144, Santos - SP, CEP: 11070-100, Brazil*

[b]*Department of Physics, Government Science College, Bangalore 560001, India*

nagabhushana.nkr@gmail.com, sonia.tatumi@gmail.com

11.1 Introduction

Nanoscience is the study of phenomena and modification of materials at atomic, molecular, and macromolecular dimensions, where properties differ significantly from those at higher scale; and nanotechnology refers to the design, characterization, production, and application of nanostructures, nanodevices, and systems by controlling shape and crystallite size at the nanoscale [1]. Nanoscience and nanotechnology deal with materials called nanomaterials having size scales within 1 to 100 nm. Literally, 1 nm is the length of ten H-atoms (Bohr radius = 0.5 Å). The nanomaterial

Luminescent Nanomaterials
Edited by Odireleng Martin Ntwaeaborwa

ISBN 978-981-4968-11-9 (Hardcover), 978-1-003-27738-5 (eBook)
www.jennystanford.com

bridges between atoms and molecules to macroscopic and bulk materials (>100 nm), it exhibits new properties, which depend on the particle size. As such, the electronic structure, conductivity, melting temperature, mechanical properties, etc. have been changed when material sizes are lower than a critical size. Indeed, surface to volume ratio of nanomaterials increases as the size decreases which makes possible new quantum mechanical effects and hence its properties change as the size changes in nano regime [2, 3]. Nanocrystals can provide a single-domain crystalline system which can be used to deliver information that can help in explaining the performance of macroscopic samples of the same materials, without muddling the presence of grain boundaries and other defects [4, 5]. Of the different interesting research areas which are encompassed by nanoscience and nanotechnology, research on luminescence properties of nanomaterials is one among them. Production of different colors from the same material by tuning the size of a semiconducting nanomaterial is well established. The research on efficient and inexpensive nanophosphors is a challenging problem for new luminescence materials [6].

Inorganic oxides are binary compounds of oxygen with an element and classified into acidic or basic oxide based on their characteristics. Inorganic oxides can be prepared by direct heating of element with oxygen, the reaction of oxygen with compounds at high temperatures, oxidation of metals and non-metals with nitric acid [7]. Inorganic materials are ionic compounds, consisting of cation and anion linked by ionic bonding and they are electrically neutral. The ions are defined by their oxidation state, and their ease of creation can be inferred from the ionization potential (cations) and/or from the electron affinity (anions) of the parent elements. The inorganic oxides are characterized by high melting point. Inorganic salts are typically poor electric and thermal conductors in the solid state.

Calcium oxide (CaO) based materials are economically sustainable and environmentally friendly. So they can apply in many fields such as radiation detection, optical storage media, night displays, direction indicators, and fiber optics [8]. CaO is a wide bandgap material (~6.1 eV), and fundamental absorption edge starts with excision transitions [9]. Nanocrystalline CaO exhibits unique enhanced luminescence, transparency, strength, and hardness properties [10]. Though many potential applications of CaO, there are not many studies of its luminescence characteristics [11] and very little work was performed on luminescence of undoped CaO nanophosphor. Nanoparticles of various scale distribution and chemical composition have been prepared by different methods [12]. Solution combustion synthesis technique (SCS) is a fast process, and the stoichiometry of the product can be controlled wherein high purity material is obtained, and uniform distribution of dopants takes place throughout the host material [7]. Cao nanophosphors used in the present work are synthesized by SCS method. The synthesized phosphors are characterized by X-ray diffraction, field emission scanning electron microscopy, energy dispersive spectroscopy, Fourier transform infrared, Raman and photoluminescence spectroscopy, and corresponding results are discussed elsewhere [13]. In this chapter, thermoluminescence (TL) and optically stimulated luminescence (OSL) properties of beta irradiated CaO nanophosphors are presented.

11.2 Thermoluminescence of CaO Nanophosphor

11.2.1 Introduction

Thermoluminescence (TL) is one of the most sensitive methods of studying radiation-induced defects. TL is the thermally stimulated emission of light from an insulator. The light energy is resulting from electron displacements within the crystal lattice of such as substance produced by previous exposure to high energy radiation. Defect centers created by ionizing radiations are directly related to the process of TL. The high energetic gamma/electron beam creates a higher density of oxygen vacancies [14]. Energy absorbed by materials during irradiation with ionizing radiation causes the production of electron and hole trap centers between valence band (VB) and conduction band (CB). The efficiency of TL increases due to an increase in the concentration of defects centers. These trapped electrons below the CB can be stimulated through various form of energies such as thermal and optical energy [15]. Subsequently thermally stimulated electrons are released from the traps, then recombine with hole centers and causes the emission of light. TL properties of materials are strongly affected by the type of ionizing radiation used. Some of the TL glow curves exhibit a broad curve due to merging of one or more individual curves and which are corresponding to a different type of trap. TL glow peak at a definite temperature is a characteristic of the material.

Generally, personal dosimeters are used for dose monitoring, in which the materials store and accumulate the absorbed radiation energy, and this accumulated energy is released in the form of light during thermal or optical stimulation. The emission intensity is proportional to the absorbed dose. Both TL and OSL are suitable for quantifying the amount of absorbed radiation dose [16, 17]. A number of commercial TL/OSL dosimeters with high sensitivity and low fading are available. However, the dose range of commercial phosphors usually is not over mGy, while higher doses need to be measured for technological applications. In this connection, interest is shown in nanophosphors because of their better radiation resistance [18, 19]. The increase in radiation resistance of nanophosphors are very significant for dosimetric applications [20]. To examine the suitability of CaO nanophosphor for TL/OSL dosimetry, the phosphor is exposed to β-rays. $^{90}Sr/^{90}Y$ beta source, which emits beta particles with a maximum energy of 2.27 MeV, is used. The source strength is about 1.48 GBq, which gives a dose rate of approximately 0.082 Gy/s at the sample position [21]. TL/OSL measurements are performed using Riso TL/OSL (DA-20) reader.

11.2.2 Effect of Annealing on TL of CaO Nanophosphor

It is well known that annealing has a significant influence on the shape of the TL glow curves when the materials are treated at different temperatures. All TL materials to be used for dosimetric purposes has to be thermally stable [22]. In the present investigation, the undoped samples are annealed at 600, 750, 900, 1050, and 1200°C to improve the sensitivity. Figure 11.1 shows TL glow curves of β-rays irradiated (50 Gy) undoped CaO and annealed at 600, 750, 900, 1050, and 1200°C. The sample annealed at 600°C showed a feeble TL response and the samples annealed at 750, 900, and 1050°C shows single prominent TL glow with a peak at 448, 456, and 453 K respectively. Whereas, 1200°C annealed sample show three prominent TL glow with peaks at 450, 585, and 708 K. These broad TL glow peaks are due to overlapping of different defects. TL response increases with the increase of annealing temperature up to 1200°C. One of the reasons for the change in the peak intensity, after annealing, is due to the cluster formation. The formation of defect clusters instead of randomly distributed defects causes the variations in the kinetics of trapping and recombination process [23]. Also, the change of annealing temperature provokes change in the crystal structure.

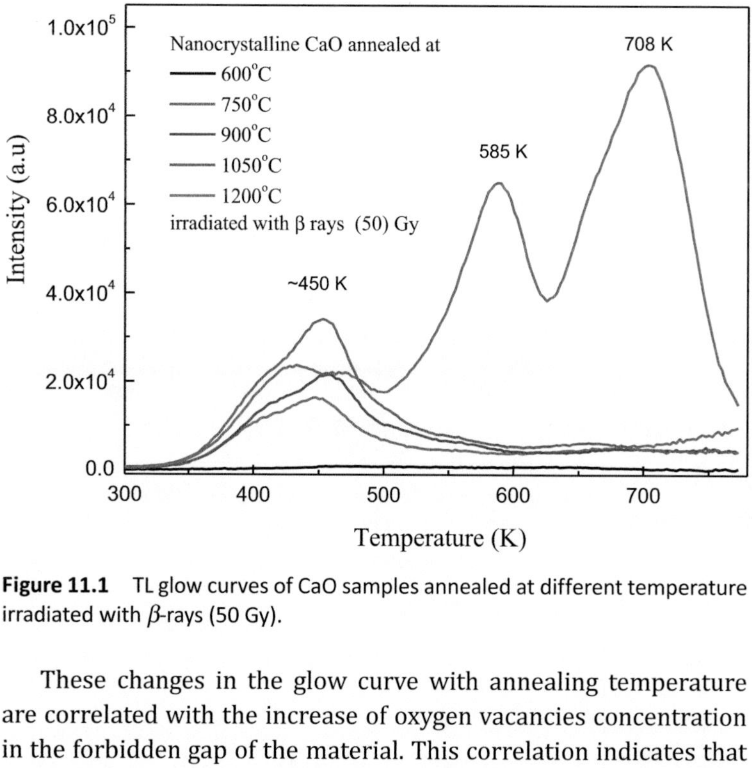

Figure 11.1 TL glow curves of CaO samples annealed at different temperature irradiated with *β*-rays (50 Gy).

These changes in the glow curve with annealing temperature are correlated with the increase of oxygen vacancies concentration in the forbidden gap of the material. This correlation indicates that even though the crystallinity in the powder improves, the presence of oxygen vacancies is significant for producing light emission. This emission is probably due to the oxygen vacancies acts as an electron trap center, responsible for the emission [24].

11.2.3 TL Dose Response of CaO Nanophosphor

Figure 11.2a shows TL glow curves of 600°C annealed and *β*-rays irradiated CaO nanophosphor. The samples are irradiated in the dose range of 10 to 100 Gy. Two TL glows a prominent one at 535 K, having a shoulder at 465 K, and the other at 675 K are observed at a higher dose. The samples do not show TL response at 10 and 20 Gy. Variation of maximum TL glow peak intensity (I_m) with dose is given Fig. 11.2b. The samples show nonlinear dose response at low doses and linear response at the high dose.

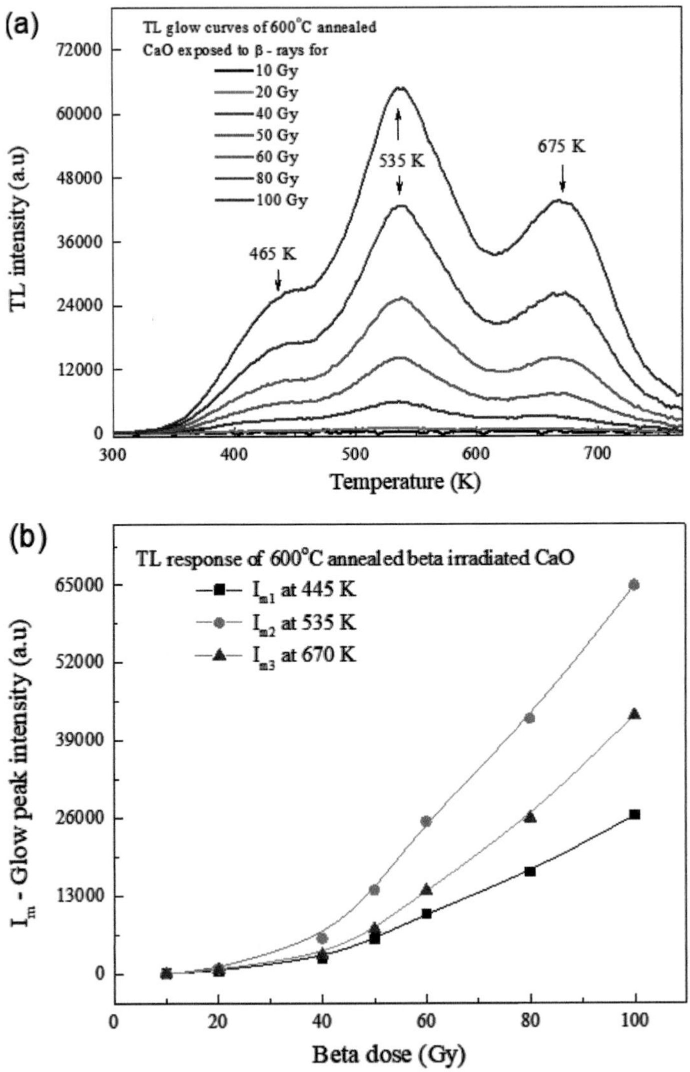

Figure 11.2 (a) TL glow curves of 600°C annealed CaO nanophosphor and (b) dose response.

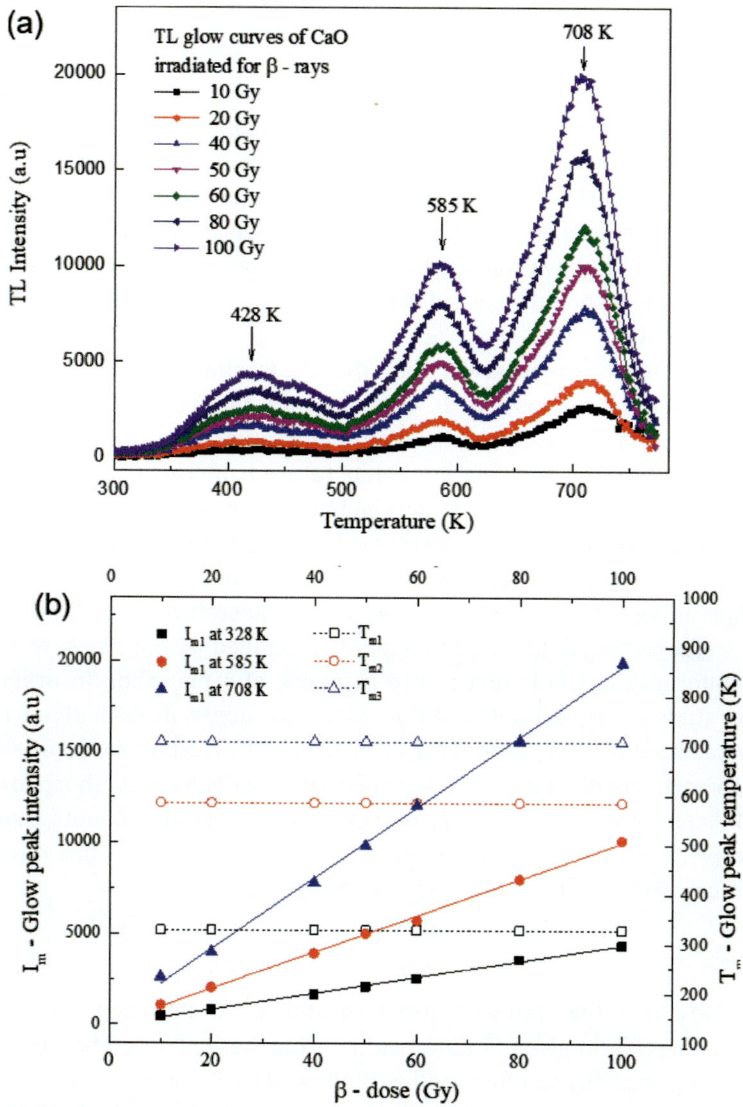

Figure 11.3 (a) TL glow curves of CaO annealed at 1200°C irradiated with *β*-rays for various doses and (b) variation of maxima TL Glow peak intensity and peak temperature with *β*-dose.

Figure 11.3a show TL glow curves of 1200°C annealed CaO phosphor and irradiated with *β*-rays in the dose range 10–100 Gy. Three prominent TL glow peaks at 428, 585, and 708 K are observed in all samples. The properties of TL materials generally depend on different parameters. The glow curve shape, position, and intensities of the peaks are related to the properties of trapping centers responsible for TL. The observed TL is explained by the tunneling recombination mechanism between electron and hole pairs in distant F^+ and V centers [25]. Variation of I_m and peak temperature (T_m) with β-dose of 1200°C annealed CaO is shown in Fig. 11.3b. It is found that TL glow peak intensity of all three peaks increases linearly with dose. It should be noted that it is a fundamental requirement of a dosimeter.

Further, it is observed that the T_m is almost constant with increasing beta dose. Also, the shape of the TL glow curve is nearly symmetric, with the high-temperature half of the curve slightly broader than the low temperature half. This is one of the characteristics of second order kinetics. In a second order reaction, significant concentrations of released electrons are retrapped before they recombine giving rise to delay in the photon emission and spreading out of the emission over wide temperature range [26].

The repeatability/reproducibility is defined as the measure of uniformity of the luminescence response of a phosphor. In order to study the repeatability of the CaO at low doses, five samples are irradiated with the same dose of 10 Gy and TL read out is recorded on each sample. Figure 11.4a shows the results for this phosphor. It is observed that T_m and I_m remains constant for all readouts. The repeatability of the TL response as a function of the absorbed dose is calculated with the following expression [27].

$$R = \frac{100 \times \sigma}{\overline{x}} \leq 7.5\% \quad (11.1)$$

where σ is the standard deviation and $\overline{x}$ is the average of all readings (I_m for given TL glow peak) during the read outs. The values for repeatability test for each peak followed by 5 read outs are found to be below 7.5%, i.e., 1.64% for T_{m1} (428 K), 0.63% for T_{m2} (588 K), and 0.58% for T_{m3} (708 K). This value fulfils the International Atomic Energy Agency (IAEA) and American National Standard Institute (ANSI) requirement [28]. In general, the three glow peaks showed very good repeatability for low dose β-rays. In Fig. 11.4b, the values for all I_m for 5 read outs are presented and include error bar of scaling factor 5 σ.

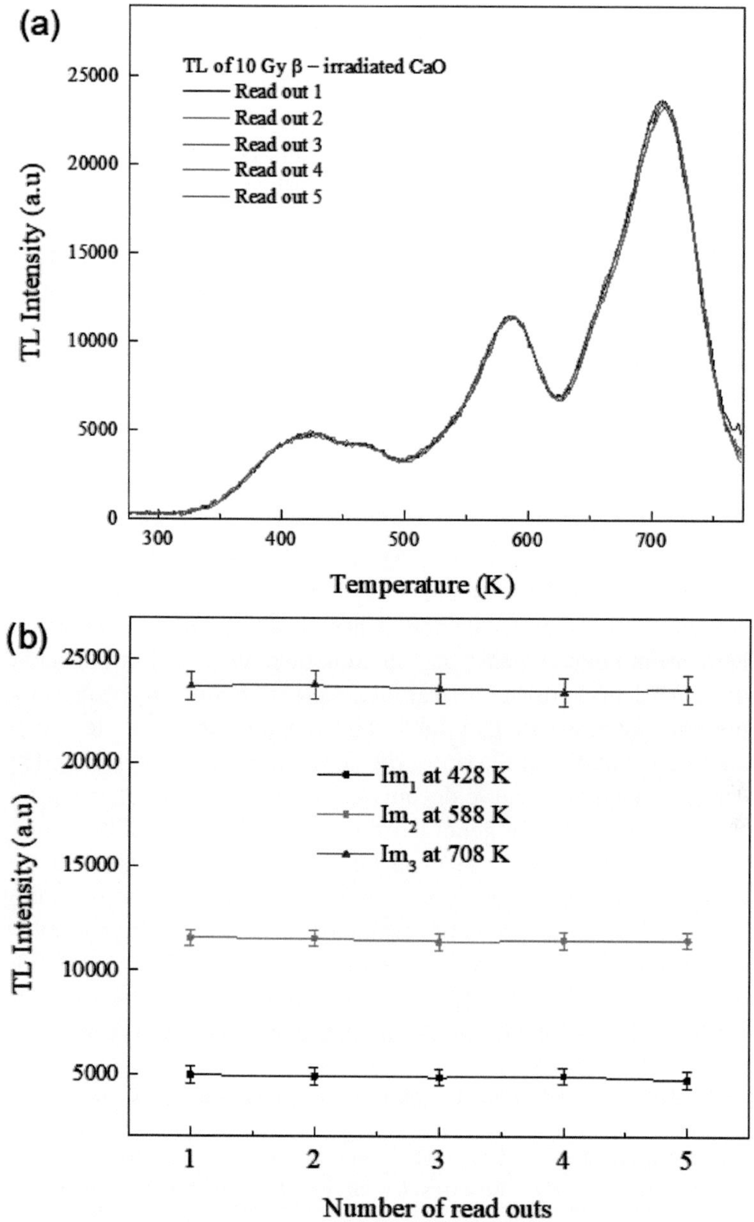

Figure 11.4 (a) TL repeatability of 1200°C annealed CaO phosphor irradiated with *β*-rays (10 Gy) and (b) glow peak intensity for different readouts.

11.2.4 Analysis of TL Glow Curves of CaO Nanophosphor

The formation of TL glow curves due to different ionizing irradiation in the CaO lattice shows the appearance of traps/defects within the material. The phosphor generally exhibits TL glow curves with one or more peaks when the charge carriers are released. The TL glow curve is characteristic of different defect lie in the energy gap of the phosphor. The defects are characterized by certain physical parameters that include activation energy (E) and frequency factor (s). Good knowledge of these kinetic parameters is essential for many applications TL material. The study of deep trap states in various solid-state materials, dosimetric behavior, TL dating, a detailed analysis of TL glow curves are indispensable [29, 30].

It is difficult to calculate the order of kinetics of individual glow peaks in complex TL glow curve. Hence, the thermal cleaning (TC) method is used to separate the glow peaks from each other. Then the order of kinetics of isolated glow peak can be calculated readily using glow curve peak shape method [29]. Figure 11.5a shows TC TL glow curves of 1200°C annealed and β-rayed (10 Gy) CaO nanophosphor. Four prominent TL peaks at 435, 500, 608, and 708 K are isolated using a TC method. The TL trap parameters obtained by peak shape method are presented in Table 11.1. Further, to get a clear idea about the position of the traps, the glow curves are deconvoluted by computerized glow curve deconvolution (CGCD) method, using the computer program for general order kinetics [31]. In CGCD method, all data points along the experimental curve are used, and the experimental results are fitted to more complex functions, including a more significant number of parameters of a curve having a number of overlapping peaks. Kitis et al. reported the TL glow curve equation of general order kinetics [32]. The best solution for a deconvolution process tends to achieve minimum figure of merit (FOM) value.

Table 11.1 TL kinetic parameters obtained from peak shape method

Temperature maxima: T_m (K)	Order of kinetics: b	Activation energy: E (eV)	Frequency factor: s (s^{-1})
435	1.00	1.61	2.07×10^8
500	2.00	1.92	1.01×10^{19}
608	2.00	1.62	6.57×10^{12}
708	2.00	1.85	2.83×10^{12}

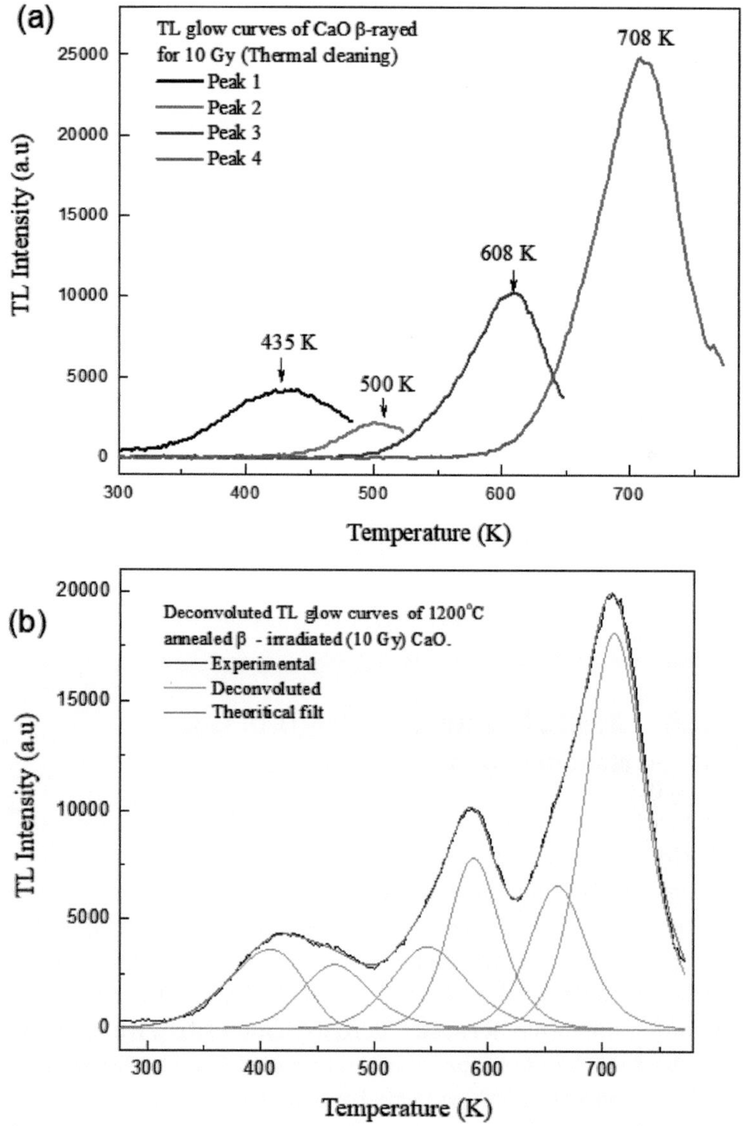

Figure 11.5 (a) Thermally cleaned TL glow curves of 1200°C annealed and 10 Gy β-rayed CaO nanophosphor and (b) deconvoluted TL glow curves by CGCD method.

The deconvoluted TL glow curves of 1200°C and annealed β-rayed (10 Gy) CaO nanophosphor is given in Fig. 11.5b. The experimental glow curve is deconvoluted to six theoretical curves with peaks at 407, 467, 546, 587, 660, and 710 K. The first glow curves at 407 K exhibit first-order kinetics and the rest of them show second order kinetics. The goodness of fit is 3.4% for the present deconvolution. The kinetic parameters obtained by CGCD method are tabulated in Table 11.2.

Table 11.2 TL kinetic parameters obtained from the glow curve deconvolution method

Temperature maxima: T_m (K)	Order of kinetics: b	Activation energy: E (eV)	Frequency factor: s (s^{-1})
407	1.00	0.41	1.74×10^8
467	2.00	0.86	4.65×10^8
546	2.00	1.08	1.80×10^9
587	2.00	1.83	1.50×10^{15}
660	2.00	2.13	4.96×10^{15}
710	2.00	2.31	6.30×10^{15}

11.2.5 Effect of Heating Rate on TL of CaO Nanophosphor

The heating rate (HR) is one of the most critical parameters in TL experiment. Variable heating rate method (VHR) is on the shift of the TL peak temperature to higher temperature side when the heating rate is increased. As the HR increases, glow peak temperature maximum shifts towards higher temperature region with a change in intensity keeping the integral area of the TL curve constant. It is found that different HR cause different TL intensities, which in turn give rise to a change in the kinetic parameter. Figure 11.6 show (a) TL glow curves of 1200°C annealed and beta irradiated (100 Gy) CaO phosphor recorded at various heating rates and (b) variation of I_m with the heating rate.

Temperature lag correction

The TL glow curve is the plot of TL light intensity as a function of temperature. The temperature is usually that of the heating element (kanthal strip) measured by the thermocouple attached. However,

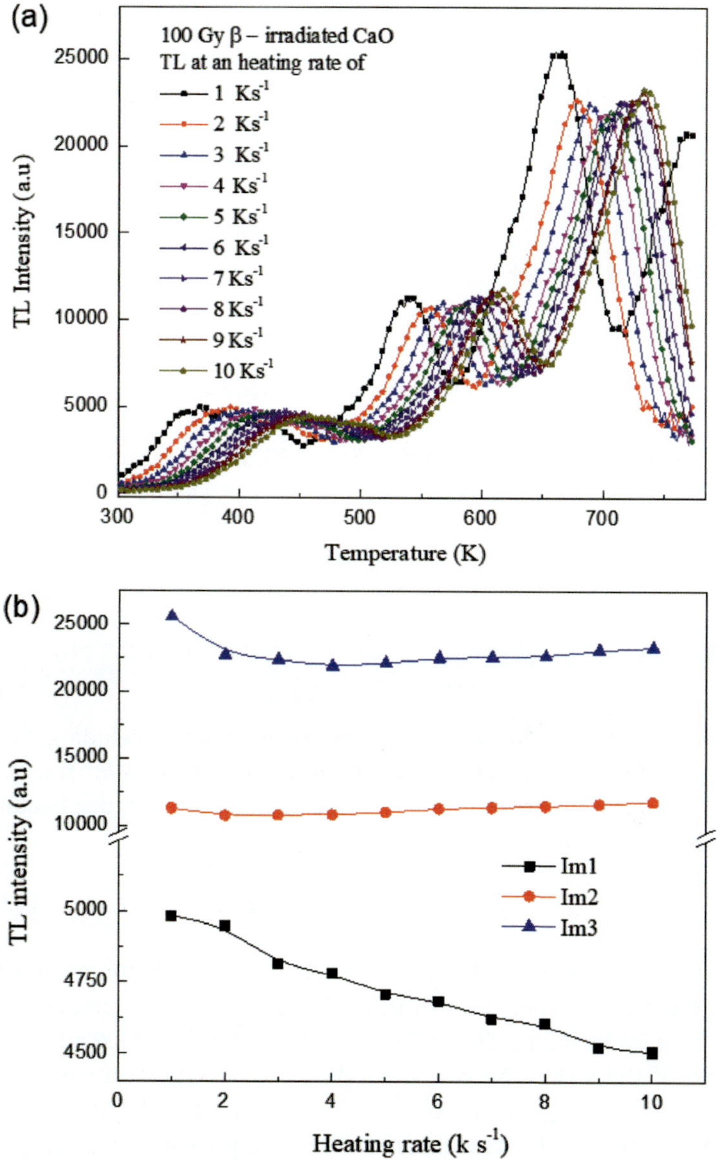

Figure 11.6 (a) TL glow curves of 1200°C annealed CaO recorded at various heating rates and (b) variation of maximum TL intensity with the heating rate.

to extract physical information from the glow curves, it is necessary to know the sample's temperature rather than a heating strip. In contact heating (most commonly applied type), the temperature of the heating element is different from the temperature of the sample. For accurate determination of the kinetic parameters from the glow curves, it is important to know the temperature of the phosphor rather than that of the heating element [33].

A simple method to correct for the temperature lag (TLA) based on TL measurements is proposed by Kitis and Tuyn [34]. The equation used to correct for the temperature lag based on TL measurements is given by

$$T_j = T_i - c\ln\left(\beta_i / \beta_j\right) \tag{11.2}$$

where T_i and T_j are the maximum temperatures of a glow peak with heating rates *i* and *j*, respectively, β_i and β_j are heating rates, *c* is a constant (calculated using Eq. 11.3), which is determined by using two lower heating rates where the TLA can be considered a negligible.

$$c = \frac{(T_{m,raw,\beta=2}) - (T_{m,raw,\beta=1})}{\ln 2} \tag{11.3}$$

The TLA (ΔT) in the position of the peak maximum is given by

$$\Delta T = T_{m,raw} - T_{m,corr} \tag{11.4}$$

Here $T_{m,raw}$ is glow peak temperature maximum including TLA, $T_{m,corr}$ is new temperatures maximum with correction for the TLA at high heating rates (using Eq. 11.2). Further, the effective heating rates due to TLA are calculated using the expression

$$\beta_{eff} = \frac{T_{m,raw} - T_o - \Delta T}{T_{m,raw} - T_o}\beta \tag{11.5}$$

where β_{eff} – effective heating rate, T_o – starting temperature (K) of the order of room temperature, $T_{m,raw}$ – glow peak temperature maximum including TLA, ΔT – temperature lag (K), β – the heating rate of the strip. The corrected glow peak temperature maxima, temperature lag, and effective heating rates are calculated using the above equation for our experimental data presented in Fig. 11.7. These parameters are tabulated in Table 11.3. The procedure adopted in the present work evaluates the temperature lag for a linear function of heating rate. Figure 11.7 show glow peak maximum temperature as a function of heating rate for CaO phosphor, curve $T_{m,raw}$ is plotted from the data including the TLA and curve $T_{m,corr}$ is plotted from the data after correction for the TLA (a) T_{m1} = 428 K, (b) T_{m1} = 585 K, and (c) T_{m1} = 708 K.

Table 11.3 Parameters obtained in temperature lag calculation of calcium oxide nanophosphor

β (K s^{-1})	Peak – 1			Peak – 2			Peak – 3			(β_{eff}) avg (K s^{-1})
	T_{m1}, raw	T_{m1}, corr	ΔT_1	T_{m2}, raw	T_{m2}, corr	ΔT_2	T_{m3}, raw	T_{m3}, corr	ΔT_3	
1	368	368	0	543	543.00	0	665	665	0	1
2	393	392.99	0.01	555	555.00	0	680	680	0	2
3	408	407.62	0.38	568	562.02	5.98	690	688.77	1.23	2.97
4	419	417.99	1.01	580	567.00	13.00	700	695.00	5.00	3.91
5	430	426.04	3.96	588	570.86	17.14	705	699.83	5.17	4.84
6	438	432.61	5.39	595	574.02	20.98	713	708.95	4.05	5.77
7	445	438.17	6.83	600	576.68	23.32	719	707.11	11.89	6.65
8	450	442.98	7.02	608	579.00	29.00	728	710.00	18.00	7.52
9	455	447.24	7.76	612	581.03	30.97	730	712.55	17.45	8.45
10	460	451.03	8.97	618	582.86	35.14	733	714.83	18.17	9.32

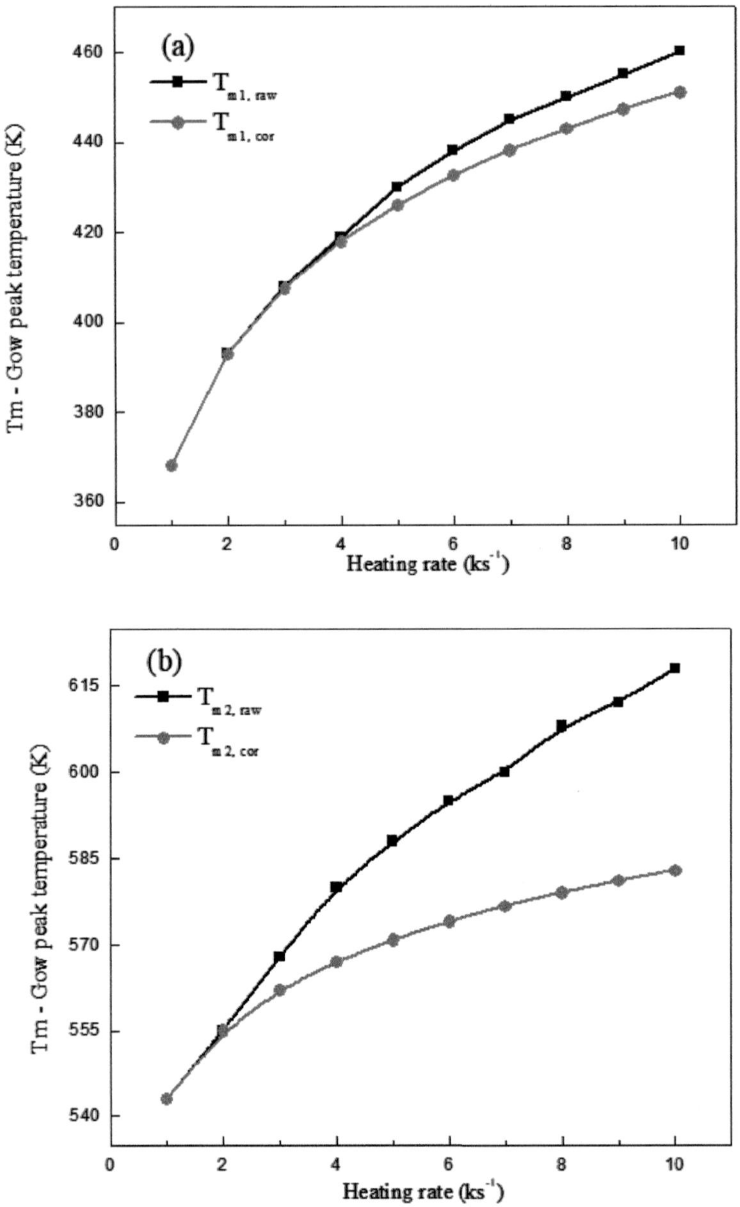
(a)
Tm - Gow peak temperature (K)
Heating rate (ks⁻¹)
360
380
400
420
440
460
0
2
4
6
8
10
(b)
Tm - Gow peak temperature (K)
Heating rate (ks⁻¹)
540
555
570
585
600
615
0
2
4
6
8
10

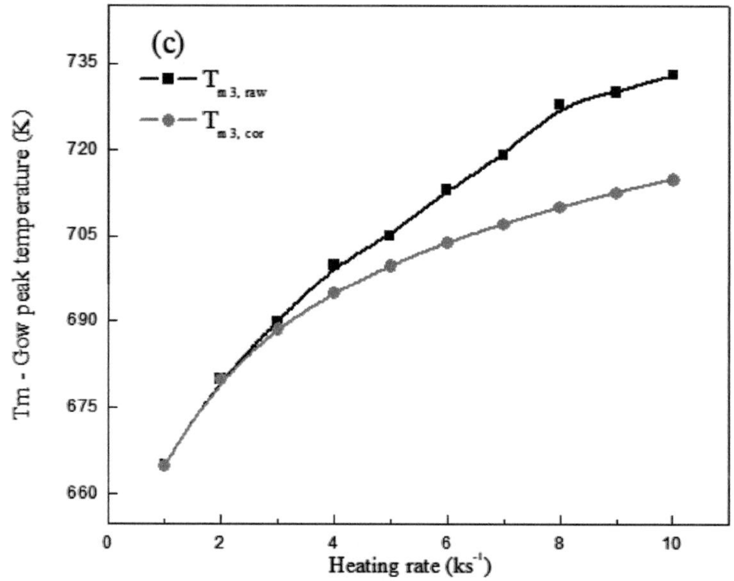

Figure 11.7 The glow-peak maximum temperature as a function of the heating rate for CaO phosphor, curve $T_{m,raw}$ is plotted from the data including the TLA and curve $T_{m,corr}$ is plotted from the data after correction for the TLA (a) T_{m1} = 428 K, (b) T_{m1} = 585 K, and (c) T_{m1} = 708 K.

The experimental heating rate is same as effective heating rate at low HR values. Whereas effective heating rate values are found to be lesser than the experimental heating rate at higher values.

Kinetic parameters by VHR method

The variable heating rate method is based on the shift in the position of the glow-peak maximum temperature as a function of the heating rate. It can be noted that, plot of $\ln(T^2{}_m/\beta)$ vs. $1/(kT_m)$ must be a straight line (method I) with slope E (activation energy), and the intercept of this plot is $\ln(E/sk)$, which allows evaluating the s (frequency factor). Also, the plot of $\ln(\beta_i/\beta_j)$ vs. $1/(kT_m)$ must be a straight line (method II) with slope E, and the intercept of this plot is $-E/(kT_m)$ which allows evaluating the maximum temperature T_m corresponding to the heating rate β_i. These plots for TLA corrected data are shown in Fig. 11.8a and b respectively. The trap parameters obtained by VHR method and given in Table 11.4.

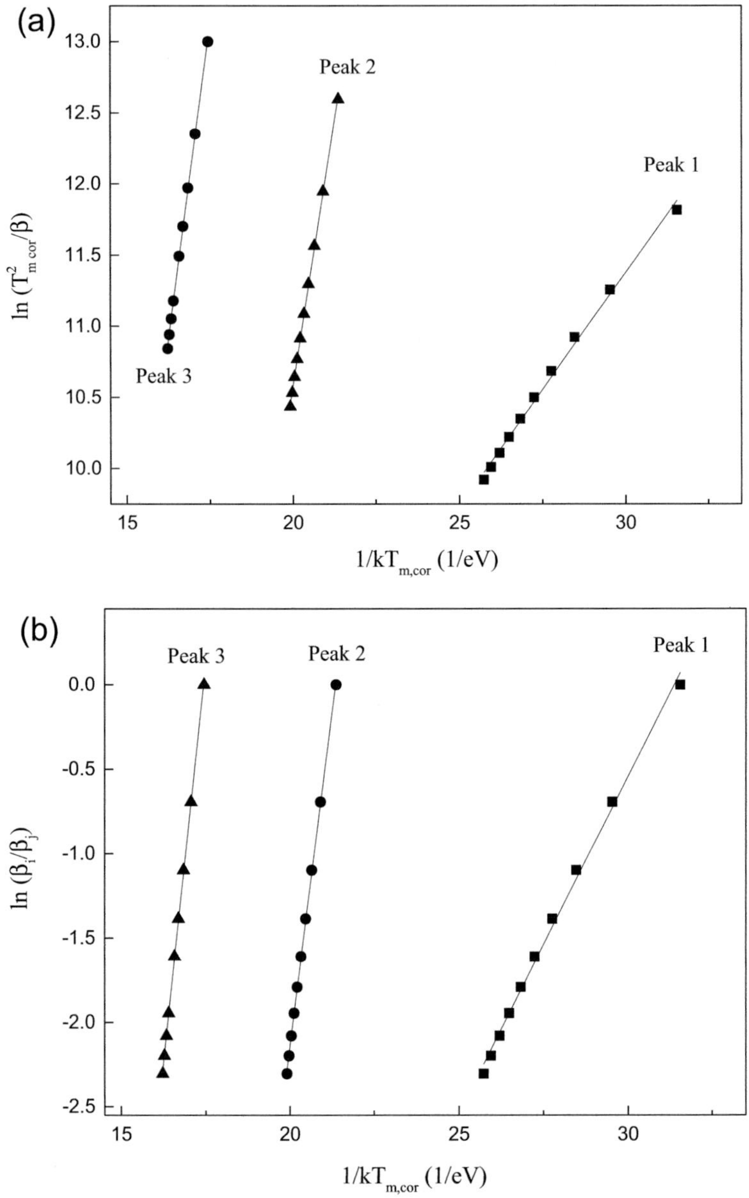

Figure 11.8 (a) Plot of $\ln(T_{m,corr}/\beta)$ against $1/(kT_{m,corr})$ for CaO nanophosphor. Curve $T_{m,corr}$ is plotted from the data after correction for the TLA. (b) Plot of $\ln(\beta_i/\beta_j)$ vs. $1/(kT_{m,corr})$.

Table 11.4 TL kinetic parameters obtained from various heating rate method

Method	Peak – 1 (428 K)		Peak – 2 (585 K)		Peak – 3 (708 K)	
	E (eV)	s (s^{-1})	E (eV)	s (s^{-1})	E (eV)	s (s^{-1})
I	0.33	1.13×10^{10}	1.48	3.48×10^{12}	1.78	1.5×10^{12}
II	0.39	—	1.58	—	1.89	—

The correct analysis of TL parameters implies that the obtained results are not in conflict with the TL theory in general and with the utilized model in particular. Generally, the correct analysis mainly tends to have activation energy and the frequency factor values in the usual limits 0.7 to 2.0 eV and 10^7 to10^{13} s^{-1} respectively [35]. In the present investigations, activation energy and frequency parameters of CaO nanophosphor values obtained by thermal cleaning, CGCD, and VHR methods are in good agreement with theoretical values.

11.3 Optically Stimulated Luminescence of CaO Nanophosphor

11.3.1 Introduction

The study of optically stimulated luminescence (OSL) properties of phosphor materials is a major field of research in radiation dosimetry. OSL is a process resulting in light emission from an insulator or semiconductor previously exposed to ionizing radiation when the material is stimulated with visible photon [36, 37]. OSL dosimetry is successfully used for monitoring of radiation of persons working in an ionizing radiation environment, nuclear accidents, cosmic radiation in space, radiation facilities in health and power sectors, geochronology (dating of sediments) and environmental radiation monitoring [38]. In particular, OSL is effectively used in medical [38], personal [39], retrospective [40], and accident dosimetry [36]. In addition to being of use for radiation dosimetry, the OSL signal contains information about the distribution of traps and recombination centers in the crystals. OSL signals can also provide insights into the charge excitation, movement of traps and their recombination in crystals. Such insights will lead to improving our understanding of the luminescence mechanism [41].

11.3.2 Types of OSL

(i) Continuous wave OSL

Continuous wave (CW) OSL is administered in which a pre-irradiated material is stimulated by a light source of constant intensity. Wavelengths of emitted luminescence signal other than the stimulation are recorded [42, 43]. CW-OSL is one of most sensitive mode because the intensity of the emitted light is the highest during stimulation. In CW-OSL, there is no restriction on the stimulation duration or recording for an optimization of the signal-to-noise ratio.

(ii) Linearly modulated OSL

Linearly modulated (LM) OSL is recorded with the intensity of the stimulating photon beam is ramped linearly. The emitted signal is observed during the ramped stimulation [44]. A graph of the light emission intensity as a function of the readout time results in the form of peak shaped curves (like TL glow peaks) instead of exponential-like decays. This is because the distribution of the traps results in different photo-ionization cross sections (PICS) for different wavelengths of stimulations. The release of electrons from the traps depends on their PICS; larger the PICS, quicker is the release. There is no correlation between LM-OSL peaks and TL glow peaks from a physics point of view. In the case of LM-OSL, the recording of luminescence is slow but peak shaped curve is obtained. Also, well separated multiple peaks can be obtained if optically active traps are having different values of PICS participating in the OSL process.

(iii) Pulsed OSL

Pulsed OSL (POSL) is a luminescence recorded sporadically following stimulation by short pulses (duration of the order of hundreds of nanoseconds) with a frequency of a few thousands of hertz [43]. The optical stimulation is thus turned on and off periodically

The OSL properties of CaO nanomaterials are characterized by an optical stimulation band that extends over a broad wavelength range from the UV region to Infrared region. Hence, a large variety of stimulation sources such as lasers, diodes, and broadband lamps can be used to record OSL curves [45].

11.3.3 Effect of Annealing on CW-OSL of CaO Nanophosphor

CW-OSL is a popular procedure for the determination of environmental radiation doses. CW-OSL is the most straight forward OSL readout approach consists of stimulating the detector with the light of constant intensity. In CW-OSL, the luminescence is recorded very fast and looks like a decay curve. The background count rate or net background is nearly constant in CW-OSL decay curves. Hence, CW-OSL is useful in dosimetric applications due to the short recording time. CaO samples annealed at various temperature are exposed to β-rays for a dose of 50 Gy. Figure 11.9a show typical OSL decay curves of CaO annealed at various temperature for 2 h and Fig. 11.9b show variation of CW-OSL intensity with annealing temperature. It is observed that the OSL intensity remains almost the same in samples annealed at 600, 750, 900, and 1050°C. Further, OSL intensity increases rapidly in samples annealed at 1200°C, this might be due to the increase of crystallite size. Hence, OSL properties of CaO annealed at 1200°C are studied for various β-dose.

11.3.4 CW-OSL Dose Response of CaO Nanophosphor

Figure 11.10a represents CW-OSL decay curves of 1200°C annealed CaO phosphor exposed to β-rays in the dose range of 10–100 Gy stimulated with blue light (470 nm). The OSL from annealed CaO sample is strong, the decay is fast, and the entire signal decays within 50 s. The CW-OSL curve of an ideal material is hardly described by a single exponential decay function. It is found that the shape of CW-OSL is not an exponential, the shape of the decay curve depends on the sample, absorbed dose, stimulation wavelength, stimulation flux, and sample temperature [46]. Also, there are many factors contributing to deviations from ideal exponential behavior, which includes (i) contribution from more than one trapping center or trapping centers, (ii) by re-trapping of the charge carriers after stimulation, and (iii) non-uniform illumination of the material.

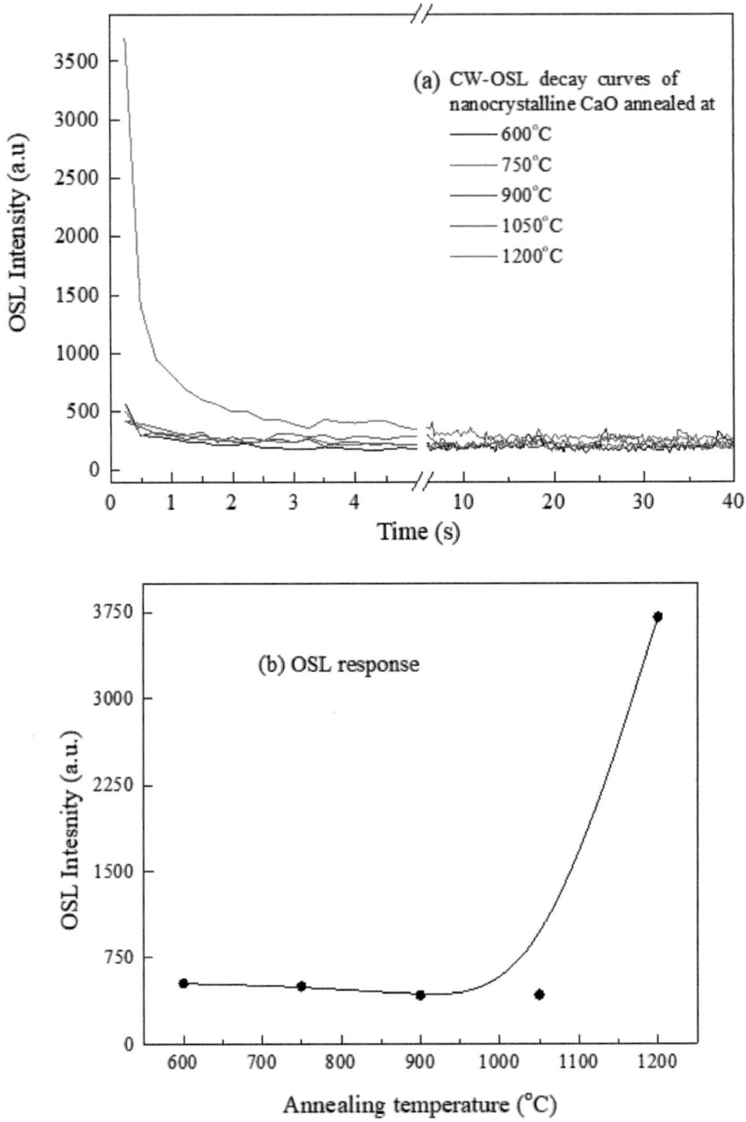

Figure 11.9 (a) CW-OSL decay curves of CaO annealed at different temperatures and (b) variation of CW-OSL intensity with annealing temperature.

Further, it is observed that the shape of CW-OSL decay curve remains same and total area under the decay curves increases with β-dose. This is due to the fact that, the rate of decay of the OSL curve depends only on photoionization cross section (σ) and incident photon flux (ϕ), but not on initial concentration of trapped charges (n_0), which means that the shape of the OSL curve does not change with the dose. On the other hand, the total area under the OSL curve depends only on n_0, but not on σ or ϕ, because

$$\int_0^\infty I_{\text{osl}}(t)dt \propto \int_0^\infty n_o\, \sigma\, \phi\, e^{-\sigma\phi t}\, dt = n_o \tag{11.6}$$

i.e., a total number of photons emitted or the total area under the OSL curve is not affected by the change in stimulation flux. These observations have a practical consequence for dosimetry.

Figure 11.10b show the variation of CW-OSL intensity with β-dose. It is found that the initial OSL intensity increases linearly with the beta dose. The increase in intensity with dose is due to the increase in a number of traps responsible for OSL [36]. An ideal dosimeter must have a signal that is linearly proportional to the absorbed dose. However, deviations from linearity is a result of the dynamics of the process of charge capture between different defects in the phosphor. The general behavior of most TL/OSL materials is to show a linear relationship between luminescence signal and absorbed dose for low doses, approaching saturation for higher doses as the trapping centers become saturated. This can be approximated by a saturating exponential function of the type:

$$S(D) = A\,(1 - e^{-BD}) \tag{11.7}$$

where A is a proportionality constant and B is a free parameter. At low doses, this function approximates the linear relation $S(D) = ABD$. At high doses the phosphor response (signal per dose) is less than the response at low doses, characterizing a sub-linear dose response. Competition between different trapping and recombination centers during either irradiation or readout can cause the material to produce more signal that would be expected based on the response at low doses. This is called supra-linear dose-response [47]. In the present work, OSL response of CaO phosphor exhibits linear dose response. This is one of the fundamental requirement of a good dosimeter and opens a new window to study other dosimetric properties such as fading, detection limits, minimum detectable dose, energy response, temperature dependence, reusability/repeatability, etc.

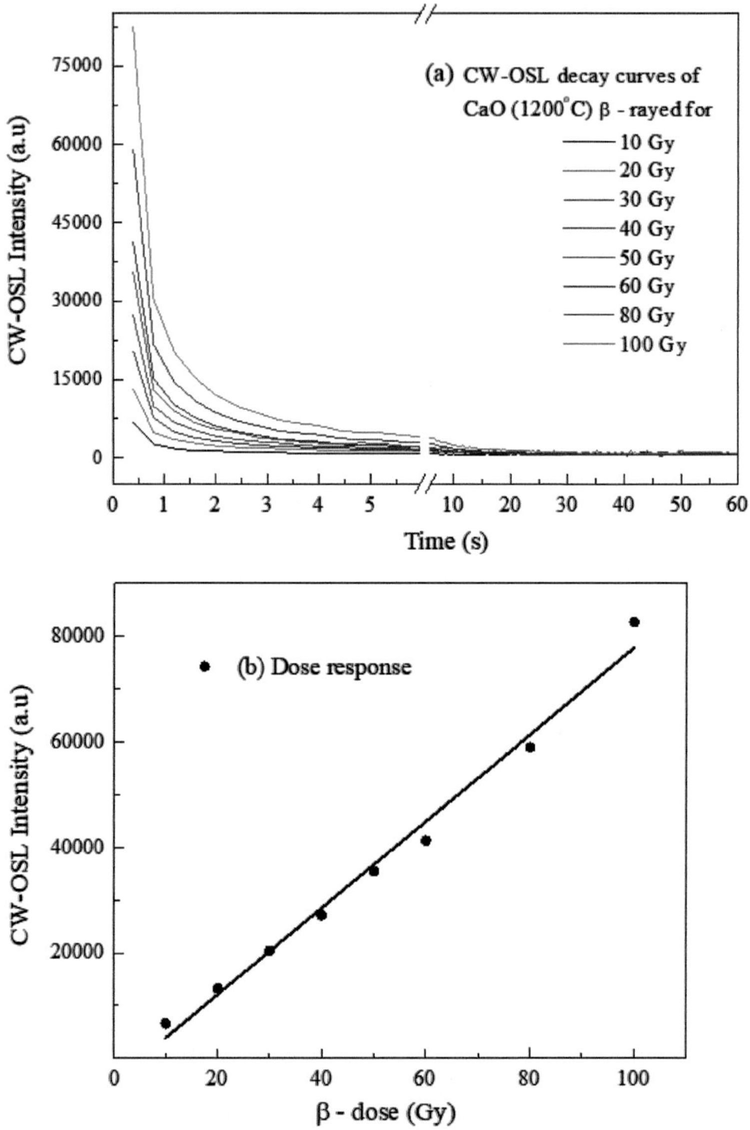

Figure 11.10 CW-OSL (a) decay curves of 1200°C annealed CaO phosphor and (b) dose-response curve.

11.3.5 LM-OSL Dose Response of CaO Nanophosphor

The peak height of LM-OSL curve is decided by the number of traps (also called as trap density) participating in OSL process and the recombination efficiency, whereas the peak position is decided by the optical stimulation rate, retrapping, and recombination cross sections and the radiation dose. LM-OSL is very useful in the characterization of traps participating in OSL process. Figure 11.11a shows LM-OSL curves of 1200°C annealed CaO phosphor exposed to beta rays in the dose range of 10–100 Gy. LM-OSL intensity for stimulation photon fluence rate (ϕ) increased linearly from zero until a maximum value ϕ_{max} during a period 'T' is given by

$$I_{LM-OSL}(t) \propto n_o \sigma \frac{\phi_{max}}{T} t\, e^{-\frac{1}{2}\sigma \frac{\phi_{max}}{T} t^2} \tag{11.8}$$

and the LM-OSL curve reaches a maximum at an instant t_{max} given by

$$t_{max} = \sqrt{\frac{T}{\sigma\, \phi_{max}}} \tag{11.9}$$

where σ is related to the photo ionization cross section. The LM-OSL maximum intensity is proportional to

$$I_{max} \propto n_o \sqrt{\frac{\sigma\, \phi_{max}}{T}}\, e^{-\frac{1}{2}} = \frac{n_o}{t_{max}} e^{-\frac{1}{2}} \tag{11.10}$$

Further, it is found that the area under the curve increases with the beta dose. Variation of LM-OSL peak intensity maxima and t_{max} as a function of dose is shown in Fig. 11.11b. This dose-response is also linear like CW-OSL dose response and t_{max} slightly shift to higher values at a higher dose. The LM-OSL curve has a reduced intensity when compared to an equivalent CW-OSL curve. However, the areas under the curves are the same [44].

11.3.6 Theoretical Fitting of OSL Curves

The PICS of a trap is an important property in detecting the appropriate wavelength for optical stimulation [37]. The PICS is defined as defined as the photoionization probability per unit time per atom (or optical center) divided by the incident photon flux (ϕ), that is:

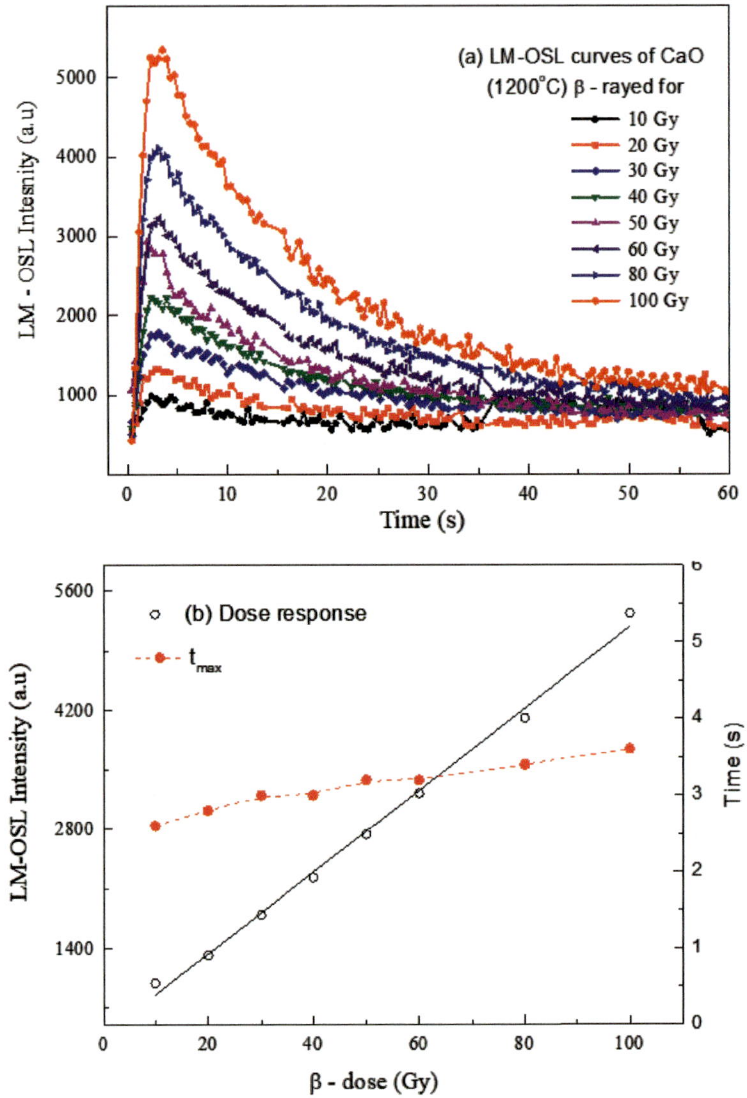

Figure 11.11 (a) LM-OSL curves of 1200°C annealed CaO phosphor and (b) dose-response curve.

$$p = \sigma\,\phi, \quad \text{here} \quad p = \frac{1}{\tau} \quad \text{thus} \quad \sigma = \frac{1}{\phi\,\tau}\,\text{cm}^2 \tag{11.11}$$

where τ is the decay constant (in s), and ϕ is the incident photon flux (photons incident per cm^2 per s). Hence, PICS has the unit 'cm^2.' Photoionization here refers to electron transition from a discrete bound defect state to the quasi-continuum density of states in the conduction band (electron emission) or to electron transition from the valence band to a localized level (hole emission) [48, 49]. Thus, PICS is related to retrapping probability (b) and maximum stimulation intensity. OSL originates from various trap levels having different PICS. These trap levels and retrapping process lead to different OSL decay components. The CW-OSL curve can be fitted with a sum of three exponentials represented by Eq. 11.12

$$I_{OSL} = C_1 e^{\frac{-t_1}{\tau_1}} + C_2 e^{\frac{-t_2}{\tau_2}} + C_3 e^{\frac{-t_3}{\tau_3}} \qquad (11.12)$$

where I_{OSL} is the initial intensity and τ_1, τ_2, and τ_3 are the decay constants of respective OSL traps. It may be assumed that the OSL in the CaO sample originates from various trap levels having different PICS [50]. These trap levels and re-rapping processes may lead to an OSL decay, which may be fitted with the sum of exponentials as given by Eq. 11.12. The coefficient in the above equation gives the idea of a number of traps responsible for each component.

Deconvoluted CW-OSL decay curves of 100 Gy beta irradiated CaO phosphor is given in Fig. 11.12a. The figure of merit and goodness of fit (GOF) for this analysis are 2.00% and 0.99% respectively which are in the acceptable limit [51]. Lifetime, the order of kinetics and PICS of three OSL components of the deconvoluted decay curves for various beta doses are calculated and tabulated in Table 11.5. Further, it is found that the lifetime associated with the traps responsible for OSL decay components are less at a higher dose. This implies that the phosphor is highly sensitive to beta particles. Further, all the three components exhibit general order kinetics, and corresponding PICS values are in theoretically acceptable range.

LM-OSL curves generally comprise of several overlapping peaks. The analysis and separation of these complex curves into their constituent components can be attained using CGCD analysis. Figure 11.12b shows the deconvoluted LM-OSL curves of beta irradiated (100 Gy) calcium oxide phosphor. The LM-OSL decay curves are deconvoluted into three components namely fast, medium, and slow using the equation below [52].

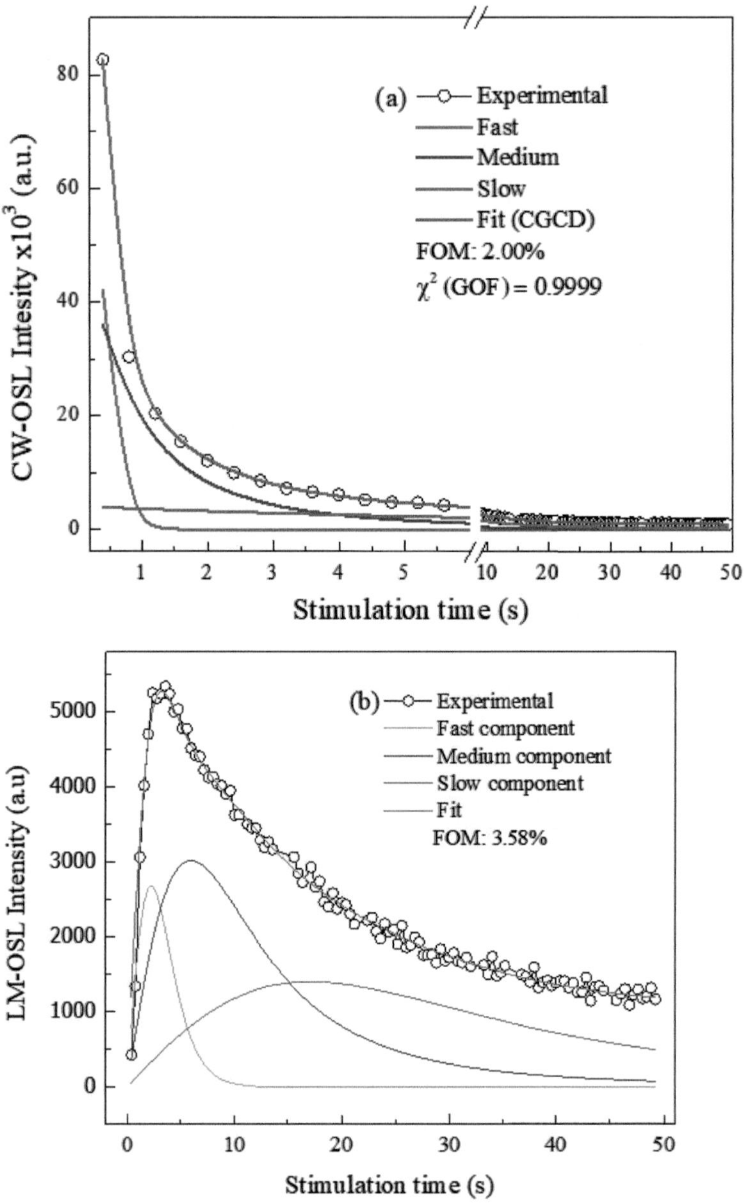

Figure 11.12 Experimental and theoretically fitted (a) CW-OSL decay curves and (b) LM-OSL curves of β-irradiated (100 Gy) CaO phosphor.

Table 11.5 CW-OSL kinetic parameters of β-rayed calcium oxide phosphors

Dose (Gy)	Decay constant/lifetime (s)			Order of kinetics			Photoionization cross section (cm^2)		
	Fast	Medium	Slow	Fast	Medium	Slow	Fast	Medium	Slow
10	0.044	1.764	41.391	1.68	1.18	1.00	1.32×10^{-16}	3.32×10^{-18}	1.42×10^{-19}
50	0.028	1.067	10.431	1.28	1.59	1.00	2.07×10^{-16}	5.48×10^{-18}	5.62×10^{-19}
100	0.037	1.000	10.000	1.29	1.44	1.88	1.60×10^{-16}	5.86×10^{-18}	5.86×10^{-19}

$$I(T) = I_m \frac{t}{t_m}\left[\frac{b-1}{2b}\frac{t^2}{t_m^2} + \frac{b+1}{2b}\right]^{b/1-b} \quad (11.13)$$

Curve fitting studies have shown that three LM-OSL components (fast, medium, and slow) are a minimum to achieve an acceptable fit. Here fast and medium components are to fit the main part of the LM-OSL curve and last one for slowly decreasing tail (see Fig. 11.12b).

FOM of this analysis is 3.58%. For each component, the time corresponding to peak maximum (lifetime/decay constant) is established which allows calculating PICS [53]. The obtained results are presented in Table 11.6. Here, fast component exhibit general order kinetics whereas medium and slow components exhibit second order kinetics. PICS is highest for the first component and decreases for the subsequent one. Further, the fast component has the smallest contribution to the total LM-OSL signal (10.85%), while light outputs of medium and slow components are comparable (44.20–44.95%).

Table 11.6 The parameters estimated from LM-OSL curves of β-rayed calcium oxide phosphors

Dose (Gy)	Components	decay constant/ lifetime (s)	Order of kinetics	Photo-ionization cross section (cm^2)	Contribution (%)
	Fast	2.29	1.2	2.56×10^{-18}	10.85
100	Medium	5.97	2.0	9.77×10^{-19}	44.20
	Slow	17.03	2.0	3.45×10^{-19}	44.95

Further, the OSL is inseparably related to the (TL) phenomenon. The relationships between OSL and TL signals are essential not only from a theoretical point of view, but also they help identify how deep, with respect to the conduction band, are the trapping centers that contribute to the OSL signal. This, in turn, leads to an estimate of the traps' thermal stability. The TL/OSL correlations also aid in understanding the influence of other trapping centers (e.g., shallow traps or deep traps) on the OSL process. Thus, a series of combined TL/OSL experiments can provide useful insight into the OSL process and its relationship with trapping centers in the material [36]. These include studies on the effects of optical bleaching on the TL curves and the effects of thermal annealing on the OSL curves. This is one of the possible scopes for future work.

11.4 TL/OSL Mechanism

The mechanism of TL/OSL phenomenon can be explained using basic concepts. The creation of free charges (electron/hole) by ionizing radiation; trapping of free charges at defect centers and emission of photons when thermally/optically stimulated trapped electron to recombine with a hole. Material irradiated with ionizing radiation led to the displacement of electrons from the valence band (VB) to the conduction band (CB) and trapped at the trapping center just below the CB. These trapped electrons are stimulated with thermal/optical energy will responsible for the TL/OSL emission [54]. The results of TL/OSL, kinetic/trap parameters, and TL emission spectra are used to propose the possible mechanism.

TL/OSL mechanism of β-ray irradiated CaO phosphor with band diagram is shown in Fig. 11.13. TL/OSL glow curve of β-ray irradiated pure CaO sample exhibit multiple overlapped glow peaks/decay curves. Energy absorbed by electrons during irradiation with ionizing radiation causes the production of free electrons and holes pairs. These free electrons move around the CB and hole move around the VB. Meanwhile, vacancies at VB and electrons at CB go through the crystal lattice until being trapped at the trapping center. These defect centers are usually called "traps." These individual electron trapping and hole trapping are due to defect centers which are called trapping and luminescent centers (TC/LC) respectively [54]. As the material stimulated with heat or optical energy, it acquires suitable energy for the liberation of electrons; these electrons can reach the recombination center. As a result of the emission of visible light through recombination of electrons with vacancies at luminescent centers. Variation TL glow curve structure with annealing temperature is due to its chemical composition and crystal structure. TL response also depends on the crystal lattice and crystal defects.

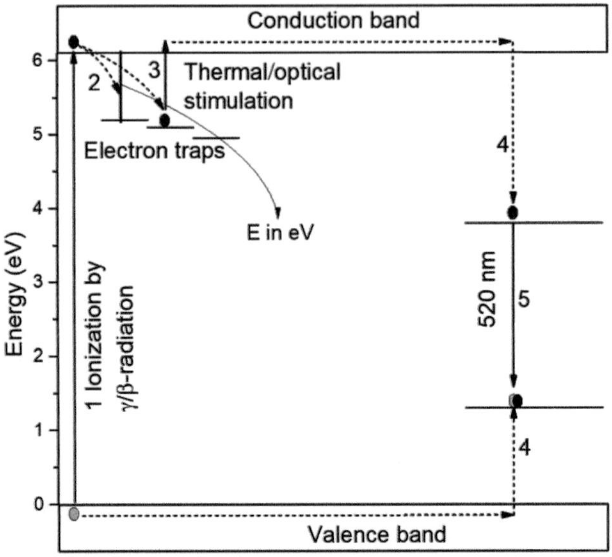

Figure 11.13 Flat band diagram representing TL/OSL mechanism of γ-/β-ray irradiated undoped CaO phosphor.

11.5 Summary

TL/OSL studies of 600, 750, 900, 1050, and 1200°C annealed undoped CaO are performed. The sample annealed at 600°C showed a feeble TL response and the samples annealed 750, 900, and 1050°C shows single prominent TL glow with a peak at 448, 456, and 453 K respectively. The samples show nonlinear dose response at low doses and liner response at high doses. TL response increases with the increase of annealing temperature. Further, 1200°C annealed sample show three prominent TL glow with peaks at 428, 585, and 708 K. TL glow peak intensity of all three peaks increases linearly with dose and T_m is almost constant. The repeatability test for each glow peak followed by 5 readouts are found to be below 7.5%, i.e., 1.64% for T_{m1} (428 K), 0.63% for T_{m2} (588 K), and 0.58% for T_{m3} (708 K).

Effect of heating rate on TL properties of 1200°C annealed undoped CaO are studied and found that glow peak temperature maxima shift towards higher temperature side with an increase in heating rate. Further, temperature lag in measuring TL glow curves and effective heating rate values are calculated. The experimental heating rate is the same as the effective heating rate at low HR values. Higher heating rate values are found to be lesser than experimental values. Kinetic parameters are calculated using VHR and CGCD methods and are in good agreement with theoretical values.

CW and LM-OSL curves of β-irradiated CaO phosphor are studied. CW-OSL intensity of samples annealed at 600, 750, 900, and 1050°C remains almost the same whereas the intensity increases rapidly in samples annealed at 1200°C. CW-OSL properties of 1200°C annealed CaO phosphor are studied under beta irradiation for various dose. The OSL curve exhibits a linear dose response over a dose range of 10 to 100 Gy. It is observed that the shape of CW-OSL decay curve remains same and total area under the decay curves increases with β-dose. The increase in intensity with dose is due to the increase in a number of traps responsible for OSL. The above observations have practical implications for dosimetry. The LM-OSL intensity graphs are peak shaped curves, instead of exponential-like decays. The peak-shaped curves are explained using theoretical equations. The area under the LM-OSL curve increase with beta dose and t_{max} slightly shift to higher values at a higher dose.

PICS and other parameters are obtained by deconvolution of OSL curves. Theoretical curve fitting is done with three OSL components (fast, medium, and slow). The FOM and GOF are in the acceptable range for both the OSL deconvolution. The kinetic parameters obtained are in theoretically acceptable values. CaO phosphor is a polycrystalline powder and has the advantages of simple preparation and low cost. These results demonstrated that the CaO material has a strong potential for OSL dosimetry. However, further work is needed in order to characterize this material for radiation dosimetry applications fully.

TL/OSL mechanism of γ/β-ray irradiated undoped CaO nanophosphor with flat band diagram are proposed using the results obtained in TL/OSL curves, TLE, and trap parameters. In undoped phosphor, the defect centers such as oxygen and cation vacancies formed during synthesis of the sample, which acts as electron traps and hole traps respectively. These trapped electron and hole will undergo recombination when stimulated with thermal/optical energy to facilitate the TL/OSL.

References

1. W.A. Goddard, *Handbook of Nanoscience, Engineering and Technology*. CRC Press, 2012.
2. K.E. Drexler, C. Peterson, G. Pergamit, *Unbounding the Future: The Nanotechnology Revolution*. Quill, 1991.
3. A.K. Geim, K.S. Novoselov, The rise of graphene, *Nat. Mater.*, vol. 6, 2007, pp. 183–191, doi: 10.1038/nmat1849.
4. H.W. Kroto, J.R.Heath, S.C. O'Brien, R.F. Curl, E. Smalley, C_{60} buckminsterfullerene, *Nature*, vol. 318 (6042), 1985, pp. 162–163, doi: 10.1038/318162a0.
5. R. Heung, X. Wang, P. Xiao, Characterisation of PSZ/Al_2O_3 composite coatings using electrochemical impedance spectroscopy, *Electrochim. Acta*, vol. 51 (8–9), 2006, pp. 1789–1796, doi: 10.1016/j.electacta.2005.02.097.
6. M. Gaft, R. Reisfeld, G. Panczer, *Luminescence Spectroscopy of Minerals and Materials*. Springer Nature, 2005.
7. K.C. Patil, M.S. Hegde, T. Rattan, S.T. Aruna, *Chemistry of Nanocrystalline Oxide Materials*. World Scientific, 2008.
8. J.A. Rodriguez, M. Fernández-García, *Synthesis, Properties, and Applications of Oxide Nanomaterials*. Wiley-Interscience, 2007.
9. R. Polat, R. Demirboąa, W.H. Khushefati, Effects of nano and micro size of CaO and MgO, nano-clay and expanded perlite aggregate on the autogenous shrinkage of mortar, *Constr. Build. Mater.*, vol. 81, 2015, pp. 268–275, doi: 10.1016/j.conbuildmat.2015.02.032.
10. V. Petit, P. Moretti, P. Camy, J.L. Doualan, R. Moncorgé, Active waveguides produced in Yb^{3+}:CaF_2 by H^+ implantation for laser applications, *J. Alloys Compd.*, vol. 451 (1–2), 2008, pp. 68–70, doi: 10.1016/J.JALLCOM.2007.04.120.
11. C. Pandurangappa, B. N. Lakshminarasappa, Optical studies of samarium-doped fluoride nanoparticles, *Philos. Mag.*, vol. 91 (35), 2011, pp. 4486–4494, doi: 10.1080/14786435.2011.610376.
12. N. Zhang, et al., Hydrothermal synthesis and electrochemical properties of alpha-manganese sulfide submicrocrystals as an attractive electrode material for lithium-ion batteries, *Mater. Chem. Phys.*, vol. 111 (1), 2008, pp. 13–16, doi: 10.1016/J.MATCHEMPHYS.2008.03.040.
13. D. Prakash, Synthesis and characterization of rare earth doped nanocrystalline oxides and their luminescence properties, Ph.D. thesis,

Visvesvaraya Technological university, Belagavi, Karnataka, India, 2018

14. N. Jiang, On the oxidation of CaF_2 in transmission electron microscope, *Micron*, vol. 43 (6), 2012, pp. 746–754, doi: 10.1016/J. MICRON.2012.02.002.
15. Y. Sun, et al., Effect of CaO on NO_x reduction by selective non-catalytic reduction under variable gas compositions in a simulated cement precalciner atmosphere, *Int. J. Environ. Res. Public Health*, vol. 14 (12), 2017, p. 1474, doi: 10.3390/ijerph14121474.
16. L. Bøtter-Jensen, S.W.S. Mckeever, A.G. Wintle, *Optically Stimulated Luminescence Dosimetry*. Elsevier, 2003.
17. R.Chen, S.W.S. McKeever, *Theory of Thermoluminescence and Related Phenomena*. World scientific, 1997.
18. V.V. Ursaki, I.M. Tiginyanu, O. Volciuc, V. Popa, V.A. Skuratov, H. Morkoç, Nanostructuring induced enhancement of radiation hardness in GaN epilayers, *Appl. Phys. Lett.*, vol. 90 (16), 2007, p. 161908, doi: 10.1063/1.2723076.
19. N. Salah, P.D. Sahare, A.A. Rupasov, Thermoluminescence of nanocrystalline LiF:Mg, Cu, P, *J. Lumin.*, vol. 124 (2), 2007, pp. 357–364, doi: 10.1016/j.jlumin.2006.04.004.
20. V.S. Kortov, Nanophosphors and outlooks for their use in ionizing radiation detection, in *Radiation Measurements*, 2010, vol. 45 (3–6), pp. 512–515.
21. http://www.usu.edu/geo/luminlab/Reader.pdf
22. J.K. Mundupuzhakal, R.H. Biswas, S. Chauhan,V. Varma, Y.B. Acharya, B.S. Chakrabarty, Effect of thermal annealing on the thermoluminescent properties of nano-calcium fluoride and its dose-response characteristics, *Radiat. Prot. Dosimetry*, vol. 167 (4), 2015, pp. 419–424, doi: 10.1093/rpd/ncu333.
23. H. Toktamiş, A.N. Yazici, Effects of annealing on thermoluminescence peak positions and trap depths of synthetic and natural quartz by means of the various heating rate method, *Chinese Phys. Lett.*, vol. 29 (8), 2012, pp. 27–30, doi: 10.1088/0256-307X/29/8/087802.
24. F.O. Ogundare, F.A. Balogun, L.A. Hussain, Heating rate effects on the thermoluminescence of fluorite, *Radiat. Meas.*, vol. 40 (1), 2005, pp. 60–64, doi: 10.1016/j.radmeas.2005.01.004.
25. Y. Yanagisawa, N. Inishi, A. Narumi, Phosphorescence and thermoluminescence of uv-irradiated CaO powders at temperatures

from 120 to 350 K, *Phys. Rev. B*, vol. 46 (17), 1992, pp. 11121–11124, doi: 10.1103/PhysRevB.46.11121.

26. K.R. Nagabhushana, D. Prakash, F. Singh, S.H. Tatumi, R.R. Rocca, S. Watanabe, Comparative studies on thermoluminescence glow curves of calcium oxide nanophosphor irradiated with various ionizing radiations, *J. Alloys Compd.*, vol. 735, 2018, pp. 1949–1954, doi: 10.1016/j.jallcom.2017.11.148.
27. S. Del Sol Fernández, et al., Thermoluminescent characteristics of LiF:MG, Cu, P and $CaSO_4$:Dy for low dose measurement, *Appl. Radiat. Isot.*, vol. 111, 2016, pp. 50–55, doi: 10.1016/j.apradiso.2016.02.011.
28. F. Sepulveda, J. Azorin, T. Rivera, C. Furetta, C. Sanipoli, Thermoluminescence (TL) characteristics of perovskite-like $KMgF_3$ activated by La ions, in *Nuclear Instruments and Methods in Physics Research, Section B: Beam Interactions with Materials and Atoms*, 2004, vol. 213, pp. 329–332.
29. K.R. Nagabhushana, B.N. Lakshminarasappa, F. Singh, Thermoluminescence studies in swift heavy ion irradiated aluminum oxide, *Radiat. Meas.*, vol. 43, 2008, pp. S651–S655, doi: 10.1016/j.radmeas.2008.04.035.
30. C. Furetta, *Handbook of Thermoluminescence*. World scientific, 2003.
31. D. Afouxenidis, G.S. Polymeris, N.C. Tsirliganis, G. Kitis, Computerised curve deconvolution of TL/OSL curves using a popular spreadsheet program, *Radiat. Prot. Dosimetry*, vol. 149 (4), 2012, pp. 363–370, doi: 10.1093/rpd/ncr315.
32. G.A. Denisenko, Luminescence enhancement of rare earth ions by metal nanostructures, *J. Rare Earths*, vol. 27 (4), 2009, pp. 641–645, doi: 10.1016/S1002-0721(08)60306-3.
33. V.T.T. Ha, N.T.Q. Hai, N.N. Long, N.N. Long, Estimation and correction for the temperature lag in thermoluminescent measurements for LiF:Mg, Cu, Na, Si phosphor, *Commun. Phys.*, vol. 21 (2), 2011, p. 145, doi: 10.15625/0868-3166/21/2/110.
34. G. Kitis, J.W.N. Tuyn, A simple method to correct for the temperature lag in TL glow-curve measurements, *J. Phys. D: Appl. Phys.*, vol. 31 (16), 1998, pp. 2065–2073, doi: 10.1088/0022-3727/31/16/017.
35. P.W. Levy, Thermoluminescence systems with two or more glow peaks described by anomalous kinetic parameters, *Nucl. Inst. Methods Phys. Res. B*, vol. 1 (2–3), 1984, pp. 436–444, doi: 10.1016/0168-583X(84)90105-8.
36. E. Yukihara, S.W.S. McKeever, *Optically Stimulated Luminescence: Fundamentals and Applications*. Wiley, 2011.

37. E.M. Yoshimura, Correlation of optically stimulated luminescence and thermoluminescence of Al_2O_3:Fe, Mg, Cr crystals, *Nucl. Instrum. Methods Phys. Res., Sect. B*, vol. 580 (1), 2007, pp. 606–609, doi: 10.1016/j.nima.2007.05.021.

38. C. Perks, C. Yahnke, Medical dosimetry using optically stimulated luminescence dots and microStar readers, *12th Int. Contress Int. Radiat. Prot. Assoc.*, 2008, p. 10.

39. P.C. Lin, S. Lin, P.C. Wang, R. Ridhar, Techniques for physicochemical characterization of nanomaterials, *Biotechnol. Adv.*, vol. 32 (4), 2014, pp. 711–26. doi: 10.1016/j.biotechadv.2013.11.006.

40. A. Soni, D.R. Mishra, G.S. Polymeris, B. C. Bhatt, M.S. Kulkarni, OSL and thermally assisted OSL response in dental enamel for its possible application in retrospective dosimetry, *Radiat. Environ. Biophys.*, vol. 53 (4), 2014, pp. 763–774, doi: 10.1007/s00411-014-0554-5.

41. C. Ankjærgaard, M. Jain, P.C. Hansen, H.B. Nielsen, Towards multi-exponential analysis in optically stimulated luminescence, *J. Phys. D: Appl. Phys.*, vol. 43 (19), 2010, p. 195501, doi: 10.1088/0022-3727/43/19/195501.

42. E.G. Yukihara, S.W.S. McKeever, Spectroscopy and optically stimulated luminescence of Al_2O_3:C using time-resolved measurements, *J. Appl. Phys.*, vol. 100 (8), 2006, p. 83512, doi: 10.1063/1.2357344.

43. E.G. Yukihara, R. Gaza, S.W.S. McKeever, C.G. Soares, Optically stimulated luminescence and thermoluminescence efficiencies for high-energy heavy charged particle irradiation in Al_2O_3:C, *Radiat. Meas.*, vol. 38 (1), 2004, pp. 59–70, doi: 10.1016/S1350-4487(03)00251-8.

44. E. Bulur, An alternative technique for optically stimulated luminescence (OSL) experiment, *Radiat. Meas.*, vol. 26 (5), 1996, pp. 701–709, doi: 10.1016/S1350-4487(97)82884-3.

45. V.H. Whitley, S.W.S. McKeever, Photoionization of deep centers in Al_2O_3, *J. Appl. Phys.*, vol. 87 (1), 2000, pp. 249–256, doi: 10.1063/1.371853.

46. E.G. Yukihara, R. Gaza, S.W.S. Mckeever, C.G. Soares, Optically stimulated luminescence and thermoluminescence effciencies for high-energy heavy charged particle irradiation in Al_2O_3:C, vol. 38, 2004, pp. 59–70, doi: 10.1016/S1350-4487(03)00251-8.

47. L. Botter-Jensen, S.W.S. McKeever, Ann G. Wintle, *Optically Stimulated Luminescence Dosimetry*. Elsevier, 2003.

48. J.C. Inkson, Deep impurities in semiconductors. II. The optical cross section, *J. Phys. C Solid State Phys.*, vol. 14 (7), 1981, pp. 1093–1101, doi: 10.1088/0022-3719/14/7/012.

49. S.T. Pantelides, The electronic structure of impurities and other point defects in semiconductors, *Rev. Mod. Phys.*, vol. 50 (4), 1978, pp. 797–858, doi: 10.1103/RevModPhys.50.797.
50. Z. Ye, M. Tan, G. Wang, J. Yuan, Novel fluorescent europium chelate-doped silica nanoparticles: Preparation, characterization and time-resolved fluorometric application, *J. Mater. Chem.*, vol. 14 (5), 2004, p. 851, doi: 10.1039/b311905j.
51. K.R. Nagabhushana, D. Prakash, S.H. Tatumi, R.R. Rocca, S. Watanabe, Beta ray induced optically stimulated luminescence properties of calcium oxide phosphor, *Indian J. Pure Appl. Phys.*, vol. 56 (0975-0959 (Online); 0301-1208 (Print)), 2018, pp. 635–638.
52. G. Kitis, V. Pagonis, Computerized curve deconvolution analysis for LM-OSL, Radiat. Meas., vol. 43, 2008, pp. 737–741, doi: 10.1016/j.radmeas.2007.12.055.
53. A. Twardak, P. Bilski, B. Marczewska, W. Gieszczyk, Analysis of TL and OSL kinetics of lithium aluminate, *Radiat. Meas.*, 2014, vol. 71, pp. 143–147.
54. S.C. Santos, C. Yamagata, L.L. Campos, S.R.H. Mello-Castanho, Processing, microstructure and thermoluminescence response of biomorphic yttrium oxide ceramics, *Ceram. Int.*, vol. 42 (11), 2016, pp. 13291–13295, doi: 10.1016/J.CERAMINT.2016.05.136.

Chapter 12

Photon Conversion in Perovskite Solar Cells

Vinod Kumar,[a,b] Leta Tesfaye Jule,[a,b] and Krishnaraj Ramaswamy[b,c]

[a]*Department of Physics, College of Natural and Computation Science, Dambi Dollo University, Ethiopia*

[b]*Centre for Excellence—Indigenous Knowledge, Innovative Technology Transfer and Enterpreneurship Dambi Dollo University, Ethiopia*

[c]*Department of Mechanical Engineering, Dambi Dollo University, Ethiopia*

vinod.phy@gmail.com

Perovskite solar cell (PSC) is one of most attractive solar cell in the third-generation solar cells, which have the ability to commercialize as a future photovoltaics technology. The active layer in PSC is absorbing only the visible photon, while the ultraviolet (UV) and near-infrared (NIR) parts in solar spectrum is promoting different kind of losses in solar cell. To utilizing the UV and NIR parts of sunlight for received the full spectral response has definitely become a significant focus to increase the power conversion efficiency (PCE) PSCs. The upconversion and downconversion materials used for

Luminescent Nanomaterials
Edited by Odireleng Martin Ntwaeaborwa

ISBN 978-981-4968-11-9 (Hardcover), 978-1-003-27738-5 (eBook)
www.jennystanford.com

PSCs have observed a considerable enhancement of PCE, which strongly depend on the power density of sunlight. The device has also significantly improved the long-term stability of PSCs in the ambient conditions with using of efficient encapsulation layer. So, lanthanide doped different host materials are used for up- and downconversion materials in solar cell. The up- and downconversion materials is incorporated in the PSC in different configuration and enhanced the PCE of the devices due to conversion of photon from UV/NIR to visible.

12.1 Introduction

Organic inorganic lead based perovskite materials have shown highly attention in research community due to their proper bandgap, carrier mobility, and high absorption coefficient. These properties in perovskite materials are making them the most attractive solar cell materials for next generation photovoltaic technology [1, 2]. The PCE of the PSCs has shown a remarkable progress in last few years [1–3]. Normally, the PSCs are combination of an electron transport layer (ETL), absorbing layer of perovskite, hole transport layer (HTL), and contact electrode. In this representative assembly of device, perovskite based light absorption materials have mostly provided response in the visible region and absorbed approximately 45% of solar spectrum. So, more than 50% of light in solar spectrum cannot be fully absorbed by the perovskite materials [4, 5]. So, enhancement in utilization of sunlight is a significant way to improve the photovoltaic performance of PSCs. In addition, ETL is responsible for collecting the photogenerated electrons from the perovskite layer and transfer the electrons into the electrode. Obviously, the limitation in the PCE of PSCs is their insensitivity of the UV and NIR regions of solar spectrum [6–8]. Therefore, making more efforts for using the UV and NIR part of solar spectrum has become an important way for growing the PCE of PSCs.

Although showing a certain degree of photoelectric response in PSCs, the high energy of UV light illumination could negatively induce defects and vacancies of oxygen. Organic $CH_3NH_3^+$ ions are the main cause of decomposition in perovskite materials and overall, dropped the performance of device [9, 10]. Therefore, decrease in the amount of UV in the solar irradiation might improve the stability of PSC devices in light. Downconversion (DC) phenomena into the PSCs can prevent the UV part of the incident solar spectrum to interact with the layer of perovskite absorber and convert UV photon into visible photon. Overall, it increased the photocurrent of solar cell as well as improved the light stability of devices in high energy UV photons [11]. On other hand, the upconversion (UC) process has been used as a promising route for utilization of NIR light in photovoltaic devices, such as silicon solar cells, dye sensitized solar cells (DSSCs), PSCs, and organic solar cells [12–15]. The NIR photons are absorbed by using UC materials and then converted into the visible photons by using anti-Stokes process through two- or multi-photon processes. The theoretical limit of upconverter modified solar cell efficiency could extent 63.2% for concentrated and 47.6% for non-concentrated sunlight, respectively [14].

Rare earth (RE) ions doped materials are used for UC and DC process in solar cell because of their sharpness and highly intense luminescent properties due to their 4f intra shell transitions. The elements in the periodic table of lanthanide series have a partially filled 4f energy level surrounded by fully occupied 5s and 5p orbitals [16, 17]. Their variable electronic structures and valence states endow RE ions with flexible redox properties, unique luminescent and electromagnetic properties. Thus, REs have been commonly used into different host materials for photon conversion from UV/NIR into visible and enhanced PCE of PSCs using these materials.

12.2 Photon Conversion

The mismatch between the conventional single-junction solar cell spectral response (SR) and spectrum of solar emission is existing. It is attributed to the limitation of absorption in the active layer of solar cell. So, different kinds of losses in the solar cells are shown as outcome. So, these spectral losses can limit the efficiency of the solar cell. In the solar cell, three kinds of losses are recorded. The modification in the solar spectrum can be of help for reducing the losses in solar cell. The first two losses are thermalization and transmission loss. Thermalization losses are dominant in the solar cell which is based on smaller band gap, while transmission losses are commonly observed in the wide band gap based solar cell. If, a semiconductor absorbed a photon of higher energy than the band gap, an electron–hole pair with higher energy is formed but the excess energy is produced in the form of heat. The third kind of loss is insufficient collection of photon-generated carriers due to recombination close to/or at the surface. When high energy photons are absorbed in this region, the result is a reduced SR at shorter wavelengths. This loss can be decreased by the conversion of photons, whereby photons are shifted into an energy range of cell. The solar spectrum of photon flux with a function of wavelength for Air Mass 1.5 Global (AM1.5G) is presented in Fig. 12.1. The wavelength conversions of photons are a capable path of decreasing the losses of the spectral mismatch. So, it is measured to be the major portion of the efficiency losses in the solar cell device. Wavelength dependent photon conversions have been discovered for improving the conversion of energy in a single-junction solar cell in recent years. QC (quantum cutting), DC, and UC are three luminescence processes. The PCE of solar cells can be enhanced by using these three approaches. One high-energy photon is able to cut into multiple low energy photons with conversion efficiency higher than 100% by DC process using QC [19]. Thus, the PCE of solar cell can be increased with minimized losses. The transformation of one absorbed high-energy photon into one lower-energy photon is observed by DS process, which is a subcategory of DC mechanism. On the other hand, two low energy photons into one high energy photon can be converted by the UC process for avoiding the transmission loss. The DS process is used for improving the efficiency of solar cell by increasing the SR of wavelength, where the silicon cell has observed high light responsivity. These UC/DC approaches of photon conversion can enable to break the Shockley-Queisser limit for single-junction solar cell [20]. The photoluminescence spectra of RE ions have shown the emission frequencies due to the shielding effects of the orbitals of 5s and 5p, which are not dependent on the host materials. RE ion doped DC and UC routes to compensate for the non-absorbable wavelength regions and enhanced the PCE of solar cell [21, 22].

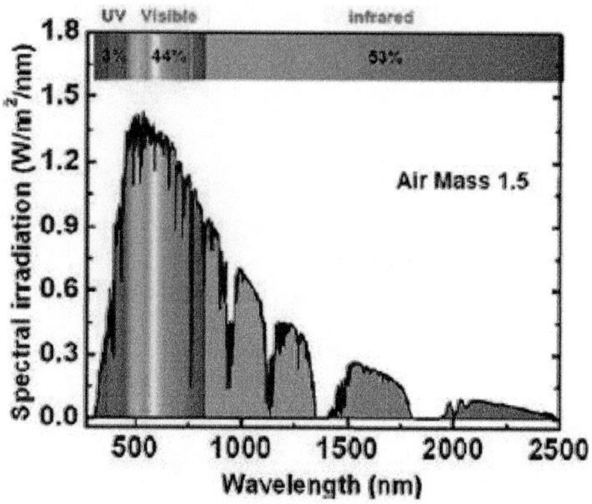

Figure 12.1 Photon flux of Air Mass 1.5 Global solar spectrum with different wavelengths [18].

12.3 Effect of DC/DS Layer on PCE of Pervoskite Solar Cell

The DC process in the solar cells was theoretically shown to lead an enlarged conversion efficiency of 36.5% in a single-junction solar cell configuration for the non-concentrated sunlight [23]. The detailed-balance calculations for single-junction c-silicon solar cells lead to a maximum efficiency of 31% [24]. These calculations were accomplished for a 6000 K blackbody spectrum with a function of band gap and refractive index of the solar cell material. The limit of the solar cell efficiency is reached for a band gap of 1.05 eV, and asymptotically approaches 39.63% for very high refractive indices larger than 10. The limit efficiency is 36.5% for c-Si solar cell with refractive index of 3.6s. Analysis of the energy content of the incident standard AM 1.5G spectrum [25] and the potential gain DC have shown that an extra amount of 32% is incident on a silicon solar cell with a DC layer [25], which can be converted at high internal quantum efficiency. So, the DC materials can be used for enhancing the PCE of solar cells like PSCs. Different RE doped host materials/ approach is used to convert the UV photon into visible photon of PSC [26–28].

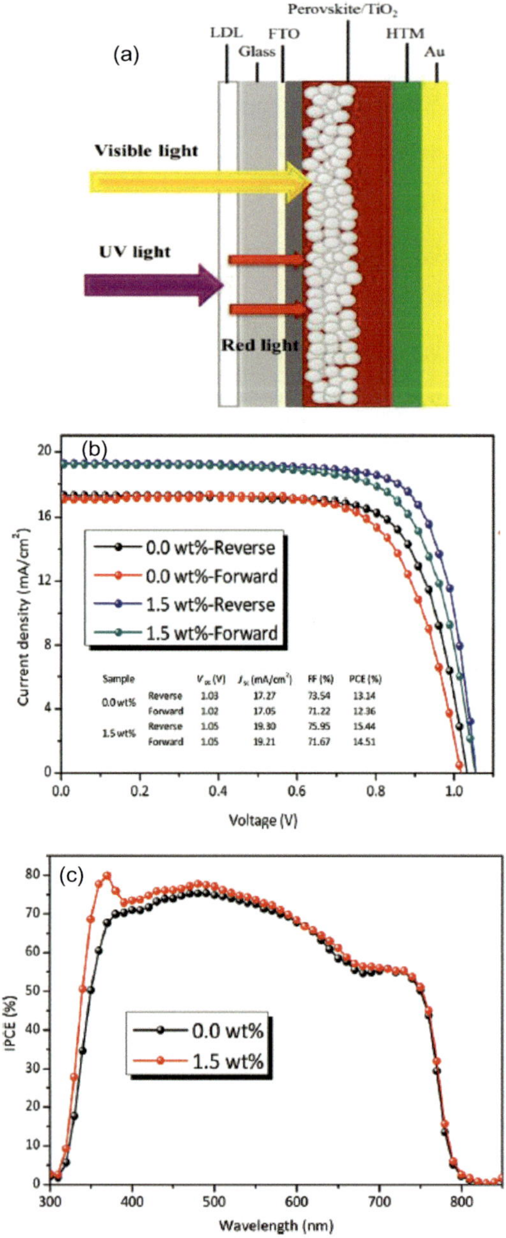

Figure 12.2 (a) Schematic diagram of PSC using the transparent Eu-complex LDL on the top, (b) *J–V* results of PSCs for reverse and forward scans at the content of Eu-complex of 0.0 and 1.5 wt%, and (c) The IPCE graph of PSCs with and without Eu-complex LDL for 0.0 and 1.5 wt% Eu-complex content [26].

DC material based on Eu-complex for improving the photovoltaic performance of PSCs was reported by Jiang et al. [26]. The transparent luminescent downconverting layer (LDL) is prepared by Eu–4,7-diphenyl-1,10-phenanthroline for improving the PCE of PSCs by utilizing high energy UV photon. The schematic configuration of Eu-complex LDL coated PSC is shown in Fig. 12.2a. When the incident light entered from the fluorine-doped tin oxide (FTO) side coated with Eu-complex LDL into the PSCs, maximum visible light of low energy photon passes from the LDL and then absorbed by the perovskite layer. Simultaneously, the high energy UV light is absorbed by the LDL and transformed into the low-energy photons of visible light. This photon conversion of UV to visible helps to increase the utilization of UV light and extending the SR range and decreasing the losses in the PSCs, which help to better stability of PSC. The current density–voltage (*J-V*) graphs for the best PSC with and without LDL are shown in Fig. 12.2b. An open-circuit voltage (V_{oc}) of 1.03 V is observed for the PSC without Eu-complex LDL with the PCE of 13.14% with a short-circuit current density (J_{sc}) of 17.27 mA/cm^2, and a fill factor of 73.54%. The PCE is obtained 17.5% with increased the Eu-complex content up to 1.5 wt% with the J_{sc} of 15.44 mA/cm^2. The enhancement in the J_{sc} is attributed to better absorption in the region of 300–380 nm wavelengths and it is confirmed by the IPCE spectra, shown in Fig. 12.2c. On the other hand, it is because of the uniform thickness and flat surface of perovskite layer in the PSC. The conversion of UV photon into visible photon using LDL is supported to avoid the damage which comes from direct interaction between UV light and perovskite materials. The PSC coated with Eu-complex LDL on the other side of the FTO glass displayed an enhancement of 11.8% in J_{sc} and 15.3% in efficiency in comparison to uncoated PSC. It is attributed to the Eu-complex LDL converting UV photon into the visible photon. It is indicated that the LDL of Eu-complex plays an important role for enhancing the PCE of PSC as well as decreasing the UV degradation in device.

Jia et al. [27] reported the improvement in the devices performance of PSCs by utilizing DC $NaYF_4$:Eu^{3+} nanophosphors prepared by hydrothermal reaction and accumulated the PSC devices according to the given configuration in Fig. 12.3a for determining the device performance of DC layer. Spin coating technique was used for preparing the DC layer of $NaYF_4:Eu^{3+}$ on the back side of the FTO for protection of the PSC devices from high energy UV photon. When sunlight entered the PSCs from the FTO side, a certain amount of UV light was absorbed by the DC layer and then converted into visible light that penetrated the FTO. It was finally absorbed by the perovskite layer. This enhancement in the consumption of UV light by extending the SR range in PSCs. Spin coating technique was used for fabrication of DC layer on the non-conductive side of the FTO. The spectra of photoluminescence emission exhibited that the UV light can be absorbed by the DC layer of $NaYF_4:Eu^{3+}$ then converted into the visible light. These low energy photons can further be absorbed by the perovskite absorption layer. As a consequence, the utilization of sunlight in the PSC was enhanced. The effect of DC layer on photovoltaic performance of device is measured by *J–V* curves as shown in Fig. 12.3b. Devices with DC layer obtained better average PCE value with respect to the PSC-blank devices. It indicates that the $NaYF_4:Eu^{3+}$ layer enhanced the device performance of the PSC. The device of PSC-6% has shown the best average PCE of 19.99%, as compared to PSC without the DC layer (16.99%). In addition, the best-behaved PSC-6% device obtained a PCE value of 20.17% with J_{sc} of 23.22 mA/cm^2, V_{oc} of 1.127 V and FF of 0.771. The DC capability of $NaYF_4:Eu^{3+}$ can be optimized with the variation of doping concentration of Eu^{3+}.

Rahman et al. [28] reported a promising DC material based on Sr_2Ce_04:Eu (SCOE) and its incorporation in the PSCs for improving the performance of device as well as stability in UV light. The schematic configuration of the PSCs is shown in Fig. 12.4a. This device consists of FTO as anode, compact TiO_2 (c-TiO_2) as ETL, a thin layer of SCOE DC material, perovskite layer of (CsFAMA)Pb(BrI)$_3$, Spiro-OMeTAD as a HTL, and Ag as cathode with the thickness of 560, 30, 150, 440, 80 and 130 nm, respectively. The *J–V* curves with and without the SCOE layer of device are shown in Fig. 12.4b. The SCOE-coated PSCs have achieved a high J_{sc} of 23.70 mA/cm^2 and a high PCE of 18.95%. It improved 9% and 14.15% respectively as compared to without SCOE layer device. Moreover, the PCE of the

SCOE based PSCs is maintained 80% of its initial value after 70 h UV-light irradiation while remained 78% after 75 days under ambient environment with 20–25% relative humidity. It is suggested that the incorporation of DC material in device is a functional component of PSCs to increase the solar SR and improve the device performance with reduced UV induced photodegradation by converting the UV light into the visible light for increasing stability of device.

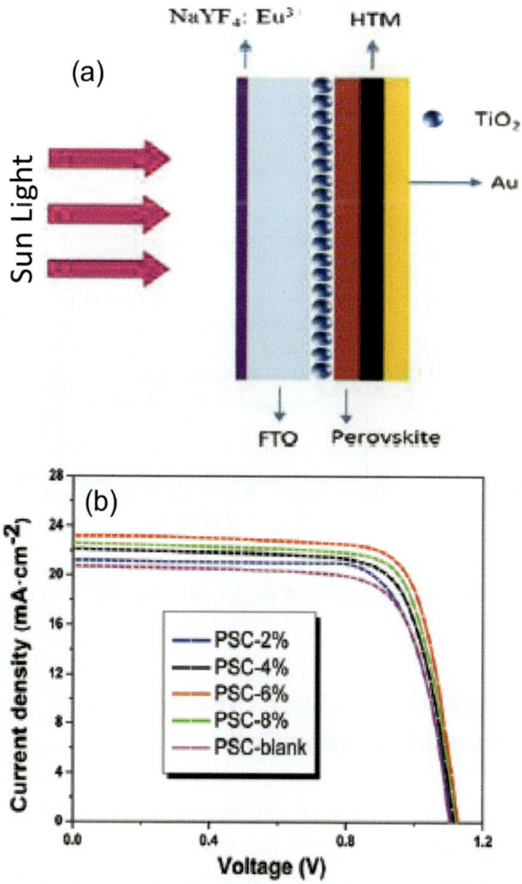

Figure 12.3 (a) Device configuration of the PSC with DC layer and (b) *J–V* graph of different PSC devices [27].

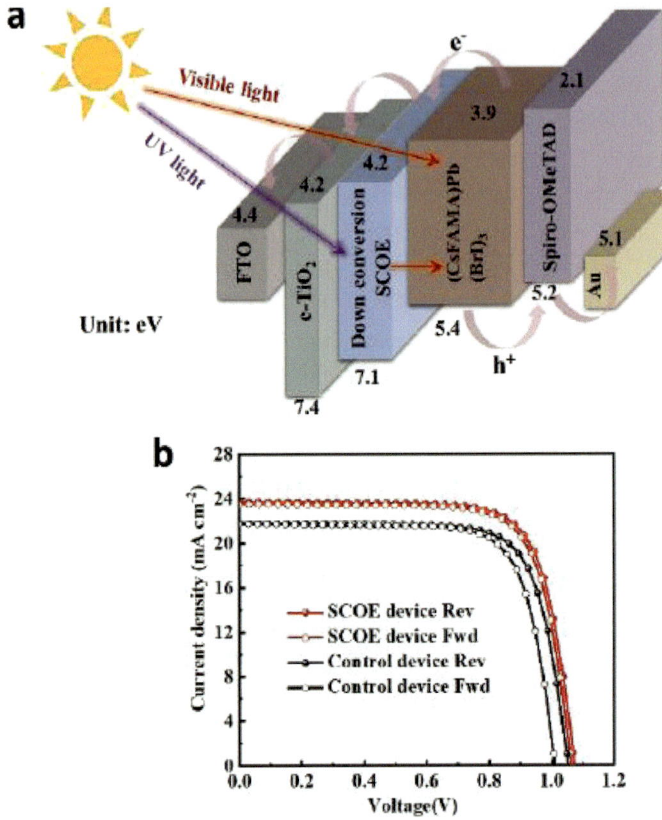

Figure 12.4 (a) Device schematic diagram of the SCOE DC material based PSCs (b) *J–V* graph of the control and SCOE DC based PSCs [28].

12.4 Effect of UC Layer on the PCE of Pervoskite Solar Cell

UC process in solar cell was calculated to potentially lead to a maximum PCE of 47.6% [29] for the non-concentrated sunlight using a 6000 K blackbody spectrum in detailed-balance calculations. This optimum is reached for a solar cell material with the band gap of ~2 eV. The PCE limit would be ~37% for silicon solar cell after using of UC materials [29]. The investigation of the energy content of the incident AM 1.5G spectrum revealed that cells with an UC layer would help from an extra amount of 35% light incident in the silicon solar cell [25]. UC process is already reported in different third-generation solar cell as well PSC for converting NIR into visible photon [30–36].

Zhang et al. reported enhancement in the PCE of PSC using the upconverting material [33]. Er^{3+}-Yb^{3+}-Li^{+} tri-doped TiO_2 UC material is synthesized through the addition of Li^{+} into Er^{3+}-Yb^{3+} co-doped TiO_2 UC materials. The addition of Li ion in UC-TiO_2 presented an enhancement in the UC emission with compared to Er^{3+}-Yb^{3+} co-doped TiO_2. This UC-TiO_2 material was used into the PSCs. The *J–V* curves of with and without UC-TiO_2 of PSC are shown in Fig. 12.5. The PCE of without UC-TiO_2 based PSC is recorded 14.0%, while the PCE of with UC-TiO_2 device is increased upto 16.5%. It showed an enhancement of 19%. This output advised that UC-TiO_2 is an effective UC material for PSC and provided a new route to increase the spectral absorption of PSC from visible light to NIR by using of UC materials with addition doping of Li^{+} ion.

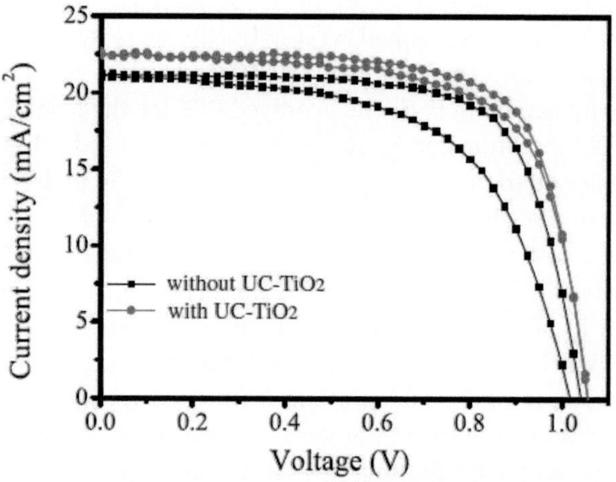

Figure 12.5 *J–V* curves of the PSC without and with UC-TiO_2 [33].

Meng et al. reported [34] that the incorporation of UC mechanism in PSCs is a facile method for spreading the spectral absorption from the visible to the NIR range and decreasing the non-absorption losses in PSCs. However, the poor compatibility between UC nanocrystals (UCNCs) with the precursor of perovskite is a big challenge for the experimental approach of UCNCs in PSCs. This compatibility issue is overcome using a ligand-exchange strategy. The β-$NaYF_4$:Yb, Er

UCNCs have been incorporated into a $CH_3NH_3PbI_3$ layer by in situ to fabricate NIR-enabled PSCs. The CH_3NH_3I capped UCNCs generated from the ligand-exchange are mixed with the perovskite precursor and served as nucleation sites for the UCNC-mediated heteroepitaxial growth of perovskite. Moreover, the in situ embedding of UCNCs into the perovskite layer are understood during the spin coating process. Finally, the UCNC incorporated perovskite layer is shown a pinhole-free uniform morphology with large crystal grains and enabled NIR absorption. It is also contributed in the energy transfer from the UCNCs to the perovskite material as well as the electron transport to the surface of the collecting electrode. The *J–V* curve of devices with schematic diagram is shown in Fig. 12.6a The UCNC-embedded perovskite film based device is recorded an average PCE of 18.60% with the 19.70% for the best device, Fig. 12.6b. The schematic diagram of UCNCs incorporated PSC device is shown in Fig. 12.6c.

Sebag et al. [35] summarized the impact of the incorporation of UC nanoparticles (UCNPs) into PSCs. Highly crystalline and uniform Yb^{3+}/Er^{3+} co-doped KY_7F_{22} UCNPs were synthesized and incorporated into the interface of FTO/perovskite or perovskite/HTL configuration in PSC. The averaged PCE for the front and rear-side UCNP insertion configuration PSC was observed 6.1 and 6.5% in comparison to the control device, respectively. To understand the contribution of UC, microscopic characterizations of PSC using the fluorescence mapping technique were carried out on the devices, where only half of the perovskite/HTL interface decorated by UCNPs (Fig. 12.7). These mapping results presented a comprehensive microscopic picture, which understand a relationship of the electrical and optical contribution of UCNPs with the morphology of device. When the laser spot goes from the part of device without UCNPs to with UCNPs, together with the appearance of UCNP fluorescence, a ~3-fold increase of J_{sc} was observed. It is directly attributed to the UC contribution. The reduced green/red ratio of the UC fluorescence spectra detected during mapping experiments were related with the higher absorbance of perovskite material for green fluorescence wavelength range. This provided a direct indication of the involvement of the UCNPs in a PSC where the UC fluorescence was harvested by the perovskite layer.

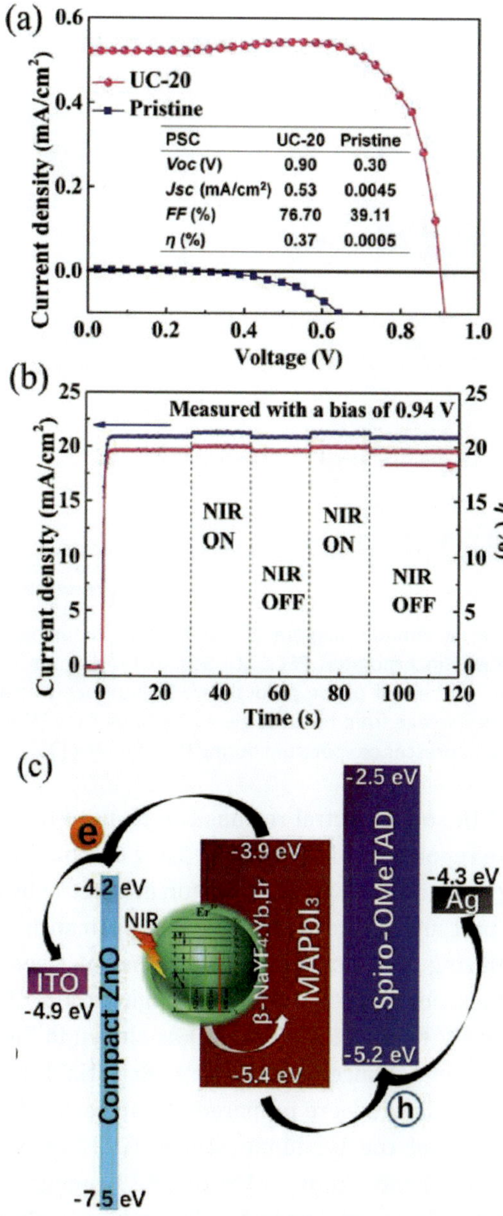

Figure 12.6 (a) *J–V* curves for pristine and UC-20 planar PSC device measured by NIR laser irradiation (980 nm). (b) The J_{sc} and PCE of UC-20 planar PSCs measured at 0.94 V under AM 1.5G with NIR laser (980 nm) switched on and off every 20 s. (c) Schematic diagram of UCNCs incorporated planar PSC device [34].

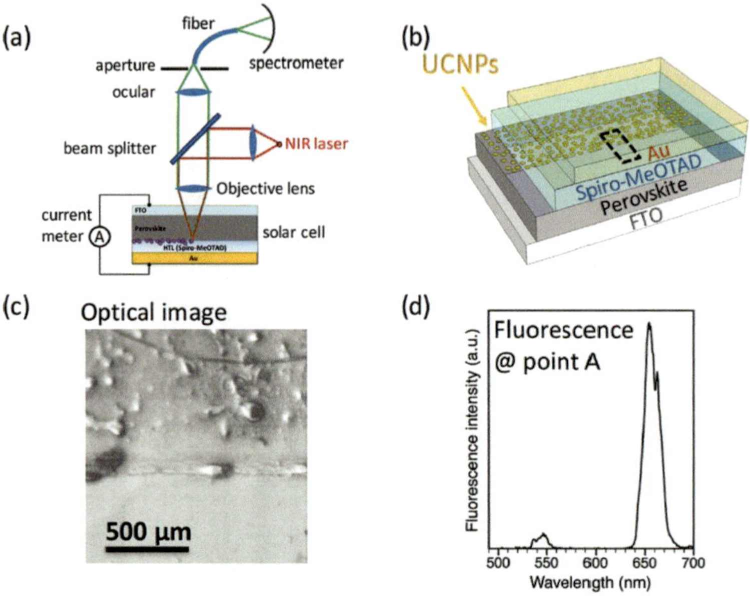

Figure 12.7 (a) Schematic diagram of fluorescence mapping experimental setup with UCNPs incorporated PSCs. (b) Schematic diagram of the UCNP-embedded PSC where half of the perovskite/HTL interface was decorated by UCNPs. (c) Optical image from the gold electrode side of the UCNP embedded PSC. (d) The UC fluorescence spectrum during the mapping [35].

The UV/NIR solar spectral regions can be used through the UC and DC route reported by Li et al. [36]. They demonstrated a novel methodology to utilize the UV/NIR photon in spectral by converting into visible part and improve the device performance of the PSCs. The PSCs utilizing only a small portion of the solar spectrum with respect to the spectral distribution of sunlight at AM 1.5G involved photons from UV to IR (280–2500 nm) as shown in Fig. 12.8. The efficient plasmon-enhanced $NaYF_4:Yb^{3+}$, $Er^{3+}/NaYF_4:Yb^{3+}$, Tm^{3+}/Ag UC composite layers were prepared on the back side of the PSC device. The PCE of the UC luminescence (UCL)-embedded PSCs is observed to enhance from 16.1% to 19.2% under the sunlight illumination of 1.5 W/cm^2. The PCE of PSC can be further enhanced up to 19.5% by incorporation of a DC layer of $(Eu(TTA)_2(Phen)MAA)$ in the incident light side of the PSCs, which converted UV to visible light. The long-term stability by moisture as well as stability in light under illumination have been significantly enhanced through the external encapsulation of the UCL and DC layers. So, this work represents an innovative method to utilize the full SR of PSCs and exhibits application potential in PSC as shown in Fig. 12.8.

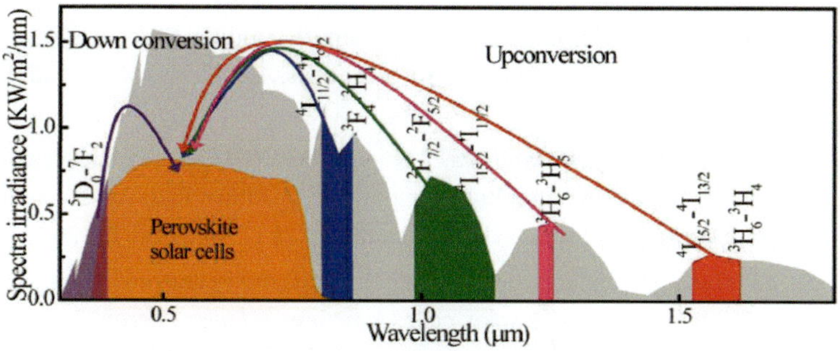

Figure 12.8 The fraction absorbed by PSCs in AM 1.5G spectrum and the spectral regions which can be used by using UC and DC method [36].

12.5 Conclusion

Organic inorganic halide perovskite solar cells (PSC) have huge prospective to influence the existing photovoltaic industry. As realized that higher power conversion efficiency of PSC is still the most crucial task but lot of mechanisms are suggested for minimizing the losses in solar cell by optimizing the electrical properties of the PSCs. In PSCs, the photon conversion processes from NIR/UV into the visible light can enhance the PCE in the devices. Research indicated that rare earth doped DC/DS/UC layers can enhance the PCE of PSCs. The combination of UC and DC photon conversion materials is a novel approach to extend the response of the UV and NIR radiation in PSC for future photovoltaic technology.

References

1. M. Ahmadi, Y.C. Hsiao, T. Wu, Q. Liu, W. Qin, B. Hu, Effect of photogenerated dipoles in the hole transport layer on photovoltaic

performance of organic-inorganic perovskite solar cells, *Adv. Energy Mater.*, 2017, 7, 1601575.

2. N.J. Jeon, J.H. Noh, W.S. Yang, Y.C. Kim, S. Ryu, J. Seo, S.I. Seok, Compositional engineering of perovskite materials for high-performance solar cells, *Nature*, 2015, 517, 476–477.
3. C. Chen, D. Liu, B. Zhang, W. Bi, H. Li, J. Jin, X. Chen, L. Xu, H. Song, Q. Dai, Carrier interfacial engineering by bismuth modification for efficient and thermos resistant perovskite solar cells, *Adv. Energy Mater.*, 2018, 170, 3659.
4. V. Kumar, O.M. Ntwaeaborwa, T. Soga, V. Dutta, H.C. Swart, Rare earth doped zinc oxide nanophosphor powder: A future material for solid state lighting and solar cell, *ACS Photonics*, 2017, 4 (11) 2613–2637.
5. A. Urbina, The balance between efficiency, stability and environmental impacts in perovskite solar cells: A review, *J. Phys.: Energy*, 2020, 2, 022001.
6. Y. Chen, T. Chen, L. Dai, Layer-by-layer growth of $CH_3NH_3PbI_{3-x}Cl_x$ for highly efficient planar heterojunction perovskite solar cells, *Adv. Mater.*, 2015, 27, 1053–1059.
7. H. Lian, Z. Hou, M. Shang, D. Geng, Y. Zhang, J. Lin, Rare earth ions doped phosphors for improving efficiencies of solar cells, *Energy*, 2013, 57, 270–283.
8. Z. He, C. Zhong, S. Su, M. Xu, H. Wu, Y. Cao, Enhanced power-conversion efficiency in polymer solar cells using an inverted device structure, *Nat. Photonics*, 2012, 6, 591–595.
9. Y. Zhang, P. Wang, X. Yu, J. Xie, X. Sun, H. Wang, J. Huang, L. Xu, C. Cui, M. Lei, D. Yang, Enhanced performance and light soaking stability of planar perovskite solar cells using an amine-based fullerene interfacial modifier, *J. Mater. Chem. A*, 2016, 4, 18509–18515.
10. S. Ito, S. Tanaka, K. Manabe, H. Nishino, Effects of surface blocking layer of Sb2S3 on nanocrystalline TiO_2 for $CH_3NH_3PbI_3$ perovskite solar cells, *J. Phys. Chem. C*, 2014, 118, 16995–17000.
11. T. Leijtens, G.E. Eperon, S. Pathak, A. Abate, M.M. Lee, H.J. Snaith, Overcoming ultraviolet light instability of sensitized TiO_2 with meso-superstructured organometal tri-halide perovskite solar cells, *Nat. Commun.*, 2013, 4, 2885.
12. V. Kumar, A. Pandey, S.K Swami, O.M. Ntwaeaborwa, H.C. Swart, V. Dutta, Synthesis and characterization of Er^{3+}-Yb^{3+} doped ZnO upconversion

nanoparticles for solar cell application, *J. Alloys Compd.*, 2018, 766, 429–435.

13. M. He, X. Pang, X. Liu, B. Jiang, Y. He, H. Snaith, Z. Lin, Monodisperse dual-functional upconversion nanoparticles enabled near-infrared organolead halide perovskite solar cells, *Angew. Chem.*, 2016, 55, 4280–4284.
14. X. Huang, S. Han, W. Huang, X. Liu, Enhancing solar cell efficiency: The search for luminescent materials as spectral converters, *Chem. Soc. Rev.*, 2013, 42, 173–201.
15. J. Wu, J. Wang, J. Lin, Z. Lan, Q. Tang, M. Huang, Y. Huang, L. Fan, Q. Li, Z. Tang, Enhancement of the photovoltaic performance of dye-sensitized solar cells by doping $Y_{0.78}Yb_{0.20}Er_{0.02}F_3$ in the photoanode, *Adv. Energy Mater.*, 2012, 2, 78–81.
16. V. Kumar, S. Som, V. Kumar, V. Kumar, E. Coetsee, O.M. Ntwaeaborwa, H.C. Swart, Tunable and white emission from ZnO:Tb^{3+} nanophosphors for solid state lighting applications, *Chem. Eng. J.*, 2014, 255, 541.
17. H. Zhou, Y. Qu, T. Zeid, X. Duan, Towards highly efficient photocatalysts using semiconductor nanoarchitectures, *Energy Environ. Sci.*, 2012, 5, 6732.
18. Q.Y. Zhang, X.Y. Huang, Recent progress in quantum cutting phosphors, *Prog. Mater. Sci.*, 2010, 55, 353–427.
19. N. Chander, A.F. Khan, V.K. Komarala, S. Chawla, V. Dutta, Enhancement of dye sensitized solar cell efficiency via incorporation of upconverting phosphor nanoparticles as spectral converters, *Prog. Photovoltaics*, 2016, 24, 692–703.
20. S. Asahi, H. Teranishi, K. Kusaki, T. Kaizu, T. Kita, Two step photon up conversion solar cells, *Nat. Commun.*, 2017, 8, 14962.
21. A.J. Nozik, J. Miller, Introduction to solar photon conversion, *Chem. Rev.*, 2010, 110 (11), 6443–6445.
22. T. Trupke, M.A. Green, P. Würfel, Improving solar cell efficiencies by down conversion of high-energy photons, *J. Appl. Phys.*, 2002, 92, 1668–1674.
23. W. Shockley, H.J. Queisser, Detailed balance limit of efficiency of p-n junction solar cells, *J. Appl. Phys.*, 1961, 32, 510.
24. ASTM 2003. Standard tables for reference solar spectral irradiances: Direct normal and hemispherical on 37° tilted surface, standard G173-

03e1 West Conshohocken, PA, USA: American Society for Testing and Materials.

25. B.S. Richards, Enhancing the performance of silicon solar cells via the application of passive luminescence conversion layers, *Sol. Energy Mater. Sol. Cells*, 2006, 90, 2329–2337.
26. L. Jiang, W. Chen, J. Zheng, L. Zhu, L. Mo, Z. Li, L. Hu, T. Hayat, A. Alsaedi, C. Zhang, S. Dai, Enhancing the photovoltaic performance of perovskite solar cells with a down-conversion Eu-complex, *ACS Appl. Mater. Interfaces*, 2017, 9, 26958–26964.
27. J. Jia, J. Dong, J. Lin, Z. Lan, L. Fan, J. Wu, Improved photovoltaic performance of perovskite solar cells by utilizing down conversion NaYF4:Eu^{3+} nanophosphors, *J. Mater. Chem. C*, 2019, 7, 937–942.
28. N.U. Rahman, W.U. Khan, S. Khan, X. Chen, J. Khan, J. Zhao, Z. Yang, M. Wu, Z. Chi, A promising europium-based down conversion material: Organic–inorganic perovskite solar cells with high photovoltaic performance and UV-light stability, *J. Mater. Chem. A*, 2019, 7, 6467–6474.
29. T. Trupke, M.A. Green, P. Würfel, Improving solar cell efficiencies by upconversion of sub-band-gap light, *J. Appl. Phys.*, 2002, 92, 4117–4122.
30. T. Morifuji, Y. Takekuma, M. Nagata, Integrated photon upconversion dye sensitized solar cell by co-adsorption with derivative of Pt–porphyrin and anthracene on mesoporous TiO_2, *ACS Omega*, 2019, 4 (6), 11271–11275.
31. J. Wild, A. Meijerink, J.K. Rath, W.G.J.H.M. van Sark, R.E.I. Schropp, Upconverter solar cells: Materials and applications, *Energy Environ. Sci.*, 2011, 4, 4835–4848.
32. M. Ran, N. Liu, H. Yang, R. Meng, M. Chen, H. Lu, Y.P. Yang, Positive effects in perovskite solar cells achieved using down-conversion $NaEuF_4$ nanoparticles, *Appl. Phys. Lett.*, 2020, 116, 113503.
33. Z. Zhang, J. Qin, W. Shi, Y. Liu, Y. Zhang, Y. Liu, H. Gao, Y. Mao, Enhanced power conversion efficiency of perovskite solar cells with an up-conversion material of Er^{3+}-Yb^{3+}-Li^{+} Tri-doped TiO_2, *Nanoscale Res. Lett.*, 2018, 13, 147.
34. F.L. Meng, J.J. Wu, E.F. Zhao, Y.Z. Zheng, M.L. Huang, L.M. Dai, X. Tao, J.F Chen, High-efficiency near-infrared enabled planar perovskite solar cells by embedding upconversion nanocrystals, *Nanoscale*, 2017, 9, 18535–18545.

35. M.S. Sebag, Z. Hu, K.O. Lima, H. Xiang, P. Gredin, M. Mortier, L. Billot, L. Aigouy, Z. Chen, Microscopic evidence of upconversion-induced near-infrared light harvest in hybrid perovskite solar cells, *ACS Appl. Energy Mater.*, 2018, 1, 3537–3543.
36. H. Li, C. Chen, J. Jin, W. Bia, B. Zhang, X. Chena, L. Xu, D. Liu, Q. Dai, H. Song, Near-infrared and ultraviolet to visible photon conversion for full spectrum response perovskite solar cells, *Nano Energy*, 2018, 50, 699–709.

Chapter 13

Structural and Optical Probe into Rare Earth Doped ZnO for Spectral Conversion in Solar Cells

Francis Otieno,[a,b,c] Mildred Airo,[a,d] Rudolph M. Erasmus,[b] Caren Billing,[a,b] Alex Quandt,[b,c] Daniel Wamwangi,[b,c] and David G. Billing[a,b]

[a]*School of Chemistry, University of the Witwatersrand, Private Bag 3, WITS 2050, South Africa*

[b]*Materials for Energy Research group, University of the Witwatersrand, Private Bag 3, WITS 2050, South Africa*

[c]*Material Physics Research Institute, School of Physics, University of the Witwatersrand, Private Bag 3, Wits, 2050, Johannesburg, South Africa*

[d]*College of Science, Engineering and Technology, University of South Africa, P.O. Box 392, Unisa, 0003, Republic of South Africa*

daniel.wamwangi@wits.ac.za

Zinc oxide continues to draw renewed interest in the semiconductor industry due to outstanding multifunctional properties with diverse

Luminescent Nanomaterials
Edited by Odireleng Martin Ntwaeaborwa

ISBN 978-981-4968-11-9 (Hardcover), 978-1-003-27738-5 (eBook)
www.jennystanford.com

application potential. These properties include high exciton binding energy, high resistance against radiation, high breakdown voltage, insensitivity to visible light, and easy wet chemical etching. The high quantum efficiency for emission by ZnO makes it a strong candidate for solid-state white lighting applications as well as transparent conducting electrode in solar cells and electronics. In order to realize efficient utilization of the multi-functional properties of ZnO for electronic and optoelectronic applications, ZnO is usually doped with different elements. Doping is aimed at enhancing and controlling its electrical, optical, and multi-functional properties. Typical dopants widely used are trivalent atoms categorized as group III in the periodic table (Al, In, Ga) and doping occurs through substitution of cation. The as-grown ZnO thin film is usually an n-type semiconductor with structural, electrical, and optical properties that can be varied depending on the growth conditions as well as post deposition treatment such as thermal annealing. RF magnetron sputtering is used for deposition of Tb^{3+} and Eu^{3+} doped ZnO films which are subsequently ex situ annealed in argon atmosphere for elucidation of quantum size effects associated with changes in nanocrystalline grains for photovoltaic applications.

ZnO doped with rare earths (REs) has exhibited electroluminescence, thus highlighting its potential for photovoltaic applications as a bi-functional layer. A doped ZnO layer is thus simultaneously utilized as transparent conducting electrode and as a spectral conversion layer. The RE doped luminescent materials can be effectively used to modify high energy and sub-band gap energy photons from the solar spectrum that would have otherwise been lost in direct band gap absorbers. In solar cells, they have been applied to reduce thermalization and transmission losses arising from the following intrinsic properties of the semiconductor material namely: (a) sub-bandgap photon loss and (b) thermalization of charge carriers resulting from absorption of high energy photons. In this chapter we present the morphological, optical, and the structural changes that occur in ZnO thin films grown using the RF magnetron sputtering technique when doped with different RE elements (Eu^{3+} and Tb^{3+}) and annealed at different temperatures.

13.1 Introduction

At ambient conditions, ZnO possesses the thermodynamically stable Wurtzite phase similar to GaN [1, 2], a factor that makes it widely applicable in high-performance optoelectronic devices. The increased research on zinc oxide is driven by its excellent properties necessary for solid-state lighting and transparent or high temperature electronics [3]. It has a direct band gap of 3.37 eV (at 300 K) that allows absorption and emission in the near-UV spectral region, a large free-exciton binding energy (60 meV) that enables excitonic emission at or even above room temperature [4–6] and a strong cohesive energy of 1.89 eV/atom [7]. This is coupled with its high optical gain (300 cm^{-1}) [5], high mechanical strength, thermal and chemical stability, ambient insensibility, and negative electron affinity [8] as well as high radiation hardness [9]. These characteristics of ZnO together with its unique combination of physical properties—optical, electrical, magnetic, piezoelectric, and ferroelectric properties make it a promising candidate material for advanced device applications. Zinc oxide, besides being one of the important materials in the classification of wide bandgap materials, is one of the important transparent conductive oxides (TCOs). Additionally, the abundance of zinc guarantees lower and sustainable costs of fabrication compared to tin and indium based oxides, the increasingly used components of TCOs. Its added advantage of being non-toxic, inert under H_2 atmosphere and low temperature fabrication makes it more suitable as front electrodes in photovoltaic applications [10–12].

Due to these advantages, the application of ZnO and Al-doped ZnO is increasing in the industry day by day. As a semiconductor, ZnO has a wide range of useful properties that greatly distinguishes it from other semiconductors or oxides such as:

- *Direct and wide band gap.* This has been found to be 3.37 eV at room temperature and 3.44 eV at low temperatures. This wide band gap as stated earlier enables ZnO to be used in optoelectronic devices in the blue/UV region, such as in light emitting diodes, laser diodes, and photodetectors [13–15].
- *Large exciton binding energy.* ZnO has a higher free-exciton binding energy of 60 meV [16, 17], even greater than GaN

(25 meV). This makes it possible for efficient excitonic emission to occur at room temperature hence making ZnO a good candidate for optical devices which are based on excitonic effects.

- *Large piezoelectric constants.* In piezoelectric materials, an applied voltage generates a deformation in the crystal and vice versa. These materials are useful for fabricating transducers, actuators and sensors to detect pressure or acceleration changes. The low symmetry of the Wurtzite crystal structure combined with a large electromechanical coupling in ZnO gives rise to strong piezoelectric and pyroelectric properties [18–21].
- *Strong luminescence.* The strong luminescence in the green-white spectral region makes it suitable for phosphor applications. These luminescence centres in ZnO are speculated to originate from oxygen vacancies or zinc interstitials [22]. This has been extensively studied in this project in subsequent chapters.
- *High sensitivity of surface conductivity to the presence of adsorbed species.* The surface conductivity of ZnO is very sensitive to exposure to a variety of gases. This enables them to be used to detect the freshness of food and drinks when used as a "smell" sensor to the trimethylamine present in the odor of sea foods, wine, and coffee [23].
- *Large non-linear optical coefficients.* Larciprete et al. [24] used second and third harmonic generation experiments to show nonlinear optical behavior in bulk crystals and thin films of ZnO. This has paved the way for integrated nonlinear optical devices. Deposition techniques capable of producing films with a strong second order nonlinear response include pulse laser deposition, reactive sputtering and spray pyrolysis.
- *High thermal conductivity.* This property makes it useful as an additive. It is a major component in tyres along with rubber which led to numerous inventions in the rubber and ceramic industries [25]. ZnO has a linear coefficient of thermal expansion (3.16×10^{-6} /°C) and melting temperature (2000 °C) close to that of silica [26]. Usually its vapor pressure reaches 760 mmHg at about 1700 °C. Leverenz [27], attributes

the high thermal conductivity to the strong directional Zn-O at the c-axis of the unit cell.

- *Large single crystal availability*. Large area single crystals ZnO are readily available as wells as commercial epi-ready substrates. Bulk crystals can be obtained by using various methods such as hydrothermal growth [28], vapor-phase transport [29], and melt growth at high pressure [30]. Thin films on the other hand can be grown by molecular-beam epitaxy, metal-organic chemical vapor deposition (MOCVD), laser ablation or sputtering [31, 32].
- *Amenability to wet chemical etching*. ZnO thin films can be easily etched using acidic, alkaline as well as mixed acids or alkali solutions. This ensures flexibility and versatility in the processing, design, and integration into electronic and optoelectronic devices. Amenability to low-temperature wet chemical etching greatly helps in device fabrication [33].
- *Radiation hardness*. Such high radiation hardness is crucial for high altitude applications or in space. According to Tuomisto, Filip, et al. ZnO exhibits exceptionally high radiation hardness enabling irradiation induced defects to fully recover upon annealing at 600 K [34]. This according to Anderson Janotti and Chris G Van de Walle is higher than that of GaN [33].

These attributes have steered intensive research on ZnO for novel applications in optoelectronics, transparent electronics, and spintronics.

13.2 Rare Earth Doping

Rare earth doping in ZnO has inspired widespread research and industrial interest especially for mechanisms that enable energy transfer between the host material (ZnO) and the RE element. These mechanisms allow for quantum cutting, up- and downconversion for the efficient and extended utilization of the solar emission spectrum in photovoltaic technology.

When doped with a rare earth metal, ZnO can be used as a sensitizer to excite RE ions such as Ce^{3+}, Er^{3+}, Ho^{3+}, Nd^{3+}, Tm^{3+}, Dy^{3+}, Eu^{3+}, and Tb^{3+}. This is made possible by its large absorption cross section and broad excitation spectrum [35]. This way, the REs can absorb the UV-blue emission from ZnO and emit in the visible or infra-red range thus enabling it to be an excellent doping host for rare earth ions with an optimal spectrum modifying matrix for diverse solar cells.

The optical properties of RE doped ZnO usually depend on dopant concentrations, fabrication process and the crystal structure of the host due to the crystal field energy and spin-orbit coupling. The emission characteristics of these rare earth ions in ZnO are sensitive to the nature of the fabrication process, the size of the ZnO nano-crystals, and their morphology. The morphology and structure of thin films can be tuned by changing the ad-atom energies and mobility through thermal treatment, either in situ or ex situ. Enhancing film crystallinity during growth due to lattice strains and thermal stresses between the ZnO film and the substrate presents one approach to spectrum modification in these films [36].

In organic solar cells, ZnO is widely used as a buffer layer since it has a low work function, excellent optical transparency, high electron mobility, and ease of fabrication [37]. Combining the positive attributes of ZnO and REs in organic solar cells would therefore produce a good spectrum modifying matrix with dual functionality as an electron transporting layer and as a photon-conversion layer. Additionally, the implementation of the rare earth doped buffer layer requires a simple architecture with no resistivity and optical losses induced by the space gap between the downconverting layer and the active layer, a feature that is consistently persistent in externally stacked phosphor based photovoltaic devices.

13.3 Photoconversion Efficiency: Enhancement

Photovoltaic (PV) technologies for solar energy conversion is a promising alternative to realizing green and renewable energy generation [38]. Although the technology is not new, solar cell production remains costly, largely due to their low power conversion efficiencies [39–41]. The main challenge to realizing high conversion efficiency of PV cells is their insensitivity to the full spectral range of the solar emission. The photon flux at Air Mass 1.5 Global (AM 1.5G) is made of photons of broad wavelength distribution ranging from ultraviolet to infrared (280–2500 nm, 0.5–4.4 eV). However

PVs utilize a relatively small fraction of the solar photon flux as determined by its energy matching with the characteristic bandgap of the active layer in the PV device [42].

Notably, the maximum theoretical photoconversion efficiency (PCE) for crystalline silicon (c-Si) with a bandgap energy (E_g) of 1.1 eV is approximately 31% as defined by the Shockley–Queisser limit [43]. Hence overcoming this drives intensive research towards minimizing energy losses attributed to transmission and thermalization. This was motivated by Trupke et al. as the theoretical extension of conversion efficiency to 38.6% after imposing material modifications for increased utility of the solar spectrum [44, 45].

In recent years, luminescent materials capable of converting a broad spectrum of light into photons of a particular wavelength have been reported, thus paving the way for reduction of these losses through energy conversion processes. RE doped TCOs have demonstrated spectral conversion through upconversion, quantum cutting, and downshifting as illustrated in Fig. 13.1.

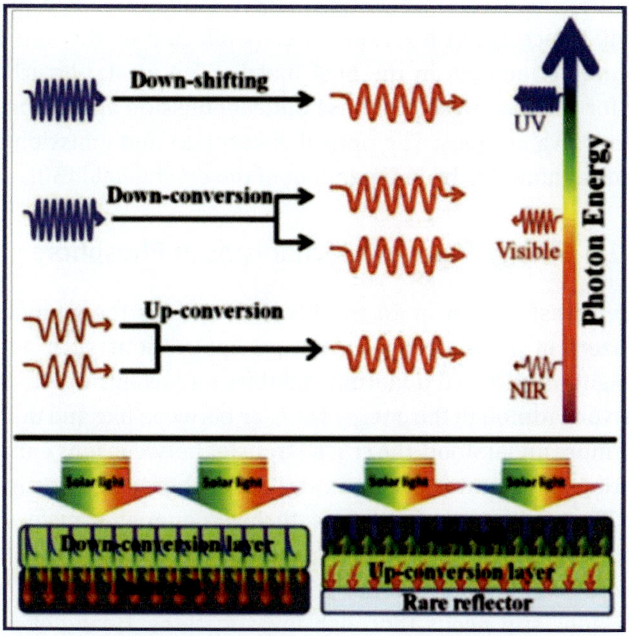

Figure 13.1 Mechanisms of photon luminescent conversion processes and an illustration of the simplified view of schemes of application into solar cells [102].

13.4 Rare Earth Metals Phosphors

Rare earth metal phosphors are isotopically stable elements with outer electronic configurations of trivalent ions having RE^{3+} as the dominant oxidation state. They can be classified into 2 groups namely actinides (Z = 89–103) or lanthanides (Z = 57–71). The lanthanides are represented by the general electronic configuration of the form [Xe] $5s^2 5p^6 4f^n$. [Xe] represents the electronic configuration of xenon and *n* the number of electrons in the respective 4f levels. The energy distribution of the partially occupied 4f orbitals favor a broad regime of electronic transitions that spans from the ultraviolet to the infra-red part of the electromagnetic spectrum. It is this electronic property within the 4f levels that facilitates their function as perfect "photon managers" for efficient energy conversion at any desired wavelength [46, 47]. The characteristic energy level diagram of rare earth ion emission is also known as the Dieke Diagram [48, 49, 86]. The partially filled 4f orbitals are completely shielded by the outer $5s^2$ and $5p^6$ electrons which have larger radial extension than the 4f shell. Hence the 4f levels are weakly affected by the ions in the surrounding medium in the host matrix. The weak perturbations allow for the narrow and intense optical emission and absorptions in these trivalent ions. The optical absorption and emission bands are thus assumed to be independent of the crystal field [50].

13.4.1 Energy Transfer Mechanisms in Phosphors

Energy transfer from or to the RE ions enables their function as sensitizers in a wide range of optical applications such as solid-state lighting, infrared quantum counters and as infrared to visible convertors. Although the energy transfer between like and unlike RE ions is more understood, the crucial transfer between ions enhancing transitions to the rare earth ions still needs thorough investigations [51]. The excitation of RE ions through energy transfers from semiconductor nanocrystal hosts such as ZnO has received lots of interest. However, the energy transfer mechanisms from ZnO to the RE ions still needs thorough investigations. Huang et al. [52] reported that the energy transfer from ZnO to RE ions, specifically Eu^{3+} ions, is predominantly due to the intrinsic defects in ZnO. Luo et al. [53] on the other hand have argued that the energy transfer is possibly due to a combination of the band edge emission from the ZnO and the defects in the ZnO. Hence there is no consensus on the single dominant mechanism of energy transfer in REs embedded in a ZnO host matrix. Many studies on ZnO-RE energy transfer have been based on steady-state photoluminescence (PL) emission from the nanocrystals to the rare earth [56].

During an excitation, ions may relax to the ground state radiatively or non-radiatively, the latter process being subdivided into internal non-radiative transitions and energy transfer. Non-radiative transfer from donors to acceptors depletes the population of the excited state of the donor and decreases the intensity and lifetime from the excited state to the lower state. In order that the transfer be significant and measurable, the rate of transfer of energy must be of the same order of magnitude as the radiative transition in the donor ion.

In general, semiconductor nanocrystals such as ZnO may be used as sensitizers to excite rare earth ions as a result of their intrinsic properties such as size-dependent luminescent properties [55], large absorption cross sections [56–58], and broad excitation spectra [59]. In this process, semiconductor nanocrystals are excited using a light source and transfer the energy to the RE ions which can then reemit light at a different wavelength [60].

13.4.2 Rare Earth Metals Used for Doping of ZnO

13.4.2.1 Terbium

Terbium was discovered in 1843 by Carl Mosander and is named after Ytterby, a village in Sweden, the place where it was mined and harvested. It has an atomic number of 65 and an atomic mass of 158.93 AMU with electronic configuration of [Xe] $4f^9 6s^2$.

The trivalent Tb^{3+} ion has been under investigation by numerous research groups [48, 61] as a result of it being a potential dopant for many solid-state devices such as light emitting diodes (LEDs), fluorescent lamps, television screens, and solar cells. When excited using ultra-violet radiation, the Tb^{3+} ions in ZnO produced during this study show narrow emission lines in the green spectral region that is plausibly attributed to ${}^5D_4 \rightarrow {}^7F_J$ (J = 6, 5, 4, 3) transitions. Additionally the ${}^5D_3 \rightarrow {}^7F_J$ (J = 6, 5, 4, 3, 2) transitions have been associated with blue spectral emissions. Figure 13.2 shows a typical absorption and emission of Tb^{3+} ions.

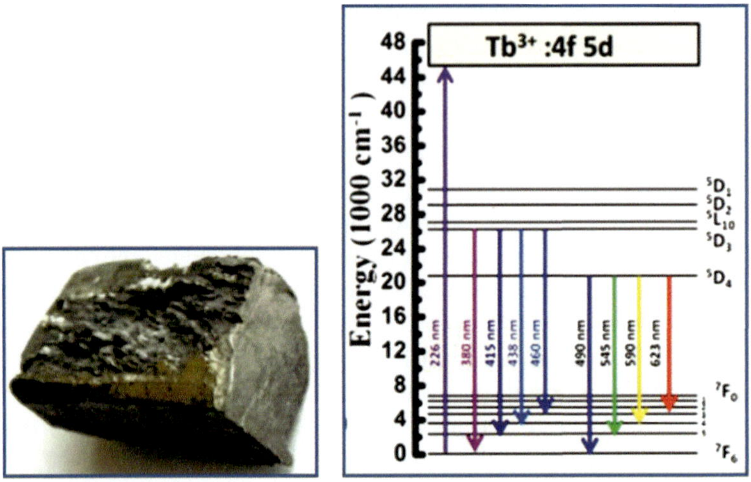

(a) (b)

Figure 13.2 (a) Terbium ore and (b) schematic energy level diagram of Tb^{3+} ion [103].

Studies on Tb doped ZnO exhibit interest on a wide variety of correlated physical properties. Work by Liu et al. [62] on Tb-doped ZnO nanocrystals have shown green luminescence due to Tb^{3+} ions. The emission was ascribed to the 4f – 4f transitions. Liu et al. [63] also looked at Tb-doped ZnO nanoparticles (NPs) synthesized through wet chemical route and reported on their PL properties. The relaxation of carriers from excited states of ZnO hosts to RE dopants was associated with the emission intensity of Tb^{3+} centers as confirmed by the emission intensity dependence. Jia et al. [64] reported a strong UV emission peak in the photoluminescence studies of Tb-doped ZnO. They attribute the red shift in the UV emission to the low energy defect levels of the Tb dopants.

13.4.2.2 Europium

Europium was discovered by Eugène-Antole Demarçay, a French chemist, in 1896. He was able to produce reasonably pure europium in 1901 from monazite sand ores. Trivalent europium ion (Eu^{3+}) is well known for red-emission as a result of $^5D_0 \rightarrow {}^7F_2$ transition [65]. It is widely used as a source of red light in preparing phosphors used in many light emitting devices. Its potential application as the source of red light in Eu-doped GaN laser was reported by Park et al. [66]. Shahroosvand, H. and Ghorbani-asl, M. further reported the [67] correlation between the red emission and the synthesis method suggesting that intrinsic defects may aid in efficient energy transfer from ZnO host to Eu^{3+} ions. The three peaks in the region of $^5D_0 \rightarrow {}^7F_J$ were observed, an indication that the Eu^{3+} is located in the surface lattice sites of the ZnO materials. They proposed an energy transfer mechanism shown in Fig. 13.3. They attributed spectroscopic properties of the Eu^{3+} ions to the transitions between the intra-4f electron energy levels. The right-hand side gives the partial energy level of Eu^{3+} ions, where the common $^{2S+1}L_J$ notation is used for the energy level, with S, L, and J referring to spin, orbital, and total angular momentum, respectively. The schematic on the left-hand side shows ZnO host with the band-edge and the surface defect emissions.

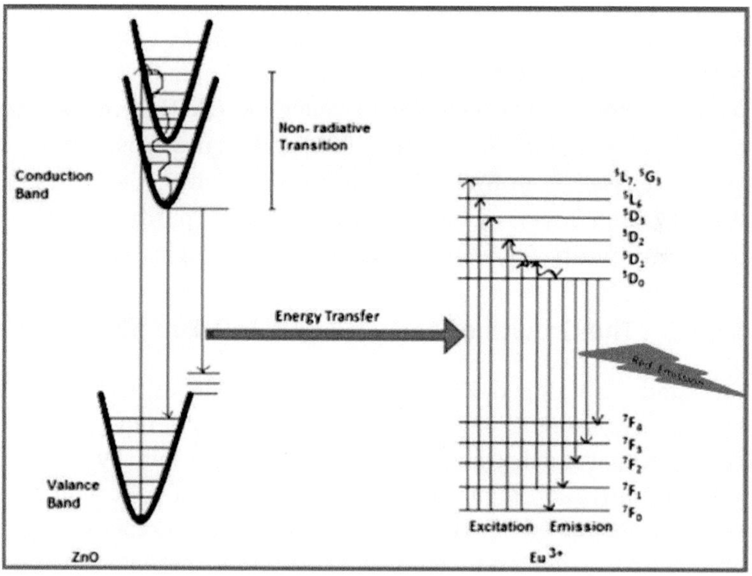

Figure 13.3 Illustration of proposed emission mechanism in Eu-doped ZnO [68].

Other previous studies on Eu-doped ZnO include: Pearton et al. [68], who reported green luminescence of the ZnO nanowires with Eu diffusion process observed in the vicinity of 515 nm and attributed it to the thermally activated diffusion of Eu ions. Additionally, the diffusion of Eu in the nanowires due to annealing led to a red-shift of the near band edge (NBE) emission. This chapter examines the relationship between structure and optoelectronic properties of rare earth doped ZnO thin films fabricated by RF magnetron sputtering for photovoltaic applications. As a proof of concept, the films are incorporated in dye sensitized solar cells for two purposes, as an electron transport layer as well as a photoactive layer capable of photon energy management through down shifting effects. In the next section, the methods for fabrication of each RE doped ZnO thin film are presented.

13.5 ZnO Growth

13.5.1 Materials Used

ZnO (97%): Tb (3%) (99.99% purity), ZnO (97%): Eu (3%) (99.99% purity) and ZnO disks targets of diameter 76 mm and thickness 6 mm sourced from Semiconductor Wafer, Inc. (SWI) were used for thin film deposition. The Si wafer (001) and quartz substrate were also purchased from SWI. ITO (surface resistivity: 30–60 Ω/sq.), PEDOT:PSS (1.3 wt% dispersion in H_2O), P3HT (purity: 99.995%) and PCBM (purity: 99.5%) were purchased from Sigma Aldrich.

13.5.2 The Growth of $ZnO:Tb^{3+}$ and $ZnO:Eu^{3+}$ Thin Films

Tb^{3+} and Eu^{3+} doped ZnO thin films were deposited at room temperature using RF magnetron sputtering onto n-type (001) Si, ITO, and quartz substrates. The substrates were initially cleaned with acetone and ethanol in an ultrasonic bath for 20 min and rinsed in deionized water before drying using a stream of dried nitrogen gas. They were then mounted on a rotating substrate holder at a distance of 6 cm from the target. The sputtering chamber was evacuated to a base pressure of about 2.5×10^{-5} mbar while sputtering was performed under argon with a working gas pressure of 3.1×10^{-3} mbar. The growth was undertaken under 13-sccm Ar flow, and 2.20 Wcm^{-2} RF power density. Post-growth annealing was carried out in an Ar filled furnace at a heating rate of 10°C/min over a temperature range of 600–900°C for 2 h.

13.6 Property Determination and Correlations

The film morphologies and structural analyses were characterized using Veeco Di-3100 atomic force microscopy (AFM) in tapping mode. Grazing incidence X-ray diffraction (XRD) measurement of the ZnO:Tb^{3+} and ZnO:Eu^{3+} thin films were taken using Cu Kα radiation from the Bruker AXS D8 Discover High Resolution X-ray Diffractometer set at 40 kV and 40 mA. The Seeman-Bohlin geometry was used with a low scanning rate of 1.2° per min at a glancing incidence angle of 1°. The transmittance spectra of the films deposited onto the quartz substrates were measured using a Cary 500 UV–visible spectrophotometer. Photoluminescence measurements were carried out using a Horiba LabRAM HR spectrometer with a 150 lines/mm grating and an excitation wavelength of 244 nm from a frequency doubled Lexel argon ion laser. Rutherford backscattering spectrometry (RBS) measurements were performed to determine the composition and Tb^{3+} ion distribution profiles in the ZnO matrix at room temperature using alpha particles ($^4He^+$) of energy 1.6 MeV at a backscattering angle of 165° (IBM geometry). The beam current was kept between 10 and 15 nA during measurements.

13.7 Experimental Results

13.7.1 Results and Discussions of ZnO:Tb^{3+} Thin Films

13.7.1.1 Morphology of ZnO:Tb^{3+}

Surface roughness in electrodes or interfacial layers contribute to device performance, through localized surface charge density or by forming short circuits, which minimize charge transport. However these effects are dependent on the thickness of the active layer and on the type of solar cell. To evaluate the magnitude of surface roughness and topography, atomic force microscopy is utilized. Figure 13.4 shows the surface topography of the ZnO:Tb^{3+} films characterized using Veeco Di3100 AFM in tapping mode after thermal annealing at select temperatures.

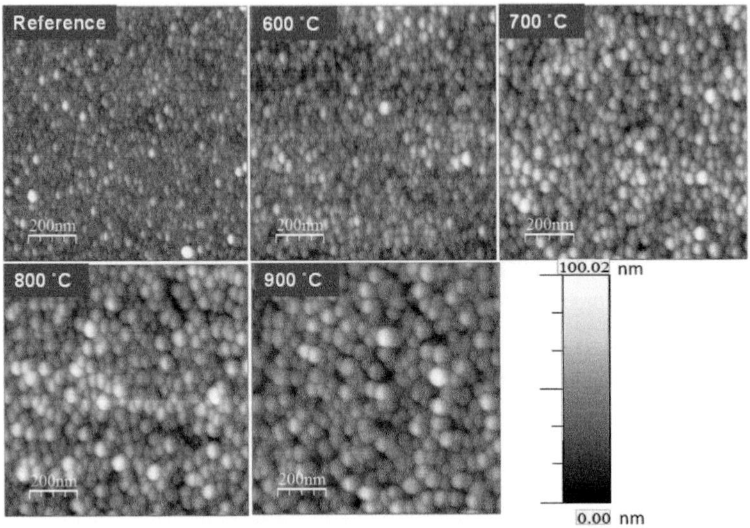

Figure 13.4 AFM images of pristine ZnO:Tb^{3+} (at room temperature) and after annealing in an Ar filled furnace for 2 h in the range between 600 and 900°C.

The average roughness of the ZnO:Tb^{3+} film surface is shown in Table 13.1 for all temperature ranges of interest. As the annealing temperature increases, the surface roughness of the film increases dramatically (from 4.9 nm for as-deposited ZnO:Tb^{3+} film to 36.9 nm when annealed at 900°C).

Table 13.1 Summary of rms roughness and grain height from taping mode AFM

Annealing temperature (°C)	Roughness rms (nm)	Grain height (nm)
RT	4.9	12.2
600	11.8	49.3
700	17.2	54.0
800	27.8	68.8
900	36.9	81.4

It is noted that the surface corrugations of ZnO:Tb^{3+} films analyzed by AFM possibly represent the particle size and could thus be larger than the grain size determined from XRD shown in Table 13.1. This is plausible since the particle size measured from AFM is the surface morphology of coalesced grains [69]. Furthermore the geometry of the XRD measurement of the grain size is normal to the surface of the film and not in-plane. Thermally activated diffusion enables the ad atoms in the film to migrate to sites with minimal potential energy and in the lattice site leading to the enhancement of film quality as shown in the XRD patterns. The increase in grain height with annealing temperature is an indication of thermally activated diffusion of atoms that coalesce to larger grains in regions with lower activation barriers for self-diffusion [70, 71]. The major grain growth also results in an increase in the surface roughness as evident by the rms values.

13.7.1.2 Rutherford backscattering spectroscopy measurements

The RBS spectrum of a ZnO:Tb^{3+} thin film on (001)Si substrate annealed at 700°C is shown in Fig. 13.5 together with the corresponding simulation obtained from XRump code. The experimental spectrum is well reproduced. The peak profiles in the RBS spectrum at around channels 180, 260, 410, and 470 is due to the non-Rutherford scattering cross section from O, Si, Zn, and Tb, respectively. The surface energy positions in the channels of these elements have been shown using the vertical arrows in the spectra. All other samples annealed at different temperatures showed insignificant change in surface energy position and concentration.

The presence of Tb^{3+} in the films is very prominent and the global atomic concentrations of the different species contained in the film can be deduced. From the XRump simulation, stoichiometry of the matrix of 48.1, 49.2 and 2.7% for ZnO and Tb, respectively was derived. The Tb^{3+} in channel 470 has a nearly flat yield suggesting a homogeneous distribution of the Tb^{3+} along the growth direction, which is independent of the annealing temperature.

13.7.1.3 Structural properties of ZnO:Tb^{3+}

Figure 13.6 shows the grazing incidence XRD patterns of pristine ZnO:Tb^{3+} and after annealing in the temperature range 600–900°C in Ar ambient. The pristine ZnO:Tb^{3+} film appears as partially crystalline with low intensity of the (hkl) reflexes. Annealing leads to preferred orientation of the ZnO reflexes. The typical strong diffraction peaks ascribed to hexagonal Wurtzite ZnO (JCP2) card number 003-0888 were observed without any extra peaks from other oxide phases at elevated temperatures. This indicates a high phase purity and that no diffraction peaks corresponding to Tb^{3+} based compounds or any other impurities were present. It is probable that the Tb^{3+} ions are incorporated in the Wurtzite lattice and/or in the grain boundaries. The diffraction peaks become gradually stronger up to 800°C, an indication of increasing crystallinity. The appearance of reduced peak intensity and dislocation density after annealing at 900°C may be attributed to the presence of porosity [72]. The dislocation density δ, defined as the length of dislocation lines per unit volume of the crystal (Williamson and Smallman) and the strain ε present in the thin film were also investigated and the results are presented in Table 13.2.

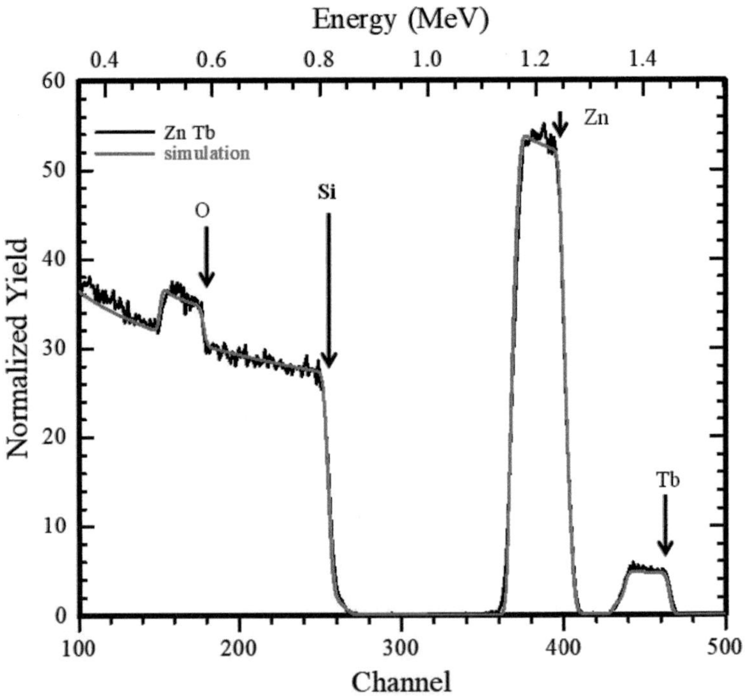

Figure 13.5 RBS spectrum of a 190 nm $ZnO:Tb^{3+}$ layer on 200 nm Si showing experimental data and computer simulation using XRump code. The sample structure and the thicknesses of the layers were derived.

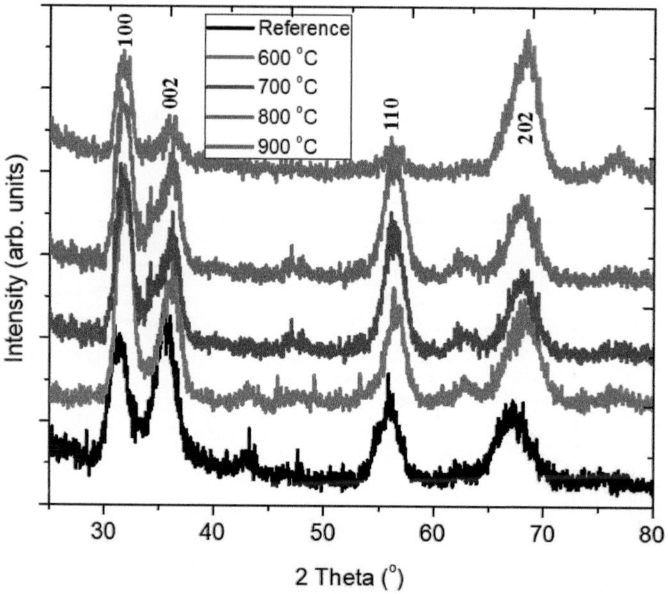

Figure 13.6 Grazing incidence X-ray diffraction of ZnO:Tb^{3+} thin films deposited by RF sputtering at room temperature then annealed different temperatures (600–900°C).

Table 13.2 Characteristics of Tb^{3+} doped ZnO estimated from XRD measurements

Temperature (°C)	FWHM (100)	2 theta	Size, D (nm)	a/(Å)	c/(Å)	δ (10^{16})	ε (10^{-3})
Reference	1.620	31.397	5.09	3.28	5.69	3.86	6.81
600	1.473	31.901	5.61	3.23	5.61	3.18	6.18
700	1.332	31.784	6.19	3.24	5.63	2.63	5.59
800	1.332	31.764	6.20	3.25	5.63	2.60	5.57
900	1.207	31.820	6.84	3.25	5.64	2.14	5.07

13.7.1.4 Optical properties

Figure 13.7 shows the transmittance of pristine and annealed ZnO:Tb^{3+} thin films on quartz substrate measured in the range between 250 and 800 nm. The film thickness for the pristine and annealed films have been determined independently to 183 and 190 nm, respectively. The sharp optical transmission edge observed around 380 nm corresponds to the inter-band transitions of the valence electrons consistent with the laser excitation energy used in the PL measurements. The contributions of the substrate have been de-convoluted from the transmission spectra through the dual beam calibration with the substrate on the reference beam.

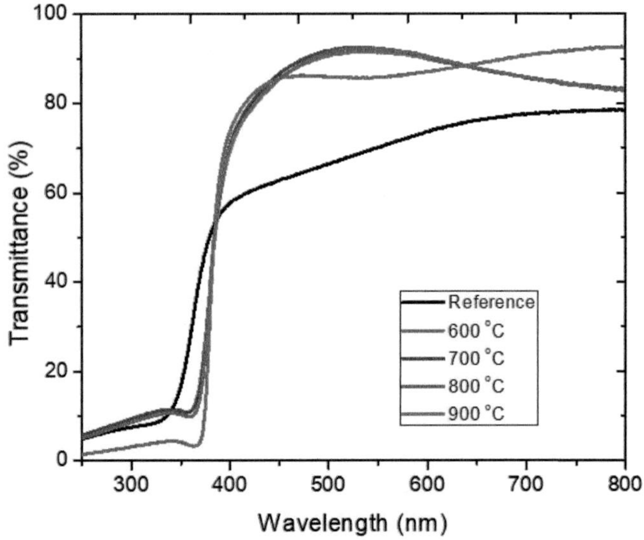

Figure 13.7 Transmission spectra of pristine ZnO:Tb^{3+} thin films and upon annealing at different temperatures, 600–900°C in an Ar filled furnace.

The average transmittance for all the deposited films is in the range between 71% and 92% within the visible region. Thus enabling the use of these ZnO:Tb^{3+} films in thin films solar cell applications [73]. At short wavelengths, the steep decrease in transmittance is attributed to the threshold absorption at the band edge. The enhanced absorption for the thin film annealed at 900°C is attributed to increased crystallinity leading to decrease in the extinction coefficient as the structural defects are minimized with increasing temperature thus reducing the absorption coefficient [74]. To correlate the effect of temperature and grain size with the optical properties of ZnO:Tb^{3+} thin films, the absorption coefficient (α) and the band gap (E_g) were determined from Tauc's formalism for direct and indirect transitions in the form given in Eq. 13.1:

$$(\alpha h\nu) = A(h\nu - E_g)^p \qquad (13.1)$$

where A is a constant, h is the Planck's constant and p is an exponent that depends on the nature of the allowed or forbidden electronic transitions. Thus, for allowed direct transition $p = 1/2$, while $p = 3/2$ describes the case for forbidden direct transition, $p = 2$ for indirect allowed transition and $p = 3$ for indirect forbidden transitions. Since $ZnO:Tb^{3+}$ is a direct band gap semiconductor with a hexagonal Wurtzite crystal structure, as confirmed in the XRD measurements, $p = 1/2$ [75–77]. $h\upsilon$ is the photon energy expressed in eV while α is the absorption coefficient α is calculated using the relation in Eq. 13.2:

$$T = \exp(-\alpha d) \quad (13.2)$$

where T is the optical transmittance and d is the thickness of $ZnO:Tb^{3+}$ films.

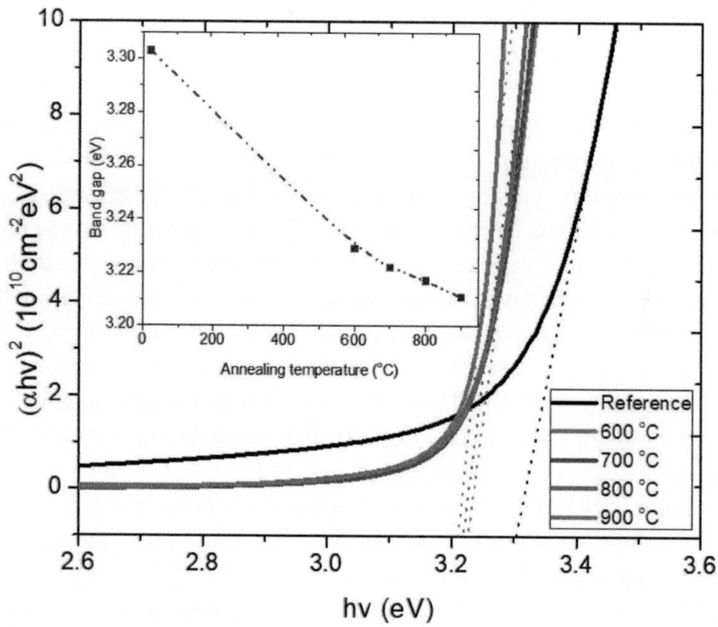

Figure 13.8 Plot of $(\alpha hv)^2$ vs. hv for the pristine $ZnO:Tb^{3+}$ thin film and after annealing at 600–900°C. The inset shows the variation of optical band gap as a function of annealing temperature.

The optical band gap of the films annealed at various temperatures was obtained by plotting $(\alpha hv)^2$ as a function of photon energy (hv). The values of the direct energy gap E_g were derived from the intercept of the extrapolation to zero absorption with the photon energy axis as shown in Fig. 13.8. The non-zero absorbance below the energy cut-off for the pristine film is attributed to the phonon contributions associated with structural defects or trap states arising from a partially crystalline pristine ZnO:Tb^{3+} film.

The sharpness of the absorption edge is found to be maximum at 800°C annealing temperature, beyond which it broadens. The change in the optical band gap is comparatively small but a minimum is shown at 900°C. The estimated values of the band gap for all the samples are summarized in the inset of Fig. 13.8. These values are within the range reported for films and single crystal [78–80] and are also in agreement with the band gap approximation derived from the PL results.

13.7.1.5 Room temperature photoluminescence studies

Figure 13.9 shows the PL emission spectra of ZnO:Tb^{3+} thin film excited at 244 nm. The spectra show emissions at 378 nm corresponding to a NBE emission of ZnO. This emission is attributed to the recombination of the excitons of granular ZnO through an exciton–exciton collision process corresponding to near-band-edge (NBE) emission of ZnO [81]. Its intensity increased with increasing annealing temperature, an indication of enhanced film crystallinity necessary for effective charge transfer (grain size commensurate to the mean free path of the electrons) in solar cell application. This is due to the decrease of the non-radiative defects and the increase in the grain size of ZnO [36]. At 900°C, this peak showed reduced peak intensity. A plausible reason is that the decrease in the length of *c*-axis of ZnO films annealed at higher temperature may induce an acute lattice distortion, which would significantly increase the polarizability in the ZnO:Tb^{3+} matrix and thus affect the optical properties of ZnO:Tb^{3+} films and quench the PL efficiency [82]. This is in agreement with XRD findings. This decrease in the emission of ZnO or the inter-band transitions could also be attributed to some of the excited electrons being transferred to the energy levels associated with oxygen vacancies while others are transferred to the energy levels of Tb^{3}. The broad emissions between 400–500 nm are attributed to deep level defect emissions arising from the recombination between the Zn_i and the valence band level [83]. ZnO has six kinds of defects classified as vacancies, interstitials, and antisites and their emission intensities can be influenced by the defect concentration and defect–defect interactions which varies with annealing temperature.

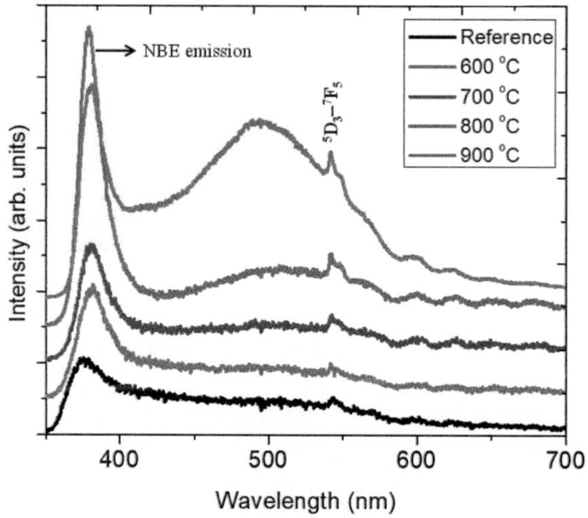

Figure 13.9 PL spectra of the ZnO:Tb^{3+} thin films annealed at different temperatures excited at 244 nm together with that of the ZnO:Tb^{3+} thin film annealed at temperature range of 600–900°C.

The sharp emission peak at 543 nm is related to Tb^{3+} emission and it corresponds to the radiative $^5D4$ – $^7F5$ transitions. Here, the electric dipole (ED) transitions between the 4f states in the free Tb^{3+} ions are parity forbidden. However the ED transitions are partially allowed and have a weak intensity whenever Tb^{3+} ions occupy the interstitials or lattice sites in ZnO [83, 84]. Hence according to Bylander et al. [85], majority of excited charge carriers trapped at Tb^{3+} centers originate from inter-band transitions in the ZnO matrix and a marginal fraction of these electrons are contributed from the 4f-4f absorption transition within the Tb^{3+} ions.

Table 13.3 shows the peak centroid and the corresponding full width at half maximum (FWHM) for the near-band-edge peak as well as the broad band peak while Fig. 13.8 gives an elaborate schematic band diagram of the optical emissions. The shifting of the peak center with annealing temperature is due to the possible relaxation of built-in strain in ZnO:Tb^{3+} thin films.

The efficiency of the energy transfer from the host ZnO to RE ions such as Tb^{3+} is usually lower due to the large mismatch in the ionic radii and the valence state of the Zn^{2+} (74 pm) and RE ions, Tb^{3+} (118 pm). Furthermore, the location of the rare earth in the host also determines the efficiency of the energy transfer, thus generally, rare earth ions can occupy three different sites in the hosts, namely: (i) in the substitution sites of Zn^{2+} in the ZnO lattice, (ii) at the ZnO grain boundaries, or (iii) at the interstitial sites. The location of Tb^{3+} on the surface or grain boundaries is not consistent with the RBS data which indicated a homogeneous distribution of Tb^{3+} in ZnO thin film. The distinction of surface and bulk emissions can thus be resolved qualitatively using the RBS profiles of the RE and constituent atoms in the ZnO film. Figure 13.10 shows a schematic diagram of the energy levels of the ZnO:Tb^{3+} thin films.

Table 13.3 Peak center positions and FWHM for the NBE PL peak

Annealing temperature (°C)	**Peak centre (nm)**	**FWHM (nm)**
Reference	477	21.9
600	381	19.1
700	382	16.6
800	384	12.2
900	398	26.3

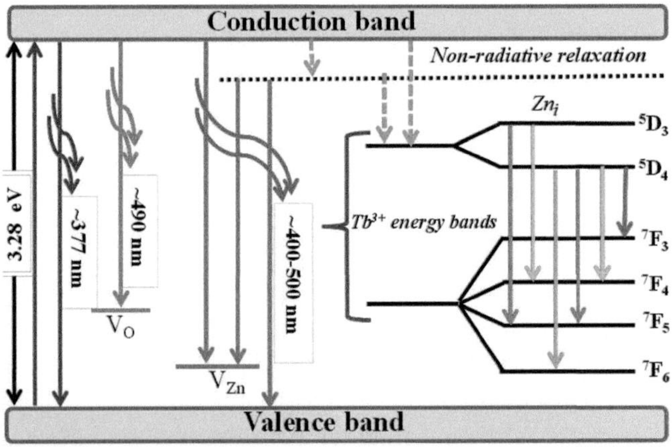

Figure 13.10 Schematic of the energy level diagram of the ZnO:Tb^{3+} thin films.

13.8 Application of $ZnO:Tb^{3+}$ Thin Films into Organic Solar Cells

13.8.1 Device Fabrication and Characterization

The photovoltaic devices were fabricated on glass coated with indium tin oxide (ITO), which had been cleaned by sonication using the procedure described in Ref. [86, 87]. Thin ZnO or $ZnO:Tb^{3+}$ (about 35 nm) films were grown on the cleaned ITO by RF magnetron sputtering using the procedure described in Section 13.5.2 before annealing at 300°C in Ar ambient for 1 h. The active layer consisting of P3HT:PCBM were blended at a mass ratio of 1:1 in chlorobenzene at a total concentration of 20 mg ml^{-1}. The mixture was subsequently spin coated at 2000 rpm for 1 min onto the hole transport layer (PEDOT:PSS) coated ZnO/ITO substrate to attain a 200 nm thick active layer. The active layers were annealed at 80°C for 10 min under inert atmosphere to remove any solvent adsorbents or moisture. Finally, an Al electrode was thermally evaporated at a pressure of 2×10^{-5} Pa. The current density–voltage (*J–V*) characteristics were obtained using a HP 4141B source measure unit (SMU) under 100 mW/cm^2 illumination (AM 1.5G). All the measurements were carried out at room temperature under standard conditions. More than 20 devices of pristine ZnO and $ZnO:Tb^{3+}$ based inverted solar cell each of area 8.4 ± 0.1 mm^2 were fabricated, characterized, and the results obtained exhibited a high degree of repeatability to within 2.5%.

13.8.2 Current Density–Voltage Characteristics for Performance Parameters Extraction

Figure 13.11 shows *J–V* characteristics curve used as a proof of concept of the influence of Tb^{3+} doping on photoactive performance in an organic solar cell with an inverted structure of ITO/$ZnO:Tb^{3+}$/P3HT:PCBM/PEDOT:PSS/Al. The inset in the figure represents the UV-visible absorption spectra of active layer (P3HT:PCBM), ZnO/P3HT:PCBM, and $ZnO:Tb^{3+}$/P3HT:PCBM films used in device fabrication. The films were annealed at 300°C to avoid the softening of the glass substrate as well as deterioration of electrical properties of the ITO film.

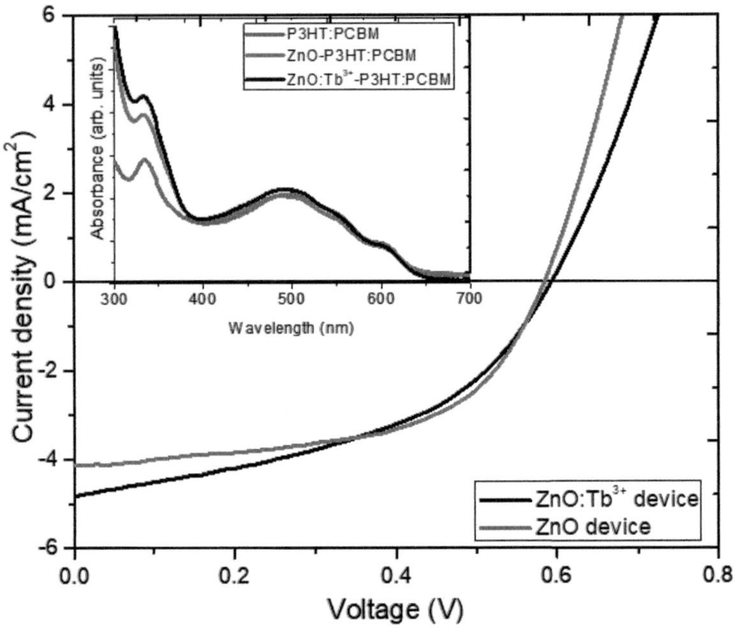

Figure 13.11 *J–V* characteristics under illumination of device formed from pristine and heat treated P3HT:PCBM blends.

ZnO and ZnO:Tb^{3+} thin films grown by RF sputtering on ITO coated glass substrate were both used as electron transport and spectra conversion layers. Table 13.4 shows the performance parameters. The two devices showed almost similar fill factor (FF) and open circuit voltage (V_{OC}) values, this shows that the energy levels in the organic photovoltaic (OPV) device were not altered by the Tb^{3+} doping. Further support to this assertion is provided by the constant values of the series and shunt resistance values of both devices. The changes upon Tb^{3+} addition in ZnO are seen in the optical emission leading to the marginal increase in J_{SC} from 4.11 to 4.82 mA cm^{-2}. This could be attributed to spectrum downconversion as a result of Tb^{3+} doping. According to the absorption measurements shown in the inset of Fig. 13.11, there is an increased absorption in the ultraviolet region from 300–400 nm with the introduction of the ZnO thin film as a layer preceding the P3HT:PCBM blend thin film which is further enhanced upon doping with Tb^{3+} leading to an increase in the number of photo-generated photons in the visible region converted from the UV region. The downconversion process altered the UV properties of the solar cell under the solar emission spectrum enabling absorption of photons in the UV range and re-emitting light at a longer wavelength in the visible region, where the organic solar cell has a considerably better response. The increase in light absorption with ZnO thin film compared to the film with P3HT:PCBM blend alone may be an indication of the emission being in resonance with the P3HT electronic structure since exciton generation occurs through the $\pi \rightarrow \pi^*$ interaction in the P3HT.

Table 13.4 Summary of the performance of the P3HT:PCBM bulk heterojunction solar cell devices with pure ZnO and ZnO:Tb^{3+} as the cathode buffer layer

Device	J_{sc} (mA/cm^2)	V_{oc} (V)	R_s (Ω cm^2)	R_{sh} (kΩ cm^2)	FF (%)	PCE (%)
ZnO	4.11 ± 0.10	0.59 ± 0.02	51.8 ± 1.2	232 ± 3	47.8 ± 1.0	1.16 ± 0.08
ZnO:Tb^{3+}	4.82 ± 0.12	0.60 ± 0.01	52.3 ± 1.0	235 ± 2	46.9 ± 1.1	1.36 ± 0.06

13.9 Results and Discussion of ZnO:Eu^{3+} Thin Films

13.9.1 Morphological Studies

As a bi-functional layer in a solar cell device the topography of the ZnO film should be relatively homogeneous and devoid of pinhole formations even at elevated temperatures. Hence investigation on morphology studies are required to determine the grain size distribution since it limits charge transport. Figure 13.12 shows the SEM images of ZnO:Eu^{3+} annealed at different temperatures ranging between 500–900°C.

The micrographs of both pristine and annealed thin films exhibit uniform and compact columnar growth of ZnO:Eu^{3+} quasi-spherical particles. As the temperature increases, the particle profile and size increase in conformity with the Zone 2 and 3 of the structural zone model [104]. Besides ad atom diffusion on the films with annealing temperature, the role of thermal stress owing to the contrasting thermal expansion coefficients of the film and the substrate can also be ascribed to changes in the columnar growth mode at the interface.

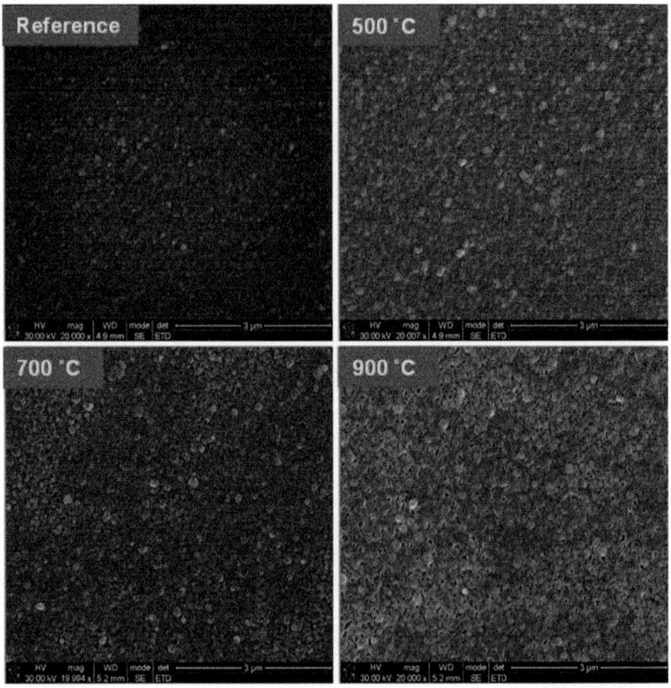

Figure 13.12 Scanning electron microscopy micrographs of Eu-doped ZnO films annealed at temperature range of 500–900°C.

13.9.2 Rutherford Backscattering Spectroscopy Measurements

The elemental and percentage composition and distribution of the Eu^{3+} ions in the ZnO thin film were established using Rutherford backscattering spectroscopy as shown in Fig. 13.13.

Figure 13.13 shows the RBS spectra of pristine ZnO:Eu^{3+} thin film. Simulation of the experimental spectrum was done using the XRump code [88]. The vertical arrows indicate the scattering energies of the elemental constituents at the surface. The black line indicates the experimental data while the red represents the simulation. Table 13.5 shows the summary of Eu^{3+} distribution in the ZnO thin film as calculated from the XRump code.

The film thickness was derived as 357 nm and this value is in agreement with the profilometer value. Small amounts of Si are observed to diffuse into the ZnO thin films as a result of the annealing effect giving the atoms adequate energy to move. The total average Eu composition is ~1% with the majority being observed at a depth of 278 nm.

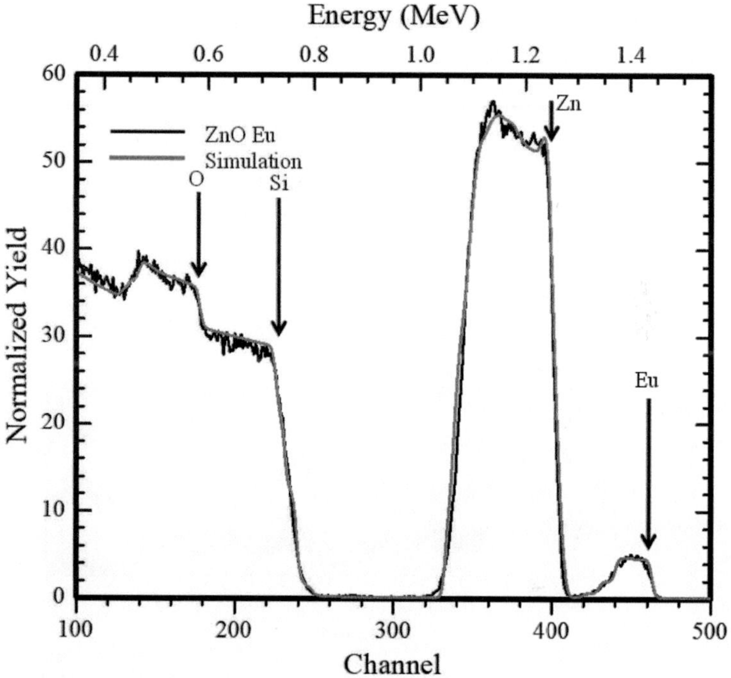

Figure 13.13 Random and simulated RBS spectra from the ZnO:Eu^{3+} thin film annealed at 900°C.

Table 13.5 Derived atomic composition at various film depths from XRump code

Layer	Thickness (nm)	Zn %	O %	Eu %	Si %
1	118	48.0	51.5	1.0	
2	50	50.0	49.5	0.4	
3	50	50.0	49.7	0.1	
4	60	50.0	46.7	0.1	3.0
5	46	26.1	25.0	0.1	46.0
6	33	15.0	12.5	0.05	70.0
7	2000				100.0

13.9.3 X-Ray Diffraction Analysis

The room temperature XRD of ZnO:Eu^{3+} thin films annealed at various temperatures are depicted in Fig. 13.14. It is seen that the XRD pattern of the films mainly consists of hexagonal ZnO (JCPDS database no: 00-003-0888).

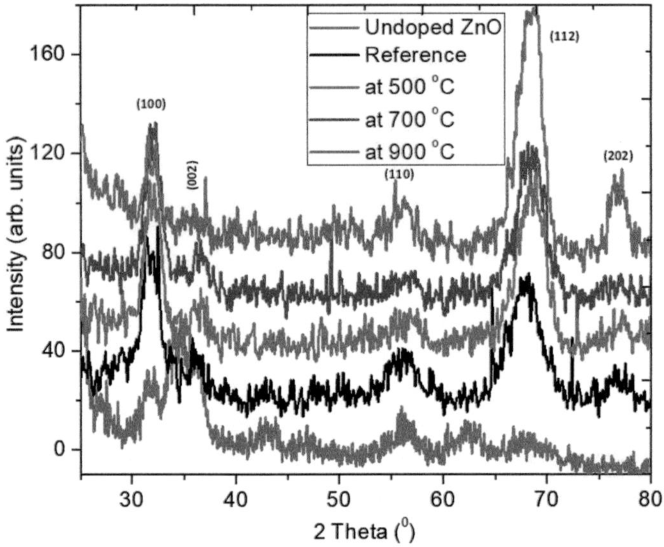

Figure 13.14 Room temperature XRD pattern of pristine and annealed ZnO:Eu^{3+} thin films.

The XRD pattern of Fig. 13.14 shows no diffraction peaks associated with europium oxides or other impurities thus indicating that the Eu^{3+} ions are probably incorporated in the ZnO lattice. The Bragg peaks in the XRD pattern of ZnO are indexed to the hexagonal ZnO phase with lattice parameters: (a) 3.249 Å and (c) 5.206 Å. These parameters are in good agreement with literature [89]. Compared to the pristine ZnO thin film, the diffraction peaks of Eu^{3+} doped ZnO samples shift by about 0.5° (2θ) to a lower angle due to the larger ionic radius of Eu^{3+} (0.95 Å) compared to that of Zn^{2+} (0.74 Å) [89].

Upon annealing the ZnO:Eu^{3+} thin films, the (100) peak which is more pronounced compared to the (002) is gradually shifted from 31.74° to 32.04° while its FWHM decreases from 2.155 to 0.999, an indication of enhanced crystallinity with annealing.

13.9.4 Optical Studies

Figure 13.15 represents transmittance (%) of the thin films as a function of wavelength. All the four films show a dip near 372 nm which is ascribed to the band edge absorption.

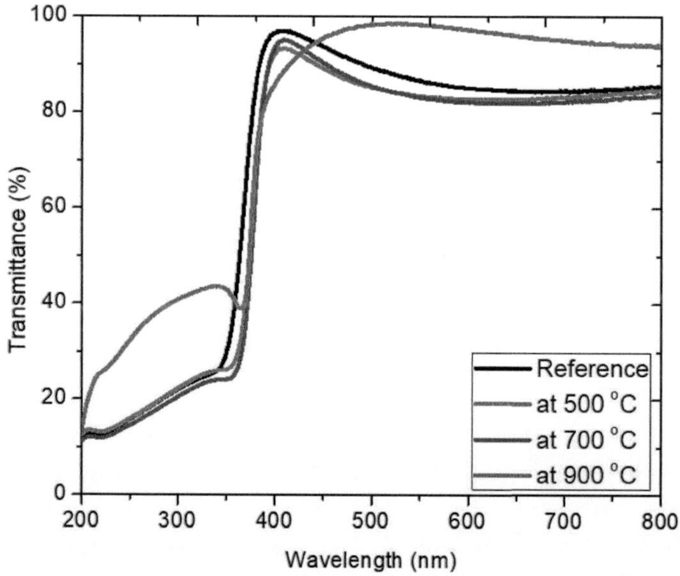

Figure 13.15 Optical transmittance of pristine and ZnO:Eu^{3+} annealed at 500–900°C.

The transmittance of Eu-doped ZnO thin films in the wavelength range of 200–800 nm are shown in Fig. 13.15. All the four films are highly transparent in the visible wavelength range and sharp ultraviolet absorption edge is observed at approximately 372 nm in the UV region. They also show good transmittance of about 95% which is even more pronounced with annealing at 900°C as a result of optical scattering caused by the surface roughness of the film. This is a very vital result as highly transparent electrodes are requisite for photovoltaic applications. In addition, a shift towards the lower energies of the absorption edge with increasing annealing temperature is evident. The incorporation of Eu^{3+} is accompanied by a systematic low-energy shift of the band gap indicating the narrowing of the optical band gap [90]. The plot of $(\alpha h\nu)^2$ against $h\nu$

in Fig. 13.16 is used to derive the optical band gap and its variation with annealing temperature. This procedure is based on the Tauc's model and Fermi-Golden rule for direct allowed transition.

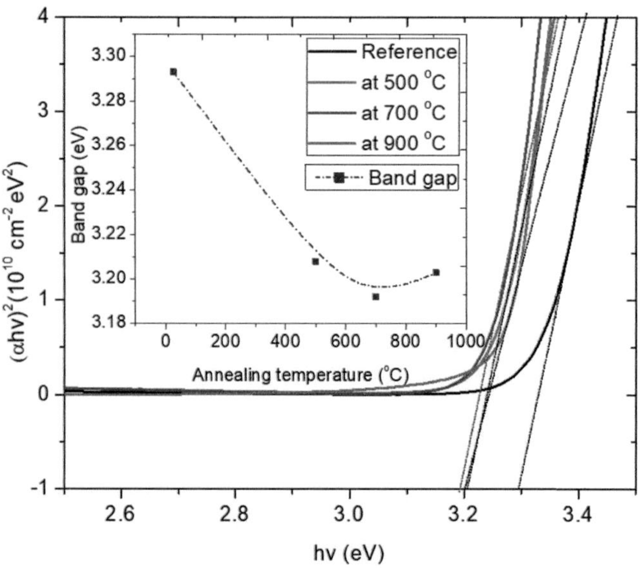

Figure 13.16 Plot of $(\alpha h\nu)^2$ vs. $h\nu$ for the pristine ZnO:Eu^{3+} thin film and after annealing at 500–900°C. The inset shows the variation of optical band gap as a function of annealing temperature.

When the films are annealed from 500–800°C, the optical bandgap is observed to decrease from 3.29 to 3.19 eV before slightly increasing upon annealing at 900°C as shown in Fig. 13.16. The tailoring of the bandgap of semiconductors could be attributed to the introduction of new energy levels in the band gap near the conduction band edge such as shallow level donor impurities [91].

13.9.5 Photoluminescence Properties

The structural and compositional changes in the films at the nanoscale level can strongly influence the luminescence properties. The correlation between the structure and the emission behavior of ZnO:Eu^{3+} thin films as a function of the annealing temperature was investigated by photoluminescence. Figures 13.17 and 13.18 show the suggested electronic transitions and the photoluminescence spectra of Eu-doped ZnO in the 350–800 nm, respectively. The observed emissions in the spectra of Fig. 13.19 represent band levels associated with transitions from the excited ^{5}D0 level to the ^{7}Fj (j = 0, 1, 2, 3, 4) levels of the Eu^{3+} $4f^6$ configuration [92] are observed. The spectrum shows that the $^5D_0 - {}^7F_2$ transitions undergo defect intensity modifications on increased thermal treatment. This indicates the presence of structural re-arrangements around the Eu^{3+} centers in agreement with the suggested transitions as indicated in Fig. 13.17.

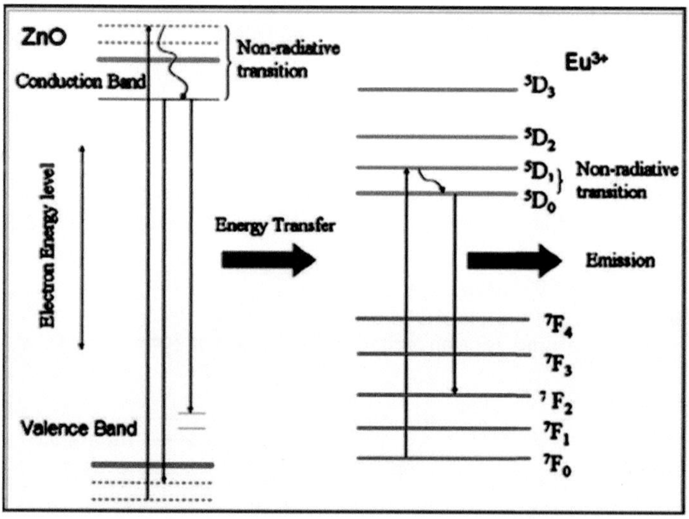

Figure 13.17 Schematic illustration of the proposed mechanism of energy transfer from the ZnO host to the Eu^{3+} ions.

From Figs. 13.18 and 13.19, the UV emission attributed to near-band-edge (NBE) exciton recombination and a broad deep level emission is depicted and discussed here. The sharp red emission peaks exhibited at 590–705 nm are attributed to Eu^{3+} ions. Figure 13.18 shows the PL spectra of films annealed at between 500–700°C while Fig. 13.19 includes films annealed at 900°C.

From Fig. 13.18, the decrease in intensity of both the near-band-edge emission and deep level defects emission at 372 nm and 536 nm, respectively, while the intense emission of the $^5D_0$ – $^7F_2$ transition at 613 nm confirms the energy transfer from ZnO to Eu^{3+} ions [93–95].

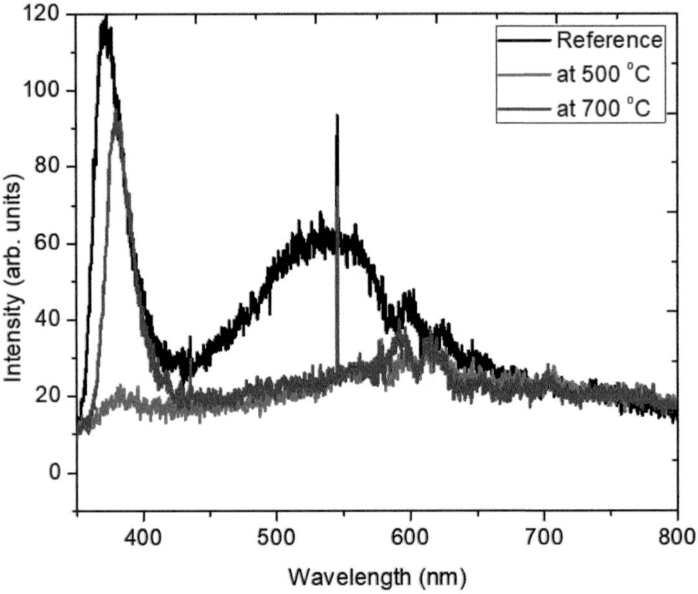

Figure 13.18 Room temperature PL spectrum of ZnO:Eu^{3+} annealed at 500–700°C measured using an excitation wavelength of 244 nm.

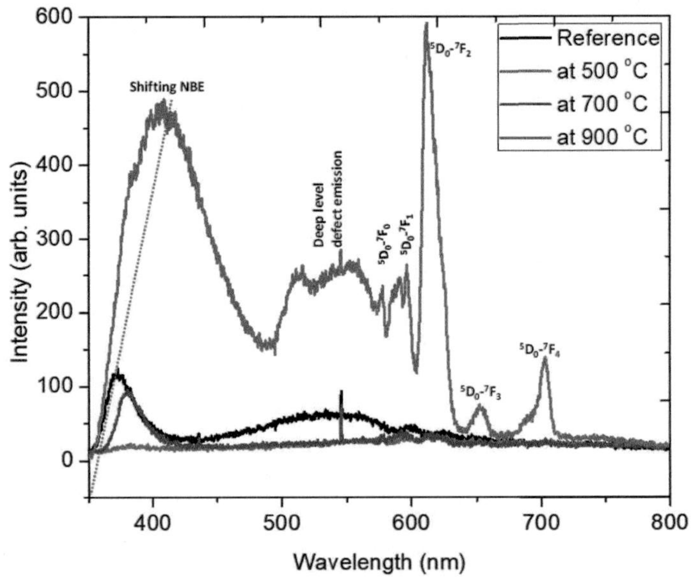

Figure 13.19 Room temperature PL spectrum of ZnO:Eu^{3+} annealed at 500–900°C measured with an excitation wavelength of 244 nm.

From Fig. 13.19, sharp peaks are observed upon annealing at 900°C at 576 nm, 594 nm, 613 nm, 651 nm, 703 nm that correspond to the intrinsic 4f $\rightarrow$ 4f transitions of Eu^{3+} in the ZnO host. The resulting intrinsic 4f $\rightarrow$ 4f transitions of Eu^{3+} were assigned to $^5D_0 - {}^7F_0$, $^5D_0 - {}^7F_1$, $^5D_0 - {}^7F_2$, $^5D_0 - {}^7F_3$, and $^5D_0 - {}^7F_4$ transitions of Eu^{3+} ions, respectively [89, 96, 97]. The peak situated at 613 nm is observed to be stronger which is a clear indication that the Eu^{3+} ions are located at the inversion anti-symmetric sides of the ZnO hosts [89]. According to the Judd–Ofelt theory, the magnetic dipole transition is permitted. However, the electric dipole transition is only permitted when the Eu^{3+} ion occupy a site without an inversion center and the intensity is significantly affected by the symmetry in local environments [98]. If the Eu^{3+} ions occupied an inversion symmetry site in the crystal lattice, the magnetic transition $^5D_0 - {}^7F_1$ (around 594 nm) would be the dominant transition. However the dominance of $^5D_0 - {}^7F_2$ transition indicates an electric dipole transition. The ratio of intensity $(^5D_0 - {}^7F_2)/(^5D_0 - {}^7F_1)$ may be used to establish the measure of distortion from the inversion symmetry of the Eu^{3+} ion local environment. A lower symmetry around Eu^{3+} ion is known to yield a larger value of the intensity ratio, referred to as the asymmetric factor or the asymmetric ratio [99]. The high intensity of the red peak at 613 nm, due to the $^5D_0 - {}^7F_2$ transition upon annealing at 900°C is a confirmation of the fact that Eu^{3+} emission is parity forbidden and observed only when the lattice environment is distorted and has non-inversion symmetry [98]. Because this is a forced electric-dipole transition, this transition is hypersensitive to the annealing temperature. Eu^{3+} ions are known to likely occupy two sites, namely; the inner and surface lattice sites of ZnO [95, 100, 101]. Theoretically, the Eu^{3+} ions are incorporated into the inner lattice site of the ZnO, or Eu^{3+} ions are at substitutional Zn site with C_{3v} symmetry, the $J = 0, J = 1$ transition would split into two crystal-field levels ($J = 0 \rightarrow J = 0, J = 0 \rightarrow J = 1$), however from Fig. 13.19, there are five peaks at $^5D_0 - {}^7F_j$, an indication that the Eu^{3+} ions have been incorporated into a lower symmetry site than C_{3v} [67].

13.10 Application of ZnO:Eu^{3+} into Dye Sensitized Solar Cells

13.10.1 Fabrication Procedure

The cleaned fluorine-doped tin oxide (FTO) coated glass substrate was used as an electrode. ZnO and Eu-doped ZnO thin films were deposited onto the FTO substrates by RF magnetron sputtering and subsequently annealed at 350 °C for 30 minutes. The aim of the heat treatment was to improve film-substrate adhesion and further increase the adsorption on ZnO and Eu-doped ZnO by the dye molecules. The thickness and area of the films was approximated to 150 nm and 0.25 cm^2, respectively. For sensitization, the ZnO films were immersed in a dye solution (3.0×10^{-4} M mixture of the ruthenizer 535-bisTBA (N-719) in methanol) and left to dry for 12 h at room temperature. The counter electrode was composed of a thin of layer of Pt (50 nm) sputter coated on a clean FTO substrate.

The electrode with ZnO thin film photoanode was placed face up on a flat surface, and the counter electrode placed on top of it. These two opposing glass plates were offset from one another so that the entire photoanode was covered by the counter electrode. The redox electrolyte solution (I^-/I_3^-), composed of 0.6 M methylhexylimidazolium iodide, 0.1 M iodine, 0.5 M tert-butylpyridine, and 0.1 M lithium iodide in 3-methoxypropionitrile, was placed at the edges of the plates, and the liquid was drawn into the space between the electrodes via capillary action. An epoxy adhesive was utilized to hold/seal the electrodes together and the *J–V* characteristics are shown in Fig. 13.20.

13.10.2 Results and Discussion

The dye sensitized solar cell (DSSC) with Eu-doped ZnO yielded a PCE of 0.56% (J_{sc} = 2.28 mA cm^{-2}, V_{oc} = 0.59 mV and FF = 42.3%), while the DSSC with pristine ZnO without Eu produced a PCE of 0.52% (J_{sc} = 2.07 mA cm^{-2}, V_{oc} = 0.57 V, FF = 43.6%) indicating 7.7% efficiency enhancement which could be attributed to spectral conversion by the Eu-doped ZnO. The values are relatively low and could be as a result of exposure of the materials to air.

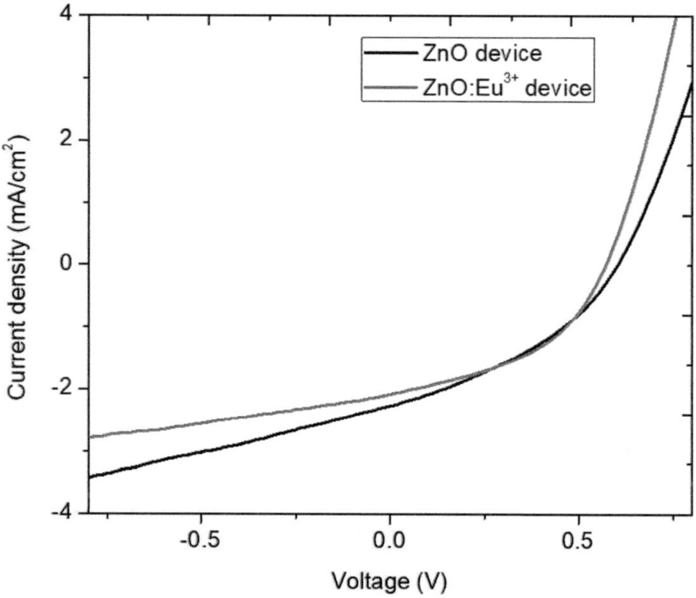

Figure 13.20 *J–V* characteristics curves of pristine and Eu-doped ZnO when used in dye sensitized solar cell as a photoanode.

13.11 Conclusion

Polycrystalline ZnO:Tb^{3+} and ZnO:Eu^{3+} thin films with high orientation were deposited on (001)Si, quartz, FTO, and ITO coated glass substrates by RF magnetron sputtering technique. The films showed excellent transparency and conductivity required for solar cell applications. X-ray diffraction (XRD), atomic force microscope (AFM) and the photoluminescence (PL) were employed to analyze the influence of the post-deposition annealing treatment on the structural properties of the ZnO:Tb^{3+} and ZnO:Eu^{3+} thin films. The films showed an increasing roughness and grain size of ZnO:Tb^{3+} with annealing temperature between 600–900°C and a shift of the diffraction peak position. As a proof of concept, hybrid OPV devices based on P3HT:PCBM blend were fabricated. Incorporation of ZnO and ZnO:Tb^{3+} thin films showed an increase in absorption within 300–400 nm wavelength range. The high absorption when ZnO:Tb^{3+}

was used showed that the films can be effectively used as a buffer layer and downconversion layer with enhancement of OPV device performance.

Down conversion has been applied to minimize thermalization losses in photovoltaic devices. In this work, Terbium doped ZnO ($ZnO:Tb^{3+}$) thin films were deposited on ITO coated glass, quartz, and silicon substrates using RF magnetron sputtering technique fitted with a high purity (99.99%) Tb-doped ZnO target (97% ZnO, 3% Tb) for use in organic solar cells as a bi-functional layer. A systematic study on the film crystallization dynamics was carried out through elevated temperature annealing in an Ar ambient. The films were characterized using grazing incidence (XRD), Rutherford backscattering spectrometry (RBS), atomic force microscopy, UV-visible transmittance and photoluminescence measurements using an excitation wavelength of 244 nm.

The tunability of the size and bandgap of $ZnO:Tb^{3+}$ nanocrystals with annealing exhibited quantum confinement effects which enable the control of emission characteristics in $ZnO:Tb^{3+}$. The energy transfer of ZnO → Tb^{3+} ($^5D_3$ - $^7F_5$) is also observed from the photoluminescence (PL) spectra. At an inter-band resonance excitation of around 300–400 nm, a typical emission band from the Tb^{3+} is obtained. The $ZnO:Tb^{3+}$ grown on ITO coated glass were then used as a bi-functional layer in an organic solar cell based on P3HT:PCBM blend as an active layer in an inverted device structure. The energy transfer through down conversion between ZnO and Tb^{3+} leads to enhanced absorption in P3HT:PCBM in the 300–400 nm range, and subsequently augments the J_{sc} of Tb^{3+} based device by 17%.

Eu-doped ZnO nanoparticles were also deposited using RF sputtering and the role of annealing on the modification of red photoluminescence was studied. From the PL spectra analysis, there is a correlation between the annealing temperature and the characteristics of the red photoluminescence, which suggests that the intrinsic defects can assist efficient energy transfer from the ZnO host to the Eu^{3+} ions. Five peaks in the region of $^5D_0$ – ^7F_J were observed, which indicate that the Eu^{3+} is located in the surface lattice sites of the ZnO materials. The DSSC with Eu-doped ZnO yielded a PCE of 0.56% compared to 0.52% obtained from DSSC with pristine ZnO without Eu produced indicating 7.7% efficiency enhancement which could be attributed to spectral conversion by the Eu-doped ZnO.

Acknowledgment

The authors would like to thank the Synchrotron techniques for Africa Research and Technology-Global Challenge Research Fund (START-GCRF), the University of the Witwatersrand, Material Physics Research Institute, School of Physics; the XRD and MMU facilities at Wits, NRF, and Material Energy Research Group (MERG) for funding.

References

1. V. Karpina, V. Lazorenko, C. Lashkarev, V. Dobrowolski, L. Kopylova, V. Baturin, S. Pustovoytov, A.J. Karpenko, S. Eremin, P. Lytvyn, Zinc oxide–analogue of GaN with new perspective possibilities, *Crystal Research and Technology*, 39 (2004) 980–992.
2. Ü. Özgür, Y.I. Alivov, C. Liu, A. Teke, M. Reshchikov, S. Doğan, V. Avrutin, S.-J. Cho, H. Morkoc, A comprehensive review of ZnO materials and devices, *Journal of Applied Physics*, 98 (2005) 11.
3. K. Nomura, H. Ohta, K. Ueda, T. Kamiya, M. Hirano, H. Hosono, Thin-film transistor fabricated in single-crystalline transparent oxide semiconductor, *Science*, 300 (2003) 1269–1272.
4. D. Thomas, The exciton spectrum of zinc oxide, *Journal of Physics and Chemistry of Solids*, 15 (1960) 86–96.
5. Y. Chen, D. Bagnall, H.-j. Koh, K.-t. Park, K. Hiraga, Z. Zhu, T. Yao, Plasma assisted molecular beam epitaxy of ZnO on c-plane sapphire: Growth and characterization, *Journal of Applied Physics*, 84 (1998) 3912–3918.
6. D. Reynolds, D.C. Look, B. Jogai, C. Litton, G. Cantwell, W. Harsch, Valence-band ordering in ZnO, *Physical Review B*, 60 (1999) 2340.
7. R.W. Chuang, R.-X. Wu, L.-W. Lai, C.-T. Lee, ZnO-on-GaN heterojunction light-emitting diode grown by vapor cooling condensation technique, *Applied Physics Letters*, 91 (2007) 231113.
8. R. Wang, C. Liu, J.-L. Huang, S.-J. Chen, Y.-K. Tseng, S.-C. Kung, ZnO nanopencils: Efficient field emitters, *Applied Physics Letters*, 87 (2005) 013110.
9. D.C. Look, D. Reynolds, J.W. Hemsky, R. Jones, J. Sizelove, Production and annealing of electron irradiation damage in ZnO, *Applied Physics Letters*, 75 (1999) 811–813.

10. S. Zafar, C. Ferekides, D. Morel, Characterization and analysis of ZnO: Al deposited by reactive magnetron sputtering, *Journal of Vacuum Science & Technology A: Vacuum, Surfaces, and Films*, 13 (1995) 2177–2182.

11. M. Kubon, E. Boehmer, F. Siebke, B. Rech, C. Beneking, H. Wagner, Solution of the ZnO/p contact problem in a-Si: H solar cells, *Solar Energy Materials and Solar Cells*, 41 (1996) 485–492.

12. T. Subramanyam, B.S. Naidu, S. Uthanna, Effect of substrate temperature on the physical properties of DC reactive magnetron sputtered ZnO films, *Optical Materials*, 13 (1999) 239–247.

13. Ü. Özgür, Y.I. Alivov, C. Liu, A. Teke, M. Reshchikov, S. Doğan, V. Avrutin, S. Cho, H. Morkoçd, A comprehensive review of ZnO materials and devices, *Journal of Applied Physics*, 98 (2005) 041301.

14. N.H. Nickel, E. Terukov, Zinc oxide - a material for micro- and optoelectronic applications, *Proceedings of the NATO Advanced Research Workshop on Zinc Oxide as a Material for Micro- and Optoelectronic Applications*, held in St. Petersburg, Russia, from 23 to 25 June 2004, Springer Science & Business Media, 2006.

15. C. Jagadish, S.J. Pearton, *Zinc Oxide Bulk, Thin Films and Nanostructures: Processing, Properties, and Applications*, Elsevier, 2011.

16. D.C. Reynolds, D.C. Look, B. Jogai, Optically pumped ultraviolet lasing from ZnO, *Solid State Communications*, 99 (1996) 873–875.

17. D. Bagnall, Y. Chen, M. Shen, Z. Zhu, T. Goto, T. Yao, Room temperature excitonic stimulated emission from zinc oxide epilayers grown by plasma-assisted MBE, *Journal of Crystal Growth*, 184 (1998) 605–609.

18. T. Yamamoto, T. Shiosaki, A. Kawabata, Characterization of ZnO piezoelectric films prepared by rf planar-magnetron sputtering, *Journal of Applied Physics*, 51 (1980) 3113–3120.

19. J. Molarius, J. Kaitila, T. Pensala, M. Ylilammi, Piezoelectric ZnO films by rf sputtering, *Journal of Materials Science: Materials in Electronics*, 14 (2003) 431–435.

20. R. Ondo-Ndong, G. Ferblantier, F. Pascal-Delannoy, A. Boyer, A. Foucaran, Electrical properties of zinc oxide sputtered thin films, *Microelectronics Journal*, 34 (2003) 1087–1092.

21. J.G. Gardeniers, Z. Rittersma, G. Burger, Preferred orientation and piezoelectricity in sputtered ZnO films, *Journal of Applied Physics*, 83 (1998) 7844–7854.

22. S. Talam, S.R. Karumuri, N. Gunnam, Synthesis, characterization, and spectroscopic properties of ZnO nanoparticles, *ISRN Nanotechnology*, 2012 (2012).

23. H. Nanto, H. Sokooshi, T. Usuda, Smell sensor using zinc oxide thin films prepared by magnetron sputtering, in: *Transducers '91: 1991 International Conference on Solid-State Sensors and Actuators. Digest of Technical Papers*, 1991, pp. 596–599.

24. M. C. Larciprete, D. Haertle, A. Belardini, M. Bertolotti, F. Sarto, P. Günter, Characterization of second and third order optical nonlinearities of ZnO sputtered films, *Applied Physics B*, 82 (2006) 431–437.

25. D.I. Florescu, L. Mourokh, F.H. Pollak, D.C. Look, G. Cantwell, X. Li, High spatial resolution thermal conductivity of bulk ZnO (0001), *Journal of Applied Physics*, 91 (2002) 890–892.

26. Z.O. Rediscovered, H. Brown, *The New Jersey Zinc Company*, New York, 1957.

27. H. Leverenz, *Luminescence of Solids*, ACS Publications, 1950.

28. K. Maeda, M. Sato, I. Niikura, T. Fukuda, Growth of 2 inch ZnO bulk single crystal by the hydrothermal method, *Semiconductor Science and Technology*, 20 (2005) S49.

29. D.C. Look, D.C. Reynolds, J. Sizelove, R. Jones, C.W. Litton, G. Cantwell, W. Harsch, Electrical properties of bulk ZnO, *Solid State Communications*, 105 (1998) 399–401.

30. D. Reynolds, C. Litton, D.C. Look, J. Hoelscher, B. Claflin, T. Collins, J. Nause, B. Nemeth, High-quality, melt-grown ZnO single crystals, *Journal of Applied Physics*, 95 (2004) 4802–4805.

31. H.K. Ardakani, Electrical conductivity of in situ "hydrogen-reduced" and structural properties of zinc oxide thin films deposited in different ambients by pulsed excimer laser ablation, *Thin Solid Films*, 287 (1996) 280–283.

32. F. Quaranta, A. Valentini, F.R. Rizzi, G. Casamassima, Dual-ion-beam sputter deposition of ZnO films, *Journal of Applied Physics*, 74 (1993) 244–248.

33. A. Janotti, C.G. Van de Walle, Fundamentals of zinc oxide as a semiconductor, *Reports on Progress in Physics*, 72 (2009) 126501.

34. F. Tuomisto, K. Saarinen, D.C. Look, G.C. Farlow, Introduction and recovery of point defects in electron-irradiated ZnO, *Physical Review B*, 72 (2005) 085206.

35. J. Muth, R. Kolbas, A. Sharma, S. Oktyabrsky, J. Narayan, Excitonic structure and absorption coefficient measurements of ZnO single crystal epitaxial films deposited by pulsed laser deposition, *Journal of Applied Physics*, 85 (1999) 7884–7887.

36. X. Teng, H. Fan, S. Pan, C. Ye, G. Li, Influence of annealing on the structural and optical properties of ZnO:Tb thin films, *Journal of Applied Physics*, 100 (2006) 053507.

37. N. Wu, Q. Luo, X. Qiao, C.-Q. Ma, The preparation of a Eu^{3+}-doped ZnO bi-functional layer and its application in organic photovoltaics, *Materials Research Express*, 2 (2015) 125901.

38. X. Huang, S. Han, W. Huang, X. Liu, Enhancing solar cell efficiency: The search for luminescent materials as spectral converters, *Chemical Society Reviews*, 42 (2013) 173–201.

39. P.K. Nayak, G. Garcia-Belmonte, A. Kahn, J. Bisquert, D. Cahen, Photovoltaic efficiency limits and material disorder, *Energy & Environmental Science*, 5 (2012) 6022–6039.

40. A.J. Nozik, J. Miller, *Introduction to Solar Photon Conversion*, ACS Publications, 2010.

41. H.A. Atwater, A. Polman, Plasmonics for improved photovoltaic devices, *Nature Materials*, 9 (2010) 205.

42. B. Richards, Enhancing the performance of silicon solar cells via the application of passive luminescence conversion layers, *Solar Energy Materials and Solar Cells*, 90 (2006) 2329–2337.

43. W. Shockley, H.J. Queisser, Detailed balance limit of efficiency of p-n junction solar cells, *Journal of Applied Physics*, 32 (1961) 510–519.

44. T. Trupke, M. Green, P. Würfel, Improving solar cell efficiencies by down-conversion of high-energy photons, *Journal of Applied Physics*, 92 (2002) 1668–1674.

45. T. Trupke, M. Green, P. Würfel, Improving solar cell efficiencies by up-conversion of sub-band-gap light, *Journal of Applied Physics*, 92 (2002) 4117–4122.

46. A. Meijerink, R. Wegh, P. Vergeer, T. Vlugt, Photon management with lanthanides, *Optical Materials*, 28 (2006) 575–581.

47. G. Blasse, B. Grabmaier, *Luminescent Materials*, Springer Science & Business Media, 2012.

48. K. Ogasawara, S. Watanabe, H. Toyoshima, M.G. Brik, First-principles calculations of $4f^n \rightarrow 4f^{n-1}5d$ transition spectra, in: *Handbook on the Physics and Chemistry of Rare Earths*, vol. 37, 2007, pp. 1–59.

49. A.L. Allenic, Structural, electrical and optical properties of p-type ZnO epitaxial films, Theses and Dissertations (2008).

50. M.M.M. Almotari, Fabrication and characterisation of zinc oxide thin films singly doped with trace amounts of rare earth materials, Theses and Dissertations (2013).

51. R. Reisfeld, Excited states and energy transfer from donor cations to rare earths in the condensed phase, in: *Rare Earths*, Springer, 1991, pp. 65–97.

52. J. Huang, S. Liu, B. Gao, T. Jiang, Y. Zhao, S. Liu, L. Kuang, X. Xu, Synthesis and optical properties of Eu^{3+} doped zno nanoparticles used for white light emitting diodes, *Journal of Nanoscience and Nanotechnology*, 14 (2014) 3052–3055.

53. L. Luo, F. Huang, G. Dong, Y. Wang, Z. Hu, J. Chen, White light emission and luminescence dynamics in Eu^{3+}/Dy^{3+} codoped ZnO nanocrystals, *Journal of Nanoscience and Nanotechnology*, 16 (2016) 619–625.

54. V. Mangalam, K. Pita, Energy transfer efficiency from ZnO-nanocrystals to Eu^{3+} ions embedded in SiO_2 film for emission at 614 nm, *Materials*, 10 (2017) 930.

55. B. Bendre, S. Mahamuni, Luminescence in ZnO quantum particles, *Journal of Materials Research*, 19 (2004) 737–740.

56. D. Chen, Y. Wang, E. Ma, F. Bao, Y. Yu, Luminescence of an Er^{3+}-doped glass matrix containing CdS quantum dots, *Scripta Materialia*, 55 (2006) 891–894.

57. J. Bang, H. Yang, P.H. Holloway, Enhanced luminescence of SiO_2:Eu^{3+} by energy transfer from ZnO nanoparticles, *The Journal of Chemical Physics*, 123 (2005) 084709.

58. M. Chong, A. Abiyasa, K. Pita, S. Yu, Visible red random lasing in Y_2O_3:Eu^{3+}/ZnO polycrystalline thin films by energy transfer from ZnO films to Eu^{3+}, *Applied Physics Letters*, 93 (2008) 151105.

59. Y. Yu, Y. Wang, D. Chen, P. Huang, E. Ma, F. Bao, Enhanced emissions of Eu^{3+} by energy transfer from ZnO quantum dots embedded in SiO_2 glass, *Nanotechnology*, 19 (2008) 055711.

60. V. Mangalam, K. Pita, C. Couteau, Study of energy transfer mechanism from ZnO nanocrystals to Eu^{3+} ions, *Nanoscale Research Letters*, 11 (2016) 73.

61. R. Verma, K. Kumar, S. Rai, Pulsed laser ablation synthesis of silver nanoparticles and their use in fluorescence enhancement of Tb^{3+}-doped aluminosilicate glass, *Solid State Communications*, 150 (2010) 1947–1950.

62. S.-M. Liu, F.-Q. Liu, H.-Q. Guo, Z.-H. Zhang, Z.-G. Wang, Correlated structural and optical investigation of terbium-doped zinc oxide nanocrystals, *Physics Letters A*, 271 (2000) 128–133.

63. S.-M. Liu, F.-Q. Liu, Z.-G. Wang, Relaxation of carriers in terbium-doped ZnO nanoparticles, *Chemical Physics Letters*, 343 (2001) 489–492.

64. T. Jia, W. Wang, F. Long, Z. Fu, H. Wang, Q. Zhang, Synthesis, characterization and luminescence properties of Y-doped and Tb-doped ZnO nanocrystals, *Materials Science and Engineering: B*, 162 (2009) 179–184.

65. S. Gao, H. Zhang, R. Deng, X. Wang, D. Sun, G. Zheng, Engineering white light-emitting Eu-doped ZnO urchins by biopolymer-assisted hydrothermal method, *Applied Physics Letters*, 89 (2006) 123125.

66. J. Park, A. Steckl, Demonstration of a visible laser on silicon using Eu-doped GaN thin films, *Journal of Applied Physics*, 98 (2005) 056108.

67. H. Shahroosvand, M. Ghorbani-asl, Solution-based synthetic strategies for Eu doped ZnO nanoparticle with enhanced red photoluminescence, *Journal of Luminescence*, 144 (2013) 223–229.

68. C. Pan, C. Chen, J. Chen, P. Huang, G. Chi, C. Chang, F. Ren, S. Pearton, Optical and structural properties of Eu-diffused and doped ZnO nanowires, *Applied Surface Science*, 256 (2009) 187–190.

69. Z. Fang, Z. Yan, Y. Tan, X. Liu, Y. Wang, Influence of post-annealing treatment on the structure properties of ZnO films, *Applied Surface Science*, 241 (2005) 303–308.

70. W.-C. Shih, M.-S. Wu, Growth of ZnO films on GaAs substrates with a SiO_2 buffer layer by RF planar magnetron sputtering for surface acoustic wave applications, *Journal of Crystal Growth*, 137 (1994) 319–325.

71. M. Puchert, P. Timbrell, R. Lamb, Postdeposition annealing of radio frequency magnetron sputtered ZnO films, *Journal of Vacuum Science & Technology A: Vacuum, Surfaces, and Films*, 14 (1996) 2220–2230.

72. F. Otieno, M. Airo, R.M. Erasmus, D.G. Billing, A. Quandt, D. Wamwangi, Structural and spectroscopic analysis of ex-situ annealed RF sputtered aluminium doped zinc oxide thin films, *Journal of Applied Physics*, 122 (2017) 075303.

73. B.-Y. Oh, M.-C. Jeong, W. Lee, J.-M. Myoung, Properties of transparent conductive ZnO:Al films prepared by co-sputtering, *Journal of Crystal Growth*, 274 (2005) 453–457.

74. V. Gupta, A. Mansingh, Influence of postdeposition annealing on the structural and optical properties of sputtered zinc oxide film, *Journal of Applied Physics*, 80 (1996) 1063–1073.

75. H. Wahab, A. Salama, A. El-Saeid, O. Nur, M. Willander, I. Battisha, Optical, structural and morphological studies of (ZnO) nano-rod thin films for biosensor applications using sol gel technique, *Results in Physics*, 3 (2013) 46–51.

76. J. Chen, J. Chen, D. Chen, Y. Zhou, W. Li, Y. Ren, L. Hu, Electrochemical deposition of Al-doped ZnO transparent conducting nanowire arrays for thin-film solar cell electrodes, *Materials Letters*, 117 (2014) 162–164.

77. B. Kang, F. Ren, Y. Heo, L. Tien, D. Norton, S. Pearton, pH measurements with single ZnO nanorods integrated with a microchannel, *Applied Physics Letters*, 86 (2005) 112105.

78. H. Yoshikawa, S. Adachi, Optical constants of ZnO, *Japanese Journal of Applied Physics*, 36 (1997) 6237.

79. T. Tansley, D. Neely, C. Foley, Optical dispersion in zinc oxide, *Physica Status Solidi (A)*, 77 (1983) 491–496.

80. G. Heiland, E. Mollwo, F. Stöckmann, Electronic processes in zinc oxide, *Solid State Physics*, 8 (1959) 191–323.

81. Y. Kong, D. Yu, B. Zhang, W. Fang, S. Feng, Ultraviolet-emitting ZnO nanowires synthesized by a physical vapor deposition approach, *Applied Physics Letters*, 78 (2001) 407–409.

82. Y.-C. Lee, S.-Y. Hu, W. Water, K.-K. Tiong, Z.-C. Feng, Y.-T. Chen, J.-C. Huang, J.-W. Lee, C.-C. Huang, J.-L. Shen, Rapid thermal annealing effects on the structural and optical properties of ZnO films deposited on Si substrates, *Journal of Luminescence*, 129 (2009) 148–152.

83. V. Kumar, O.M. Ntwaeaborwa, H.C. Swart, Deep level defect correlated emission and Si diffusion in ZnO:Tb^{3+} thin films prepared by pulsed laser deposition, *Journal of Colloid and Interface Science*, 465 (2016) 295–303.

84. S.-M. Liu, F.-Q. Liu, Z.-G. Wang, Relaxation of carriers in terbium-doped ZnO nanoparticles, *Chemical Physics Letters*, 343 (2001) 489–492.

85. E. Bylander, Surface effects on the low-energy cathodoluminescence of zinc oxide, *Journal of Applied Physics*, 49 (1978) 1188–1195.

86. F. Otieno, B.K. Mutuma, M. Airo, K. Ranganathan, R. Erasmus, N. Coville, D. Wamwangi, Enhancement of organic photovoltaic device performance via P3HT:PCBM solution heat treatment, *Thin Solid Films*, 625 (2017) 62–69.

87. F. Otieno, M. Airo, K. Ranganathan, D. Wamwangi, Annealed silver-islands for enhanced optical absorption in organic solar cell, *Thin Solid Films*, 598 (2016) 177–183.

88. E. Kótai, Computer methods for analysis and simulation of RBS and ERDA spectra, *Nuclear Instruments and Methods in Physics Research Section B: Beam Interactions with Materials and Atoms*, 85 (1994) 588–596.

89. P.P. Pal, J. Manam, Structural and photoluminescence studies of Eu^{3+} doped zinc oxide nanorods prepared by precipitation method, *Journal of Rare Earths*, 31 (2013) 37–43.

90. R. Swapna, T. SrinivasaReddy, K. Venkateswarlu, M.S. Kumar, Effect of post-annealing on the properties of Eu doped ZnO nano thin films, *Procedia Materials Science*, 10 (2015) 723–729.

91. V. Kumar, S. Som, V. Kumar, V. Kumar, O. Ntwaeaborwa, E. Coetsee, H. Swart, Tunable and white emission from $ZnO{:}Tb^{3+}$ nanophosphors for solid state lighting applications, *Chemical Engineering Journal*, 255 (2014) 541–552.

92. L. Armelao, F. Heigl, S. Brunet, R. Sammynaiken, T. Regier, R.I. Blyth, L. Zuin, R. Sankari, J. Vogt, T.K. Sham, The origin and dynamics of soft X-ray-excited optical luminescence of ZnO, *ChemPhysChem*, 11 (2010) 3625–3631.

93. M. Wang, C. Huang, Z. Huang, W. Guo, J. Huang, H. He, H. Wang, Y. Cao, Q. Liu, J. Liang, Synthesis and photoluminescence of Eu-doped ZnO microrods prepared by hydrothermal method, *Optical Materials*, 31 (2009) 1502–1505.

94. X. Zeng, J. Yuan, Z. Wang, L.-d. Zhang, Nanosheet-based microspheres of Eu^{3+}-doped ZnO with efficient energy transfer from ZnO to Eu^{3+} at room temperature, *Advanced Materials*, 19 (2007) 4510–4514.

95. Y. Liu, W. Luo, R. Li, G. Liu, M.R. Antonio, X. Chen, Optical spectroscopy of Eu^{3+} doped ZnO nanocrystals, *The Journal of Physical Chemistry C*, 112 (2008) 686–694.

96. Y.-K. Park, J.-I. Han, M.-G. Kwak, H. Yang, S.-H. Ju, W.-S. Cho, Time-resolved spectroscopic study of energy transfer in $ZnO{:}EuCl_3$ phosphors, *Journal of Luminescence*, 78 (1998) 87–90.

97. T.-H. Fang, Y.-S. Chang, L.-W. Ji, S.D. Prior, W. Water, K.-J. Chen, C.-F. Fang, C.-N. Fang, S.-T. Shen, Photoluminescence characteristics of ZnO doped with Eu^{3+} powders, *Journal of Physics and Chemistry of Solids*, 70 (2009) 1015–1018.

98. V. Natarajan, M. Bhide, A. Dhobale, S. Godbole, T. Seshagiri, A. Page, C.-H. Lu, Photoluminescence, thermally stimulated luminescence and electron paramagnetic resonance of europium-ion doped strontium pyrophosphate, *Materials Research Bulletin*, 39 (2004) 2065–2075.

99. P. Dorenbos, L. Pierron, L. Dinca, C. Van Eijk, A. Kahn-Harari, B. Viana, 4f–5d spectroscopy of Ce^{3+} in $CaBPO_5$, $LiCaPO_4$ and Li_2CaSiO_4, *Journal of Physics: Condensed Matter*, 15 (2003) 511.

100. S. Lima, F. Sigoli, M.R. Davolos, M. Jafelicci Jr, Europium (III)-containing zinc oxide from Pechini method, *Journal of Alloys and Compounds*, 344 (2002) 280–284.

101. Y. Liu, W. Luo, R. Li, X. Chen, Spectroscopic evidence of the multiple-site structure of Eu^{3+} ions incorporated in ZnO nanocrystals, *Optics Letters*, 32 (2007) 566–568.

102. H.-Q. Wang, M. Batentschuk, A. Osvet, L. Pinna, C.J. Brabec, Rare-earth ion doped up-conversion materials for photovoltaic applications, *Advanced Materials*, 23 (2011) 2675–680.

103. R. Nikifor, B.L. Whualkuer, F.L. Edilson, G.B. Renato, M.S. Glauco, de A.B. Cid, Three- and four-photon excited up-conversion luminescence in terbium doped lutetium silicate powders by femtosecond laser irradiation, *Optical Materials Express*, 3 (2013) 1803–1809.

104. O. Kluth, G. Schlöpe, J. Hüpkes, C. Agashe, J. Müller, B. Rech, Modified Thornton model for magnetron sputtered zinc oxide: film structure and etching behavior, *Thin Solid Films*, 442 (2003) 80–85.

Chapter 14

The Analysis of Collocated Glow-Peaks in the Thermoluminescence of Superluminous Phosphors

Makaiko Chithambo
Department of Physics and Electronics, Rhodes University, Makhanda, South Africa
m.chthambo@ru.ac.za

The thermoluminescence glow-curves of many superluminous phosphors are usually dominated by an apparently-single high intensity peak. We consider the analysis of such peaks by using examples drawn from the kinetic analysis of thermoluminescence of beta-irradiated phosphors including $SrAl_2O_4:Eu^{2+}$, Dy^{3+}, $Sr_4Al_{14}O_{25}:Eu^{2+}$, Dy^{3+}, and $Ca_5(PO_4)_3OH:Gd^{3+}$, Pr^{3+} although much of the discussion centres on the first case in this list. Using a number of tests including partial dynamic-heating, isothermal heating, phosphorescence, and the effect of fading, one can show that the main peak in particular and the glow-curve in general consists of a set of closely spaced peaks. Information from the experimental separation and analysis of the peaks can be complemented by data

Luminescent Nanomaterials
Edited by Odireleng Martin Ntwaeaborwa

ISBN 978-981-4968-11-9 (Hardcover), 978-1-003-27738-5 (eBook)
www.jennystanford.com

obtained by studying the prominent peak as though it were a *bona-fide* single one. We show how a model based on density of energy states can be helpful in accounting for any differences between the qualitative and quantitative results.

14.1 Introduction

Thermoluminescence (TL) is a sensitive method for monitoring changes in concentration of point-defects in insulators. The basis of TL is that intrinsic or extrinsic defects can act as charge traps or luminescence centres. Ionizing radiation can move electrons or holes to these trapping sites. The subsequent release of the trapped electrons by heating at a controlled rate can be recorded as a temperature-dependent set of luminescence signals termed a glow curve.

The preceding summary of thermoluminescence assumes the presence of localised and discrete energy levels within the band gap. Because the study of thermoluminescence relies on kinetic analysis, a common practical concern is the analysis of glow-peaks for kinetic parameters. One of these is the activation energy or the energy depth of a localized energy level (serving as an electron trap) below the conduction band. Most methods developed for this purpose only apply for isolated single peaks. It follows then that if a method applicable to an isolated peak is used on a peak that is not, the results may be unreliable. Indeed, simply associating a prominent peak with a single discrete electron trap potentially disregards other peaks within the peaks-profile and may lead to incomplete or erroneous conclusions about the mechanisms involved in the thermoluminescence. It is thus essential to develop and understand procedures that can be used in cases where multiple secondary peaks are concealed within a dominant peak to such an extent that the collection appears singular.

The question of how to deal with collocated glow-peaks is the subject of this chapter. Although there are many examples of such experimental structure, for example in $SrAl_2O_4:Eu^{2+}$, Dy^{3+} [1, 2], $Sr_2SiO_4:Eu^{2+}$ [3], $Ca_5(PO_4)_3OH:Gd^{3+}$, Pr^{3+} [4], and a number of other phosphors including $CaAl_2O_4:Eu^{2+}$, Nd^{3+} [5], we will mainly be concerned with the thermoluminescence of $SrAl_2O_4:Eu^{2+}$, Dy^{3+}, a well-known phosphor first reported on by Matsuzawa et al. [6].

This chapter is limited to the kinetic analysis of the prominent peak in $SrAl_2O_4:Eu^{2+}$, Dy^{3+}. We discuss methods for separating closely-spaced peaks as a means to analyse a composite peak. For comparison, we also show results obtainable when such a peak is analysed as if it were a genuinely single one.

14.2 Experimental Methods

Thermoluminescence reported was measured on a RISØ TL/OSL DA-20 Luminescence Reader. The luminescence was detected by an EMI 9235QB photomultiplier tube through a stack of a BG-39 bandpass filter (transmission band ~335–610 nm FWHM) and a BG-3 bandpass filter (transmission band ~250–600 nm FWHM). Samples of $SrAl_2O_4:Eu^{2+}$, Dy^{3+} in powder form were placed on stainless steel discs of diameter 10 mm and thickness 0.3 mm and irradiated to required dose using a $^{90}Sr/^{90}Y$ beta source at 0.10 Gy s^{-1}.

Radioluminescence (RL) where the luminescence is recorded as a function of the emission wavelength, was used to identify emission bands of stimulated luminescence in $SrAl_2O_4:Eu^{2+}$, Dy^{3+}. RL spectra were measured at ~1 mbar vacuum by heating to 400°C at 10 K min^{-} using a high sensitivity spectrometer [7]. Samples were irradiated using a Philips MCN-101 X-ray tube at a dose rate of 1.8 Gy min^{-1}.

14.3 Radioluminescence

Figure 14.1 shows RL spectra of $SrAl_2O_4:Eu^2$, Dy^{3+} [2]. The isometric plot is shown in Fig. 14.1a and the corresponding contour map in Fig. 14.1b. Two emissions stand out, one at 475 nm and a more intense one at 575 nm. Using the intensity scale as a guide, one notes evidence of increasing non-radiative transitions up to ~75°C in the emission at 475 nm and up to 125°C at 575 nm. There are also less intense emissions near 405, 510, 600, 610, and 660 nm. What is shown in Fig. 14.1 is stimulated luminescence which should be distinguished from spontaneously emitted emissions such as fluorescence.

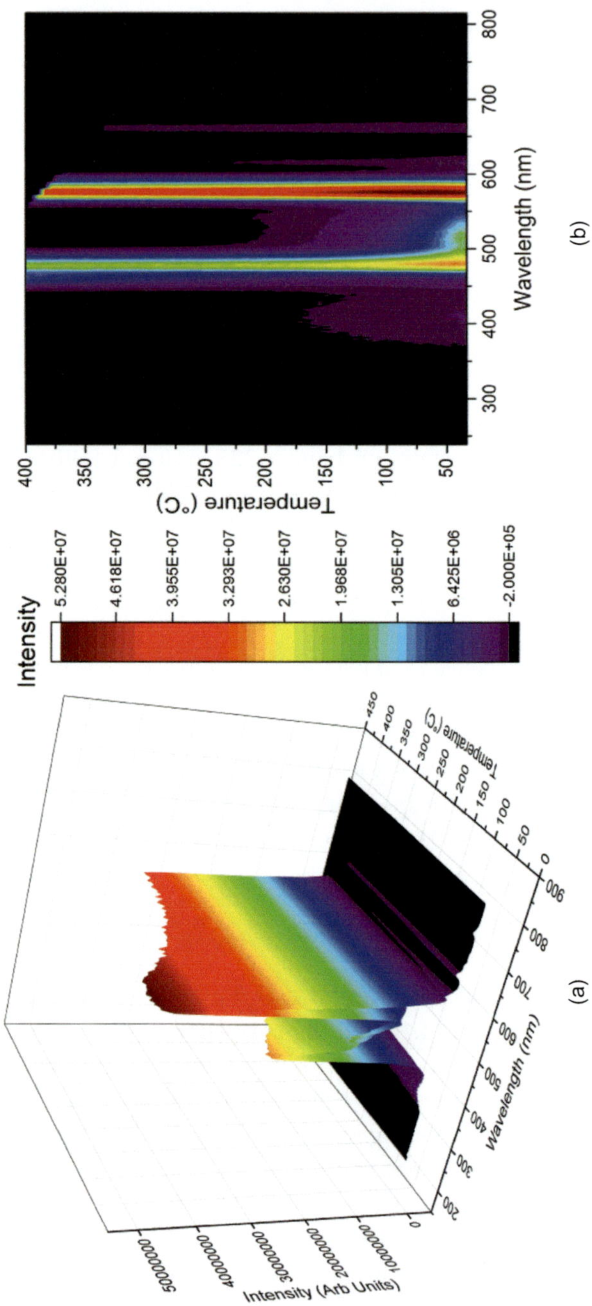

Figure 14.1 Temperature-dependence of RL spectra (a) and its contour map (b). Reproduced from Ref. [1] with permission from Elsevier.

14.4 Features of the Thermoluminescence Glow Curve

14.4.1 Glow Curve Features

Figure 14.2 shows a glow-curve measured at 1°C s^{-1} after irradiation to 20 Gy. There is a prominent and apparently single peak at 34°C. The signal before the onset of thermoluminescence is phosphorescence whereas that beyond 400°C is due to blackbody radiation.

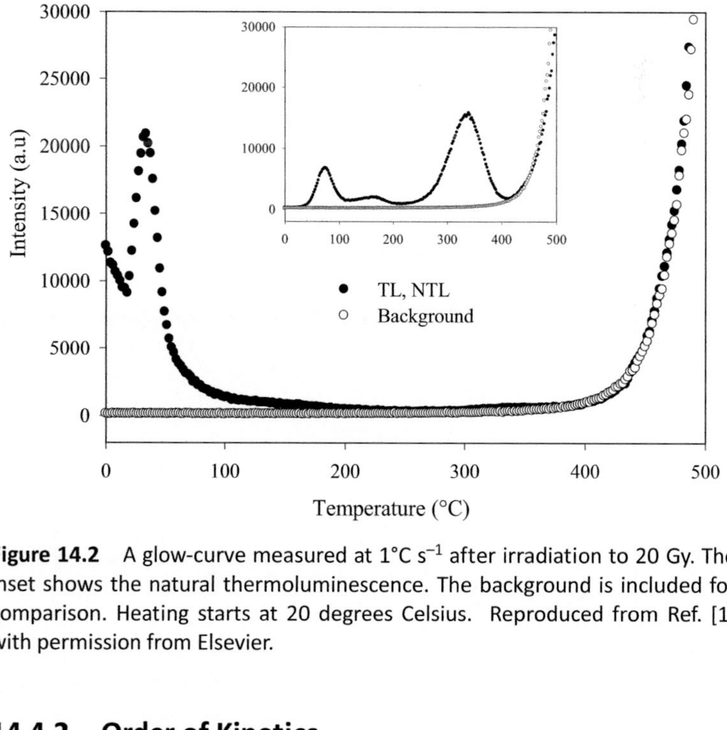

Figure 14.2 A glow-curve measured at 1°C s^{-1} after irradiation to 20 Gy. The inset shows the natural thermoluminescence. The background is included for comparison. Heating starts at 20 degrees Celsius. Reproduced from Ref. [1] with permission from Elsevier.

14.4.2 Order of Kinetics

An important first step in kinetic analysis is to determine the order of kinetics of the peak(s) in question. For the apparently single peak of Fig. 14.2, the T_m – T_{stop} procedure was used to do so. Three sets of measurement corresponding to irradiation to 20, 60, and 100 Gy were made. An irradiated sample was partially heated from 20 to 22°C and after cooling to room temperature, re-heated to 500°C to obtain the complete glow-curve. The position T_m of the resultant peak was noted. The procedure was repeated with T_{stop} incremented in turn by 2°C up to 60°C.

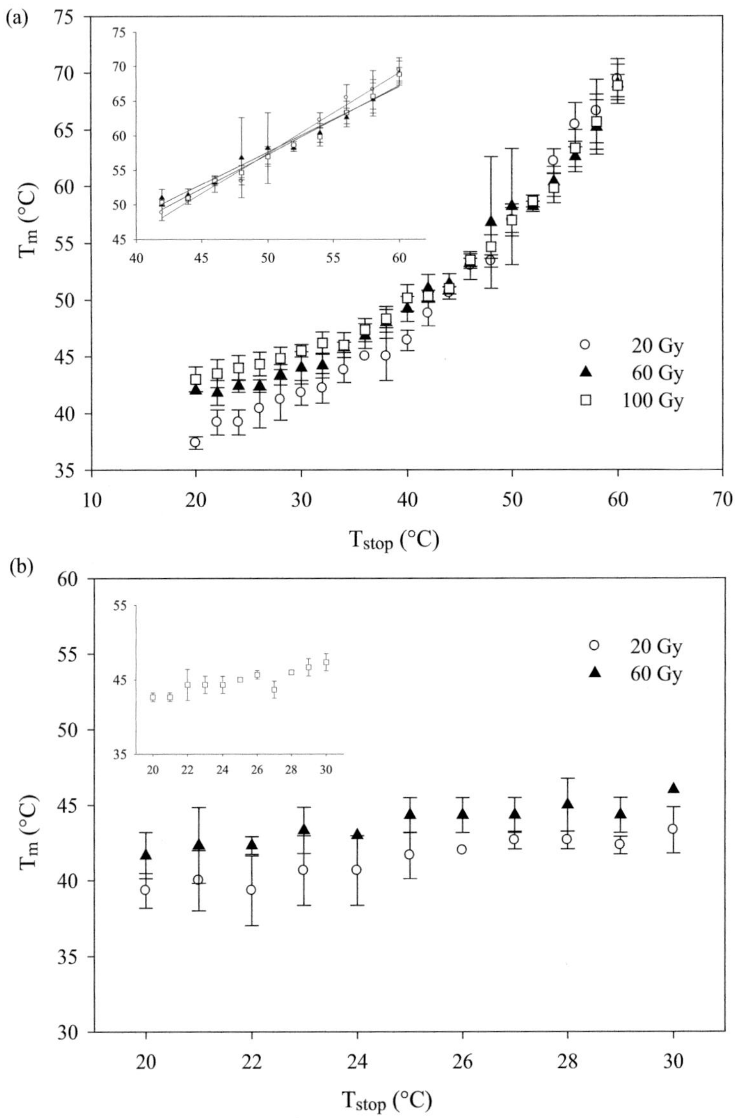

Figure 14.3 T_m against T_{stop} for measurements corresponding to beta irradiation to 20, 60, and 100 Gy. Each data point is an average of five and the margin of error is the standard deviation of the set (a) similar measurements corresponding to 20, 60, and 100 Gy (inset) between 20 and 30°C made at 1°C intervals (b). Reprinted from Ref. [1] with permission from Elsevier.

Figure 14.3 shows the peak position T_m against T_{stop} for the three sets of measurement. The position T_m increases with T_{stop} but does so non-monotonically. Between 20 and 50°C, the change of T_m with T_{stop} is glacial. Indeed, for each dose, there are regions where T_m is independent of T_{stop}. This is exemplified by the portion from 20 through 26°C for measurements corresponding to 60 Gy (solid triangles). It is instructive to consider the detailed dependence of T_m on T_{stop} between 20 and 30°C as illustrated in Fig. 14.3b for measurements at 1°C intervals on another test sample. This shows that as T_m increases with T_{stop}, it initially passes through regions in which it is independent of T_{stop}. On the basis of TL theory [8, 9], each such region corresponds to a separate TL peak.

Closer examination by linear regression of the dependence of T_m on T_{stop} between 42 and 60°C (Fig. 14.3a, inset), reveals that the slopes corresponding to 20, 60, and 100 Gy are 1.16 ± 0.04, 0.93 ± 0.05, and 1.02 ± 0.05. A linear increase with a slope close to 1, as found, is expected for a quasi-continuum set of first or second-order peaks [9]. The increase of T_m with T_{stop} in this example therefore suggests that the apparently single peak is a composite and consists of multiple components. However, since this result is consistent with either of first or second-order kinetics, further tests are necessary to settle this question. These will be described.

14.4.3 Influence of Irradiation on Peak Position

The dependence of T_m on dose is also characteristic of the order of kinetics and complements the T_m – T_{stop} procedure described in the preceding section. The peak temperature of a first-order peak is independent of irradiation dose whereas that of a second-order decreases with dose [8].

Figure 14.4 shows a plot of T_m against dose. The peak position increases with dose but through regions in which it is uninfluenced by dose. This is further evidence that the apparently single peak of Fig. 14.2 consists of multiple first-order peaks which individually become prominent at specific irradiation doses.

The increase of peak position with dose, which has also been reported in various cases, for example in $Ca_5(PO_4)_3OH:Gd^{3+}, Pr^{3+}$ [4] or in γ-irradiated $SrAl_2O_4:Eu^{2+}, Dy^{3+}$ [10] is somewhat of an anomaly as it cannot be accounted for by either first or second-order kinetics.

The increase of T_m with T_{stop} in Fig. 14.3 shows that the main peak being a composite. Now, since a first-order peak is independent of dose, the presence of regions where T_m is independent of dose, as also observed in $Sr_4Al_{14}O_{25}:Eu^{2+}, Dy^{3+}$ [11], suggests that the peak is made up of multiple first-order components each of which becomes dominant at specific doses. This leads to a plot where T_m appears to increase with dose. For some doses, the glow-curve is dominated by a particular component with a particular T_m and the result is a graph where the position appears to be independent of dose until the onset of another higher dose range. The interpretations of Figs. 14.3 and 14.4 are qualitative and are built on standard theories of thermoluminescence. It is necessary to go beyond this and experimentally examine the conclusion drawn that the apparently single peak of Fig. 14.2 consists of several overlapping components.

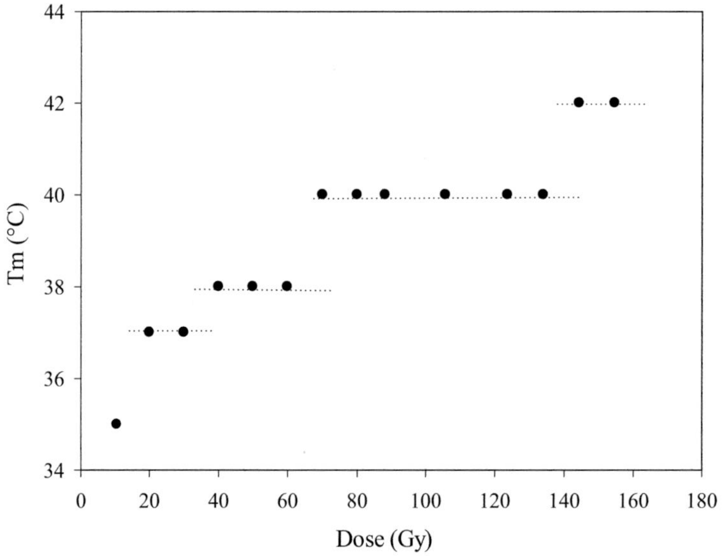

Figure 14.4 The dependence of T_m on irradiation dose. The dashed lines are only visual guides. Reprinted from Ref. [1] with permission from Elsevier.

14.5 Glow-Curve Resolution

The dependence of peak position on partial heating as well as on irradiation dose as presented thus far implies that the prominent peak of Fig. 14.2 consists of collocated components. By this, we mean several components embedded within each other to the extent that the product appears as single. We assume that the characteristics of the peak corresponding to a particular dose represent the most dominant component. We proceed to explore this idea experimentally using three methods namely, partial heating, isothermal heating, and by monitoring the influence of fading on the peak.

14.5.1 Glow-Curve Resolution by Partial Heating

The aim of this approach is to investigate the presence of any components of the main peak and to isolate any other (not necessarily to find *all*) throughout the glow-curve. To do this, a sample irradiated to 20 Gy was partially heated to 40°C at 1.0°C/s to remove the dominant peak at 34°C. After cooling to room temperature, the sample was heated again to 400°C to monitor the presence and position of the first remnant peak. The measurements were repeated with the preheating temperature increased by 5°C each time up to 270°C.

The results of the partial heating experiments are shown in Fig. 14.5. Figure 14.5a shows peaks observed after partial heating as described. As each resultant peak is removed by partial heating, previously concealed ones appear. Each new peak after partial heating is itself made up of collocated components.

Figure 14.5b shows all peaks obtained corresponding to 5°C intervals in the partial heating. What appears as a single peak in Fig. 14.2 may indeed be a collection of closely spaced ones. This is a collocated peak. In addition, the presence of any peaks beyond 240°C is not obvious. However, the same procedure reveals distinct peaks at 190°C and 330°C after preheating to 180 and 270°C (inset). Figure 14.5 experimentally verifies that both the main peak in particular and the glow-curve as a whole consist of closely-spaced peaks.

14.5.2 Glow-Curve Resolution Using Isothermal Heating

Where a peak is collocated, its components cannot be disaggregated using thermal cleaning since this method relies on heating beyond an identified peak-maximum to reveal the adjacent peak. Instead, the glow-curve was also resolved into its components using isothermal heating. The basis of this method is that if a sample is isothermally heated for a time at a temperature corresponding to a position on the rising edge of a peak, the heating essentially removes the first peak in the series to reveal the next one it is collocated with [3].

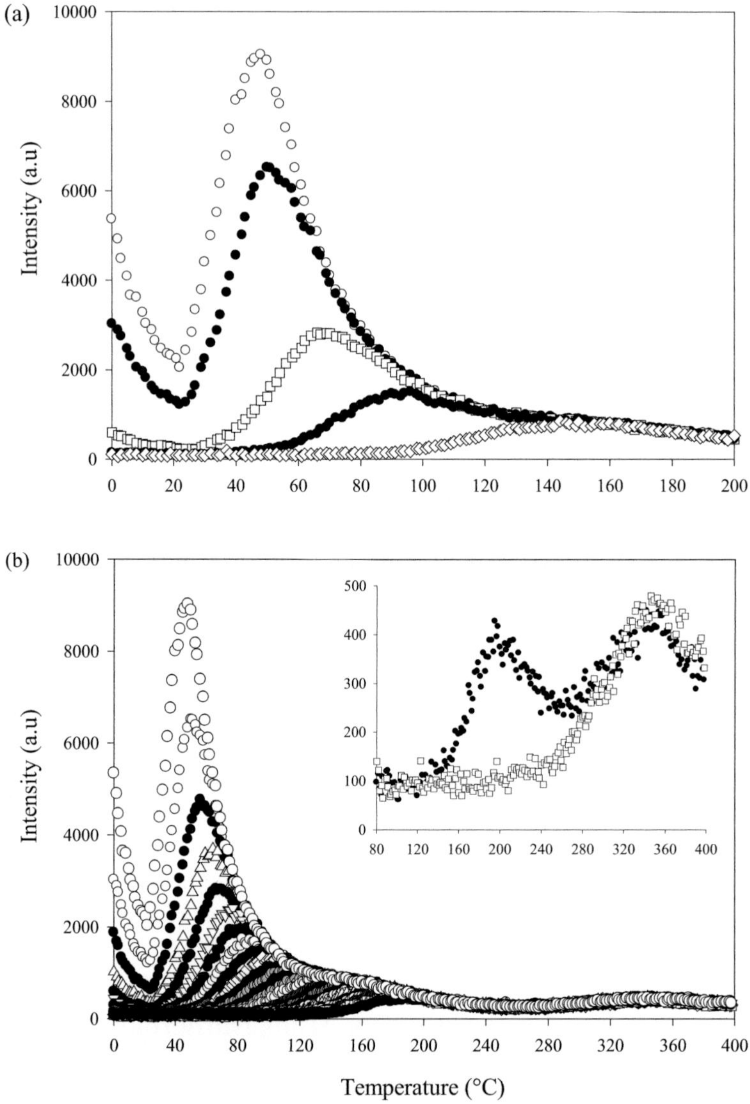

Figure 14.5 Glow-peaks observed after partial heating to 40, 45, 60, 80, and 145°C (a) the complete set of such glow-peaks and, in the inset, ones found following preheating to 180 and 270°C (b). Reproduced from Ref. [1] with permission from Elsevier.

Figure 14.6 shows examples of peaks obtained using isothermal-heating at 80, 130, 240, 330, and 380°C for 100 s respectively after irradiation to 100 Gy in each case. This is further evidence that the glow-curve consists of closely-spaced peaks.

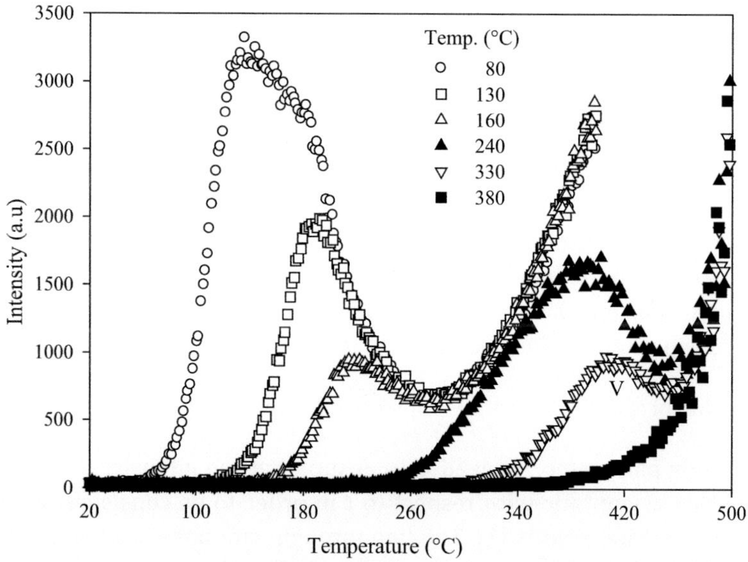

Figure 14.6 Some glow-curve components obtained using isothermal-heating at 80, 130, 240, 330, and 380°C for 100 s respectively with each corresponding to irradiation to 100 Gy. Reprinted from Ref. [1] with permission from Elsevier.

14.5.3 Fading as a Means for Glow-Curve Resolution

The influence of delay between irradiation and measurement on the thermoluminescence intensity can be used to study the possibility that an apparently single peak consists of multiple closely-spaced components.

Figure 14.7a shows the influence of delayed thermal stimulation on the main peak. Only the portion up to 120°C is shown to better illustrate the result. The luminescence intensity fades with the delay and the apparent peak-position increases (inset). Such fading cannot be accounted for by thermally-induced loss of charge from one electron trap. If this were so, the intensity would decrease but the peak-position would either be unaffected or increase. The shift of as much as 40°C here is improbable for a single peak. We ascribe the change to loss of signal from multiple electron traps decaying at different rates. What happens then is that as a less stable peak at a lower temperature fades, a subsequent one at a higher temperature in the series appears. In this way, Fig. 14.7a is additional evidence that the glow-curve consists of multiple closely-set peaks.

The peak fades due to phosphorescence. Phosphorescence is the emission of thermoluminescence as a function of time at a constant temperature. Its intensity at a particular temperature is exponentially proportional to time for first order kinetics but scales inversely as $[1+(b-1)p''n_o^{b-1}t]^{b/(b-1)}$ for general-order kinetics. Here b is the order of kinetics and p'' is the decay constant. Figure 14.7b shows the phosphorescence measured after irradiation to 100 Gy. The phosphorescence here does not follow first- or general-order kinetics as may be expected neither could it be described forms of the experimental Becquerel decay law. Instead, an empirical model of the form $I(t) = A\exp\left(\frac{\kappa}{t+\gamma}\right)$ sufficed where A is a constant and the scaling parameter γ is needed to ensure that the intensity $I(t)$ does not go asymptotic with respect to t in order to be consistent with experimental results [1]. The line through data points in Fig. 14.7b is the best fit of this expression which like the Becquerel law, is empirical. The relevant point here is that the phosphorescence that causes fading of the apparently-single peak cannot be modelled on the basis of a single electron trap.

Figure 14.7c shows how the glow-curve fades owing to phosphorescence by comparing a glow-curve recorded immediately after irradiation (open circles) with one measured after 30000 s. For ease of comparison, the intensity of the initial glow-curve has been scaled down by 96%. The main peak fades completely by 30000 s after irradiation. In contrast, peaks near and beyond 100°C are less affected. Thus the phosphorescence of Fig. 14.7b is due to loss of electrons from electron traps responsible for the main peak. Such charge depletion from multiple electron traps explains why the phosphorescence cannot be modelled as a simple exponential decay. To examine this assumption further, phosphorescence was measured from a sample where the main peak had been removed by preheating to 100°C after irradiation to 100 Gy. The inset to Fig. 14.7c shows the result where it is apparent that no phosphorescence at all is observed in the absence of the main peak. Figures 14.7b and c is further incidental evidence that the main peak consists of several closely spaced components.

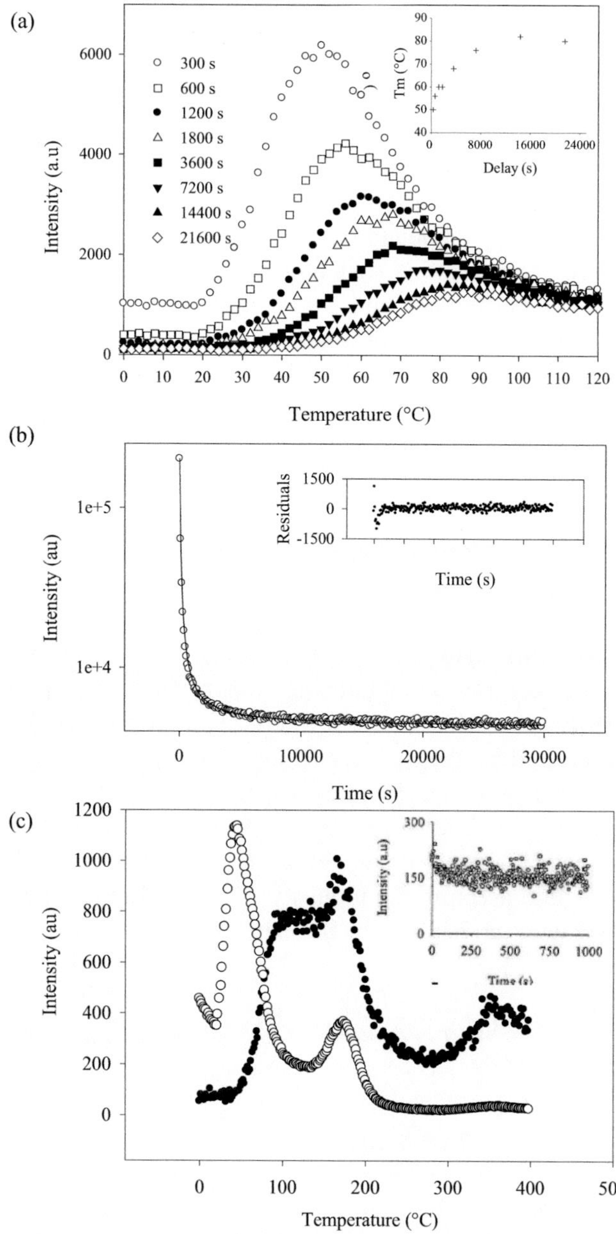

Figure 14.7 The influence of delayed stimulation on the thermoluminescence (a) Phosphorescence measured for 30000 s following beta irradiation to 100 Gy. (b) Glow-curves measured immediately after irradiation (×0.04) and after 3000 s; the inset shows a measurement after preheating to 100°C intended to remove the dominant peak (c). Reproduced from Ref. [1] with permission from Elsevier.

14.6 Kinetic Analysis

We have shown that the prominent thermoluminescence peak consists of closely-spaced components. Conventional methods for dealing with thermoluminescence peaks apply to either isolated peaks or ones with adjacent peak maxima. Apart from extracting and analysing the components separately, the question that needs to be addressed is whether such methods can reliably be applied to analyse a composite peak and if so how this can be done and the result interpreted. In view of this, we apply kinetic analysis in two ways; using methods that totally avoid or minimise the influence of one peak on another for example peak-position techniques and, for comparison, methods that exploit the whole peak thereby assuming that the peak is genuinely single.

14.6.1 Curve-Fitting of the Main Peak as a Supposed Single Peak

To start with, we analyse the peak as though it is simple and single to demonstrate results obtainable under such an assumption. Preparatory tests described elsewhere [1] showed that the main peak is affected by thermal quenching. The background theory is omitted but is discussed elsewhere e.g. Refs. [8, 9].

The thermoluminescence was analysed on the basis of general-order kinetics using the model of Kitis [12]. Taking thermal quenching into account, the intensity $I(T)$ of the apparently-single main peak with can be fitted by the expression

$$I(T)=\left\{I_{\mathrm{m}}b^{\frac{b}{b-1}}\exp\left(\frac{E}{kT}\frac{T-T_{\mathrm{m}}}{T_{\mathrm{m}}}\right)\left[(b-1)(1-\Delta)\frac{T^2}{T_{\mathrm{m}}^2}\exp\left(\frac{E}{kT}\frac{T-T_{\mathrm{m}}}{T_{\mathrm{m}}}\right)\right.\right.$$

$$\left.\left.+Z_{\mathrm{m}}\right]^{\frac{-b}{b-1}}\right\}\eta(T)+\sum_{i=1}^{i=3}f_i(T), \qquad (14.1)$$

where $f(T)$ is the same term within the braces, I_{m} is the peak maximum, T_{m} the corresponding peak temperature, $\Delta = 2kT/E$, $\Delta_{\mathrm{m}} = 2kT_{\mathrm{m}}E$, and $Z_{\mathrm{m}} = 1 + (b - 1)\Delta_{\mathrm{m}}$ and $n(T)$ is the luminescence efficiency given by $n(T) = 1/[1 + C\exp(-\Delta E/kT)]$. The best fit of Eq. (14.1) for a sample irradiated to 60 Gy gave $E = 0.89 \pm 0.01$ eV and $b = 1.51 \pm 0.17$. The frequency factor was evaluated as $3.0 \times 10^{13}\text{s}^{-1}$.

To determine whether the activation energy has any systematic dependence on dose, the curve fitting was carried out on glow-curves corresponding to a range of doses (Fig. 11 in Ref. [1]). The activation energy is of the order of 0.90 eV and $b \sim 1.5$ without any systematic dependence on dose. Thus if the peak is analysed using curve-fitting assuming that it is a simple single peak, the order of kinetics is found to be general-order rather than first-order as predicted by the $T_m - T_{stop}$ and T_m – dose methods.

14.6.2 Curve-Fitting of Digitally Mustered Components

Components of the main peak recovered by digital mustering [3] can likewise be analysed using curve fitting as explained. The purpose here is to digitally separate peaks and obtain their complete shape to facilitate kinetic analysis. A peak at a lower temperature is found by digitally subtracting a glow-peak obtained after partial heating from the initial glow-peak. The order of kinetics for the first two components are 1.58 ± 0.08 and 1.41 ± 0.09 and the activation energy, 0.96 ± 0.04 and 0.88 ± 0.04 respectively [1]. The results are consistent with those in the preceding section but more importantly, these results also show that when the components of the peak are analysed as explained, the order of kinetics is found as general rather than first order.

14.6.3 Kinetic Analysis Using the Temperature-Dependence of the Area Underneath an Isothermal Decay-Curve

The phosphorescence measured at temperatures coincident with those of the main peak can be analysed using the temperature-dependence of the area underneath an isothermal decay-curve [3]. Here, a series of data points corresponding to the area Φ under an isothermal decay-curve measured at a specific temperature for a specific time are plotted against the measurement temperature.

This produces a pseudo-peak which can then be used for kinetic analysis. The basis of the technique is that since phosphorescence is thermoluminescence measured at a constant temperature, each data point Φ resembles a data point at that temperature were such data obtained contiguously at a specific heating rate. This method cannot isolate components of a composite peak but it helps to also illustrate results obtainable when the peak is analysed as a *bona fide* single peak.

We refer to the dependence of the area Φ corresponding to 5 s on temperature between 28 and 60°C was studied for measurements corresponding to 20 and 100 Gy [1]. The runs were made on the same aliquot. Each data point is obtained between irradiation and heating to 400°C, the latter step included to remove residual charge in electron traps. A fit of the Kitis general-order equation gave $b = 2.07 \pm 0.22$ and $E = 0.87 \pm 0.09$ eV for phosphorescence corresponding to 20 Gy and $b = 2.01 \pm 0.39$ and $E = 0.84 \pm 0.10$ eV for the 100 Gy data. The activation energy values are consistent with those from curve-fitting in earlier sections but notably the order of kinetics is found to be second-order.

14.7 Mechanisms of Thermoluminescence in $SrAl_2O_4:Eu^{2+}, Dy^{3+}$

14.7.1 Explanation Using a Model Based on a Continuum of Energy Levels

The qualitative and quantitative discussion thus far has been based on the premise that the electron traps form discrete single-valued localized states within the energy band gap. However, we have demonstrated, the TL from $SrAl_2O_4:Eu^{2+}, Dy^{3+}$ can be explained in terms of a quasi-continuum set of electron traps. Although both $T_m - T_{stop}$ and T_m – dose procedures imply that the main peak is a collection of first-order peaks, whole-peak methods produce general or second-order kinetics for the same peak. It is clear that the seemingly simple glow peaks of superluminous phosphors are anything but and it is necessary to reconcile these observations.

Since the thermoluminescence of $SrAl_2O_4:Eu^{2+}$, Dy^{3+} suggests the presence of closely-spaced peaks, its electron trap should not be modelled as single-valued but rather as a spread of energy values about the putative energy position. The outlines of such an approach are discussed elsewhere e.g. [8]. If we label the density of states as $D(E)$, then the number of energy levels between E and $E + dE$ is $D(E)\,dE$. Therefore, the concentration of trapped electrons n is,

$$n = \int_0^{E_c} f(E)g(E)dE, \tag{14.2}$$

where $f(E)$ is the probability of occupancy of an energy level. The concentration of electrons in the conduction band n_c is increased by stimulation from electron traps and decreased owing to re-trapping and loss to recombination centres. Thus

$$\frac{dn_c}{dt} = \int_0^{\infty} pf(E)D(E)dE - n_c A_n \int_0^{\infty} (1 - f(E))D(E)dE - n_c m A_{mn}, \tag{14.3}$$

where p is thermal stimulation probability, A_n the retrapping probability, m the concentration of recombination centres and A_{mn} the recombination probability. Taking charge neutrality into account, the TL intensity I can be expressed as

$$I(t) = -\eta \frac{dm}{dt}, \tag{14.4}$$

where η is the luminescence efficiency. When the distribution of energy levels is continuous over a finite range $\Delta E = E_f - E_i$ under the assumption of first-order kinetics, relevant in this discussion, the thermoluminescence intensity $I(T)$ becomes

$$I(T) = \frac{n_o s}{\Delta E} \int_{E_i}^{E_f} \exp\left(-\frac{E}{kT}\right) \times \exp\left[-\frac{s}{\beta} \int_{T_o}^{T} \exp\left(-\frac{E}{kT'}\right) dT'\right] dE, \tag{14.5}$$

where all symbols are as defined before.

In practise, the thermoluminescence intensity for a quasi-continuum of electron traps as in $SrAl_2O_4:Eu^{2+}$, Dy^{3+} requires that the form of its density of states $D(E)$ be known. This is however not known at this time. Therefore, when we analyse a selected glow-peak corresponding to a particular dose, we are choosing the most dominant component of a composite because this is based on reducing the number of energy levels to one. In this sense, analyses of an experimental or digitally-mustered component peak by curve-fitting are approximations.

It is relevant to point out studies by [13] who used expression (14.5) to determine the order of kinetics of glow-peaks. The reported that the qualitative features of the peak changed from first to second-order kinetics when the value of ΔE increased. This is consistent with our demonstration that when the main peak is analysed using a limited range e.g. the $T_m - T_{stop}$ method where data points are chosen from a limited portion of the peak, the order of kinetics found as first-order. On the other hand, when the peak is analysed as a simple single one, the procedure subsumes multiple components and increases the range of ΔE. This explains why the order of kinetics appears to increase from first to general or second-order kinetics.

In summary, this chapter has looked at the kinetic analysis of thermoluminescence from superluminous phosphors. We have shown, using various experimental tests including partial heating, isothermal heating, phosphorescence and fading, that the prominent peak in a TL glow-curve of the example studies consist of closely-spaced peaks. The question we addressed is how one is to analyse such a peak and how to make sense of the results. Analysis of the peak for its order of kinetics using qualitative methods such as the $T_m - T_{stop}$ one implies that it is a first-order peak. However, when whole peak methods are applied, the order of kinetics determined is other than first order. We invoked a model based on density of energy states to reconcile the differences. In brief, qualitative and quantitative analysis show that it is the presence of collocation within peaks that accounts for the inconclusive nature of its order of kinetics as deduced from experimental results. The cautionary statement from this work is that in analyzing TL from phosphors, a less naïve model that dispenses with the presumption that electron trapping energy levels are discrete may be required. A useful first step would be to determine the form of the density of states. Therein lies the next obstacle.

References

1. Chithambo, M.L., Wako, A.H., Finch, A.A. (2017). Thermoluminescence of $SrAl_2O_4$:Eu^{2+}, Dy^{3+}: Kinetic analysis of a composite-peak. *Radiat. Meas.* **97**, pp. 1–13.
2. Chithambo, M.L. (2017). Thermoluminescence of the main peak in $SrAl_2O_4$:Eu^{2+}, Dy^{3+}: Spectral and kinetics features of secondary emission detected in the ultra-violet region. *Radiat. Meas.* **96**, pp. 29–41.
3. Chithambo, M.L., Niyonzima, P. (2014). On isothermal heating as a method of separating closely collocated thermoluminescence peaks for kinetic analysis. *J. Lumin.* **155**, pp. 70–78.
4. Mokoena, P.P., Chithambo, M.L., Vinay Kumar, H.C. Swart, H.C., Ntwaeaborwa, O.M. (2015). Thermoluminescence of calcium phosphate co-doped with gadolinium and praseodymium. *Radiat. Meas.* **77**, pp. 26–33.
5. Bos, A.J.J., van Duijvenvoorde, R.M., van der Kolk, E., Drozdowski, W. (2011). Thermoluminescence excitation spectroscopy: A versatile technique to study persistent luminescence phosphors. *J. Lumin.* **131**, pp. 1465–1471.
6. Matsuzawa, T., Aoki, Y., Takeuchi, N., Murayama, Y. (1996). A new long phosphorescent phosphor with high brightness, $SrAl_2O_4$:Eu^{2+}, Dy^{3+}. *J. Electrochem. Soc.*, **143**, pp. 2670–2673.
7. Luff, B.J., Townsend, P.D. (1993). High sensitivity thermoluminescence spectrometer. *Meas. Sci. Technol.* **4**, pp. 65–71.
8. Chen, R., McKeever, S.W.S. (1997). *Theory of Thermoluminescence and Related Phenomena* (World Scientific, Singapore).
9. Pagonis, V., Kitis, G., Furetta, C. (2006). *Numerical and Practical Exercises in Thermoluminescence* (Springer, Berlin).
10. Bedyal, A.K., Vinay, K., Lochab, S.P., F. Singh, Ntwaeaborwa, O.M., Swart, H.C. (2013). Thermoluminescence response of gamma irradiated $SrAl_2O_4$:Eu^{2+}/Dy^{3+} nanophosphor. *Int. J. Mod. Phys.*, **22**, pp. 365–373.
11. Chithambo, M.L. (2021). Thermal assistance in the optically stimulated luminescence of superluminous $Sr_4Al_{14}O_{25}$:Eu^{2+}, Dy^{3+}. *Physica B: Condens. Matter.* **603**, p. 412722.
12. Kitis, G., Gomez-Ros, J.M., Tuyn, J.W.N. (1998). Thermoluminescence glow-curve deconvolution functions for first, second and general orders of kinetics. *J. Phys. D: Appl. Phys.* **31**, pp. 2636–2641.

13. Hornyak, W.F., Chen, R. (1989). Thermoluminescence and phosphorescence with a continuous distribution of activation energies. *J. Lumin.* **44**, pp. 73–81.

Chapter 15

Tunable Luminescence from Dy^{3+} and Pr^{3+} Doped Mixed Rare-Earth Oxyorthosilicate Phosphors

Simon N. Ogugua, Robin E. Kroon, and Hendrik C. Swart
Department of Physics, University of the Free State, Bloemfontein ZA9300, South Africa
swarthc@ufs.ac.za

Varying the molar ratios of La, Gd, and Y in mixed rare-earth oxyorthosilicates (R_2SiO_5) will manipulate the unit cell of the host and hence change the chemical nature of the crystals. This can influence the luminescence properties of a dopant in these host matrices and can lead to the generation of tunable colour emission. The structure of the phosphors was analysed using X-ray diffraction. It is demonstrated here that the emitted colour from $La_{2-x}Gd_xSiO_5{:}Dy^{3+}$ (x = 0, 0.5, 1.0, 1.5, and 2.0) and $La_{2-x}Y_xSiO_5{:}Pr^{3+}$ (x = 0, 0.5, 1, 1.5, and 2) could be tuned to generate multicoloured and white light for potential light emitting diode application. The tunable nature of the phosphors was also demonstrated by lifetime results and the band gaps obtained from the diffuse reflectance data.

Luminescent Nanomaterials
Edited by Odireleng Martin Ntwaeaborwa

ISBN 978-981-4968-11-9 (Hardcover), 978-1-003-27738-5 (eBook)
www.jennystanford.com

15.1 Introduction

Rare-earth oxyorthosilicates are represented by the general formula R_2SiO_5. They are created by the mixture of equal ratio (1:1) of rare-earth silicate (R_2O_3-SiO_2) compounds. They are classified into small (La-Tb) and large (Y, Sc, Dy-Lu) ionic radius rare-earth ions based on the size of their ionic radius [1]. Among these trivalent rare-earth ions, only Sc^{3+}, Y^{3+}, La^{3+}, Lu^{3+} and Gd^{3+} are usually used as hosts since they are optically inert ions and cannot show photoactive behaviour. This is because they possess stable, empty or filled states. Gd^{3+} however, has a half-empty 4f shell and possesses an $^8S_{7/2}$ ground state with excitation energy around 275 nm and hence Gd^{3+} can give emission in the ultraviolet wavelength (<318 nm) from the $^6P \rightarrow {}^8S$ transition [2, 3]. Because most rare-earth ions and some host materials absorb in the ultraviolet region, there is easy energy transfer from the ^6P energy level of Gd^{3+} to other rare-earth ions or host lattice groups. Hence, Gd^{3+} is considered generally as an inert ion [2]. Our focus in this chapter, however, is on R_2SiO_5 (R = La, Gd, or Y) doped rare-earth ions. These set of host matrices belong to the monoclinic crystallographic point group. They can be classified based on their space group as *P21/c* or *I2/a*, and *C2/c* or *B2/b* depending on their phase. The space groups *P21/c* and *C2/c* belong to the low temperature (*XI*) monoclinic phase, while the *I2/a* and *B2/b* belong to the high temperature (*X2*) monoclinic phase [1, 4–6]. Table 15.1 summarized the crystallographic properties of La_2SiO_5, Gd_2SiO_5, and Y_2SiO_5 based on their point groups, space group, lattice parameters and ionic radii. R_2SiO_5 have two non-equivalent crystallographic sites namely R1 and R2 coordinated by different numbers of oxygen atoms in the low and high temperature phases. The R1 site is coordinated by nine and R2 site is coordinated by seven oxygen atoms in the low temperature phase. In the high temperature phase however, the R1 site and the R2 site are coordinated by seven and six oxygen atoms, respectively [1, 6–8]. Figure 15.1a and b show the crystal structures of the low and high temperature phases of R_2SiO_5, respectively, using the Y_2SiO_5 as an example. In each phase the two R sites are linked by a tetrahedral Si (SiO_4) site [9]. Phase transformation temperatures of 1190°C and 1250°C from X1 to X2 phase have been reported for Y_2SiO_5 [1, 10]. Table 15.2 summarized the bond lengths between La–O, Y–O, and Gd–O in the La1 and La2 sites in La_2SiO_5, Y1 and Y2 sites in Y_2SiO_5, and Gd1 and Gd2 sites in Gd_2SiO_5 [6, 11].

Table 15.1 Lattice parameters and ionic radii of La_2SiO_5, Gd_2SiO_5, and Y_2SiO_5

Compound	Crystallographic point group	Space group	Lattice parameters	Ionic radius (nm)	Ref.
La_2SiO_5	Monoclinic	*X1* = *P21/c*, *X2* = *I2/a*	a = 0.93329 nm, b = 0.75087 nm, c = 0.70337 nm, β = 108.675°, V = 0.4669 nm^3	R1 = 0.1216 R2 = 0.110	[4]
Y_2SiO_5	Monoclinic	*X1* = *C2/c*, *X2* = *B2/b*	a = 1.44300 nm, b = 1.04100 nm, c = 0.67330 nm, β = 122.130°, V = 0.3974 nm^3	R1 = 0.0960 R2 = 0.090	[4–6]
Gd_2SiO_5	Monoclinic	*X1* = *P21/c*, *X2* = *I2/a*	a = 0.91105 nm, b = 0.69783 nm, c = 0.68544 nm, β = 107.141°, V = 0.4164 nm^3	R1 = 0.1107 R2 = 0.100	[1, 4]

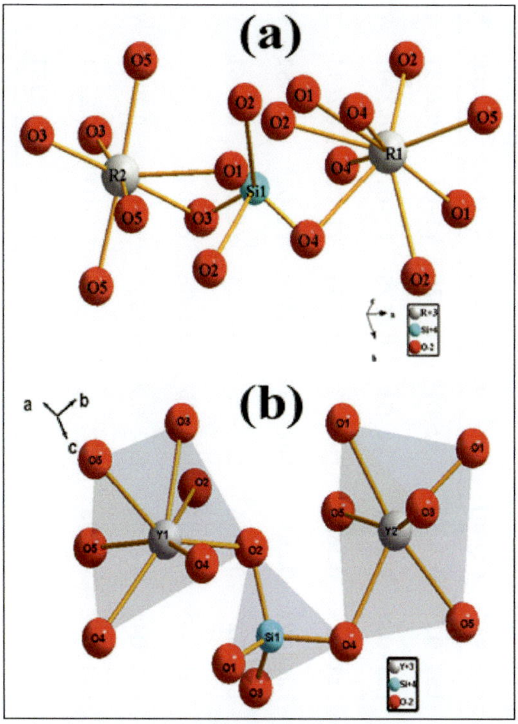

Figure 15.1 The crystal structure of R_2SiO_5 showing the (a) *X1* phase and (b) *X2* phase.

Table 15.2 The La–O, Y–O, and Gd–O distances for La1 and La2 sites in La_2SiO_5, Y1 and Y2 sites in Y_2SiO_5, and Gd1 and Gd2 sites in Gd_2SiO_5

Nuclear	*d* (nm)	Nuclear	*d* (nm)	Nuclear	*d* (nm)	Nuclear	*d* (nm)	Nuclear	*d* (nm)	Nuclear	*d* (nm)
La1–O1	0.2654	La2–O1[f]	0.2676	Y1–O1	0.2391	Y2–O1[f]	0.2421	Gd1–O1	0.2272	Gd2–O5	0.2295
La1–O1[a]	0.2370	La2–O3	0.2575	Y1–O1[a]	0.2201	Y2–O3	0.2541	Gd1–O2	0.2337	Gd2–O5	0.2300
La1–O2	0.2666	La2–O3[f]	0.2656	Y1–O2	0.2611	Y2–O3[f]	0.2562	Gd1–O5	0.2351	Gd2–O5	0.2303
La1–O2[b]	0.2618	La2–O3[g]	0.2489	Y1–O2[b]	0.2451	Y2–O3[g]	0.2342	Gd1–O4	0.2412	Gd2–O3	0.2382
La1–O2[c]	0.2516	La2–O5	0.2324	Y1–O2[c]	0.2461	Y2–O5	0.2342	Gd1–O2	0.2495	Gd2–O1	0.2399
La1–O4[d]	0.3055	La2–O5[h]	0.2506	Y1–O4[d]	0.2561	Y2–O5[h]	0.2302	Gd1–O1	0.2510	Gd2–O3	0.2502
La1–O4[e]	0.2570	La2–O5[i]	0.2358	Y1–O4[e]	0.2951	Y2–O5[i]	0.2231	Gd1-O2	0.2557	Gd2–O3	0.2532
La1–O4[b]	0.2695			Y1–O4[b]	0.2251			Gd1–O4	0.2689		
La1–O5[c]	0.2509			Y1–O5[c]	0.2231			Gd1–O4	0.2761		
Ave.	**0.2628**	Ave.	**0.2512**	Ave.	**0.2456**	Ave.	**0.2390**				
Ref.	[11]				[11]				[6]		

Table 15.3 Some applications of R_2SiO_5 (R = La, Gd or Y) doped rare-earth ions

Phosphor material	Excitation Source	Emission wavelength (nm)	Application(s)	Ref.
Gd_2SiO_5:Er	940 nm continuous wave diode laser	1550	Near infrared laser	[19]
Gd_2SiO_5:Yb	940 nm continuous wave diode laser	1089	Near infrared laser	[20]
Y_2SiO_5:Eu^{3+}	80 MHz dye laser	—	Storage of information	[21]
Y_2SiO_5:Pr^{3+}	4 GHz dye laser	—	High-resolution nonlinear imaging processing	[22]
Y_2SiO_5:Tm^{3+}, Er^{3+}, Yb^{3+}	975 nm continuous wave diode laser	475, 550, and 650	Optical display	[23]
Y_2SiO_5:Pr^{3+}, Li^{+}	488 and 515 nm argon laser, 447 nm diode-pumped solid-state laser, 589 nm diode-pumped solid-state laser	277 and 492	Light-utilizing technologies	[24]
Y_2SiO_5:Sm^{3+}	401 nm xenon lamp	656	Advance forensic	[25]
Y_2SiO_5:Tb^{3+}	147 nm deuterium lamp	542	Plasma display panel	[26]
Y_2SiO_5:Dy^{3+}	388 nm xenon short arc lamp	575	Cool white light emitting diode	[27]
Y_2SiO_5:Sm^{3+}	403 nm xenon lamp	600	White light emitting diode	[28]

(Continued)

Table 15.3 (*Continued*)

Phosphor material	Excitation Source	Emission wavelength (nm)	Application(s)	Ref.
$Y_2SiO_5:Pr^{3+}$	425 nm Xenon lamp, 52 W Filament lamp	360	Environmental pollutant degradation	[29]
$TiO_2/Y_2SiO_5:Pr^{3+}$, Li(I)	488 nm Xenon lamp, 100 W triphosphor tube light	312	Pollutant degradation	
Y_2SiO_5:Ce, Sm	85 KV and 20 mA X-ray source	670	X-ray storage	[30]
$Gd_2SiO_5:Ce^{3+}$, Tb^{3+}, Eu^{3+}	345 nm Xenon lamp	440, 545, and 614	Optical processing, memory and laser frequency stabilization	[31]
Gd_2SiO_5:Ce	Gamma-ray	—	Scintillation	[32]
$LaGdSiO_5:Dy^{3+}$	425 and 475 nm Xenon lamp	425, 485, and 575	Light emitting diode	[33]
$LaGdSiO_5:Dy^{3+}$	325 nm He-Cd laser	415, 485, and 573	Light emitting diode	[34]
$La_{2-x}Y_xSiO_5:Pr^{3+}$	235 nm Xenon lamp	490, 611, and 650	Light emitting diode	[11]
$La_{2-x}Gd_xSiO_5:Dy^{3+}$	325 nm He-Cd laser	415, 485, and 573	Light emitting diode	[35]

Compared to some top luminescence host materials such as bismuth germanium oxide (BGO) and yttrium aluminium garnet (YAG), R_2SiO_5 has shown a quantum yield of twice and lifetime five times faster than that of BGO [12, 13] and better luminescence than YAG [14]. They have been used widely as host matrices for rare-earth dopants used in various applications due to their high chemical and thermal stability [15–18]. Because of these qualities, R_2SiO_5 doped rare-earth ions phosphors are used in numerous technological applications as given in Table 15.3, including light emitting devices, scintillation, pollutant degradation, lasers, etc.

In this chapter, we focused on the synthesis and characterizations of single and mixed hosts R_2SiO_5 (R = La, Gd, or Y) doped with Dy^{3+} or Pr^{3+} phosphors. We studied the effect of the molar ratios of La, Gd, and Y on the properties of the phosphors. The idea here is to design sets of phosphor materials with tunable properties, for example tunable emission, by tuning the crystal field of the host matrices. The crystal field plays an important role in the chemical nature of the optical centre of a phosphor material. Crystal field engineering of a host material can lead to the shifting of the wavelength ranges of optical transitions, increasing the rates of radiative transitions and minimizing loss by non-radiative decay and excited state absorption. These objectives can be accomplished by manipulating the unit cell containing the optical centre. One of the ways to manipulate the unit cell is by changing its chemical composition. By varying the molar ratio of La, Gd, and Y in mixed R_2SiO_5 hosts (R = La, Gd, or Y) the unit cell of the host can be manipulated and hence tuning the crystal field strength. These, to a large extent, can influence the luminescence properties of Pr^{3+} and Dy^{3+} in these host matrices and can lead to the generation of tunable colour and white light for LED application. We demonstrated that the emitted colour from $La_{2-x}Gd_xSiO_5{:}Dy^{3+}$ ($x = 0, 0.5, 1.0, 1.5$, and 2.0) and $La_{2-x}Y_xSiO_5{:}Pr^{3+}$ ($x = 0, 0.5, 1, 1.5$, and 2) could be tuned.

15.2 Experimental

15.2.1 Preparation

The phosphors were prepared by the solution combustion method. The starting materials $La(NO_3)_3{\cdot}6H_2O$ (99.8%), $Gd(NO_3)_3{\cdot}6H_2O$

(99.8%), $Y(NO_3)_3 \cdot 6H_2O$ (99.8%) were used and either $Pr(NO_3)_3 \cdot 6H_2O$ (99.9%) or $Dy(NO_3)_3 \cdot 6H_2O$ (99.9%) were used depending on the dopant. For the synthesis of $La_{2-x}Gd_xSiO_5$:Dy^{3+} 1.0 mol% (x = 0, 0.5, 1.0, 1.5, 2.0), the starting materials $La(NO_3)_3 \cdot 6H_2O$ (99.8%), $Gd(NO_3)_3 \cdot 6H_2O$ (99.8%) and $Dy(NO_3)_3 \cdot 6H_2O$ (99.9%) were used, while $La(NO_3)_3 \cdot 6H_2O$ (99.8%), $Y(NO_3)_3 \cdot 6H_2O$ (99.8%) and $Pr(NO_3)_3 \cdot 6H_2O$ (99.9%) were used for the synthesis of $La_{2-x}Y_xSiO_5$:Pr^{3+} 1.0 mol% (x = 0, 0.5, 1, 1.5, and 2). The compositions x = 0 and 2 represent La_2SiO_5, Gd_2SiO_5 or Y_2SiO_5, while x = 0.5, 1, and 1.5 are the admixture of these hosts. In all the synthesis, sicilic acid (SiO_2H_2O, 99.9%) was used as source of silicon, while urea ($CO(NH_2)_2$, 98%) was used as fuel. All the chemicals were bought from Sigma-Aldrich. The starting materials were weighed according to the required stoichiometric compositions (with Dy^{3+} or Pr^{3+} replacing La^{3+}, Gd^{3+} or Y^{3+}) for each sample and were dissolved in 2 ml of distilled water contained in a glass beaker. After vigorous stirring for about 15 min at 100°C using a magnetic stirrer on a hot plate, the homogeneous solution obtained (which looks transparent through the glass beaker) was transferred to an alumina crucible boat and instantly into a muffle furnace pre-heated to 600 ± 10°C. A combustion reaction took place within 2 min by releasing of flame and a puff sound, resulting in the emission of gases like N_2, H_2O, and CO_2. The foam-like product was removed from the oven and allowed to cool naturally to room temperature and ground into a fine powder using an agate mortar and pestle. Each sample was stored in a separate sample holder for characterization. No emission was observed in the Pr^{3+} doped samples after preparation, hence they were annealed at 950°C in air for 1 h. The procedure for the synthesis of the samples is illustrated schematically in Fig. 15.2.

15.2.2 Characterization

The crystal structures of the samples were analysed using a Bruker D8 Advance X-ray diffractometer. The photoluminescence (PL) spectra were measured using an FLS980 fluorescence spectrometer (Edinburgh Instruments) equipped with a 450 W xenon lamp as steady-state excitation source and xenon flashlamp for lifetime measurements. The ultraviolet visible (UV-Vis) diffuse reflectance measurements were taken using a Lambda 950 UV-Vis spectrometer from PerkinElmer.

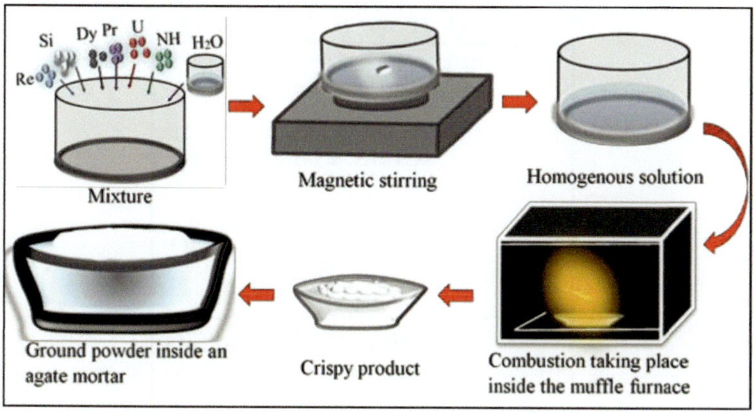

Figure 15.2 A schematic illustrating the synthesis procedure of R_2SiO_5 doped Dy^{3+} and Pr^{3+} (Re = La = $La(NO_3)_3 \cdot 6H_2O$, Gd = $Gd(NO_3)_3 \cdot 6H_2O$ or $Y(NO_3)_3 \cdot 6H_2O$, Dy = $Dy(NO_3)_3 \cdot 6H_2O$, Si = SiO_2H_2O, U = $CO(NH_2)_2$, NH = NH_4NO_3, and H_2O = distilled water) powder phosphors.

15.3 Results and Discussion

The XRD patterns of La_2SiO_5, Gd_2SiO_5, and Y_2SiO_5 phosphors compared with the respective Joint Committee on Powder Diffraction Standards (JCPDS) files are shown in Fig. 15.3. The patterns of Gd_2SiO_5 and Y_2SiO_5 matched with the monoclinic phases referenced in the JCPDS files 40-0287 and 21-1456, respectively, while La_2SiO_5 patterns is a mixture of a monoclinic phase of L_2SiO_5 and hexagonal phase of $La(OH)_3$ (marked with *) referenced in the JCPDS files 40-0234 and 36-1481, respectively. The peak marked with triangle (Δ) in the Y_2SiO_5 XRD pattern has been observed in many reports but was not assigned to any crystallographic plane [36–38]. Looking at Fig. 15.3, it is obvious that the crystalline structure of Gd_2SiO_5 and Y_2SiO_5 are similar and are very different from the structure of La_2SiO_5. This is because of the presence of the hexagonal $La(OH)_3$ phase in the La_2SiO_5 sample as can be seen in Fig. 15.3.

Figure 15.4a and b depicts the XRD patterns of $La_{2-x}Gd_xSiO_5:Dy^{3+}$ and $La_{2-x}Y_xSiO_5:Pr^{3+}$ (x = 0, 0.5, 1, 1.5, and 2) phosphors, respectively. Compare to the XRD patterns of the undoped La_2SiO_5 and Gd_2SiO_5 in Fig. 15.3, incorporation of Dy^{3+} did not cause any apparent change in the patterns of the equivalent doped samples (x = 0 and 2) in

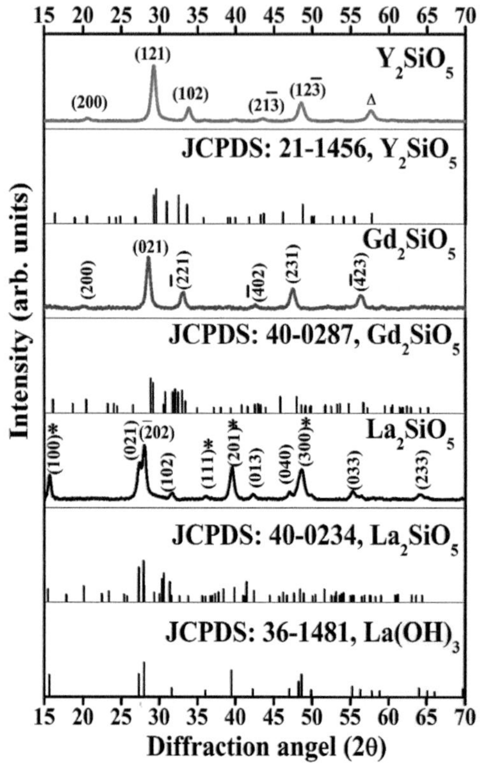

Figure 15.3 The XRD patterns of La_2SiO_5, Gd_2SiO_5, and Y_2SiO_5 compared to their respective standard JCPDS files.

Fig. 15.4a. This is an indication that the dopant is well incorporated into the La and Y lattice sites. Akin to the undoped samples, the patterns matched well with the standard JCPDS files of 40-0234 (La_2SiO_5), 36-1481 ($La(OH)_3$), and 40-0287 (Gd_2SiO_5). The mixed hosts, $x = 0.5$, 1 and 1.5, exhibited XRD patterns with diffraction peaks from La_2SiO_5, $La(OH)_3$, and Gd_2SiO_5, suggesting that admixtures of these three phases were crystallized. As a result, there is an obvious shift in the diffraction peak positions as the molar ratio of La to Gd changes in the matrices. For instance, the $(\bar{2}02)$ and (033) peaks shifted to higher diffraction angle (2θ) as the ratio of Gd increases (the amount of La decreases). The shifting of the peaks could be ascribed to the ion substitution induced lattice strains. The observed behaviour of the XRD patterns with the variation in the ratios of La and Gd is an indication that the matrices either crystallized in a mixed

La_2SiO_5 and $La(OH)_3$ phase ($x = 0$) or pure Gd_2SiO_5 phase ($x = 2$) or an admixture of these phases ($x = 0.5$, 1, and 1.5). In Fig. 15.4b, more peaks become prominent in the $x = 0$ sample, which is an evidence that the samples become more crystalline after the post annealing treatment at 950°C. For example, the (210), (121), and (123) planes were visible in the $x = 0$ samples in Fig. 15.4b. Similar to what was observed in Fig. 15.4a, as the ratio of La is decreased (from $x = 0$ to $x = 1.5$), the pattern of the samples assumed the crystal nature of Y_2SiO_5. When $x = 2$, the XRD pattern assumed the monoclinic point group and the *C2/c* space group of Y_2SiO_5. Also no visible change was observed between the undoped Y_2SiO_5 (see Fig. 15.3) and the Y_2SiO_5:Pr^{3+} ($x = 2$ in Fig. 15.4b).

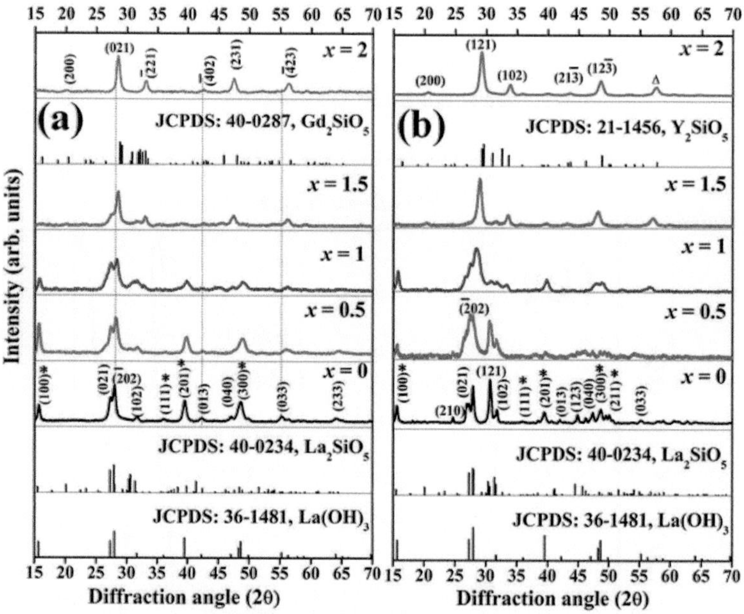

Figure 15.4 The XRD patterns of (a) $La_{2-x}Gd_xSiO_5$:Dy^{3+} and (b) $La_{2-x}Y_xSiO_5$:Pr^{3+}, ($x = 0$, 0.5, 1, 1.5, and 2) compared to their respective standard JCPDS files.

Figure 15.5a is the spectra of diffuse reflectance as a function of wavelength of undoped La_2SiO_5, Gd_2SiO_5, and Y_2SiO_5 hosts. The band-to-band transitions were observed at the absorption edges of the spectra around 207 nm for La_2SiO_5 and Y_2SiO_5, and 224 nm for Gd_2SiO_5. The spectra show that La_2SiO_5 is more transparent (reflects less light) than Gd_2SiO_5, while Y_2SiO_5 has the highest reflectance. The absorption deep levels around 275 and 313 nm are assigned to the $^8S_{7/2} \rightarrow {}^6I_J$ and $^8S_{7/2} \rightarrow {}^8S_{3/2}$ electronic transitions of Gd^{3+} [39], respectively.

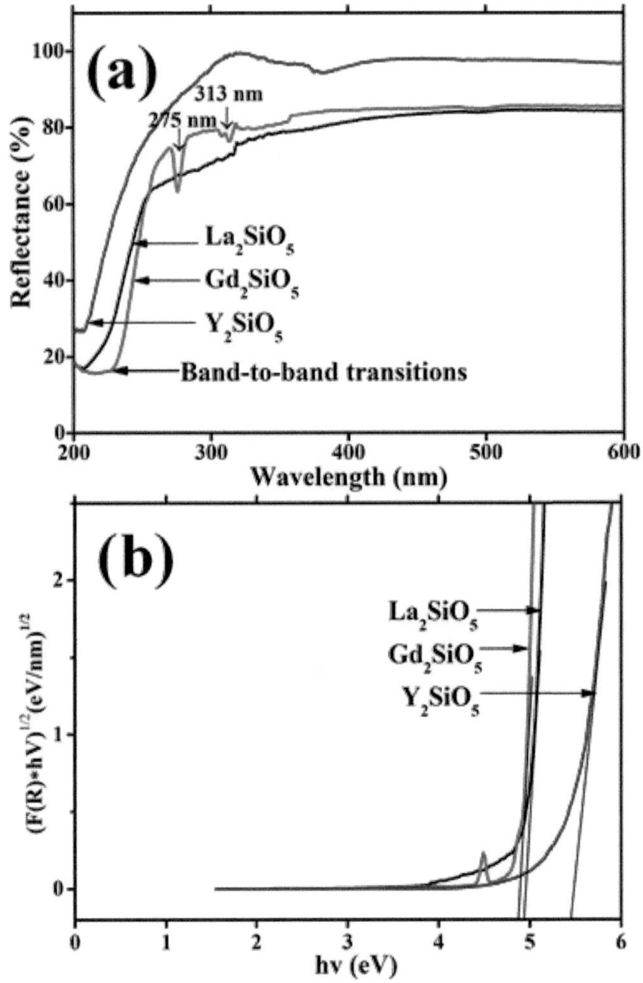

Figure 15.5 (a) The diffuse reflectance and (b) band gaps of La_2SiO_5, Gd_2SiO_5, and Y_2SiO_5.

Figure 15.5b shows the plot of the hv versus $[F(R_{100})hv]^{1/2}$ constructed from the diffuse reflectance data of the undoped La_2SiO_5, Gd_2SiO_5, and Y_2SiO_5 using the Kubelka-Munk equation [40, 41]

$$F(R_\infty)=\frac{(1-R_\infty)^2}{2R_\infty} \tag{15.1}$$

where R_∞ is the reflectance of a semi-infinite medium. $F(R_\infty)$ is proportional to the absorption coefficient (α) of any material which is related to the band gap (E_g) of the material by

$$(\alpha h v)^{1/n} = A(hv - E_g) \tag{15.2}$$

where h is the Planck's constant, v is the frequency of vibration, n can take values ½, 3/2, 2, or 3 for direct allowed, direct forbidden, indirect allowed and indirect forbidden transitions, respectively, and A is a proportionality constant. $R_\infty = R_{100}$ for a finite medium with 100% reflectance. $F(R_\infty)$ is proportional to α, hence comparing Eqs. (15.1) and (15.2) yields

$$(F(R_{100})hv)^{1/n} = A(hv - E_g). \tag{15.3}$$

Setting $n = 2$ in Eq. (15.3), the band gaps of the phosphors can be determined by extrapolating the point of intersection of a tangent line and the inflection point of the curve to the hv axis as shown in the graph of hv versus $[F(R_{100})hv]^{1/2}$ in Fig. 15.5b. The band gaps are 4.87, 4.94, and 5.46 eV for Gd_2SiO_5, La_2SiO_5, and Y_2SiO_5, respectively.

Figure 15.6a shows the diffuse reflectance curves of $La_{2-x}Gd_xSiO_5$:Dy^{3+} (x = 0, 0.5, 1, 1.5, and 2). The band-to-band transition of the phosphors varies from 207–224 nm. The reflectance of the samples decreased initially with increasing x, but started increasing from x = 1 to 2. The initial decrease in the reflectance could be ascribed to the strains induced by the lattice mismatch between the two mixed hosts. The absorption deep levels around 275 and 313 nm are assigned to the $^8S_{7/2} \rightarrow {}^6I_J$ and $^8S_{7/2} \rightarrow {}^8S_{3/2}$ electronic transitions of Gd^{3+} [39], respectively, while the one at 351 nm is assigned to the $^6H_{15/2} \rightarrow {}^6P_{7/2}$ electronic transition of Dy^{3+} [42]. The band gaps of the samples shown in Fig. 15.6b decreased from 4.94 to 4.51 eV for x = 0 and x = 0.5, respectively. For further increasing x values, the band gap increased to 4.65, 4.71, and 4.77 eV for x = 1, 1.5, and 2 eV, respectively. Similar to the reflectance curve, the decrease in the band gap could be due to the strain associated with the lattice mismatch [43] between the La_2SiO_5 and Gd_2SiO_5 host matrices. Comparing the band gap of undoped La_2SiO_5 and Gd_2SiO_5 in Fig. 15.5b to those of the Dy^{3+} doped samples in Fig. 15.6b

(x = 0 and 2), the band gap of La_2SiO_5 did not change, while the band gap of Gd_2SiO_5 increased by only 0.1 eV. This is evidence that the incorporation of Dy^{3+} into these hosts has no significant effect on the lattice, probably due to the similarities between the ionic radii of these ions. In their nine and seven coordination, La^{3+} has ionic radii of 1.216 and 1.1 nm, while Gd^{3+} has ionic radii of 1.107 and 1 nm, respectively. These can easily be substituted by Dy^{3+} ions with ionic radii of 1.083 and 0.97 nm for nine and seven coordination, respectively [1, 4].

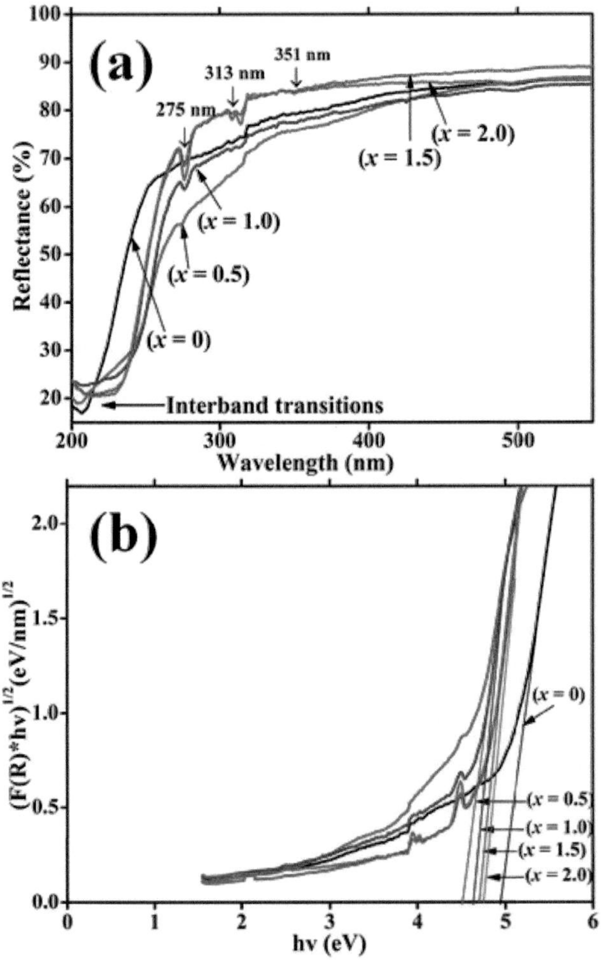

Figure 15.6 (a) The diffuse reflectance and (b) band gaps of $La_{2-x}Gd_xSiO_5$:Dy^{3+} (x = 0, 0.5, 1, 1.5, and 2).

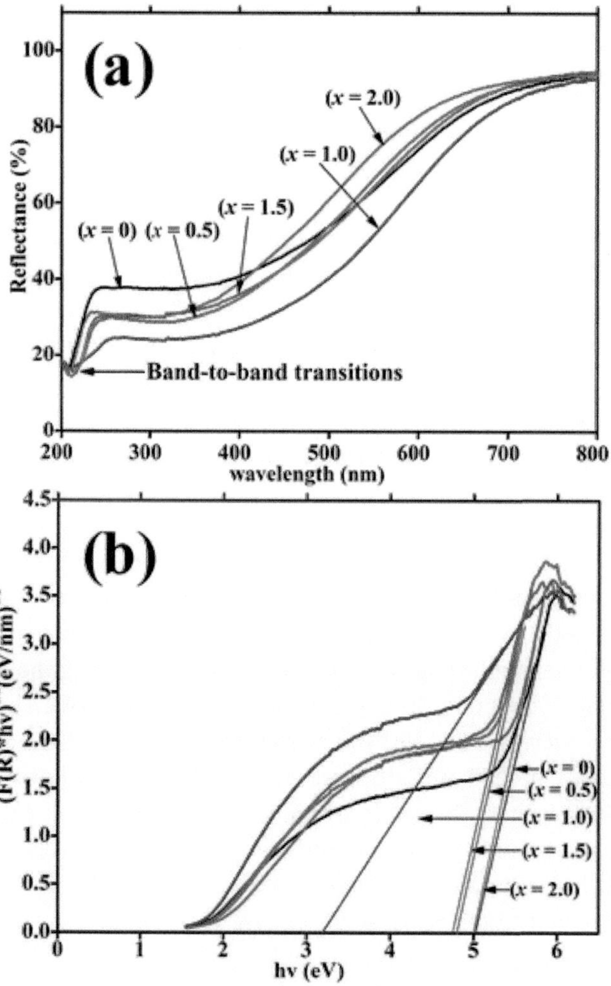

Figure 15.7 (a) The diffuse reflectance and (b) band gaps of $La_{2-x}Y_xSiO_5:Pr^{3+}$ (x = 0, 0.5, 1, 1.5, and 2).

The diffuse reflectance curves of $La_{2-x}Y_xSiO_5:Pr^{3+}$ (x = 0, 0.5, 1, 1.5, and 2) is shown in Fig. 15.7a. The reflectance curves showed strong UV absorption and the band-to-band transitions were observed around 207 nm. The UV absorption which started around 620 nm with maximum around 376 nm is ascribed to the $O^{2-} \rightarrow Pr^{4+}$ charge transfer band (CTB). Notice that the UV absorption was not observed in the diffuse reflectance curves of the undoped

La_2SiO_5 and Y_2SiO_5 in Fig. 15.5 or in that of Dy^{3+} doped La_2SiO_5 in Fig. 15.6a. The CTB is detrimental to the quality of the luminescence of the materials since it suppresses the absorption of Pr^{3+} and it is accompanied by no luminescence [44, 45]. However, other highly charged cations such U^{6+}, Nd^{5+}, and W^{6+} have been reported to display luminescence from the transition between the metal ion orbitals and the ligand molecular orbitals [46–48]. Ogugua et al. [49] has shown that annealing Pr^{3+} doped hosts in reducing atmosphere can convert Pr^{4+} to Pr^{3+} and hence reducing CTB absorption intensity and improving the luminescence property of the material. The band gaps of the $La_{2-x}Y_xSiO_5:Pr^{3+}$ ($x = 0$, 0.5, 1, 1.5, and 2) phosphors are shown in Fig. 15.7b with estimated values of 5.0, 4.8, 3.16, 4.75, and 5.0, respectively. Comparing the band gaps of the undoped La_2SiO_5 (4.94 eV) and Y_2SiO_5 (5.46 eV) and those of Pr^{3+} doped counterparts ($x = 0$ and 2), it could be seen that the band gap of La_2SiO_5 increased by 0.06 eV, while that of Y_2SiO_5 decreased by 0.46 eV after doping.

The photoluminescence (PL) excitation spectra of the $La_{2-x}Gd_xSiO_5:Dy^{3+}$ ($x = 0$, 0.5, 1, 1.5, and 2) phosphors were recorded in the wavelength range of 200–500 nm by monitoring the 573 nm emission of Dy^{3+} and the spectra are shown in Fig. 15.8a. The broad bands covering 200–250 nm wavelength range with maximums between 230–235 nm are host related and are attributed to the transition from O (2p) valence bands to the La (5d6s) and Gd (5d6s) conduction bands (band-to-band transition) of $La_{2-x}Gd_xSiO_5$ hosts [50, 51]. The peaks at 247 and 254 nm, 275 nm, 307 and 313 nm are assigned to the $^8S_{7/2} \rightarrow {}^6D_{J=9/2,\ 7/2}$, $^8S_{7/2} \rightarrow {}^6I_{J=15/2,\ 9/2,\ 7/2}$, and $^8S_{7/2} \rightarrow {}^6P_{J=7/2,\ 5/2,\ 3/2}$ f → f manifold transitions of Gd^{3+} [39, 52, 53], respectively. Excitation transitions from Dy^{3+} f → f energy levels were observed at 295 nm ($^6H_{15/2} \rightarrow {}^4K_{13/2}$), 325 nm ($^6H_{15/2} \rightarrow {}^4K_{15/2}$, $^6P_{3/2}$), 349 nm ($^6H_{15/2} \rightarrow {}^6P_{7/2}$, $^4M_{15/2}$), 364 nm ($^6H_{15/2} \rightarrow {}^6P_{5/2}$), 387 nm ($^6H_{15/2} \rightarrow {}^4I_{13/2}$), 424 nm ($^6H_{15/2} \rightarrow {}^4G_{11/2}$), 454 nm ($^6H_{15/2} \rightarrow {}^4I_{15/2}$) and 474 nm ($^6H_{15/2} \rightarrow {}^4F_{9/2}$) [42, 54–56]. The plot of PL intensity as a function of *x*, the ratio of La:Gd, is shown in Fig. 15.8b. The band-to-band transitions of the host, the 275 nm transitions of Gd^{3+} (349 nm transition of Dy^{3+} for the sample $x = 0$) were compared and the highest luminescence intensity was observed when $x = 2$.

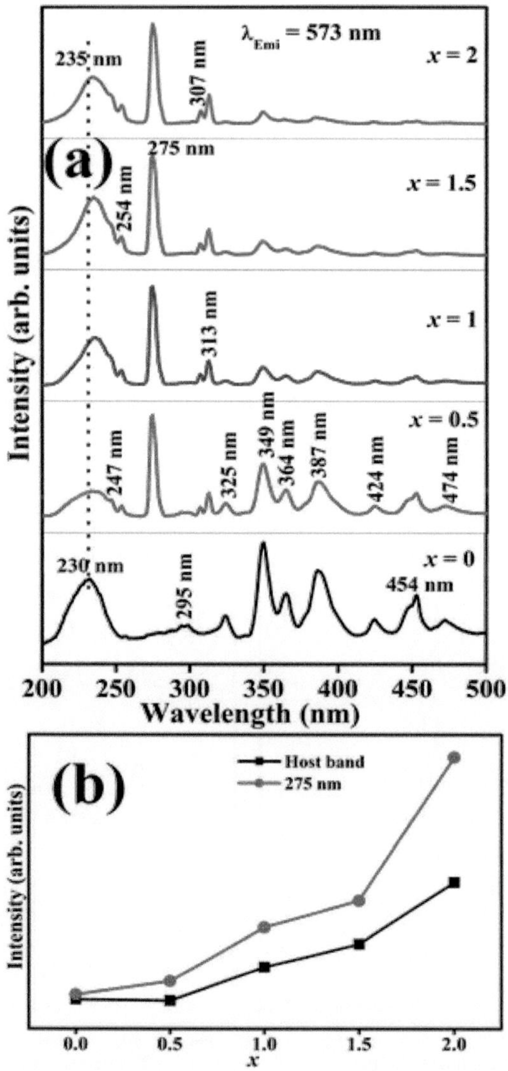

Figure 15.8 (a) The PL excitation spectra and (b) plot of PL excitation intensity as a function x for $La_{2-x}Gd_xSiO_5:Dy^{3+}$ (x = 0, 0.5, 1, 1.5, and 2).

The PL emission spectra of $La_{2-x}Gd_xSiO_5:Dy^{3+}$ (x = 0, 0.5, 1, 1.5, and 2) were recorded by exciting the host band-to-band excitation transition, 230 nm for sample $x = 0$ and 235 nm for $x = 0.5$, 1, 1.5, and 2, and the results are shown in Fig. 15.9a. The spectra showed two prominent emission peaks at 485 and 573 nm which are assigned to the ${}^4F_{9/2} \rightarrow {}^6H_{15/2}$ and ${}^4F_{9/2} \rightarrow {}^6H_{13/2}$ transitions of Dy^{3+} [6], respectively. The other peaks at 611, 668, and 756 nm are attributed to the ${}^5D_0 \rightarrow {}^7F_2$ transition of Eu^{3+} impurity from the $Gd(NO_3)_3 \cdot 6H_2O$) precursor [57], ${}^4F_{9/2} \rightarrow {}^6H_{11/12}$ and ${}^4F_{9/2} \rightarrow {}^6H_{9/2}$ f → f transitions of Dy^{3+} [6], respectively. The broad band covering 400–460 nm region with maximums at 415 and 438 nm was assigned to the recombination of hole with an electron occupying oxygen vacancies within the matrix and the self-trapped excitons (STE) in SiO_2, respectively [58, 59]. The PL emission spectra of the samples recorded by exciting the 275 nm excitation line of Gd^{3+} (349 nm line of Dy^{3+} for the sample $x = 0$) are shown in Fig. 15.9b. The spectra showed transitions comparable to those in Fig. 15.9a when the samples were excited using the band-to-band transition.

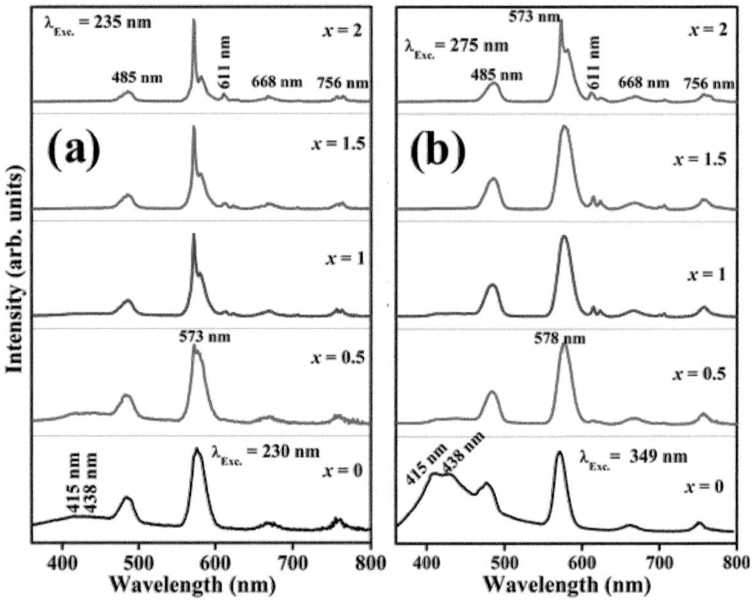

Figure 15.9 The PL emission spectra of $La_{2-x}Gd_xSiO_5:Dy^{3+}$ (x = 0, 0.5, 1, 1.5, and 2) phosphors when exciting the (a) host band-to-band excitation, 230 to 235 nm, and (b) 275 nm excitation line of Gd^{3+} (349 nm line of Dy^{3+} for the sample $x = 0$).

The PL excitation spectra of $La_{2-x}Y_xSiO_5:Pr^{3+}$ (x = 0, 0.5, 1, 1.5, and 2) phosphors were studied by monitoring the 611 nm transition of Pr^{3+} and the result is presented in Fig. 15.10. The spectra showed prominent broad bands in the UV region around 235, 237, 237, 237, and 238 nm for x = 0, 0.5, 1, 1.5, and 2, respectively. These bands are assigned to the transitions from the O 2p valence bands to the La (5d6s) and Y (4d5s) conduction bands of La_2SiO_5 and Y_2SiO_5 [3]. The encircled region of Fig. 15.10 is Pr^{3+} related and is assigned to the 4f → 5d electronic transition of Pr^{3+} [60, 61]. The other three excitation peaks in the visible region of the spectra around 447, 472, and 486 nm are assigned to the $^3H_4 \rightarrow {}^3P_2$, $^3H_4 \rightarrow {}^3P_1$, $^1I_6$, and $^3H_4 \rightarrow {}^3P_0$ 4f → 4f electronic transitions of Pr^{3+} [62], respectively.

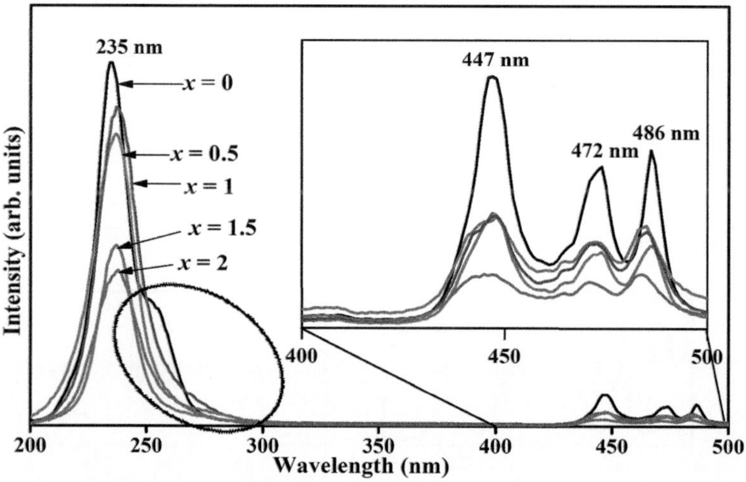

Figure 15.10 The PL excitation spectra of $La_{2-x}Y_xSiO_5$:Pr^{3+} (x = 0, 0.5, 1, 1.5, and 2) phosphors when monitoring the 611 nm emission line of Pr^{3+}. The encircled region is the excitation in the wavelength range of 245–290 nm, while the inset is the enlarge area of the excitation in the wavelength range of 400–500 nm.

The PL emission spectra of $La_{2-x}Y_xSiO_5$:Pr^{3+} (x = 0, 0.5, 1, 1.5, and 2) phosphors are shown in Fig. 15.11a–e. Three major peaks were observed from the spectra at 490, 611, and 650 nm, with the first and the second peak having shoulders at 500 and 630 nm, respectively. The 490 and 500 nm, and 611 and 630 nm peaks are ascribed to the manifolds of the $^3P_0 \rightarrow {}^3H_5$, and $^1D_2 \rightarrow {}^3H_4$ transitions of Pr^{3+} [63–65], respectively. Meanwhile, the peak at 650 nm is assigned to the $^3P_0 \rightarrow {}^3F_2$ electronic transition of Pr^{3+} [66]. The three peaks between 500 and 600 nm positioned at 528, 540, and 555 nm are ascribed to the manifolds of the $^3P_0 \rightarrow {}^3H_5$ electronic transitions of Pr^{3+} [66], whereas the peak at 730 nm with shoulder at lower and higher wavelengths is assigned to the manifolds of the

$^3P_0 \rightarrow {}^3F_4$ transition of Pr^{3+} [67]. The broad band in the wavelength range of 400–460 nm, which splits into two peaks with maximums at 415 and 430 nm, is assigned to the recombination of hole with an electron occupying oxygen vacancies within the matrix and the self-trapped excitons (STE) in SiO_2 [58, 59], respectively. Figure 15.11f shows how the emission intensities of the 490 nm ($^3P_0$), 611 nm ($^3P_0$) and 650 nm ($^1D_2$) peaks varied with *x* (La: Y ratio). It may be observed that the intensities of the $^3P_0$ and $^1D_2$ lines decreased linearly as *x* is increased, except for *x* = 1 where there is a deviation. Furthermore, it is clear that the branching ratio of the $^1D_2$ transitions dominates the spectra as the value of *x* is increased. This could be due to the non-radiative $^3P_0 \rightarrow {}^1D_2$ energy transfer through the 4f5d level of Pr^{3+} [64]. Hence from Fig. 15.11a–e, it is obvious that rate of $^3P_0 \rightarrow {}^1D_2$ transition is minimum when $x = 0$ ($La_2SiO_5:Pr^{3+}$) and increased as the value of *x* is increased, attaining maximum when $x = 2$ ($Y_2SiO_5:Pr^{3+}$).

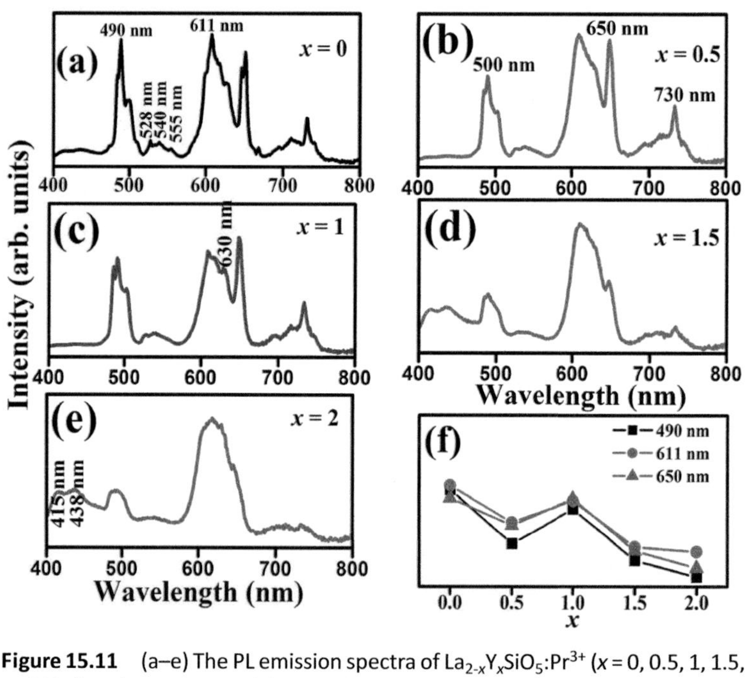

Figure 15.11 (a–e) The PL emission spectra of $La_{2-x}Y_xSiO_5:Pr^{3+}$ (x = 0, 0.5, 1, 1.5, and 2) phosphors when exciting at 235, 237, 237, 237, and 238 nm, respectively. (f) Variation of the intensities of the 490, 611, and 650 nm emission lines with *x*.

Fluorescence lifetime (τ) defines the average time an electron stays in the excited state before emitting a photon (recombination). For a mono-exponential decay, the luminescent intensity $I(t)$ after a time (t) is defined by

$$I(t) = I_0 e^{-pt} \tag{15.4}$$

where I_0 is the luminescent intensity at $t = 0$ and $p = 1/\tau$ is the transition probability of a radiative transition constant. For n^{th}–order exponential decay, Eq. (15.4) can be replaced by

$$I(t) = I_1 e^{-t/\tau_1} + I_2 e^{-t/\tau_2} + I_3 e^{-t/\tau_3} + \ldots I_n e^{-t/\tau_n} \tag{15.5}$$

where I_1, I_2, I_3, and I_n are the luminescent intensities with corresponding lifetimes τ_1, τ_2, τ_3, and τ_n [68]. The decay curves of the $La_{2-x}Gd_xSiO_5:Dy^{3+}$ ($x = 0, 0.5, 1, 1.5$, and 2) exhibited a tri-exponential behaviour (Fig. 15.12a). In Fig. 15.12b, the different exponents of the lifetime components (τ_1, τ_2, and τ_3) of a tri-exponential decay curve are shown. The curve was obtained by plotting the graph of intensity as a function of time for the sample $x = 0$, which was measured by exciting the $^4F_{9/2} \rightarrow {}^6H_{13/2}$ (573 nm) transition of Dy^{3+} at 220 nm. Considering Eq. (15.5), the luminescent intensity-time relation for a tri-exponential decay curve can be expressed mathematically by Eq. (15.6)

$$I(t) = I_1 e^{-t/\tau_1} + I_2 e^{-t/\tau_2} + I_3 e^{-t/\tau_3}. \tag{15.6}$$

The lifetimes are also measured by exciting the 573 nm transition of Dy^{3+} using the Gd^{3+} line at 275 nm and Dy^{3+} at 349 nm and the values of the lifetimes are shown in Table 15.4. The lifetime component with short lifetimes, τ_1, could be assigned to transitions related to SiO_x [35], whereas, the long lifetimes components, τ_2 and τ_3, could be related to Dy^{3+} occupying different La and Gd sites in the host matrices [35].

The decay curve of the $La_{2-x}Y_xSiO_5:Pr^{3+}$ ($x = 0, 0.5, 1, 1.5$, and 2) phosphors were excited at 235 nm and the decay time of the $^3P_0$ (490 nm) and $^1D_2$ (611) emission lines of Pr^{3+} were measured. Figure 15.12c shows the decay curve of the 611 nm emission line of Pr^{3+}. Similar to the Dy^{3+} doped samples, the decay curves show a tri-exponential behaviour and the results obtained from the calculated lifetime are shown in Table 15.5. Notice that the $^1D_0$ transition have longer lifetime compared to the $^3P_0$ transition in all the samples, which could be attributed to the spin-forbidden nature of the transition originating from the former [69, 70].

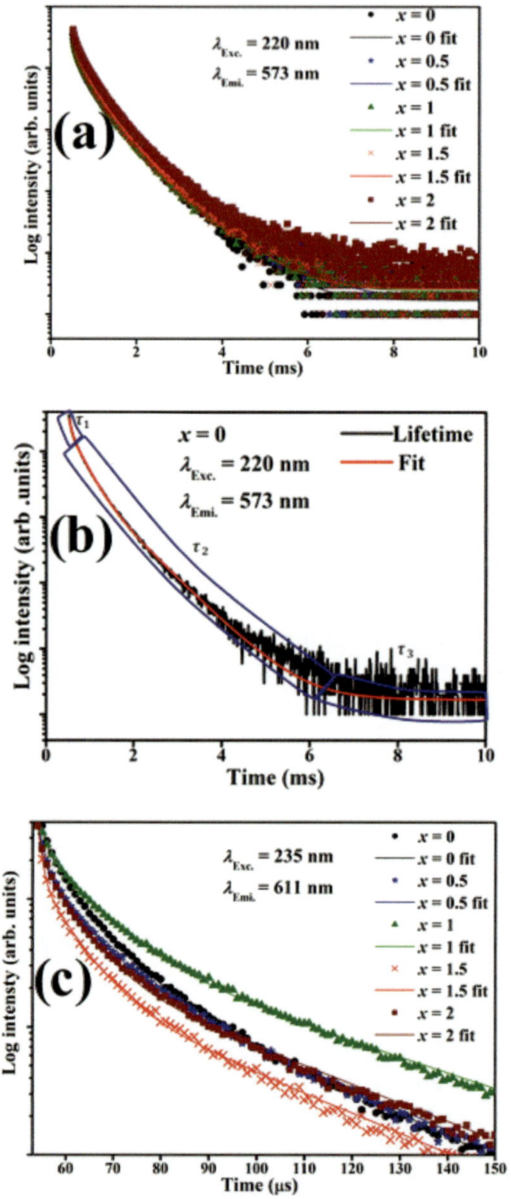

Figure 15.12 The decay of (a) $La_{2-x}Gd_xSiO_5$:Dy^{3+} (x = 0, 0.5, 1, 1.5, and 2) for the 573 nm emission of Dy^{3+} excited at 220 nm, (b) The three component of tri-exponential lifetimes (τ_1, τ_2, and τ_3) taking from the decay curve of La_2SiO_5:Dy^{3+} (sample x = 0), and (c) The decay curves of $La_{2-x}Y_xSiO_5$:Pr^{3+} (x = 0, 0.5, 1, 1.5, and 2) for the 611 nm emission of Pr^{3+} excited at 235 nm.

Table 15.4 Lifetime (in μs) of $La_{2-x}Gd_xSiO_5$:Dy^{3+} when monitoring the $^4F_{9/2} \rightarrow$ $^6H_{13/2}$ (573 nm) emission of Dy^{3+} under the excitation wavelengths of 220, 275, and 349 nm

	$\lambda_{Exc\cdot}$ = 220 nm, $\lambda_{Emi\cdot}$ = 573 nm			$\lambda_{Exc\cdot}$ = 275 nm, $\lambda_{Emi\cdot}$ = 573 nm			$\lambda_{Exc\cdot}$ = 349 nm, $\lambda_{Emi\cdot}$ = 573 nm		
	τ_1 (μs)	τ_2 (μs)	τ_3 (μs)	τ_1 (μs)	τ_2 (μs)	τ_3 (μs)	τ_1 (μs)	τ_2 (μs)	τ_3 (μs)
$x = 0$	44	276	682	—	—	—	16	234	666
$x = 0.5$	43	282	781	28	358	875	20	293	752
$x = 1.0$	39	268	716	19	335	807	12	297	746
$x = 1.5$	51	286	722	20	296	756	26	300	732
$x = 2.0$	69	336	798	45	361	820	32	298	687

Table 15.5 The values of lifetime (τ) for the $^3P_0$ (490 nm) and $^1D_2$ (611 nm) transitions of Pr^{3+} in $La_{2-x}Y_xSiO_5$ (x = 0, 0.5, 1, 1.5, and 2)

	$^3P_0$ (490 nm) emission			$^1D_2$ (611 nm) emission		
$La_{2-x}Y_xSiO_5$:Pr^{3+}	τ_1 (μs)	τ_2 (μs)	τ_3 (μs)	τ_1 (μs)	τ_2 (μs)	τ_3 (μs)
$x = 0$	3.8	11	27	4.6	28	100
$x = 0.5$	2.9	11	100	3.7	29	110
$x = 1$	2.5	14	110	3.7	35	140
$x = 1.5$	2.9	13	110	3.3	35	110
$x = 2$	1.5	11	61	2.7	24	120

Figure 15.13a,b shows the CIE chromaticity coordinates diagrams of $La_{2-x}Gd_xSiO_5$:Dy^{3+} samples when excited at 230–235 nm and 275 nm (from blue to yellow through white). The tuning nature of the CIE coordinates (from to yellow through to white) is evidence that the branching ratios of the luminescence emission lines of the Dy^{3+} doped samples changed with variation in the ratio of La:Gd in the host matrices. Table 15.6 summarises the CIE coordinates of $La_{2-x}Gd_xSiO_5$:Dy^{3+} (x = 0, 0.5, 1, 1.5, and 2). The CIE coordinated of $La_{2-x}Y_xSiO_5$:Pr^{3+} (x = 0, 0.5, 1, 1.5, and 2) (Fig. 15.13c) showed red colour and it is summarized in Table 15.6.

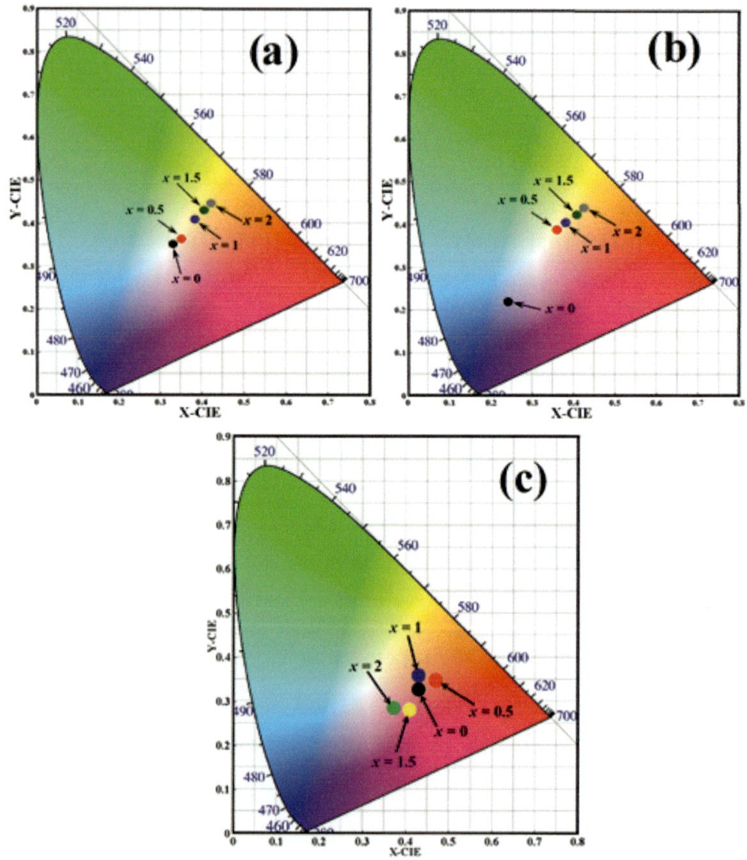

Figure 15.13 CIE coordinates diagram of (a), (b) $La_{2-x}Gd_xSiO_5:Dy^{3+}$ (x = 0, 0.5, 1, 1.5, and 2) for the 573 nm emission of Dy^{3+} excited at 230–235 nm and 275 nm (349 nm for sample x = 0), respectively, and (c) $La_{2-x}Y_xSiO_5:Pr^{3+}$ (x = 0, 0.5, 1, 1.5, and 2) for the 611 nm emission of Pr^{3+} excited at 235–238 nm.

15.4 Conclusion

The unit cells of La_2SiO_5, Gd_2SiO_5, and Y_2SiO_5 could be manipulated by mixing the doped hosts matrices in the following manner: $La_{2-x}Gd_xSiO_5:Dy^{3+}$ and $La_{2-x}Y_xSiO_5:Pr^{3+}$ (x = 0, 0.5, 1.0, 1.5, and 2.0). Properties of the phosphors such as the band gap, emitted luminescence colours and lifetimes exhibited tunable nature. The luminescence colours of $La_{2-x}Gd_xSiO_5:Dy^{3+}$ phosphors demonstrated multicolours and white colour emission. Furthermore, the luminescence the $La_{2-x}Y_xSiO_5Pr^{3+}$ phosphors demonstrated a non-radiative $^3P_0 \rightarrow {}^1D_2$ energy transfer through the 4f5d level of Pr^{3+}. The rate of $^3P_0 \rightarrow {}^1D_2$ transition is a minimum when $x = 0$ ($La_2SiO_5:Pr^{3+}$) and increased as the value of x increased, attaining a maximum when $x = 2$ ($Y_2SiO_5:Pr^{3+}$). The $La_{2-x}Y_xSiO_5:Pr^{3+}$ phosphors exhibited red emission with tunable property.

Table 15.6 CIE coordinates of $La_{2-x}Gd_xSiO_5:Dy^{3+}$ phosphors for the 230–235 and 275/349 excitation and $La_{2-x}Y_xSiO_5:Pr^{3+}$ phosphors for the 235–238 nm excitation

	$La_{2-x}Gd_xSiO_5:Dy^{3+}$		$La_{2-x}Y_xSiO_5:Pr^{3+}$
Ratio of hosts	**Excitation wavelengths (nm)**		
	230–235	**275/349**	**235–238**
	CIE Coordinates		
$x = 0$	(0.356, 0.357)	(0.259, 0.234)	(0.464, 0.341)
$x = 0.5$	(0.371, 0.369)	(0.386, 0.395)	(0.501, 0.350)
$x = 1.0$	(0.414, 0.416)	(0.414, 0.407)	(0.462, 0.355)
$x = 1.5$	(0.435, 0.436)	(0.439, 0.428)	(0.432, 0.286)
$x = 2.0$	(0.447, 0.447)	(0.447, 0.436)	(0.411, 0.290)

Acknowledgment

This research was sponsored by the Department of Science and Technology South Africa, under the South African National Nanoscience Postgraduate Teaching and Training Platform, the competitive programme for rated researchers (CPRR) (Grant No. CPR20110724000021870) of the National Research Foundation (NRF), and the rental pool programme of the National Laser Centre (NLC) (Grant No. NLC-LREGM00-CON-001). The South African Research Chairs Initiative of the Department of Science and Technology and the National Research Foundation of South Africa (84415). This research is also based on the research supported in part by the National Research Foundation of South Africa (Grant Number 93214, R.E. Kroon) for photoluminescence measurements.

References

1. Felsche, J. (1973). *Structure and Bonding*, **13**, eds. Penneman, R.A. Ryan, R.R. Rosenzweig, A. Reisfeld, R. Felsche, J. and Jørgensen, C.K. The crystal chemistry of the rare-earth silicates (Springer, Berlin) pp. 99–197.
2. Yan, B. (2017). *Photofunctional Rare Earth Hybrid Materials* (Springer Nature, Singapore) p. 6.
3. Ogugua, S.N. Shaat, S.K.K. Swart, H.C. and Ntwaeaborwa, O.M. (2015). Optical properties and chemical composition analyses of mixed rare earth oxyorthosilicate (R_2SiO_5, R = La, Gd and Y) doped Dy^{3+} phosphors prepared by urea-assisted solution combustion method, *J. Phys. Chem. Sol.* **83**, pp. 109–116.
4. Shannon, R.D. (1976). Revised effective ionic radii and systematic studies of interatomic distances in halides and chaleogenides, *Acta Crystallogr. A* **32**, pp. 751–767.
5. Bosak, A.A. Dubourdieu, C. Sénateur, J.P. Yu. O. Gorbenko, and Kaul, A.R. (2002). Hexagonal rare earth (R = Eu-Dy) manganites: XRD study of epitaxially stabilized films, *Cryst. Eng.* **5**, pp. 355–364.
6. Lisiecki, R. Dominiak-Dzik, R. Solarz, G. Ryba-Romanowski, P. Berkowski, W. and Glowacki, M. (2010). Optical spectra and luminescence dynamics of the Dy-doped Gd_2SiO_5 single crystal, *Appl. Phys. B* **98**. pp. 337–346.
7. Smolin, Y.I. and Tkachev, S.P. (1969). Determination of the structure of gadolinium oxyorthosilicate Gd_2O_3 SiO_2, *Kristallografiya* **14**, p. 22.
8. Fukuda, K. and Iwata, T. (2006). Crystal structure of lanthanum oxyorthosilicate, La_2SiO_5, *Powder Diffr.* **21**, pp. 300–303.
9. Wen, J. Duan, C.K. Ning, L. Huang, Y. Zhan, S. Zhang, J. and Yin, M. (2014). Spectroscopic distinctions between two types of Ce^{3+} Ions in X2-Y_2SiO_5: A theoretical investigation, *J. Phys. Chem. A* **118**, pp. 4988–4994.
10. Rabinovich, E.M. Shmulovich, J. Fratello, V.J. and Kopylov, N.J. (1987). Sol-gel deposition of Tb^{3+}:Y_2SiO_5 cathodoluminescent layers, *Am. Ceram. Soc. Bull.* **66**, pp. 1505–1509.
11. Ogugua, S.N. Swart, H.C. and Ntwaeaborwa, O.M. (2017). The dynamics of the photoluminescence of Pr^{3+} in mixed lanthanum yttrium oxyorthosilicate hosts, *Sens. Actuators B Chem.* **250**, pp. 285–299.
12. Takagi, K. and Fukazawa, T. (1983). Cerium-activated Gd_2SiO_5 single crystal scintillator, *Appl. Phys. Lett.* **42**, pp. 43–45.

13. Ishibashi, H. Shimizu, K. and Susa, K. (1989). Cerium doped GSO scintillators and its application to position sensitive detectors, *IEEE Trans. Nucl. Sci.* **36**, pp. 170–172.

14. Pejchal, J. Nikl, M. Mihóková, E. Mareš, J.A. Yoshikawa, A. Ogino, H. Schillemat, K.M. Krasnikov, A. Vedda, A. Nejezchleb, K. and Múčka, V. (2009). Pr^{3+}-doped complex oxide single crystal scintillators, *J. Phys. D: Appl. Phys.* **42**, p. 055117-10.

15. Yokota, H. Yoshida, M. Ishibashi, H. Yano, T. Yamamoto, H. and Kikkawa, S. (2011). Cathodoluminescence of Ce-doped Gd_2SiO_5 and $Gd_{9.33}(SiO_4)_6O_2$ phosphor under continuous electron irradiation, *J. Alloys Compd.* **509**, pp. 800–804.

16. Kobayashi, M. and Ishii, M. (1993). Effect of cerium doping on the radiation hardness of gadolinium silicate Gd_2SiO_5, *Nucl. Instrum. Methods Phys. Res. B* **82**, pp. 85–90.

17. Coetsee, E. Terblans, J.J. and Swart, H.C. (2005). Degradation of Y_2SiO_5: Ce phosphor powders, *J. Lumin.* **126**, pp. 37–42.

18. Cannas, C. Mainas, M. Musinu, A. Piccaluga, G. Speghini, A. and Bettinelli, M. (2005). Nanocrystalline luminescent Eu^{3+}-doped Y_2SiO_5 prepared by sol–gel technique, *Opt. Mater.* **27**, pp. 1506–1510.

19. Xu, X. Zhao, G. Wu, F. Xu, W. Zong, Y. Wang, X. Zhao, Z. Zhou, G. and Xu, J. (2008). Growth and spectral properties of Er:Gd_2SiO_5 crystal, *J. Cryst. Growth* **310**, pp. 156–159.

20. Yan, C. Zhao, G. Zhang, L. Xu, J. Liang, X. Juan, D. Li, W. Pan, H. Ding, L. and Zeng, H. (2006). A new Yb-doped oxyorthosilicate laser crystal: Yb:Gd_2SiO_5, *Solid State Commun.* **137**, pp. 451–455.

21. Equall, R.W. Sun, Y. and Cone, R.L. (1994). Ultraslow optical dephasing in Eu^{3+}:Y_2SiO_5, *Phys. Rev. Lett.* **72**, pp. 2179–2182.

22. Ham, B.S. Hemmer, P.R. and Shahriar, M.S. (1997). Efficient electromagnetically induced transparency in a rare-earth doped crystal, *Opt. Commun.* **144**, pp. 227–230.

23. Rakov, N. Maciel, G.S. Guimarães, R.B. and Carvalho, I.C.S. (2010). Upconversion emission from Tm^{3+}:Yb^{3+} and Tm^{3+}:Er^{3+}:Yb^{3+} doped Y_2SiO_5 powders prepared by combustion synthesis, *Mater. Chem. Phys.* **123**, pp. 199–202.

24. Cates, E.L. and Kim, J-H. (2013). Upconversion under polychromatic excitation: Y_2SiO_5:Pr^{3+}, Li^+ converts violet, cyan, green, and yellow light into UVC, *Opt. Mater.* **35**, pp. 2347–2351.

25. Gowri, M.M. Darshan, G.P. Naik, Y.V. Premkumar, H.B. Kavyashree, D. Sharma, S.C. and Nagabhushana, H. (2019). Phase dependent

photoluminescence and thermoluminescence properties of Y_2SiO_5:Sm^{3+} nanophosphors and its advanced forensic applications, *Opt. Mater.* **96**, p. 109282-14.

26. Zhang, Z.H. Wang, Y.H. Hao, Y. and Liu, W.J. (2007). Synthesis and VUV photoluminescence of green-emitting X2–Y_2SiO_5:Tb^{3+} phosphor for PDP application, *J. Alloys Compd.* **433**, pp. L12–L14.

27. Ramakrishna, G. Nagabhushana, H. Hareesh, K. and Sunitha, D.V. (2017). Driving the photoluminescent and structural properties of X2-Y_2SiO_5 by varying the dopant Dy^{3+} concentration towards cool WLED applications, *Solid State Sci.* **69**, pp. 56–63.

28. Ramakrishna, G. Nagabhushana, H. Prashantha, S.C. Sharma, S.C. and Nagabhushana, B.M. (2015). Role of flux on morphology and luminescence properties of Sm^{3+} doped Y_2SiO_5 nanopowders for WLEDs, *Spectrochim. Acta A: Mol. Biomol. Spectrosc.* **136**, pp. 356–365.

29. Yang, Y. Liu, C. Mao, P. and Wang, L-J. (2013). Upconversion luminescence and photodegradation performances of Pr doped Y_2SiO_5 nanomaterials, *J. Nanomater.* **2013**, p. 6.

30. Meijerink, A. Schipper, W.J. and Blasse, G. (1991). Photostimulated luminescence and thermally stimulated luminescence of Y_2SiO_5-Ce, Sm, *J. Phys. D: Appl. Phys.* **24**, pp. 997–1002.

31. Zhang, X. Cheng, Y. Zhou, L. Pang, Q. and Gong, M. (2014). Synthesis of a broad-band excited and multicolor tunable phosphor Gd_2SiO_5:Ce^{3+}, Tb^{3+}, Eu^{3+} for near-ultraviolet light-emitting diodes, *Ind. Eng. Chem. Res.* **53**, pp. 6694–6698.

32. Kamae, T. Fukazawa, Y. Isobe, N. Kokubun, M. Kubota, A. Osone, S. Takashashi, T. Tsuchida, N. and Ishibashi, H. (2002). Improvement on the light yield of a high-Z inorganic scintillator GSO (Ce), *Nucl. Instrum. Methods Phys. Res. A* 490, pp. 456–464.

33. Ogugua, S.N. Swart, H.C. and Ntwaeaborwa, O.M. (2016). White light emitting $LaGdSiO_5$:Dy^{3+} nanophosphors for solid state lighting applications, *Physica B* **480**, pp. 131–136.

34. Ogugua, S.N. Shaat, S.K.K. Swart, H.C. and Ntwaeaborwa, O.M. (2016). The influence of Dy^{3+} ions concentration and annealing on the properties of $LaGdSiO_5$:Dy^{3+} nanophosphors, *J. Lumin.* **179**, pp. 154–164.

35. Ogugua, S.N. Shaat, S.K.K. Swart, H.C. Kroon, R.E. and Ntwaeaborwa, O.M. (2019). Structure and optical properties of $La_{2-x}Gd_xSiO_5$:Dy^{3+} phosphors, *J. Alloys Compd.* **775**, pp. 950–968.

36. Singh, S. and Singh, D. (2020). Synthesis and spectroscopic investigations of trivalent europium-doped M_2SiO_5 (M = Y and Gd) nanophosphor for display applications, *J. Mater. Sci.: Mater. Electron.* **31**, pp. 5165–5175.

37. Ghosh, P. Sadhu, S. and Patra, A. (2016). Preparation and photoluminescence properties of $Y_2SiO_5:Eu^{3+}$ nanocrystals, *Phys. Chem. Chem. Phys.* **8**, pp. 3342–3348.

38. Qin, X. Ju, Y. Bernhard, S. and Yao, N. (2007). Europium-doped yttrium silicate nanophosphors prepared by flame synthesis, *Mater. Res. Bull.* **42**, pp. 1440–1449.

39. Li, Y-C. Chang, Y-H. Chang, Y-S. Lin, Y-J. and Laing, C-H. (2007). Luminescence and energy transfer properties of Gd^{3+} and Tb^{3+} in $LaAlGe_2O_7$, *J. Phys. Chem. C* **111**, pp. 10682–10688.

40. Kortum, G. (1969). *Reflectance Spectroscopy* (Springer-Verlag, New York) pp. 103–169.

41. Tauc, J. (1966). *The Optical Properties of Solids* (Academic Press, New York).

42. Liu, X. Xiang, W. Chen, F. Hu, Z. and Zhang, W. (2013). Synthesis and photoluminescence characteristics of Dy^{3+} doped $NaY(WO_4)_2$ phosphors, *Mater. Res. Bull.* **48**, pp. 281–285.

43. Fox, M. and Ispasoiu, R. (2017), *Springer Handbook of Electronic and Photonic Materials*, 2nd edn., eds. Kasap, S. and Capper, P., Chapter 40 - Quantum wells, superlattices, and band-gap engineering (Springer, Cham) pp. 1037–1057.

44. Boutinaud, P. Pinel, M. Dubois, E. Vink, A.P. and Mahiou, R. (2005). UV-to-red relaxation path ways in $CaTiO_3:Pr^{3+}$, *J. Lumin.* **111**, pp. 69–80.

45. Hoefdraad, H.E. (1975). Charge-transfer spectra of tetravalent lanthanide ions in oxides, *J. Inorg. Nucl. Chem.* **37**, pp. 1917–1921.

46. de Hair, J.T.W. and Blasse, G. (1973). Luminescence of hexavalent uranium in ordered perovskites, *J. Lumin.* **8**, pp. 97–100.

47. Alemi, A. Khademinia, S. Joo, S.W. Dolatyari, M. Bakhtiari, A. Moradi, H. Saeidi, S. and Esmaeilzadeh, A. (2014). Hydrothermal synthesis and investigation of optical properties of Nb^{5+}-doped lithium silicate nanostructures, *Int. Nano Lett.* **4**, pp. 100–110.

48. Rajeshwaran, P. Sivarajan, A. Raja, G. Madhan, D. and Rajkumar, P. (2016). Effect oftungsten (W^{6+}) metal ion dopant on structural, optical and photocatalytic activity of SnO_2 nanoparticles by a novel microwave method, *J. Mater. Sci: Mater. Electron.* **27**, pp. 2419–2425.

49. Ogugua, S.N. Hendrik, C.S. and Ntwaeaborwa, M.O. (2018). The influence of post-preparation annealing atmospheres on the optical properties and energy transfer between Pr^{3+} and Dy^{3+} in mixed lanthanum-yttrium oxyorthosilicate hosts, *Opt. Mater.* **76**, pp. 125–140.

50. Li, G. Li, C. Zhang, C. Cheng, Z. Quan, Z. Peng, C. and Lin, J. (2009). Tm^{3+} and/or Dy^{3+} doped LaOCl nanocrystalline phosphors for field emission displays, *J. Mater. Chem.* **19**, pp. 8936–8943.

51. Jia, M. Zhang, J. Lu, S. Sun, J. Luo, Y. Ren, X. Song, H. and Wang, X. (2004). UV excitation properties of Eu^{3+} at the S_6 site in bulk and nanocrystalline cubic Y_2O_3, *Chem. Phys. Lett.* **384**, pp. 193–196.

52. Tamboli, S. Rajeswari, B. and Dhoble, S.J. (2016). Investigation of UV-emitting Gd^{3+}-doped $LiCaBO_3$ phosphor, *Luminescence* **31**, pp. 551–556.

53. Chen, Y. Liu, B. Shi, C. Kirm, M. True, M. Vielhauer, S. and Zimmerer, G. (2005). Luminescent properties of Gd_2SiO_5 powder doped with Eu^{3+} under VUV–UV excitation, *J. Phys. Condens. Matter.* **17**, pp. 1217–1224.

54. Pålsson, L.O. and Monkman, A.P. (2002). Measurements of solid-state photoluminescence quantum yields of films using a fluorimeter, *Adv. Mater.* **14**, pp. 757–758.

55. Reddy, G.V. Moorthy, L.R. Chengaiah, T. and Jamalaiah, B.C. (2013). Enhanced white light emission and energy transfer studies of Dy^{3+}/Ce^{3+} co-doped $YAl_3(BO_3)_4$ phosphors for white light emitting diodes, *Adv. Mater. Lett.* **4**, pp. 841–848.

56. Camal, W.T. Fields, P.R. and Rajnak, K. (1968). Electronic energy levels in the trivalent lanthanide aquo ions. I. Pr^{3+}, Nd^{3+}, Pm^{3+}, Sm^{3+}, Dy^{3+}, Ho^{3+}, Er^{3+}, and Tm^{3+}, *J. Chem. Phys.* **49**, pp. 4424–4442.

57. Wegh, R.T. and Meijerink, A. (1996). First observation of visible luminescence from trivalent gadolinium, *Acta Phys. Pol. A* **90**, pp. 333–337.

58. Hu, C.G. Liu, H. Dong, W. Zhang, Y. Bao, G. Lao, C. and Wang, Z.L. (2007). $La(OH)_3$ and La_2O_3 nanobelts—synthesis and physical properties, *Adv. Mater.* **19**, pp. 470–474.

59. Itoh, C. Tanimura, K. and Itoh, N. (1988). Optical studies of self-trapped excitons in SiO_2, *J. Phys. C Solid State Phys.* **21**, pp. 4693–4702.

60. Zhuravleva, M. Novoselov, A. Yoshikawa, A. Pejchal, J. Nikl, M. and Fukuda, T. (2007). Crystal growth and scintillation properties of Pr-doped $YAlO_3$, *Opt. Mater.* **30**, pp. 171–173.

61. Zych, A. de Lange, M. de Mello, C.D. and Meijerink, A. (2012). Analysis of the radiativelifetime of Pr^{3+} d-f emission, *J. Appl. Phys.* **112**, pp. 1–12.

62. Li, Y.C. Chang, Y.H. Lin, Y.F. Chang, Y.S. and Lin, Y.J. (2007). Luminescent properties of trivalent praseodymium-doped lanthanum aluminum germanate $LaAlGe_2O_7$, *J. Phys. Chem. Sol.* **68**, pp. 1940–1945.

63. Reut, E.G. and Ryskin, A.I. (1973). Virtual recharge: mechanism of radiationless transition in scheelite and fergusonite type crystals doped with rare-earth ions, *Phys. Stat. Sol.* **17**, pp. 47–57.

64. Aumuller, G.C. Kostler, W. Grabmailer, B.C. and Frey, R. (1994). Luminescence properties of Pr^{3+} in cubic rare earth oxide, *J. Phys. Chem. Solids.* **55**, pp. 767–772.

65. Annapurna, K. Chakrabarti, R. and Buddhudu, S. (2007). Absorption and emission spectraanalysis of Pr^{3+}: Tellurite glasses, *J. Mater. Sci.* **42**, pp. 6755–6761.

66. Lee, K.H. Chae, K.W. Cheon, C.I. and Kim, J.S. (2011). Effect of crystal structural environment of Pr^{3+} on photoluminescence characteristics of double tungstates, *J. Korean Ceram. Soc.* **48**, pp. 183–188.

67. Ji, Y. Cao, J. Zhu, Z. Li, J. Wang, Y. and Tu, C. (2011). Luminescence properties and white emission in Pr^{3+} doped hexagonal $YAlO_3$ nanophosphors, *Mater. Express* **1**, pp. 231–236.

68. Dutta, S. Som, S. and Sharma, S.K. (2015). Excitation spectra and luminescence decay analysis of K^+ compensated Dy^{3+} doped $CaMoO_4$ phosphors, *RSC Adv.* **5**, pp. 7380–7387.

69. Mahlik, S. Malinowski, M. and Grinberg, M. (2011). High pressure luminescence and time resolved spectra of $La_2Be_2O_5:Pr^{3+}$, *Opt. Mater.* **34**, pp. 164–168.

70. Oskam, K. Houtepen, A. and Meijerink, A. (2002). Site selective 4f5d spectroscopy of $CaF_2:Pr^{3+}$, *J. Lumin.* 97, pp. 107–114.

Index

For Product Safety Concerns and Information please contact our EU representative GPSR@taylorandfrancis.com Taylor & Francis Verlag GmbH, Kaufingerstraße 24, 80331 München, Germany